Statistical Sensor Fusion

FREDRIK GUSTAFSSON

The photos on page 4 and 76 published with permission from
Lantmäteriverket. From Ortofoto – Svartvitt
© Lantmäteriverket, Gävle, Sweden 2010.
Medgivande MEDGIV-2010-24724.

⚠ **Copying prohibited**

All rights reserved. No part of this publication may be reproduced or transmitted in any form or by any means, electronic or mechanical, including photocopying, recording, or any information storage and retrieval system, without permission in writing from the publisher.

The papers and inks used in this product are eco-friendly.

Art. No 33373
ISBN 978-91-44-07732-1
Edition 2:1

© Fredrik Gustafsson and Studentlitteratur 2012
www.studentlitteratur.se
Studentlitteratur AB, Lund

Cover design by Jakob Meijling

Printed by Graficas Cems S.L., Spain 2012

Contents

Preface   **xi**

1 Introduction   **1**
    1.1 Sensor Networks 3
    1.2 Inertial Navigation 6
    1.3 Situational Awareness 7
    1.4 Statistical Approaches 10
    1.5 Software Support 12
    1.6 Outline of the Book 14

Part I   Fusion in the Static Case   **19**

2 Linear Models   **21**
    2.1 Introduction 22
    2.2 Least Squares Approaches 24
    2.3 Fusion 29
    2.4 The Maximum Likelihood Approach 36
    2.5 Cramér-Rao Lower Bound 39
    2.6 Summary 41

3 Nonlinear models   **45**
    3.1 Introduction 46

	3.2	Nonlinear Least Squares	47
	3.3	Linearizing the Measurement Equation	52
	3.4	Inversion of the Measurement Equation	57
	3.5	A General Approximation Strategy	65
	3.6	Conditionally Linear Models	65
	3.7	Implicit Measurement Equation	68
	3.8	Summary	70

4 Sensor Networks 73
	4.1	Typical Observation Models	74
	4.2	Target Localization	76
	4.3	NLS and SLS Solutions	84
	4.4	Dedicated Least Squares Solutions	89
	4.5	Extended Estimation Problems	92
	4.6	Summary	97

5 Detection and Classification Problems 101
	5.1	Detection	101
	5.2	Classification	111
	5.3	Association	118
	5.4	Summary	121

Part II Fusion in the Dynamic Case 123

6 Filter Theory 125
	6.1	Introduction	126
	6.2	The Fusion–Diffusion Approach to Filtering	129
	6.3	The Classical Approach to Nonlinear Filtering	131
	6.4	Grid Based Methods	140
	6.5	Nonlinear Filtering Bounds	143
	6.6	Summary	149

7 The Kalman Filter 153
	7.1	Kalman Filter Algorithms	154
	7.2	Practical Issues	162
	7.3	Computational Aspects	168
	7.4	Smoothing	175
	7.5	Square Root Implementation	181
	7.6	Filter Monitoring	186
	7.7	Examples	188
	7.8	Summary	192

8 The Extended and Unscented Kalman Filters — 195
- 8.1 DARE-based Extended Kalman Filter 197
- 8.2 Riccati-Free EKF and UKF 201
- 8.3 Target Tracking Examples 205
- 8.4 Summary 208

9 The Particle Filter — 211
- 9.1 Introduction 212
- 9.2 Recapitulation of Nonlinear Filtering 214
- 9.3 The Particle Filter 218
- 9.4 Tuning 224
- 9.5 Choice of Proposal Distribution 226
- 9.6 Theoretical Performance 233
- 9.7 Complexity Bottlenecks 236
- 9.8 Marginalized Particle Filter Theory 239
- 9.9 Particle Filter Code Examples 250
- 9.10 Summary 258

10 Kalman Filter Banks — 261
- 10.1 General Solution 262
- 10.2 On-Line Algorithms 264
- 10.3 Off-Line Algorithms 272
- 10.4 Summary 276

11 Simultaneous Localization and Mapping — 279
- 11.1 Introduction 280
- 11.2 Kalman Filter Approach 288
- 11.3 The FastSLAM Algorithm 302
- 11.4 Marginalized FastSLAM 306
- 11.5 Summary 310

Part III   Practice — 313

12 Modeling — 315
- 12.1 Discretizing Linear Models 316
- 12.2 Discretizing Nonlinear Models 318
- 12.3 Discretizing State Noise 321
- 12.4 Linearization Error and Choice of State Coordinates 323
- 12.5 Sensor Noise Modeling 326
- 12.6 Choice of Sampling Interval 330
- 12.7 Calibration of Dynamical Systems 333
- 12.8 Summary 340

13 Motion Models — 343
- 13.1 Translational Kinematics 344
- 13.2 Rotational Kinematics 346
- 13.3 Rigid-Body Kinematics 351
- 13.4 Constrained Kinematic Models 352
- 13.5 Odometric Models 356
- 13.6 Vehicle Models 358
- 13.7 Aircraft Dynamics 362
- 13.8 Underwater Vehicle Dynamics 364
- 13.9 Summary 369

14 Sensors and Sensor Near Processing — 373
- 14.1 Ranging Sensors 373
- 14.2 Physical Sensors 384
- 14.3 Wheel Speed Sensors 387
- 14.4 Wireless Network Measurements 402
- 14.5 Summary 410

15 Filter and Model Validation — 413
- 15.1 Parametric Uncertainty 413
- 15.2 Ground Truth Data 416
- 15.3 Sensor Calibration Issues 422
- 15.4 Summary 424

16 Applications — 427
- 16.1 Sensor Networks 427
- 16.2 Kalman Filtering 431
- 16.3 Particle Filter Positioning Applications 439

Appendices — 453

A Statistics Theory — 455
- A.1 Selected Distributions 455
- A.2 Conjugate Priors 457
- A.3 Nonlinear Transformations 457

B Sampling Theory — 471
- B.1 Generating Samples from Uniform Distribution 471
- B.2 Accept-Reject Sampling 472
- B.3 Bootstrap 475
- B.4 Resampling 475

	B.5	Stochastic Integration by Importance Sampling	476
	B.6	Markov Chain Monte Carlo	479
	B.7	Gibbs Sampling	480

C Estimation Theory   483
    C.1 Basic Concepts 483
    C.2 Cramér-Rao Lower Bound 484
    C.3 Sufficient Statistics 488
    C.4 Rao-Blackman-Lehmann-Scheffe's Theorem 490
    C.5 Maximum Likelihood Estimation 490
    C.6 The Method of Moments 491
    C.7 Bayesian Methods 492
    C.8 Recursive Bayesian Estimation 495

D Detection Theory   497
    D.1 Notation 497
    D.2 The Likelihood Ratio Test 498
    D.3 Detection of Known Mean in Gaussian Noise 499
    D.4 Eliminating Unknown Parameters 502
    D.5 Nuisance Parameters 504
    D.6 Bayesian Extensions 505
    D.7 Linear Model 506

E Least Squares Theory   509
    E.1 Derivation of Least Squares Algorithms 509
    E.2 Matrix Notation and QR Factorizations 512
    E.3 Comparing On-Line and Off-Line Expressions 513
    E.4 Asymptotic Expressions 518
    E.5 Derivation of Marginal Densities 519

Index   538

Preface

The objective of this book is to explain state of the art theory and algorithms for estimation, detection and nonlinear filtering with applications to localization, navigation and tracking problems. The book starts with a review of the theory on linear and nonlinear estimation, with a focus on sensor network applications. Then, general nonlinear filter theory is surveyed, with a particular focus on particle filtering. The application perspective is an important theme in the book, and common sensors and models are surveyed.

The content of this book is influenced by two key stimuli. The first one comes from the industrial needs reflected in more than 150 master thesis projects at some twenty different companies I have examined during the last fifteen years. The selection of topics is guided by these applications, and the book also contains many results and conclusions of this collective effort corresponding to 75 man years work. Only some of these theses are explicitly references, but they have all together guided my selection of topics for this book.

The second stimulus comes from applied and theoretical research projects over the last fifteen years in the sensor fusion area. Many of these are industrial collaborations, where long term collaborations with Volvo Car, Saab Aero, NIRA Dynamics, Ericsson, Xsens are in particular acknowledged. These projects involve the contributions of 24 former and current PhD students in the sensor fusion area.

The applications in this book come from numerous applied projects with grants from several funding agencies, including Vinnova, SSF, VR, IVSS, NFFP, NRFP, EU. However, this book project was first mentioned in an application to the Swedish Research Council (VR) in 2004, where the key idea was to complement the applied projects with theoretical research. The resulting six year major grant from this agency made this book mission possible.

The kind of complex algorithms described in this book can hardly be fully comprehended without hands-on exercises. In the end, the practioner has to code the algorithms to really understand all details. A lot of intuition, understanding and quick results can, however, be obtained from high-level programming environments. A lot of effort has been spent on developing accompanying software that illustrates all theory in the book in a user-friendly way. The algorithms and examples are accessible in *Signals and Systems Lab*. A demo version is available at

http://www.control.isy.liu.se/~fredrik/sigsyslab.

Complete and reproducible code examples are provided in the text, and chapter summaries overview the functionality. The lab contains a rich database of benchmark trajectories and measurement data, and pre-defined models, which should provide a good basis for applied work.

A lot of people have contributed to the book, including Dr. Fredrik Gunnarsson, Dr. Rickard Karlsson, Dr. Thomas Schön, Dr. Christina Grönwall, Dr. Per-Johan Nordlund, Dr. Andreas Eidehall, Dr. Gustaf Hendeby, Dr. David Törnqvist, Dr. Christian Lundquist, Dr. Jeroen Hol, Lic. Per Skoglar, Dr. David Lindgren, Dr. Umut Orguner, Dr. Gabriele Bleser, Kjell Magne Fauske, Dr. Mussa Bshara, Fredrik Lindsten and Niklas Wahlström. The contributions of these individuals as well as others who have provided constructive feedback are gratefully acknowledged. Special thanks go to Gustaf Hendeby for 24/7 LaTeX support. The second edition of this book includes mainly corrections based on feedback from the two years as an undergraduate course, and in particular valuable suggestions from Prof. Bo Bernhardsson.

The book emanates from a series of courses in sensor fusion, and in particular the master level course in *Sensor fusion* at Linköping University. However, there will certainly be future printings and editions of the book, and these rely on constructive feedback from the readers. More information, such as errata list, exercises, links etc., is found at the local book homepage:

http://www.control.isy.liu.se/books/sensorfusion
www.studentlitteratur.se.

1

Introduction

"Fusion can refer to combining two or more distinct things", according to Wikipedia, and the word is used in many different contexts ranging from nuclear power to cuisine. The same source defines sensor fusion as

> "Sensor fusion is the combining of sensory data or data derived from sensory data from disparate sources such that the resulting information is in some sense better than would be possible when these sources were used individually."

This is indeed an excellent crisp clear definition that reveals the main objective with sensor fusion. The raw data may be pre-processed before the fusion process, and the goal is to get as accurate information as possible. The literature mentions three different levels of fusion: *information fusion*, *sensor fusion* and *data fusion*. The highest level of information fusion is often used in artificial intelligence contexts for fusing information that cannot always be represented with real numbers. In contrast, sensor fusion and data fusion merge numerical data from multiple sources. The distinction is not always clear, but data fusion is considered to take place closer to the sensors often on raw sensor data, and sensor fusion is the next level of fusion.

This book restricts the sensor fusion problem to a probabilistic approach using statistical inference from multiple observations. This requires a probabilistic framework on sensor observations, where sensor models relate what is measured to the parameter or state to be estimated. The theory covers many areas in modern model-based signal processing.

There are two chapters dedicated to newcomers, practitioners or researchers who want a quick start into the sensor fusion world without an excessive amount of equations:

- The first Chapter 1 explains how to "speak" the sensor fusion language. Here, the main concepts and notation are presented, and three architypical applications are explained in general terms without going into details.

- The final Chapter 16 describes a number of applications. The appropriate algorithms and models are given by reference to the theory chapters, and summaries of user experience are provided. All of these applications have been evaluated on real data and in many cases on-line using real-time implementations.

All in all, these two chapters are intended to be a broad survey over the field for the impatient readers, who perhaps need a better motivation to dig into the details, and also to illustrate *"the magic of sensor fusion"*.

The sensor information to be fused can be classified as one or more of the following categories:

- Sensors with different *modality*, meaning that they measure physically different things. This requires models for how the different physical quantities relate to the sensor observations.

- Spatially distributed sensors. This requires spatial models for how the sensor observation depends on its position.

- Temporal information from sensors. This requires dynamical models describing how the parameter/state evolves in time.

Thus, models are central for our approach of sensor fusion. The reader is in the sequel provided with some illustrative examples of typical sensor combinations that illustrate the fundamental sensor fusion problems. The dual purpose of the following sections is to overview the "language" of sensor fusion:

- Section 1.1 describes all the different problem areas covered in this book in terms of a sensor network with randomly distributed sensors of the same kind measuring distance to an object. Besides providing a glossary of terminology, it also introduces the thematic application in the first part of the book.

- Section 1.2 discusses the temporal fusion of two sensors of different modality: a *GPS* (*global positioning system*) and an *IMU* (*inertial measurement unit*). GPS provides position with low sampling rate and good accuracy while the IMU provides position (second) derivatives with high sampling rate and with good accuracy. Together, they have the potential to provide both position and position derivatives with high sampling rate and high accuracy.

- Section 1.3 describes how radar and camera are combined in automotive target tracking applications. Each sensor delivers, after extensive pre-processing in the sensor, complementary information about the target. The camera provides width and height of each tracked vehicle with high resolution, and also color and shape information. The radar gives rough horizontal angular resolution and even worse vertical angular resolution, but on the other hand it gives a very accurate range and range rate. Sensor fusion algorithms can here give an overall accurate description of each vehicle in view.

The GPS/IMU and radar/camera sensor combinations are perfect illustrations of the sensor fusion principle: sensory data from disparate sources are combined such that the resulting information is better than would be possible when these sources were used individually.

Section 1.4 overviews the main concepts from the statistical approaches. Section 1.5 gives the purpose and a summary of the software support in the Signal and Systems Lab. Section 1.6 gives a chapter by chapter summary of the book.

1.1 Sensor Networks

Most of the problems covered by this book can be illustratred in terms of two-dimensional sensor networks, which provide a geometrical intuition for the problems that will hopefully facilitate generalizations to other applications.

Figure 1.1 illustrates first a simulation environment where four sensors and four targets are randomly placed in the unit square. Figure 1.2 illustrates a more realistic setup where twelve sensors are placed in the terrain, and a vehicle is driven through the network.

The type of sensors falls into one of two categories:

- *Range measurements*, such as *time of arrival* (*TOA*). Active radar, lidar and sonar sensors emit a pulse and measure the round trip time for the returned pulse, which can be converted to range. Passive radar and sonar sensors listen to the target's own emissions, and then only the range difference between pairs of sensors can be computed. Stereo cameras also provide range measurements.

  In synchronized wireless networks, time of arrival gives range information. The power of radio, acoustic, magnetic and seismic waves decays exponentially with distance, which provides a kind or range measurements.

- *Bearing measurements*, such as *direction of arrival* (*DOA*). Radar and similar sensors with directional antennas give bearing to detected targets. Vision sensors such as cameras provide two angles to objects, but no range. Antenna arrays and lobe forming techniques also give bearing.

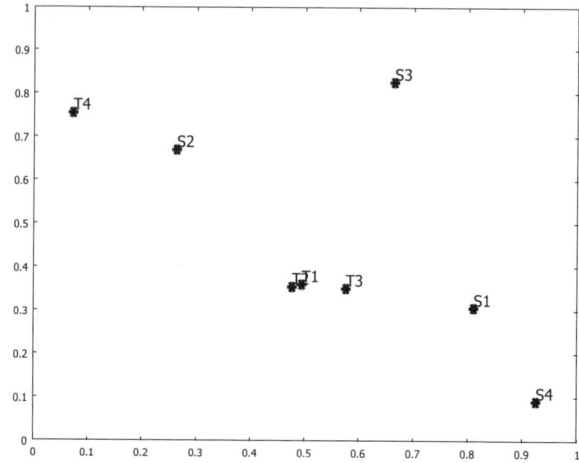

Figure 1.1: Example of sensor network with four sensors (S) and four targets (T).

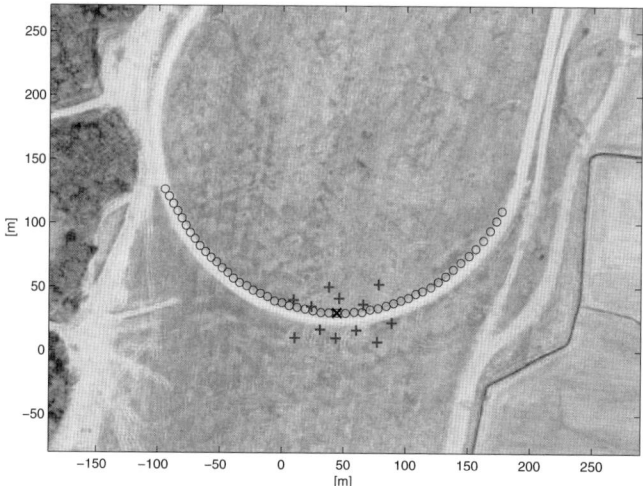

Figure 1.2: Example of deployment of a multiple sensor single target scenario, where the target is moving. The sensor positions are marked with plus signs and the target trajectory with circles.

1.1 Sensor Networks

This sensor network example can be extended in many ways. There can be a multitude of sensor modalities. The network in Figure 1.2 consists of acoustic, seismic and magnetic field sensors in each node.

Consider now the sensor network in Figure 1.1, consisting of $M = 4$ sensors at positions $p_m^s = (X_m^s, Y_m^s)^T$, $m = 1, 2, \ldots, M$, and $N = 4$ targets at positions $p_n^t = (X_n^t, Y_n^t)^T$, $n = 1, 2, \ldots, N$. The following problems can be formulated:

- *Detection* of one target ($N = 1$) from M sensors to decide whether there is a target present or not, is logically the first problem.

- *Localization* of a target ($N = 1$) from M sensors, which involves estimation of p^t. The observations can be categorized into *range* and *bearing* measurements. The main principles are *triangulation* of bearing measurements and *trilateration* of range measurements by computation circle intersections.

- *Tracking* of a target ($N = 1$) from M sensors, which extends localization to adaptive estimation of the target's time-varying position.

- *Navigation* is a dual problem to tracking. The difference is mainly conceptual, is the operator in control of the target or not? The target becomes the own platform in navigation, and the sensors can be interpreted as landmarks, or beacons. The mathematical framework differs essentially only when inertial sensor information is merged to landmark information.

- *Single Target* ($N = 1$) *Multiple Sensor* ($M > 1$) tracking. Using a multitude of sensors at different locations, the potential performance improves over the single sensor case. The basic algorithms are the same, while communication constraints, synchronization and distributed algorithms face the challenges.

- *Multiple Target* ($N > 1$) tracking poses the challenging *data association* problem: to which target does a bearing (or range) observation belong to? In Figure 1.1, each sensor delivers four measurements each ideally. For a single sensor, ordering is not a problem, but when information from other sensors is going to be merged, ordering becomes important. Already in this small example, there are $(4!)^3 = 13824$ combinations to consider at each time instant, theoretically.

- *Calibration* (or *gray-box identification*) of the sensor positions, which concerns estimation of p_n^s. Calibration is an important task that has to be done before all other tasks above. It often involves dedicated experiments. In the sensor network setup, one can let a target equipped with an accurate satellite navigation system move around the network, and apply a single sensor multiple target algorithm. Here, when the sensors observe the target, there is no association problem.

- *Simultaneous localization and mapping (SLAM)* is the counterpart to joint calibration and localization. The target starts with a blank map and while it moves around it detects landmarks and updates a landmark map on the fly. This is the most complex problem which includes all aspects above.

Practical questions for tracking applications with collaborating sensor nodes include:

- Who is responsible for the fusion, each node or a central node? That is, should centralized or de-dentralized fusion be applied? This leads to estimation algorithms that apply batch or sequential data processing.

- Communication protocols, controlling what information is transmitted, when and to whom.

- Sensors with different sampling frequencies (*multi-rate sampling*).

- Communication constraints (bandwidth, battery power) may restrict information exchange. Each sensor must decide if its information is worth transmitting.

- How to avoid double-counting information that is circulating among the sensor nodes?

1.2 Inertial Navigation

One of the most classical sensor fusion applications is to integrate inertial measurements with absolute position information. This has been done for a long time in surface ship and aircraft navigation, where dead-reckoning is used for long-term navigation. Historically, *visual triangulation* was used for short-term corrections of the position, and now landmark based triangulation or trilateration is used by most systems.

We will here illustrate this principle by the modern application of fusing data from a *GPS* (*global positioning system*) and an *IMU* (*inertial measurement unit*). This is an illustrative example of sensors with different modality and also different sampling rates:

- GPS provides slow-rate (1–10 Hz) accurate three-dimensional position measurements.

- IMU provides fast-rate (100–1000 Hz) accurate six-dimensional motion as three-dimensional acceleration and three-dimensional angular rates.

1.3 Situational Awareness

Figure 1.3: Satellite view of a highway intersection.

Figure 1.4 shows an example of measurements recorded inside a car during a route back and forth on a highway, which is visualized in the aerial image in Figure 1.3. The GPS gives an accurate trajectory of the actual test drive, while the IMU illustrates the forces acting on the vehicle.

Now, if the road was perfectly flat, the vehicle had no roll or pitch dynamics and the sensors were perfect, then the data in Figure 1.4(a) could be derived from the data in 1.4(b), and *vice versa*. This is not the case here. Figure 1.4(c) illustrates the large drifts associated with dead-reckoning inertial measurements. Similarly, Figure 1.4(d) shows that the GPS data are too noisy to give any useful motion information after double differentiation.

There are two causes for the drift in dead-reckoning. One is sensor offsets and the other one is the combination of banked roads and vehicle dynamics which gives leakage from the gravity field into the measured horizontal acceleration. All these effects cause quadratic drift over time.

1.3 Situational Awareness

In the automotive area, many *advanced driver assistant systems* (*ADAS*) have been released during the past twenty years, and there are many more to come. The performance of these hinges on available sensor information, and sensor fusion can here both increase performance, enable new application and reduce costs Gustafsson (2009).

Figure 1.5 summarizes one sensor fusion vision for future automotive safety systems. All available sensor information is communicated to the centralized or distributed sensor fusion algorithm. This includes vision sensors (visible or infrared light), distance measuring equipment (radar, lidar, sonar), an in-

8   Introduction

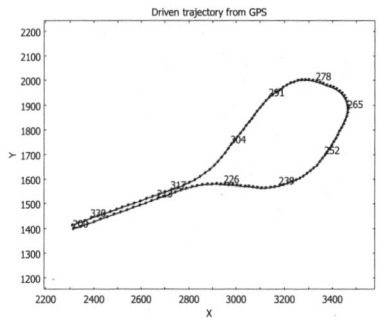

(a) Position from GPS measured at 1 Hz.

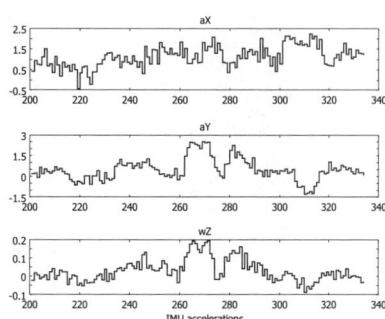

(b) Horizontal acceleration and angular rate from an IMU measured at 100 Hz.

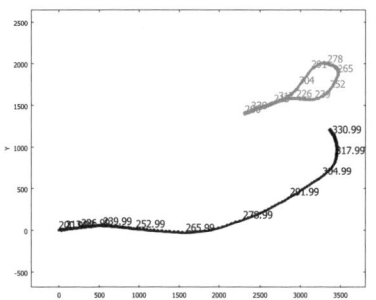

(c) GPS position (gray) and dead-reckoned position (black) from IMU.

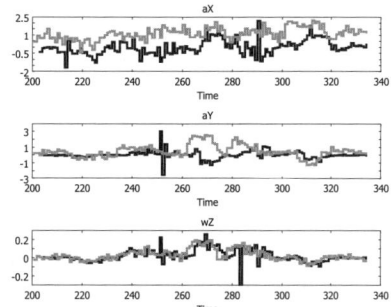

(d) IMU acceleration and angular rate (dark) and double differentiated GPS position (black)

Figure 1.4: Measurements from a test drive with on the highway in Figure 1.3.

ertial measurement unit (IMU) with up to three accelerometers and three gyroscopes, satellite navigation, wheel speeds and steering angle. The output from the sensor fusion includes virtual sensor signals, navigation information, tracking information and road geometry for collision avoidance systems.

We will here focus on the situational awareness problem. Consider the traffic scenario in Figure 1.6. Suppose each vehicle in front of the host car is

1.3 Situational Awareness

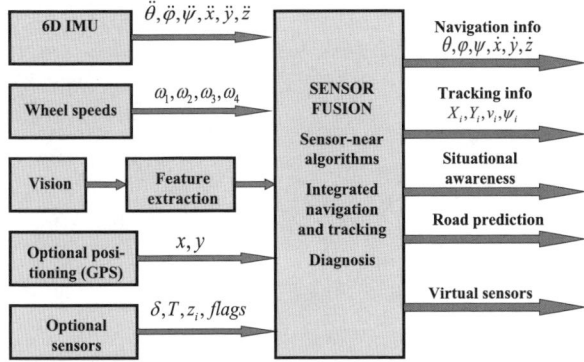

Figure 1.5: Sensor fusion structure for automotive applications as a middleware between sensors and active driver assistant systems.

represented with the following parameter vector:

$$x = \begin{pmatrix} \text{range} \\ \text{range rate} \\ \text{heading} \\ \text{width} \\ \text{height} \\ \text{color} \end{pmatrix}. \qquad (1.1)$$

Range and heading describe the location of the vehicle relative to the own vehicle. Range rate can be used to predict future risk of conflicts. The other parameters can be used for several purposes: object identification, classification of risks, association between different time instants and sensor modalities, etc.

It is clear that the union of information from the camera and radar covers the whole parameter vector. The camera provides angular width and height of each tracked vehicle with high resolution, and the actual width and height can be computed by multiplying with the range. The radar gives range and range rate but poor angular information.

It is, however, equally important that the intersection of information is non-empty. Otherwise, it would be hard or even impossible to associate the objects to each other. In the snapshot view in Figure 1.6, the computer vision algorithm delivers two rectangles. If the radar only delivers two pairs of range and range rate values, one would have to guess or use heuristics as "the larger

Figure 1.6: Low-resolution camera front view with rectangle approximations of detected vehicles from the computer vision system.

rectangle is probably closer to the own vehicle and corresponds for that reason most likely to the smallest observed radar range". Now, the angular resolution of the radar is good enough to resolve this data association problem.

Target classification, association and tracking is only one sub-task in *situation awareness*, where also stationary obstacles and road geometry are needed. Figure 1.7 illustrates one example from Lundquist and Schön (2009) on high-level information that can be obtained from sensor fusion of only a camera and a radar. Figure 1.7(b) shows a situation map that indicates the road geometry ahead, the own vehicles position, tracked vehicles and drivable area that can be used for emergency maneuvers.

1.4 Statistical Approaches

To overview the basic approaches, some brief notation is needed. Let y denote the vector of sensor observations, and x the unknown target position(s). A *sensor model* is used to compute a prediction of the sensor observation $\hat{y}(x; \theta)$ for each x. This is typically a range value or bearing angle, which is a function of the sensor positions $\theta = p_m^s$ and target position(s) x. The basic approaches at hand are:

- The least squares (LS) principle, that finds the x that minimizes the sum of squared prediction errors $V(x; \theta) = (y - \hat{y}(x; \theta))^T (y - \hat{y}(x; \theta))$ with respect to x for given θ. The solution is written $\hat{x}^{LS} = \arg\min_x V(x; \theta)$. Calibration is here a dual problem where the minimization is over θ for

1.4 Statistical Approaches

(a) (b)

Figure 1.7: (a) Camera view of a traffic situation. (b) Situation awareness map from sensor fusion of camera and radar, where the own vehicle (circle), tracked vehicles (black rectangles), road geometry (solid and dashed lines), stationary obstacles (dots) and drivable area (white regions) are marked.

a given x. This is also refered to as *system identification*. The main difference is that in sensor fusion applications, the model structure is mostly known and nominal values of the parameters are also at hand. The weighted least squares (WLS) estimate, where the squared prediction errors are weighted with the sensor (co-)variance. The least squares based methods deliver a covariance matrix to each estimate, that can be illustrated as confidence bounds. The nonlinear least squares (NLS) estimate differs from LS and WLS in that it uses a sensor model $\hat{y}(x;\theta)$ which is nonlinear in x (and where linearization is not used).

- The maximum likelihood (ML) method, where the likelihood function $p(y|x,\theta)$ is maximized with respect to x: $\hat{x}^{ML} = \arg\min_x p(y|x,\theta)$.

- The Bayesian approach, where x and θ are considered as stochastic variables with a distribution. The Bayesian approach computes the posterior distribution $p(x|y,\theta)$. This distribution contains a more detailed description than just an estimate with a covariance matrix, which is one advantage. Another flexibility is that prior knowledge can naturally be incorporated.

Fundamental estimation bounds as the Cramér-Rao lower bound (CRLB) gives a performance bound for a given problem independent on which estimation approach is used. The bound states that $\text{Cov}(\hat{x}) \geq P^{\text{CRLB}}$ for an unbiased estimator.

In general, Monte Carlo evaluation is recommended whenever possible to assess the performance of a specific approach. The general idea is that the estimation algorithms are evaluated at different realizations of the sensor readings, or over different realization of prior distributions of certain parameters. This can for instance be used to find the sensitivity of an uncertain sensor location, or to investigate if an estimator is unbiased.

1.5 Software Support

The software support in *Signals and Systems Lab* is object oriented, and the basic structure relevant at this stage is as follows:

- Models, signals and probability distributions are represented as objects defined in class constructors. The constructor takes a minimal number of arguments to define a meaningful object, and default values of other properties can then be changed. The most central classes are:

    - The SENSORMOD class that defines sensor signal models. It is a child of the more general NL class for nonlinear dynamic systems.
    - The SIG class for representing signals, and their uncertainties. Monte Carlo (MC) realizations are in many cases automatically generated and stored into the signal object, and these MC samples are used for illustrating uncertainty in various plots. The MC samples are further propagated in subsequent processing of the signals.
    - The set of PDFCLASS classes, which contains a number of named distributions, as the normal distribution NDIST.

    One key idea of creating classes for rather simple data structures, such as a signal vector and a signal model, is that all logical and sanity checks are done once for all when the object is created. The methods associated to an object can then trust that the objects have all required properties, and focus on the core algorithm.

- There are databases with standard predefined objects. For instance, `exsensor` contains standard sensor models, `exmotion` contains motion models, and `extraj` contains benchmark trajectories.

- The algorithms are hidden in methods of the classes. These are called as any other function, and the object is in most cases the first argument. Examples include:

1.5 Software Support

- `display` for displaying text information for an object.
- `simulate` computes a realization of a signal from a sensor model.
- `estimate` computes an estimate of x from a sensor model and a measurement signal.
- `crlb` computes the fundamental lower bound for the estimation error.
- `plot` illustrates a sensor network when a SENSORMOD is the first input argument, and the signal when a SIG object is the first input. There is also an `xplot` method for plotting the target location x and `xplot2` for a two-dimensional plot.

Note that the actual code that is executed for a method with a certain name depends on the first input argument. To get help for a method, the object has to be specified, for instance `help sensormod.plot`, `help sig.plot` and `help ndist.plot` give different help texts.

For illustration, consider a synthetic network with two sensors. The network is summarized in a SENSORMOD object, where target location is the state and sensor position the parameter vector. In the first example, the two sensors measure range. Simulation at an arbitrary time instant, here 1, means evaluation of the ranges and adding noise according to the specified distribution.

```
M=2; N=1;
th=[0.4 0.1 0.6 0.1];
x0=[0.5 0.5];
stoa=exsensor('toa',M,N);
stoa.th=th;
stoa.x0=x0;
stoa.pe=0.005*eye(M);
y=simulate(stoa,1)
SIG object with continuous time stochastic state space data (no input)
  Sizes:       N = 1,   ny = 2,  nx = 2
  MC is set to: 30
  #MC samples:  0
stoa
SIGMOD object: TOA
          / sqrt((x(1,:)-th(1)).^2+(x(2,:)-th(2)).^2) \
   y  = \ sqrt((x(1,:)-th(3)).^2+(x(2,:)-th(4)).^2) / + e
   x0' = [0.5        0.5]
   th' = [0.4        0.1          0.6          0.1]

   States:    x1        x2
   Outputs:   y1        y2
y.y
ans =
        0.4685      0.4319
plot(stoa)
hold on
lh2(stoa,y);
hold off
axis([0 1 0 1])
```

The likelihood for the target position, conditioned on these two ranges and that the sensor positions are given, is shown in Figure 1.8(a). The particular realization of the two random noise numbers influences where the actual estimate is located, while the banana shape of the level curves comes from the geometry of the problem. The two range circles are almost parallel at the target, so the bearing cannot be accurately determined.

In the next example, the two range sensors are replaced with bearing sensors, but the geometry is kept.

```
sdoa=exsensor('doa',M,N);
sdoa.th=th;
sdoa.x0=x0;
sdoa.pe=0.005*eye(M);
y=simulate(sdoa,1)
SIG object with continuous time stochastic state space data (no input)
  Sizes:         N = 1,   ny = 2, nx = 2
  MC is set to: 30
  #MC samples:   0
plot(sdoa)
sdoa
SIGMOD object: DOA
        / atan2(x(1,:)-th(1),x(2,:)-th(2)) \
    y = \ atan2(x(1,:)-th(3),x(2,:)-th(4)) / + e
    x0' = [0.5        0.5]
    th' = [0.4        0.1        0.6        0.1]

    States:  x1        x2
    Outputs: y1        y2
y.y
ans =
      0.2351     -0.2604
hold on
lh2(sdoa,y);
hold off
axis([0 1 0 1])
```

Figure 1.8(b) shows the level curves of the likelihood function in this case. Here, the geometry implies that the range is more uncertain than the bearing.

The level curves in Figure 1.8(a,b) are translated depending on the noise realization, while the shape reflects the underlying geometry and estimation potential of the network. A more fundamental issue is how the geometry affects the performance in an average sense.

Figure 1.8(c,d) shows the Cramér-Rao lower bound (CRLB) for the range and bearing networks. This bound is a function of the model only, and does not depend on the observations. It shows a fundamental estimation accuracy for this network.

1.6 Outline of the Book

The book consists of four parts, where the outline is as follows.

- Estimation and detection.

1.6 Outline of the Book

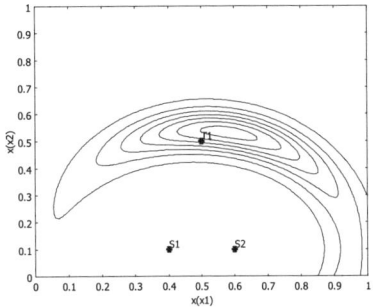

(a) Trilateration from two noisy range measurements.

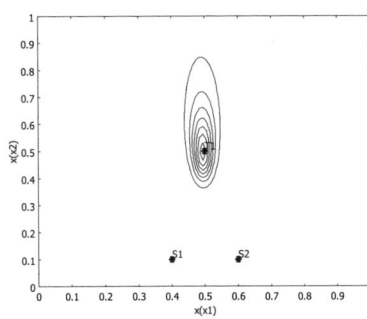

(b) Triangulation from two noisy bearing measurements.

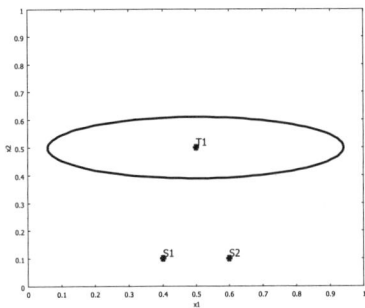

(c) CRLB for range measurements.

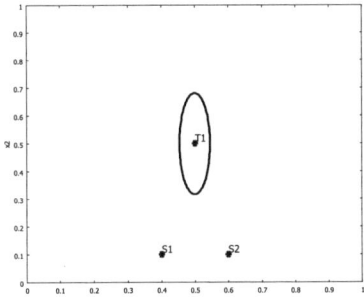

(d) CRLB for bearing measurements.

Figure 1.8: *Likelihood level curves computed over a uniform grid (a,b) and the corresponding CRLB (c,d) for a sensor network.*

- Chapter 2 describes methods to estimate the parameters in the linear model $y_k = H_k x + e_k$.
- Chapter 3 describes methods to estimate the parameters in the nonlinear model $y_k = h_k(x) + e_k$, or more generally in implicit models of the form $h_k(y_k, x, e_k) = 0$.
- Chapter 4 applies the methods in Chapters 2 and 3 to sensor network applications, and describes the background and particular solutions in more detail.
- Chapter 5 overviews a number of discrete problems as detection, diagnosis and association that are closely related to estimation.

• Nonlinear filtering.

- Chapter 6 describes a general approach to nonlinear filtering, where

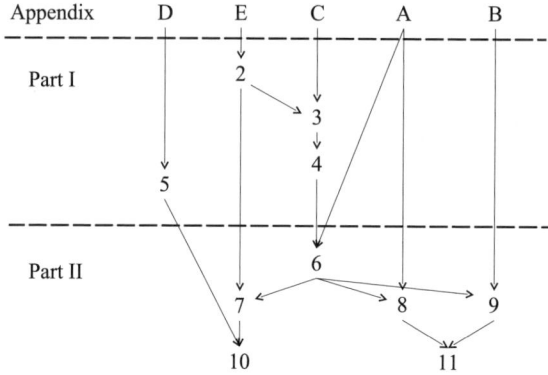

Figure 1.9: *Logical relations between the chapters.*

the parameter x is now a time-varying state x_k with a dynamic prediction model $x_{k+1} = f(x_k) + v_k$.

- Chapter 7 treats the special case of linear Gaussian models leading to the Kalman filter (KF), and the particular problems encountered in sensor networks.
- Chapter 8 presents the extended and unscented Kalman filters (EKF and UKF, respectively) for nonlinear and non-Gaussian systems.
- Chapter 9 describes the general particle filter (PF) approach to nonlinear filtering. It also extends the particle filter framework to larger systems, where linear substructures are utilized to get feasible algorithms. This is referred to as the marginalized particle filter (MPF).
- Chapter 11 describes an important application of EKF and MPF to sensor network applications where both target and sensor locations are unknown, known as the simultaneous localization and mapping (SLAM) problem.

• Practice.

- Chapter 12 describes in general terms the task of deriving motion and signal models. The focus is on discretizing continuous time models and linearizing nonlinear models.
- Chapter 13 lists a number of motion models that are suitable for describing target motion.

1.6 Outline of the Book

- Chapter 14 describes the fundamental functionality of various sensors useful in sensor networks.
- Chapter 15 gives some practical user guidelines for how to validate the models and design a filter.
- Chapter 16 presents applications where the theory in the preceding chapters are combined.

• Appendices. Here background material is provided on statistics (A), sampling (B), estimation (C), detection (D) and least squares (E) theory.

Figure 1.9 shows the relations between the chapters, and provides some implicit reading advice for readers only interested in some chapters.

The chapters can be approached in the following way. The preamble describes the core problem addressed in that chapter, the first section gives an introduction and often an illustrative example used throughout the chapter, while the last section contains a summary. The chapter summaries are divided into two parts: one theory summary and one software summary. The latter can be seen as a mini-manual to the sensor fusion part of Signals and Systems Lab.

Part I
Fusion in the Static Case

2

Linear Models

This chapter summarizes estimation and fusion principles for linear models. The general estimation theory in Appendix C, and in particular the least squares theory in Appendix E, is applied to a linear model. The problem and algorithms are illustrated with simple sensor network examples.

The basic batch formulation of a linear sensor model is

$$\mathbf{y} = \mathbf{H}x + \mathbf{e}. \tag{2.1}$$

The unknown state or parameter x has dimension n_x. This vector model can be related to N sensor readings in $\mathbf{y}$ (over time or space), each of dimension n_y:

$$y_k = H_k x + e_k, \quad k = 1, 2, \ldots, N. \tag{2.2}$$

Sequential models are indexed with k, and the boldface notation is throughout this part reserved for the batch model consisting of stacked measurements, noises and models. The index k is primarily thought of as the sensor index in this part, but it could also be considered a time index which is typically the case in the next part about filtering. For the estimation theory, though, it makes no difference if the observations come from different sensors or the same sensor at different times. The sensor network example, as motivated in Section 1.1, is used for numerical illustrations in this chapter. Here, all sensors are of the same kind (modality) measuring the position x of a target with an accuracy that depends on the sensor location.

As an alternate illustration, one could study the problem in Section 1.3. The parameter vector x could here include the range, width and height (the latter two represented as angles) of a tracked vehicle. The observations come primarily from two sensors of different modality. A radar delivers y_1 with range, width and coarse height, while the camera measures width and height but not range.

This chapter is organized as follows:

- Section 2.2 provides the formulas for how to estimate x from $\mathbf{y}$ using the least squares (LS) method. Both batch formulas suitable for centralized solutions and sequential forms for decentralized implementations are given.

- Section 2.3 discusses general sensor fusion formulas in a LS framework.

- Section 2.4 applies the maximum likelihood (ML) method, with a similar structure as for the LS method.

- Section 2.5 gives expressions for estimation bounds.

2.1 Introduction

The following example will be used as a motivating case study throughout this section.

---Example 2.1: Two sensors with good range resolution---

The location of two sensors S1 and S2 are known. Each one measures range and bearing to a target located at the unknown position $x = (x_1, x_2)^T$. We assume here that the range and bearing observations can be converted to Cartesian coordinates. The observation model is then

$$y_1 = x + e_1, \quad \text{Cov}(e_1) = R_1 \tag{2.3a}$$

$$y_2 = x + e_2, \quad \text{Cov}(e_2) = R_2 \tag{2.3b}$$

$$\mathbf{y} = \mathbf{H}x + \mathbf{e}, \quad \text{Cov}(\mathbf{e}) = \mathbf{R} \tag{2.3c}$$

$$\mathbf{H} = \begin{pmatrix} I \\ I \end{pmatrix} \tag{2.3d}$$

$$\mathbf{R} = \begin{pmatrix} R_1 & 0 \\ 0 & R_2 \end{pmatrix} \tag{2.3e}$$

The sensor fusion problem is here to estimate the position x as accurately as possible.

```
p1=[0;0];
p2=[2;0];
x0=[1;1];
Y1=ndist(x0,0.05*[1 -0.8;-0.8 1]);
Y2=ndist(x0,0.05*[1 0.8;0.8 1]);
y1=rand(Y1,20);
y2=rand(Y2,20);
plot2(Y1,Y2,'conf',90,'legend','off','markersize',18)
ans =
```

2.1 Introduction

```
   171.0089   172.0084   175.0084   176.0084
hold on
plot2(empdist(y1),empdist(y2),'conf',90,'legend','off','linewidth',2)
ans =
     []
axis([-0.5 2.5 -0.5 1.5])
hold on
plot(p1(1),p1(2),'*b',p2(1),p2(2),'*g','linewidth',2)
text(p1(1),p1(2),'S1')
text(p2(1),p2(2),'S2')
hold off
```

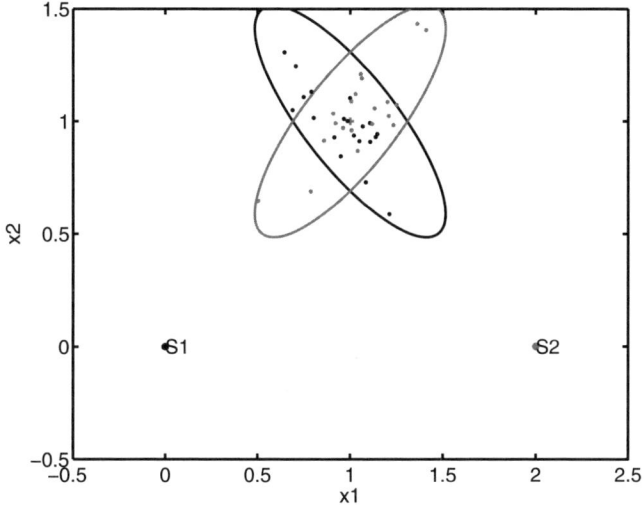

Figure 2.1: *Two sensors S1 and S2 measure a target with good range resolution but poor angle resolution. The dots illustrate random samples from the sensors, and the ellipsoids indicate an area where 90% of these measurements will occur.*

The following questions arise:

- How to merge information from many different sensors over space and time? This is the basic sensor fusion problem.

- How to communicate a sufficient representation of the sensor information? This leads to sequential algorithms.

- How to avoid double-counting information? That is, the information from one sensor may go around the network and return.

The questions above can all be addressed in the least squares framework.

2.2 Least Squares Approaches

This section provides the condensed results in Appendix E applied to the linear model.

2.2.1 Batchwise Least Squares

The *least squares (LS)* estimate is defined as the solution to the following optimization problem:

$$\hat{x}^{LS} = \arg\min_x V^{LS}(x), \tag{2.4a}$$

$$V^{LS}(x) = \sum_{k=1}^{N} (y_k - H_k x)^T (y_k - H_k x) = (\mathbf{y} - \mathbf{H}x)^T (\mathbf{y} - \mathbf{H}x). \tag{2.4b}$$

Here, $V^{LS}(x)$ is the LS loss function, and $\arg\min_x$ means the minimizing argument. Direct differentiation and setting the result to zero gives the estimate

$$\hat{x}^{LS} = \left(\sum_{k=1}^{N} H_k^T H_k\right)^{-1} \sum_{k=1}^{N} H_k^T y_k = \left(\mathbf{H}^T \mathbf{H}\right)^{-1} \mathbf{H}^T \mathbf{y}. \tag{2.5}$$

If $\text{Cov}(e_k) = R_k$ and $\mathbf{R} = \text{diag}(R_1, \ldots, R_N)$, then the *weighted least squares estimate (WLS)* is defined as the minimizing argument of the following loss function

$$\hat{x}^{WLS} = \arg\min_x V^{WLS}(x), \tag{2.6a}$$

$$V^{WLS}(x) = \sum_{k=1}^{N} (y_k - H_k x)^T R_k^{-1} (y_k - H_k x) \tag{2.6b}$$

$$= (\mathbf{y} - \mathbf{H}x)^T \mathbf{R}^{-1} (\mathbf{y} - \mathbf{H}x), \tag{2.6c}$$

with the solution

$$\hat{x}^{WLS} = \left(\sum_{k=1}^{N} H_k^T R_k^{-1} H_k\right)^{-1} \sum_{k=1}^{N} H_k^T R_k^{-1} y_k \tag{2.7a}$$

$$= \left(\mathbf{H}^T \mathbf{R}^{-1} \mathbf{H}\right)^{-1} \mathbf{H}^T \mathbf{R}^{-1} \mathbf{y}. \tag{2.7b}$$

If the data are indeed generated by the model for some true value x^o of the parameters,

$$\mathbf{y} = \mathbf{H}x^o + \mathbf{e}, \quad \text{Cov}(\mathbf{e}) = \mathbf{R}, \tag{2.8}$$

we have

$$\hat{x}^{LS} = x^o + \left(\mathbf{H}^T \mathbf{H}\right)^{-1} \mathbf{H}^T \mathbf{e}, \tag{2.9a}$$

$$\hat{x}^{WLS} = x^o + \left(\mathbf{H}^T \mathbf{R}^{-1} \mathbf{H}\right)^{-1} \mathbf{H}^T \mathbf{R}^{-1} \mathbf{e}. \tag{2.9b}$$

2.2 Least Squares Approaches

The estimates have mean

$$\mathrm{E}(\hat{x}^{LS}) = \mathrm{E}(\hat{x}^{WLS}) = x^o, \qquad (2.10)$$

and covariance

$$\mathrm{Cov}(\hat{x}^{LS}) = \left(\mathbf{H}^T\mathbf{H}\right)^{-1}\left(\mathbf{H}^T\mathbf{R}\mathbf{H}\right)\left(\mathbf{H}^T\mathbf{H}\right)^{-1} \triangleq P^{LS}, \qquad (2.11\mathrm{a})$$

$$\mathrm{Cov}(\hat{x}^{WLS}) = \left(\mathbf{H}^T\mathbf{R}^{-1}\mathbf{H}\right)^{-1} \triangleq P^{WLS}, \qquad (2.11\mathrm{b})$$

respectively. Thus, both estimates are *unbiased*. WLS is the *best linear unbiased estimator* (*BLUE*) for the linear problem, so we must have $P^{WLS} \leq P^{LS}$, which is perhaps not obvious from the algebraic expressions.

Further, if the distribution of the noise is assumed Gaussian, the estimates become Gaussian as well

$$\mathbf{e} \sim \mathcal{N}(0, \mathbf{R}) \Rightarrow \qquad (2.12\mathrm{a})$$

$$\hat{x}^{WLS} \sim \mathcal{N}(x^o, P^{WLS}), \qquad (2.12\mathrm{b})$$

$$\hat{x}^{LS} \sim \mathcal{N}(x^o, P^{LS}) \qquad (2.12\mathrm{c})$$

In the Gaussian case, WLS is also the *minimum variance estimator* (*MV*).

—— **Example 2.2: Two sensors with good range resolution** ——

Consider the setup in Example 2.1. First, a sensor model has to be defined. This is done by copying the information in the two Gaussian variables in Example 2.1. Then, one measurement is simulated (Example 2.1, 20 realizations were generated) and this is used to compute the LS and WLS estimate, respectively. The complete code for this is given below.

```
sm=sensormod('[x;x]',[2 0 4 0]);  % nx=2, ny=4
sm.x0=x0;
sm.pe=blkdiag(cov(Y1),cov(Y2));   % R=diag(R1,R2) Gaussian
y=simulate(sm,1);
xls=ls(sm,y);
xwls=wls(sm,y);
xplot2(xls,xwls,'conf',90,'linewidth',2)
```

Figure 2.2 illustrates the result. Note that the actual result is stochastic and depends on the noise realization in y_1, but the relative size of the ellipsoids reflect the accuracy.

2.2.2 Estimation of Noise Variance

For the case of unknown and constant noise variance $R_k = \lambda I_{n_y}$, that is $\mathbf{R} = \lambda I_{Nn_y}$, the LS loss function can be used to get an unbiased estimate,

$$\hat{\lambda} = \frac{1}{Nn_y - n_x} V^{LS}(\hat{x}^{LS}). \qquad (2.13)$$

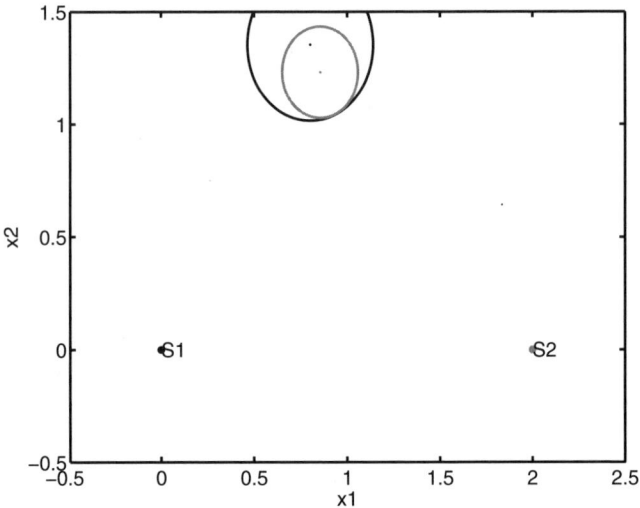

Figure 2.2: The LS and WLS estimates of the sensor information in Figure 2.1. The ellipsoids indicate a 90% confidence interval for the estimates from LS (black) and WLS (gray), respectively.

This follows from

$$V^{LS}(\hat{x}) = (\mathbf{y} - \mathbf{H}\hat{x}^{LS})^T(\mathbf{y} - \mathbf{H}\hat{x}^{LS}) \tag{2.14a}$$

$$= \left((\mathbf{H}x^o + \mathbf{e}) - \mathbf{H}(x^o + (\mathbf{H}^T\mathbf{H})^{-1}\mathbf{H}^T\mathbf{e})\right)^T$$
$$\times \left((\mathbf{H}x^o + \mathbf{e}) - \mathbf{H}(x^o + (\mathbf{H}^T\mathbf{H})^{-1}\mathbf{H}^T\mathbf{e})\right) \tag{2.14b}$$

$$= \mathbf{e}^T\left(I_{Nn_y} - \mathbf{H}\left(\mathbf{H}^T\mathbf{H}\right)^{-1}\mathbf{H}^T\right)\mathbf{e} \tag{2.14c}$$

$$= \operatorname{tr}\left(\mathbf{e}\mathbf{e}^T\left(I_{Nn_y} - \mathbf{H}\left(\mathbf{H}^T\mathbf{H}\right)^{-1}\mathbf{H}^T\right)\right). \tag{2.14d}$$

Here, the trace rule $\operatorname{tr}(AB) = \operatorname{tr}(BA)$ is used. Taking the expected value of both sides using $\mathbf{R} = \operatorname{E}(\mathbf{e}\mathbf{e}^T) = \lambda I_{Nn_y}$, we get

$$\operatorname{E}(V^{LS}(\hat{x})) = \lambda \operatorname{tr}\left(I_{Nn_y} - \mathbf{H}\left(\mathbf{H}^T\mathbf{H}\right)^{-1}\mathbf{H}^T\right) \tag{2.15a}$$

$$= \lambda\left(Nn_y - \operatorname{tr}\left(\mathbf{H}^T\mathbf{H}\left(\mathbf{H}^T\mathbf{H}\right)^{-1}\right)\right) \tag{2.15b}$$

$$= \lambda\left(Nn_y - \operatorname{tr} I_{n_x}\right) \tag{2.15c}$$

$$= \lambda\left(Nn_y - n_x\right). \tag{2.15d}$$

Here, the trace rule is used once more together with linear properties of the trace operator: $\operatorname{tr}(A + B) = \operatorname{tr}(A) + \operatorname{tr}(B)$ and $\operatorname{E}(\operatorname{tr}(A)) = \operatorname{tr}(\operatorname{E}(A))$. This shows that (2.13) is an unbiased estimate of λ.

2.2 Least Squares Approaches

A similar case occurs when the noise covariance contains an unknown scaling, so the true covariance is $\bar{R}_k = \lambda R_k$. The estimate (2.13) can then be generalized to

$$\hat{\lambda} = \frac{1}{Nn_y - n_x} V^{WLS}(\hat{x}^{WLS}). \tag{2.16}$$

2.2.3 Information Form of Least Squares

The WLS estimate can conveniently be computed as

$$\mathcal{I}_N = \sum_{k=1}^{N} H_k^T R_k^{-1} H_k = \mathbf{H}^T \mathbf{R}^{-1} \mathbf{H}, \tag{2.17a}$$

$$\iota_N = \sum_{k=1}^{N} H_k^T R_k^{-1} y_k = \mathbf{H}^T \mathbf{R}^{-1} \mathbf{y}, \tag{2.17b}$$

$$\hat{x}^{WLS} = \mathcal{I}_N^{-1} \iota_N. \tag{2.17c}$$

Here, ι_N is the *sufficient statistics* for the observations, which represents the minimal amount of data that is needed to form the estimate. This will also be referred to as the *information state*. Further, $\mathcal{I}_N$ denotes the *information matrix*, which is a measure of how much information that is available in the model. This will be referred to as the *information form of WLS*.

2.2.4 Sequential Least Squares

If the estimate is updated sequentially in space or time over the network, the information form (2.17) suggests one suitable sequential algorithm

$$\mathcal{I}_k = \mathcal{I}_{k-1} + H_k^T R_k^{-1} H_k, \tag{2.18a}$$

$$\iota_k = \iota_{k-1} + H_k^T R_k^{-1} y_k, \tag{2.18b}$$

$$\hat{x}_k = \mathcal{I}_k^{-1} \iota_k. \tag{2.18c}$$

This form is suitable for information messaging, where each sensor transmits its information state and matrix together with an identification number to all other nodes. However, when the parameter vector is large, the needed matrix inversion may cause unnecessary computational burden.

Appendix E.1 applies the *matrix inversion lemma* to get the following sequential form, initialized with $\hat{x}_0$ and P_0:

$$\hat{x}_k = \hat{x}_{k-1} + P_{k-1} H_k^T \left(H_k P_{k-1} H_k^T + R_k \right)^{-1} (y_k - H_k \hat{x}_{k-1}), \tag{2.19a}$$

$$P_k = P_{k-1} - P_{k-1} H_k^T \left(H_k P_{k-1} H_k^T + R_k \right)^{-1} H_k P_{k-1}. \tag{2.19b}$$

The least squares loss function can be computed sequentially using the following relation,

$$\min_x V(x) = \sum_{k=1}^{N} (y_k - H_k \hat{x}_N)^T R_k^{-1} (y_k - H_k \hat{x}_N) \tag{2.20a}$$

$$= \sum_{k=1}^{N} (y_k - H_k \hat{x}_{k-1})^T (H_k P_{k-1} H_k^T + R_k)^{-1} (y_k - H_k \hat{x}_{k-1})$$
$$- (\hat{x}_0 - \hat{x}_N)^T P_0^{-1} (\hat{x}_0 - \hat{x}_N). \tag{2.20b}$$

The sequential algorithm needs to be initialized in $\hat{x}_0$ and P_0, which gives rise to the last correction term in (2.20). Initialization is algebraically necessary to get a non-singular information matrix so the matrix inversion lemma can be applied.

2.2.5 Square-Root Implementations

We will here study the numerical problems to compute the inverse of $\mathbf{HH}^T$ in (2.5). The same arguments hold for the inverse of $\mathbf{HR}^{-1}\mathbf{H}^T$ in (2.7).

The condition number (ratio of largest and smallest singular value) of $\mathbf{HH}^T$ is the square of the condition number of $\mathbf{H}$, so the inversion $\mathbf{HH}^T$ may become ill-conditioned. One common solution (for instance implemented in the backslash operator in MATLAB™) is to compute the QR factorization of $\mathbf{H}$ as (note that R is not a covariance in (2.21) for this QR context)

$$\mathbf{H} = Q \begin{pmatrix} R \\ 0 \end{pmatrix} \tag{2.21a}$$

where Q is unitary $QQ^T = Q^T Q = I$ and R is upper triangular. We then get

$$\begin{pmatrix} \bar{\mathbf{y}} \\ \bar{\mathbf{e}} \end{pmatrix} \triangleq Q^T \mathbf{y} = Q^T (\mathbf{H}x + \mathbf{e}) \tag{2.21b}$$

$$= Q^T Q \begin{pmatrix} R \\ 0 \end{pmatrix} x + Q^T \mathbf{e} = \begin{pmatrix} R \\ 0 \end{pmatrix} x + Q^T \mathbf{e}. \tag{2.21c}$$

From this we get the triangular system of equations for the LS estimate

$$R\hat{x} = \bar{\mathbf{y}}, \tag{2.21d}$$

with the minimizing loss function

$$V(\hat{x}) = \bar{\mathbf{e}}^T \bar{\mathbf{e}}. \tag{2.21e}$$

This is one example from the class of *square-root algorithms*, where the numerical condition number is the square root of the straightforward implementation.

There are square root algorithms for the sequential problem as well, for instance the *Bierman's UD factorization algorithm*.

2.3 Fusion

This section describes how different data sets or estimates can be combined to yield one single estimate.

2.3.1 Fusion of Independent Measurements

Suppose, as in Example 2.1, that there are two direct measurements of the parameter available.

$$y_1 = x + e_1, \quad e_1 \sim \mathcal{N}(0, R_1), \tag{2.22a}$$
$$y_2 = x + e_2, \quad e_2 \sim \mathcal{N}(0, R_2). \tag{2.22b}$$

This is a fundamental sensor fusion problem. Suppose further to start with that the noises e_i are independent. Then, the sensor fusion problem can be seen as a special case of WLS. The WLS estimate (2.7) gives directly

$$\hat{x}^{WLS} = \left(R_1^{-1} + R_2^{-1}\right)^{-1}\left(R_1^{-1}y_1 + R_2^{-1}y_2\right). \tag{2.22c}$$

This will be referred to as the *sensor fusion formula*. Note that the LS estimate (2.5) gives the average $\hat{x}^{LS} = 0.5y_1 + 0.5y_2$ as the result in this case. A first application of the sensor fusion formula for computing the WLS estimate is given below.

---**Example 2.3: Two sensors with good range resolution**---

Consider the setup in Example 2.1. The WLS and LS estimates, respectively, are computed with

```
xhatwls=fusion(Y1,Y2);   % WLS
xhatls=0.5*Y1+0.5*Y2;    % LS
plot2(xhatwls,xhatls,'conf',90,'legend','off','linewidth',2)
ans =
  539.0020   540.0015   543.0015   544.0015
```

and illustrated in Figure 2.3. Note that `fusion` computes the WLS estimate, and the algebraic average corresponds to the least squares estimate. The WLS estimate corresponds to the smaller ellipsoid.

The example illustrates that the difference between LS and WLS can be quite significant. In the sequel, we will mostly just consider the WLS estimate because of its optimality properties (BLUE generally and MV for Gaussian noise), and commonly refer to it as just the least squares estimate, where the optimal data weighting is implicitly used.

2.3.2 Fusion of Independent Estimates

The fusion formula (2.22) assumes direct measurements $y_k = x + e_k$ of the parameters. In case of a linear model $y_k = H_k x + e_k$, each observation can

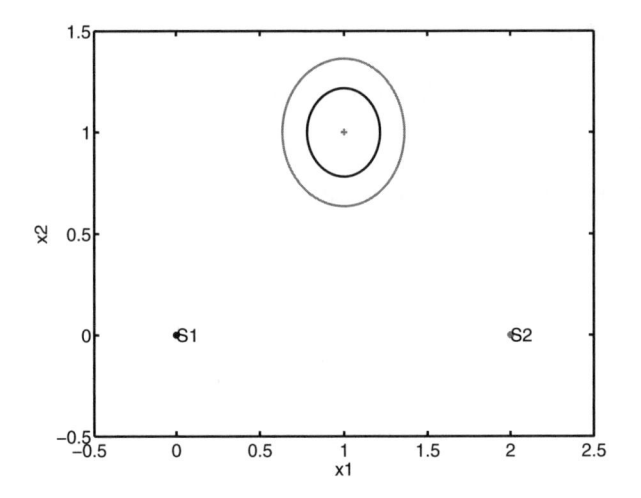

Figure 2.3: *The expected performance of LS and WLS estimates of the sensor information in Figure 2.1. Compare with the estimate from one realization in Figure 2.2. The ellipsoids indicate a 90% confidence interval for the estimates from LS (outer ellipsoid) and WLS (inner elipsoid), respectively.*

be converted to an *equivalent measurement* using a local WLS estimate $\hat{x}_k = (H_k^T R_k^{-1} H_k)^{-1} H_k^T R_k^{-1} y_k$ with the properties

$$\mathrm{E}(\hat{x}_1) = \mathrm{E}(\hat{x}_2) = x, \tag{2.23a}$$
$$\mathrm{Cov}(\hat{x}_1) = P_1, \tag{2.23b}$$
$$\mathrm{Cov}(\hat{x}_2) = P_2. \tag{2.23c}$$

The general *fusion formula* extends the special case in (2.22) to this case of two (or more) unbiased estimates with arbitrary covariance matrices.

Since the estimates can be interpreted as measurements in a linear model,

$$\hat{x}_1 = x + \tilde{x}_1, \quad \mathrm{Cov}(\tilde{x}_1) = P_1, \tag{2.23d}$$
$$\hat{x}_2 = x + \tilde{x}_2, \quad \mathrm{Cov}(\tilde{x}_2) = P_2, \tag{2.23e}$$

the WLS estimate (2.7) gives

$$P = \left(P_1^{-1} + P_2^{-1}\right)^{-1}, \tag{2.23f}$$
$$\hat{x} = P\left(P_1^{-1}\hat{x}_1 + P_2^{-1}\hat{x}_2\right). \tag{2.23g}$$

This formula requires that the two estimates are independent, the case with correlated estimated is discussed in Section 2.3.5.

2.3 Fusion

Example 2.4: Two sensors with good range resolution

Consider the setup in Example 2.1. Each sensor measures the target location, and thus has a local estimate. When one sensor receives another local estimate, it can fuse this using the fusion formula (2.23). In this case when $y_k = x + e_k$, the fusion formula (2.22) coincides with the general WLS fusion formula (2.23), and the result will be the same as in Figure 2.3.

2.3.3 Fusion on Information Form

The general fusion formula (2.23) is conveniently expressed in the information $\mathcal{I}_i = P_i^{-1}$ form as

$$\mathcal{I} = \mathcal{I}_1 + \mathcal{I}_2, \tag{2.24a}$$

$$\hat{x} = \mathcal{I}^{-1}\Big(\underbrace{\mathcal{I}_1 \hat{x}_1}_{\iota_1} + \underbrace{\mathcal{I}_2 \hat{x}_2}_{\iota_2}\Big) \tag{2.24b}$$

The information can thus be seen as a weighting of the estimates. The sensor fusion formula can be extended to more than two terms by recursive application of the formulas above.

Note that (2.24) holds even if the information matrices are rank deficient, if only their sum has full rank. This implies that fusion formula (2.23) applies even when the covariance is singular, if the inverse is replaced by pseudo-inverse.

Information can be used as a decision criterion. In some applications, the sensors can be controlled (directed to different targets, turned on and off, and so on), and this can be done with the objective to maximize the total information.

Another case is in sensor networks with communication constraints. Consider a sensor node that receives information from near-by sensors. It can use this information to decide whether the sensor's own information is worth transmitting, or if it is better to remain silent and save battery power or communication bandwidth. This is a local decision, which sometimes is referred to as *censoring sensors*, where each sensor censors its own information.

2.3.4 Defusion

The reverse operation of fusion can be called *defusion*, and it is sometimes needed to get rid of old information that is obsolete, or used multiple times to form a fused estimate. To remove the information $\mathcal{I}_2$ from $\mathcal{I}_1$, apply

$$\mathcal{I} = \mathcal{I}_1 - \mathcal{I}_2, \tag{2.25a}$$

$$\hat{x} = \mathcal{I}^{-1}\Big(\underbrace{\mathcal{I}_1 \hat{x}_1}_{\iota_1} - \underbrace{\mathcal{I}_2 \hat{x}_2}_{\iota_2}\Big) \tag{2.25b}$$

2.3.5 Safe Fusion of Dependent Estimates

A key problem in fusion is to avoid double counting information in observations. Consider for instance a sensor network, where all sensors communicate their observations in an ad-hoc network. Eventually there will be information loops.

One solution is to tag each observation with an identity number. Then, each user is responsible to compare this list with the own observations, and apply the fusion and defusion formula consequently. However, after a while this list might be too long in the end. That is, such a solution is not scalable.

An alternative solution in such a case is to apply a safe fusion formula outlined below. Consider the two ellipsoids in Figure 2.1 corresponding to P_1 and P_2. There are two immediate alternatives for how to choose P, leading to two different algorithms:

- The *covariance intersection algorithm* designs P such that it corresponds to the *smallest* ellipsoid *containing* the intersection of P_1 and P_2. This was the first approach proposed in Julier and Uhlmann (1997, 2001).

- The *largest ellipsoid algorithm* designs P such that it corresponds to the *largest* ellipsoid *inside* the intersection of P_1 and P_2. The algorithm is described in Benaskeur (2002); Zhou and Li (2008). It was called the *internal ellipsoid approximation* in Zhou and Li (2008).

Both algorithms share the appealing property that the same information can be included indefinitely many times without affecting the result, thus avoiding double-counting information. Generally, the covariance intersection algorithms give a more conservative covariance than the largest ellipsoid algorithm. We will focus on the former, and refer to it as *safe fusion*.

For the intuition, consider a scalar x where one wants to fuse two estimates $\hat{x}_1$ and $\hat{x}_2$ with variances $\text{Cov}(\hat{x}_1) = P_1$ and $\text{Cov}(\hat{x}_2) = P_2$, respectively. The only sensible thing to do if nothing is known about the underlying information structure, is to take the best one, that is, the one with the smallest P_i. In the multivariate case, the trick is to make a change of variable so that the components in the transformed state become independent for both estimates. This is done in Algorithm 2.1.

The algorithm is illustrated in Figure 2.4. Geometrically, the transformation converts the ellipsoid from S1 into a circle by the first SVD. The second SVD rotates the coordinate system so that the second ellipsoid becomes aligned to the coordinate axes, leaving the circle unaffected. Then the fifth step in the algorithm checks whether each semi-axis of the ellipsoid from S2 is larger than one or not.

Algorithm 2.1 Safe fusion

Given two unbiased estimates $\hat{x}_1$ and $\hat{x}_2$ with information $\mathcal{I}_1 = P_1^{-1}$ and $\mathcal{I}_2 = P_2^{-1}$ (pseudo-inverses if singular covariances), respectively. The estimate and covariance assuming worst case correlation $P_{12} = \mathrm{E}\big((x_1 - \hat{x}_1)(x_2 - \hat{x}_2)^T\big)$ are computed as follows:

1. SVD: $\mathcal{I}_1 = U_1 D_1 U_1^T$.

2. SVD: $D_1^{-1/2} U_1^T \mathcal{I}_2 U_1 D_1^{-1/2} = U_2 D_2 U_2^T$.

3. Transformation matrix: $T = U_2^T D_1^{1/2} U_1$.

4. State transformation: $\hat{\bar{x}}_1 = T\hat{x}_1$ and $\hat{\bar{x}}_2 = T\hat{x}_2$. The covariances of these are $\mathrm{Cov}(\hat{\bar{x}}_1) = I$ and $\mathrm{Cov}(\hat{\bar{x}}_2) = D_2^{-1}$, respectively.

5. For each component $i = 1, 2, \ldots, n_x$, let

$$\hat{\bar{x}}^i = \begin{cases} \hat{\bar{x}}_1^i, & \text{if } D_2^{ii} < 1, \\ \hat{\bar{x}}_2^i, & \text{if } D_2^{ii} \geq 1, \end{cases} \quad (2.26a)$$

$$\bar{D}^{ii} = \begin{cases} 1, & \text{if } D_2^{ii} < 1, \\ D_2^{ii}, & \text{if } D_2^{ii} \geq 1. \end{cases} \quad (2.26b)$$

where $\bar{D}$ is a diagonal matrix.

6. Inverse state transformation:

$$\hat{x} = T^{-1} \hat{\bar{x}}, \quad (2.26c)$$

$$P = T^{-1} \bar{D}^{-1} T^{-T} \quad (2.26d)$$

─── **Example 2.5: Two sensors with good range resolution** ───

Consider the setup in Example 2.1. Suppose that we transmit our observation in S1 to S2, who applies the fusion formula (X3 below), and then transmits this fused estimate back to us. If there is no list of information tags attached to this estimate, we can never know if our own information is already counted or not. If we apply the fusion formula (X4 below), the range uncertainty becomes too small. The safe fusion formula that avoids double-counting information (X5 below), however, gives a correct estimate in this case, as illustrated in Figure 2.5.

```
xhat12=fusion(Y1,Y2);
xhat121=fusion(Y1,xhat12);
xhatsafe=safefusion(Y1,xhat12);
plot2(xhat12,xhat121,xhatsafe,'conf',90,'legend','off','linewidth',2)
ans =
```

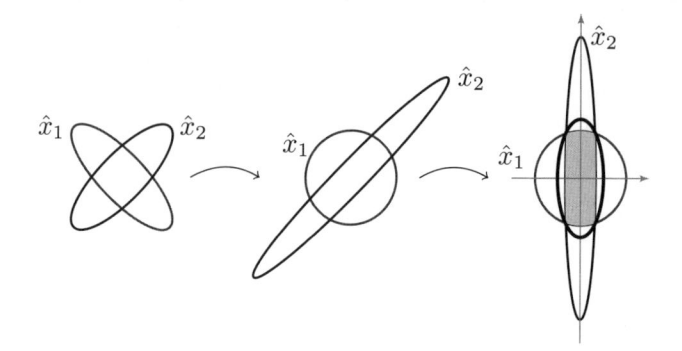

Figure 2.4: *Illustration of the steps in the covariance intersection fusion algorithm. First, both estimates are transformed such that one covariance becomes the identity matrix. Then, both transformed estimates are rotated so the other ellipsoid is aligned to the coordiate axes. The fusion covariance is now given by the ellipsoid that passes the intersections of the circle and ellipsoid. Finally, the result has to be rotated and transformed back to the original coordinates again (not illustrated in the figure). The largest ellipsoid algorithm gives a smaller ellipsoid, where the semi-axes are taken as the minimum of the other two ellipsoids' semi-axes.*

```
   722.0020    723.0015    726.0015    727.0015    728.0015    729.0015
axis([-0.5 2.5 -0.5 1.5])
hold on
plot(p1(1),p1(2),'*b',p2(1),p2(2),'*g','linewidth',2)
text(p1(1),p1(2),'S1')
text(p2(1),p2(2),'S2')
hold off
```

The figure shows that double counting the information from S1 gives a too optimistic covariance with respect to the distance to S1. The safe fusion algorithm, however, can include the same information infinitely many times without changing the shape of the covariance ellipsoid.

An important aspect is that the the safe fusion algorithm is not invariant to the order of incoming information. The following example illustrates this problem.

─── **Example 2.6: Processing order in safe fusion** ───

Suppose we have three different observations of the unknown x, as illustrated with the thin ellipsoids in Figure 2.6. We can now apply the safe fusion algorithm first to sensors (1,2), (1,3) or (2,3), respectively, and then to the remaining one. The perfect information processing algorithm should be invariant to this order. However, the safe fusion algorithm is not. The thick

2.3 Fusion

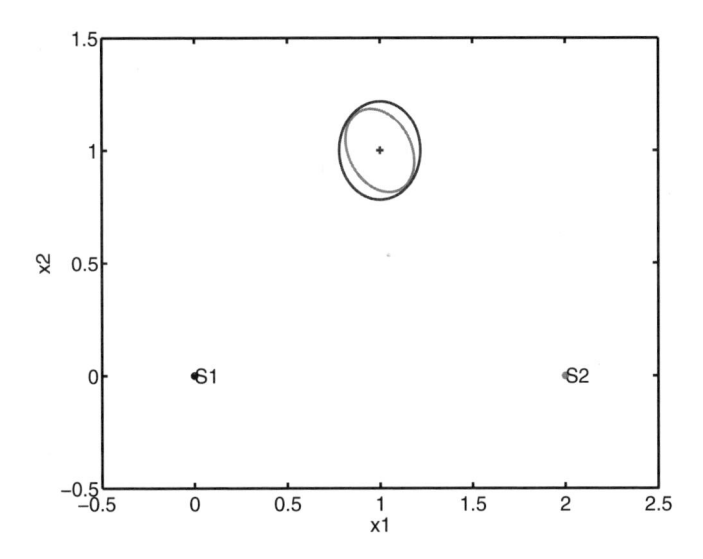

Figure 2.5: *The WLS estimate X3 (black), the estimate X4 (gray) obtained by double-counting the information in S1, and the result of safe fusion X5 (black) (which coincides with X3 from WLS here), using the sensor information in Figure 2.1. Here, X4 gives too small range uncertainty.*

ellipsoids in Figure 2.6 illustrate the different results. The code for this example is given below.

```
X1= ndist([1 ; 1],0.1*[1.5 0;0 3]);
X2= ndist([2 ; 2],0.1*[2 1; 1 2]);
X3= ndist([3 ; 3],0.1*[3 0; 0 1.5]);
Y1=safefusion(safefusion(X1,X2),X3);
Y2=safefusion(safefusion(X1,X3),X2);
Y3=safefusion(safefusion(X2,X3),X1);
plot2(Y1,Y2,Y3,'conf',90,'legend','off','linewidth',3)
ans =
   907.0020   908.0015   911.0015   912.0015   913.0015   914.0015
hold on
plot2(X1,X2,X3,'conf',90,'legend','off','linewidth',1)
ans =
   915.0015   916.0015   917.0015   918.0015   919.0015   920.0015
hold off
hold off
```

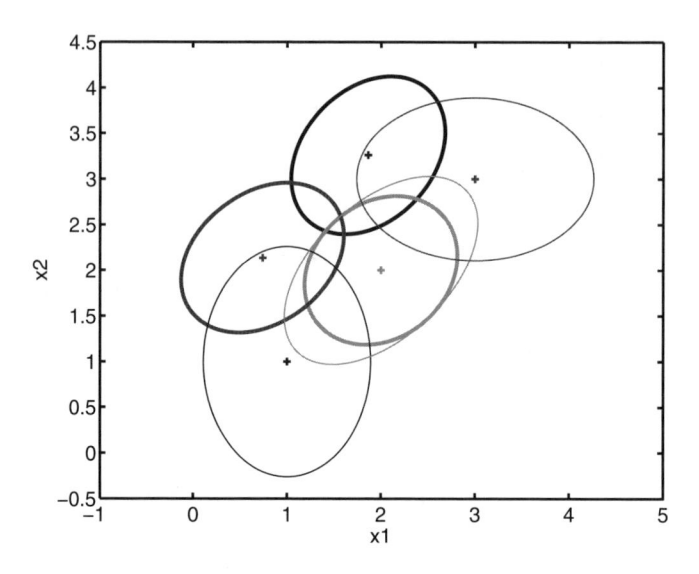

Figure 2.6: *Three possibly dependent observations of x, represented with thin ellipsoids. The thick ellipsoids represent the results of successive application of the safe fusion algorithm to the ordering (1,2,3), (1,3,2) and (2,3,1), respectively.*

2.4 The Maximum Likelihood Approach

The *likelihood* is the conditional probability density function $p(\mathbf{y}|x)$ given the parameter x. The *maximum likelihood* (ML) estimate is defined as

$$\hat{x}^{ML} = \arg\max_{x} p(\mathbf{y}|x). \tag{2.27}$$

It is a general property of the ML estimate that it is asymptotically Gaussian distributed

$$\sqrt{N}\big(\hat{x}^{ML} - x^o\big) \in \text{As}\mathcal{N}(0, P), \quad N \to \infty. \tag{2.28}$$

In a less formal way, this should be intepreted as $\hat{x}^{ML}$ is approximately distributed as $\mathcal{N}(x^o, P/N)$ for large N for some P that is independent of N.

We will use the notation $y_{1:N} = \mathbf{y}$ to clearly indicate the order of processing in the sequential algorithms. Only the case of Gaussian noise is treated below in order to get explicit results.

2.4.1 Batchwise Maximum Likelihood

For a linear Gaussian model, the likelihood is given by the Gaussian probability density function (PDF)

$$p(y_{1:N}|x) = \frac{1}{(2\pi)^{Nn_y/2}\prod_{k=1}^{N}\sqrt{\det(R_k)}} e^{-\frac{1}{2}\sum_{k=1}^{N}(y_k-H_kx)^T R_k^{-1}(y_k-H_kx)} \qquad (2.29a)$$

$$= \frac{1}{(2\pi)^{Nn_y/2}\prod_{k=1}^{N}\sqrt{\det(R_k)}} e^{-\frac{1}{2}V^{WLS}(x)}. \qquad (2.29b)$$

Maximizing the likelihood gives the same result as minimizing the negative log likelihood,

$$-2\log\big(p(y_{1:N}|x)\big) = Nn_y\log(2\pi) + \sum_{k=1}^{N}\log\big(\det(R_k)\big) + V^{WLS}(x). \qquad (2.30)$$

That is, the ML estimate coincides with the WLS estimate when the noise is Gaussian. Further, if there is a true value x^o of the parameters, then

$$\hat{x}^{ML} = \hat{x}^{WLS} \in \mathcal{N}(x^o, P). \qquad (2.31)$$

2.4.2 Inference for Fusion of Measurements

The fusion formula for two (sets of) measurements y_1 and y_2 in terms of the likelihood function is implied by Bayes formula

$$p(y_1, y_2|x) = p(y_2|y_1, x)p(y_1|x). \qquad (2.32)$$

This is in statistical theory referred to as *inference*. Note that this is not the same as multiplying the individual likelihood functions, since $p(y_1, y_2|x) \neq p(y_2|x)p(y_1|x)$.

2.4.3 Sequential Maximum Likelihood

Recursive application of Bayes' formula directly provides the sequential form of the likelihood function,

$$p(y_{1:N}|x) = p(y_1|x)\prod_{k=2}^{N} p(y_k|y_{1:k-1}, x). \qquad (2.33)$$

For Gaussian measurements, this can be expressed as

$$p(y_{1:N}) = \prod_{k=1}^{N} \mathcal{N}(y_k; H_k\hat{x}_{k-1}, H_k P_{k-1} H_k^T + R_k) \qquad (2.34)$$

using the sequential WLS algorithm to compute $\hat{x}_k$ and P_k starting with $\hat{x}_0$ and P_0. Here, $\mathcal{N}(x; \mu, P)$ denotes the Gaussian PDF. This shows that the sequential form comes as a direct parallel to the sequential form of WLS.

The maximum likelihood of data $p(y_{1:N}|\hat{x}^{ML}) = \max_x p(\mathbf{y}|x)$ is an important quantity in model selection and validation, change detection and diagnosis problems. However, as Theorem E.2 shows, the batch and sequential forms are related as

$$p(y_{1:N}|x) = \mathcal{N}(\hat{x}_N; x_0, P_0)\sqrt{\det(P_N)} \prod_{k=1}^{N} \mathcal{N}(y_k; H_k \hat{x}_N, R_k) \qquad (2.35a)$$

$$= \prod_{k=1}^{N} \mathcal{N}(y_k; H_k \hat{x}_{k-1}, H_k P_{k-1} H_k^T + R_k). \qquad (2.35b)$$

This shows how the initial conditions, which are necessary for applying the sequential form, influence the maximum of the likelihood function.

2.4.4 Marginalization of Noise Variance

We have seen above that the ML expressions are explicit and easily derived, and closely related to the WLS formulas. If the measurement noise is unknown, one has a much more challenging problem. For scalar measurements y_k, let the noise distribution be $\mathcal{N}(0, \lambda)$. As a more general problem formulation, let the multivariate measurement noise be distributed as $\mathcal{N}(0, \lambda R_k)$, where R_k is known and λ unknown. We will here apply the theory in Appendix C.7.1 for the parameter vector $x = (x, \lambda)$.

There are two conceptually different approaches:

1. Eliminate the nuisance parameters by estimation, which leads to maximization of a *generalized likelihood* and the *generalized maximum likelihood estimate* (*GML*):

$$\hat{x}^{GML} = \arg\max_x \max_\lambda p(y_{1:N}|x, \lambda) \qquad (2.36)$$

$$= \arg\max_x p(y_{1:N}|x, \hat{\lambda}(x)). \qquad (2.37)$$

For the linear Gaussian model, this does not affect the estimate, but the covariance has to be scaled with the minimized loss function as

$$\hat{\lambda} = \frac{1}{N n_y - n_x} V^{WLS}(\hat{x}^{WLS}), \qquad (2.38a)$$

$$P = \left(\mathbf{H}\mathbf{R}^{-1}\mathbf{H}^T\right)^{-1} \hat{\lambda}. \qquad (2.38b)$$

2. Eliminate the nuisance parameters by marginalization, which leads to maximization of a *marginalized likelihood* and the *marginalized maximum likelihood estimate* (*MML*):

$$\hat{x}^{MML} = \arg\max_x p(y_{1:N}|x) \qquad (2.39)$$

$$= \arg\max_x \int p(y_{1:N}|x, \lambda)p(\lambda|x)\, d\lambda \qquad (2.40)$$

The nuisance parameter λ is here considered to be stochastic with a prior distribution $p(\lambda|x)$ that may depend on the parameters x. The quite complicated but explicit formulas are given in Appendix E.5.

2.5 Cramér-Rao Lower Bound

This section summarizes the essential material in Section C.2. The *Fisher Information Matrix* (*FIM*) is for one observation y_k in the linear model defined as

$$\mathcal{I}_k(x) = \mathrm{E}\left[\left(\frac{d\log p(y_k|x)}{dx}\right)\left(\frac{d\log p(y_k|x)}{dx}\right)^T\right]. \qquad (2.41)$$

For Gaussian noise and a linear model, the information is easily shown to be

$$\mathcal{I}_k(x) = H_k^T R_k^{-1} H_k. \qquad (2.42)$$

This form coincides with the previously definition of information in Section 2.3.3. Information is additive for independent observations, so

$$\mathcal{I}_{1:N}(x) = \sum_{k=1}^N \mathcal{I}_k(x) = \sum_{k=1}^N H_k^T R_k^{-1} H_k, \qquad (2.43)$$

where the latter expression holds for a linear Gaussian model.

The *Cramér-Rao lower bound* (*CRLB*) states that any unbiased estimate must have a covariance matrix larger than or equal to the inverse FIM,

$$\mathrm{Cov}(\hat{x}) \geq \mathcal{I}_{1:N}^{-1}. \qquad (2.44)$$

2.5.1 Gaussian Case

For Gaussian noise, the FIM matrix is identical to the information matrix $\mathcal{I}_N$ defined earlier. That is, the asymptotic result (2.28) holds also in the finite sample case,

$$\hat{x}^{ML} \sim \mathcal{N}(x^o, \mathcal{I}^{-1}) = \mathcal{N}(x^o, P). \qquad (2.45)$$

The conclusion is that ML, and thus also WLS, is the MVU (minimum variance unbiased) estimate for the linear Gaussian model.

2.5.2 Non-Gaussian Case

For non-Gaussian noise, there might exist estimators that are nonlinear functions of data that outperform WLS, which still is the best linear combination of data (BLUE). The CRLB shows the potential benefit of using a nonlinear estimator.

In the scalar case $y_k = H_k x + e_k$ with i.i.d. (independent and identically distributed) measurement noise, the CRLB contains a correction factor that depends solely on the PDF $p_e(e_k)$ for e_k. Define the so called *intrinsic accuracy* of the noise distribution as the Fisher information for the problem to estimate the mean μ of the noise,

$$\mathcal{I}(\mu) = \mathrm{E}\left(\frac{d \log p_e(y|\mu)}{d\mu}\right)^2. \tag{2.46}$$

If only one observation is present, the most natural estimator $\hat{\mu} = y_1$ has variance $\mathrm{Var}(\hat{\mu}) = \mathrm{Var}(e) = R$. However, the CRLB says that $\mathrm{Var}(\hat{\mu}) = R \geq \mathcal{I}^{-1}(\mu)$, or expressed in another way

$$\mathcal{I}(\mu) \geq R^{-1} = \frac{1}{\mathrm{Var}(e)}, \tag{2.47}$$

We know that equality hold if e is Gaussian. Now, it can be shown that for any other distribution, the inequality is strict. A formal proof is found in Kay (1993); Lehmann (1991a).

The factor $\bigl(R\mathcal{I}(\mu)\bigr)^{-1} \geq 1$ can be defined as the *relative accuracy* for the noise distribution. The more non-Gaussian distribution, the larger relative accuracy and the larger potential benefits in using nonlinear estimators.

This conclusion carries over to the general linear model. It can be shown that

$$\mathrm{Cov}(\hat{x}) \geq \frac{P_N^{WLS}}{R\mathcal{I}(\mu)} \tag{2.48}$$

2.6 Summary

2.6.1 Theory

This chapter described methods to estimate the parameters x in the linear model

$$y_k = H_k x + e_k, \quad \text{Cov}(e_k) = R_k, \quad k = 1, \ldots, N,$$
$$\mathbf{y} = \mathbf{H}x + \mathbf{e}, \quad \text{Cov}(\mathbf{e}) = \mathbf{R}.$$

WLS minimizes the loss function

$$V^{WLS}(x) = \sum_{k=1}^{N}(y_k - H_k x)^T R_k^{-1}(y_k - H_k x) = (\mathbf{y} - \mathbf{H}x)^T \mathbf{R}^{-1}(\mathbf{y} - \mathbf{H}x).$$

The WLS solution is in batch form

$$\hat{x} = \left(\sum_{k=1}^{N} H_k^T R_k^{-1} H_k\right)^{-1} \sum_{k=1}^{N} H_k^T R_k^{-1} y_k = \left(\mathbf{H}^T \mathbf{R}^{-1} \mathbf{H}\right)^{-1} \mathbf{H}^T \mathbf{R}^{-1} \mathbf{y},$$
$$P = \left(\sum_{k=1}^{N} H_k^T R_k^{-1} H_k\right)^{-1} = \left(\mathbf{H}^T \mathbf{R}^{-1} \mathbf{H}\right)^{-1}.$$

and sequential form

$$\hat{x}_k = \hat{x}_{k-1} + P_{k-1} H_k^T \left(H_k P_{k-1} H_k^T + R_k\right)^{-1}(y_k - H_k \hat{x}_{k-1}),$$
$$P_k = P_{k-1} - P_{k-1} H_k^T \left(H_k P_{k-1} H_k^T + R_k\right)^{-1} H_k P_{k-1},$$

initialized with some $\hat{x}_0$ and P_0 that can be interpreted as a prior. The LS estimate is covered by the special case $R_k = I$ and $\mathbf{R} = I$, respectively. The LS estimate can be computed in a numerically more stable way by the backslash operator $\hat{x} = \mathbf{H} \backslash \mathbf{y}$, and the WLS estimate by $\hat{x} = \left(\mathbf{R}^{-1/2}\mathbf{H}\right) \backslash \left(\mathbf{R}^{-1/2}\mathbf{y}\right)$.

The minimizing loss function is given in batch and sequential form, respectively, by

$$\min_x V(x) = \sum_{k=1}^{N}(y_k - H_k \hat{x}_N)^T R_k^{-1}(y_k - H_k \hat{x}_N)$$
$$= \sum_{k=1}^{N}(y_k - H_k \hat{x}_{k-1})^T (H_k P_{k-1} H_k^T + R_k)^{-1}(y_k - H_k \hat{x}_{k-1})$$
$$- (\hat{x}_0 - \hat{x}_N)^T P_0^{-1}(\hat{x}_0 - \hat{x}_N).$$

If the scale of the noise distribution is unknown, so $R_k^o = \lambda R_k$, then the minimized loss function can be used to estimate the scale factor λ and compensate the covariance as

$$\hat{\lambda} = \frac{1}{Nn_y - n_x} V^{WLS}(\hat{x}^{WLS}),$$
$$P = \left(\mathbf{H}\mathbf{R}^{-1}\mathbf{H}^T\right)^{-1}\hat{\lambda}.$$

The sequential forms can be used here as well.

The fusion formula for two independent estimates is

$$\mathrm{E}(\hat{x}_1) = \mathrm{E}(\hat{x}_2) = x, \quad \mathrm{Cov}(\hat{x}_1) = P_1, \quad \mathrm{Cov}(\hat{x}_2) = P_2 \Rightarrow$$
$$\hat{x} = P\left(P_1^{-1}\hat{x}_1 + P_2^{-1}\hat{x}_2\right),$$
$$P = \left(P_1^{-1} + P_2^{-1}\right)^{-1}.$$

If the estimates are not independent, the safe fusion (or covariance intersection algorithm) provides a pessimistic lower bound accounting for worst case correlation.

2.6.2 Software

The following three objects relate to the theory in this chapter:

ndist is a class for Gaussian distributions, where many useful methods are over-loaded. This class can be used to represent the result of estimation as $x \sim \mathcal{N}(\hat{x}, P_x)$ for linear Gaussian problems.

sensormod defines a *sensor model*

$$y_k = Hx_k + e_k, \quad e_k \sim p_e(e)$$

where $p_e(e)$ is one object from the pdfclass family (see list(pdfclass)). Default is ndist(0,R).

sig defines a signal with fields y,t,u,x and representation of uncertainty in terms of covariances or Monte Carlo samples.

All objects store high-level information as descriptions and labels, which are inherited in subsequent processing and shown in display functions and plots.

The most important methods for these classes for linear estimation are

- X=ndist(xhat,P)

    - Transformations A1*X1+A2*X2+b.
    - Fusion X=fusion(X1,X2).
    - Safe fusion by upper bounding the covariance with covariance intersection X=safefusion(X1,X2).

2.6 Summary

- `ysig=sig(y)`, or `ysig=sig(y,t,u,x,Py,Px)`, converts a signal vector to a signal object.
    - Plot of signals `plot(ysig)`.
    - Plot of states `xplot(ysig)`.
    - Two-dimensional plot of states with confidence ellipsoids `xplot2(ysig,'conf',90)`.
- `s=sensormod(h,[nx nu ny nth])`; where `h` is a string or inline object in the variables `t,x,u,th`. The default values in the constructor are the simplest possible (deterministic models, zero initial parameter and so on). Use `s.x0=x0` to change the initial parameter, `s.px0=P0` to make it stochastic with covariance P_0, and `s.pe=R` to set the measurement noise to $\mathcal{N}(0,R)$. For the linear model, the simplest definition is `s=sensormod([mat2str(H),'*x'],[nx,0,ny,0]);`
    - Simulation `y=simulate(s,t)`, returns a SIG object `y` from a signal model object `s` and time vector `t`.
    - LS estimation `xhat=ls(s,y)`, returns a SIG object `xhat` from a signal model object `s` and signal object `y`.
    - WLS estimation `xhat=wls(s,y)`, returns a SIG object `xhat` from a signal model object `s` and signal object `y`.
    - ML estimation `xhat=ml(s,y)`, returns a SIG object `xhat` from a signal model object `s` and signal object `y`.

3

Nonlinear models

This chapter extends the results in Chapter 2 to nonlinear models. Primarily, the models are assumed to have explicit additive noise,

$$y_k = h_k(x) + e_k, \qquad (3.1a)$$
$$\mathbf{y} = \mathbf{h}(x) + \mathbf{e}. \qquad (3.1b)$$

The following approaches and cases are treated in this chapter:

- Section 3.2 discusses the nonlinear least squares (NLS) method, which minimizes $V(x) = (\mathbf{y} - \mathbf{h}(x))^T (\mathbf{y} - \mathbf{h}(x))$. This section also discusses the maximum likelihood estimate and its relation to NLS.

- Section 3.3 describes alternative approaches based on approximative observation models, as for instance the Taylor expansion $\mathbf{y} = \mathbf{h}(x) + \mathbf{e} \approx \mathbf{h}(\bar{x}) + \mathbf{H}(x - \bar{x}) + \mathbf{e}$.

- Section 3.4 applies the principle of inverse mappings $x = \mathbf{h}^{-1}(\mathbf{y} - \mathbf{e})$, when such an inverse exists.

- Section 3.5 describes a direct approach utilizing a formula from multivariate statistics.

- Section 3.6 discusses the important class of models $\mathbf{y} = \mathbf{h}^n(x^n) + \mathbf{h}^l(x^n)x^l + \mathbf{e}$ that are partly linear, partly nonlinear, in the parameter vector $x = (x^{n,T}, x^{l,T})^T$.

- The final Section 3.7 generalizes the model (3.1) to implicit measurement noise $\mathbf{h}(\mathbf{y}, x, \mathbf{e}) = 0$.

As in Chapter 2, the first Section 3.1 below defines an intuitive sensor network related example suitable for illustrating the main concepts.

3.1 Introduction

The example that will be used for analytic applications and simulations is a variation of the sensor network example in Example 2.1, but with a more realistic sensor model with independent range and bearing observations:

$$y = (r, \varphi)^T = h(x_1, x_2) + e, \qquad (3.2a)$$

$$r = \sqrt{x_1^2 + x_2^2} + e_r, \qquad (3.2b)$$

$$\varphi = \arctan2(x_1, x_2) + e_\varphi. \qquad (3.2c)$$

Here $\arctan2(x_1, x_2)$ denotes the four quadrant inverse tangent function. Note that (3.2) can be inverted to

$$x = h^{-1}(y), \qquad (3.3a)$$

$$x_1 = y_1 \cos(y_2), \qquad (3.3b)$$

$$x_2 = y_1 \sin(y_2). \qquad (3.3c)$$

Thus, there is a one to one relation between a noise-free measurement and the parameter x. This example will be used to illustrate all approaches in this chapter.

┌──── **Example 3.1: Ranging sensors: problem setup** ────┐

It is a standard assumption in radar theory that range and bearing are independent. We will in the simulations emphasize the bearing uncertainty. In this case, the problem falls into the category of ranging sensors, where the bearing information stems from the sensor direction sensitivity (antenna lobe for instance).

Both range and bearing errors are assumed Gaussian. To compare the measurements from the two sensors, a large number of observations are collected and transformed into Cartesian coordinates using the inverse transformation (3.3).

```
R1=ndist(100*sqrt(2),5);       % range distribution of sensor 1
Phi1=ndist(pi/4,0.1);          % bearing distribution of sensor 1
p1=[0;0];                      % position of sensor 1
hinv=inline('[p(1)+R*cos(Phi);p(2)+R*sin(Phi)]','R','Phi','p');
                               % inverse mapping
p2=[200;0];
R2=ndist(100*sqrt(2),5);
Phi2=ndist(3*pi/4,0.1);
```

Each sensor equation can be inverted and Monte Carlo samples of the measurements can be transformed by the inverse mapping to form an uncertainty region for each sensor.

```
xhat1=hinv(R1,Phi1,p1);
xhat2=hinv(R2,Phi2,p2);
plot2(xhat1,xhat2,'legend','')
ans =
    []
```

Figure 3.1 shows how the two "banana shaped" observation clouds intersect around the true target location at $x^o = (100, 100)^T$.

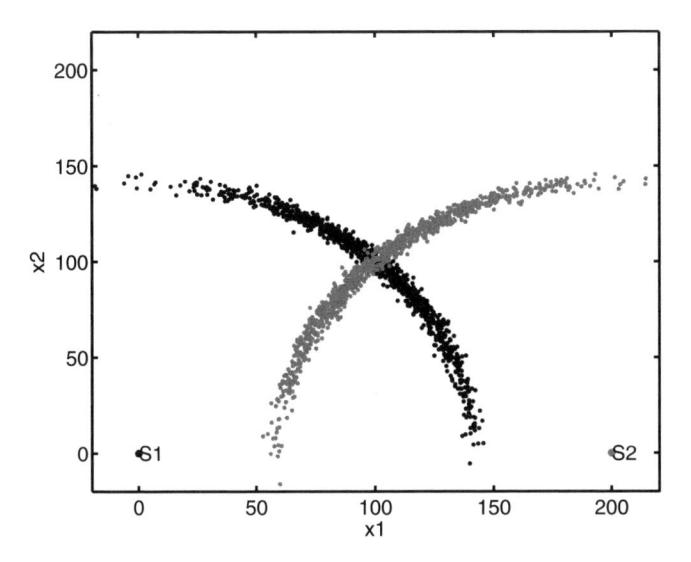

Figure 3.1: *Monte Carlo samples of two ranging sensors transformed to Cartesian coordinates.*

The example raises the natural question of how to fuse observations from these two sensors into one estimate with an associated covariance matrix (or, even better, the complete posterior distribution).

3.2 Nonlinear Least Squares

This section will give the general formulation of the *nonlinear least squares* (*NLS*) problem.

3.2.1 Scalar Observations

When the observations are scalar, the NLS cost function has the following form

$$V^{NLS}(x) = \frac{1}{2} \sum_{k=1}^{N} \varepsilon_k^2(x), \qquad (3.4)$$

where the residual $\varepsilon_k(x) = y_k - h_k(x)$ is a smooth function from $\mathbb{R}^{n_x}$ to $\mathbb{R}$, $x \in \mathbb{R}^{n_x}$ and $N \geq n_x$. Define the vector

$$\varepsilon(x) = \begin{pmatrix} \varepsilon_1(x) & \varepsilon_2(x) & \cdots & \varepsilon_N(x) \end{pmatrix}^T. \tag{3.5}$$

Using this notation (3.4) can be written

$$V^{NLS}(x) = \frac{1}{2} \|\varepsilon(x)\|_2^2 = \frac{1}{2} \varepsilon^T(x) \varepsilon(x). \tag{3.6}$$

The derivatives of $V^{NLS}(x)$ are collected in the Jacobian $J(x) \in \mathbb{R}^{n_x \times N}$, which is a matrix of first order partial derivatives of the residuals,

$$J(x) = \begin{pmatrix} \frac{\partial \varepsilon_1}{\partial x_1} & \frac{\partial \varepsilon_2}{\partial x_1} & \cdots & \frac{\partial \varepsilon_N}{\partial x_1} \\ \frac{\partial \varepsilon_1}{\partial x_2} & \frac{\partial \varepsilon_2}{\partial x_2} & \cdots & \frac{\partial \varepsilon_N}{\partial x_2} \\ \vdots & \vdots & \ddots & \vdots \\ \frac{\partial \varepsilon_1}{\partial x_{n_x}} & \frac{\partial \varepsilon_2}{\partial x_{n_x}} & \cdots & \frac{\partial \varepsilon_N}{\partial x_{n_x}} \end{pmatrix} = \frac{\partial \varepsilon^T(x)}{\partial x} = -\frac{\partial h^T(x)}{\partial x}.$$

Hence, the Jacobian is constructed by stacking the gradients of the residuals one after another. The gradient and Hessian of the cost function $V^{NLS}(x)$ are given as follows:

$$\frac{dV^{NLS}(x)}{dx} = \sum_{k=1}^{N} \varepsilon_k(x) \frac{d\varepsilon_k(x)}{dx} = J(x) \varepsilon(x), \tag{3.7a}$$

$$\frac{d^2 V^{NLS}(x)}{dx^2} = \sum_{k=1}^{N} \frac{d\varepsilon_k(x)}{dx} \left(\frac{d\varepsilon_k(x)}{dx} \right)^T + \sum_{k=1}^{N} \varepsilon_k(x) \frac{d^2 \varepsilon_k(x)}{dx^2}$$

$$= J(x) J^T(x) + \sum_{k=1}^{N} \varepsilon_k(x) \frac{d^2 \varepsilon_k(x)}{dx^2}. \tag{3.7b}$$

3.2.2 Multivariable Residuals

When the residuals are multivariable, the cost function takes the following form,

$$V^{NLS}(x) = \frac{1}{2} \sum_{k=1}^{N} \varepsilon_k^T(x) \varepsilon_k(x) = \frac{1}{2} \varepsilon^T(x) \varepsilon(x), \tag{3.8}$$

where

$$\varepsilon(x) = \begin{pmatrix} \varepsilon_1^T(x) & \varepsilon_2^T(x) & \cdots & \varepsilon_N^T(x) \end{pmatrix}^T, \tag{3.9}$$

which implies that the problem is in the same form as for the scalar case, where the total residual $\varepsilon(x)$ is constructed by stacking the individual residuals $\varepsilon_k(x)$

on top of each other. In other words, the problem is transformed into a scalar problem by applying the vectorisation operator to the individual residuals. The Jacobian $J(x) \in \mathbb{R}^{n_x \times N n_y}$ is given by,

$$J(x) = \frac{\partial \boldsymbol{\varepsilon}^T(x)}{\partial x} = \begin{pmatrix} \frac{\partial \varepsilon_{11}}{\partial x_1} & \frac{\partial \varepsilon_{12}}{\partial x_1} & \cdots & \frac{\partial \varepsilon_{Nn_y}}{\partial x_1} \\ \frac{\partial \varepsilon_{11}}{\partial x_2} & \frac{\partial \varepsilon_{12}}{\partial x_2} & \cdots & \frac{\partial \varepsilon_{Nn_y}}{\partial x_2} \\ \vdots & \vdots & \ddots & \vdots \\ \frac{\partial \varepsilon_{11}}{\partial x_{n_x}} & \frac{\partial \varepsilon_{12}}{\partial x_{n_x}} & \cdots & \frac{\partial \varepsilon_{Nn_y}}{\partial x_{n_x}} \end{pmatrix},$$

where the individual residuals are indexed in the following way

$$\varepsilon_k(x) = \begin{pmatrix} \varepsilon_{k1}(x) \\ \varepsilon_{k2}(x) \\ \vdots \\ \varepsilon_{kn_y}(x) \end{pmatrix}. \tag{3.10}$$

Using the objects introduced above the gradient and Hessian of the cost function (3.8) are given by

$$\frac{dV(x)}{dx} = \sum_{k=1}^{N} \sum_{i=1}^{n_y} \varepsilon_{ki}(x) \frac{d\varepsilon_{ki}(x)}{dx} = J(x)\boldsymbol{\varepsilon}(x), \tag{3.11a}$$

$$\frac{d^2 V(x)}{dx} = \sum_{k=1}^{N} \sum_{i=1}^{n_y} \frac{d\varepsilon_{ki}(x)}{dx} \left(\frac{d\varepsilon_{ti}(x)}{dx} \right)^T + \sum_{k=1}^{N} \sum_{i=1}^{n_y} \varepsilon_{ki}(x) \frac{d^2 \varepsilon_{ki}(x)}{dx^2}$$

$$= J(x)J^T(x) + \sum_{k=1}^{N} \sum_{i=1}^{n_y} \varepsilon_{ki}(x) \frac{d^2 \varepsilon_{ki}(x)}{dx^2}. \tag{3.11b}$$

3.2.3 Brief Algorithm Overview

Numerical optimization methods are usually iterative, where the parameter estimates are updated according to

$$\hat{x}^{(i+1)} = \hat{x}^{(i)} + \alpha^{(i)} f^{(i)}, \tag{3.12}$$

where $f^{(i)}$ is the search direction and $\alpha^{(i)}$ denotes the step length. Depending on how the search direction is obtained we can roughly group the algorithms into three categories:

- Methods using function values only.
- Methods using function values and gradients.
- Methods using function values, gradients, and Hessians.

A simple *Gauss-Newton algorithm* basically works as in Algorithm 3.1. The basic approximation here, is to neglect the second term in (3.11b). If the estimate candidate $\hat{x}^{(i)}$ is good, the residual $\varepsilon_k(\hat{x}^{(i)})$ should be close to e_k, and the central limit theorem indicates that this second term grows an order of magnitude slower than the first term, which is a quadratic form. For large residuals or small number of data, this might be a crude approximation, and there are other algorithms available.

Algorithm 3.1 Gauss-Newton

---

1. Given initial value $\hat{x}^{(0)}$, the function $\mathbf{h}(x)$ and its gradient $J(x) = -\frac{\partial \mathbf{h}^T(x)}{\partial x}$. Set $i = 0$.

2. Set $\alpha^{(i)} = 1$.

3. Compute
$$\hat{x}^{(i+1)} = \hat{x}^{(i)} + \alpha^{(i)} \left(J(x) J^T(x) \right)^{-1} J(x) \left(\mathbf{y} - \mathbf{h}(x) \right).$$

4. If the cost $V(\hat{x}^{(i+1)}) > V(\hat{x}^{(i)})$ has increased, set $\alpha^{(i)} = \alpha^{(i)}/2$ and repeat from step 3.

5. Terminate if the change in cost, the change in estimate, or the size of the gradient is small enough, or if the number of iterations has reached an upper limit.

6. Otherwise, set $i := i + 1$ and repeat from 2.

Algorithm 3.1 can be extended and generalized in the following ways:

- The symbolic form of the gradient $J(x)$ can be replaced by a numeric gradient $J(i,:)^T(x) \approx \left(\mathbf{h}(x + \epsilon e_i) - \mathbf{h}(x - \epsilon e_i) \right)/(2\epsilon)$, where e_i denotes the i'th column of the identity matrix. Here ϵ is a critical step size. It should be small enough to find the linear region of $\mathbf{h}(x)$, but still large enough to avoid numerical ill-conditioning.

- If the noise covariance $R_k = \text{Cov}(e_k)$ is known and not identical to the identity matrix, the *nonlinear weighted least squares* (*NWLS*) can be used by replacing step 3 with

$$\hat{x}^{(i+1)} = \hat{x}^{(i)} + \alpha^{(i)} \Big(\sum_{k=1}^{N} J_k(x) R_k^{-1} J_k^T(x) \Big)^{-1} \sum_{k=1}^{N} J_k(x) R_k^{-1} \left(y_k - h_k(x) \right).$$

3.2 Nonlinear Least Squares

It solves the optimization problem

$$\hat{x}^{NWLS} = \arg\min_x V^{NWLS}(x)$$

$$= \arg\min_x \frac{1}{2} \sum_{k=1}^{N} (y_k - h(x_k))^T R_k^{-1} (y_k - h(x_k)).$$

- The NWLS estimate corresponds to the maximum likelihood (ML) estimate in case of Gaussian measurement noise with covariance R_k. The likelihood function is

$$p(\mathbf{y}|x) = \frac{1}{(2\pi)^{Nn_y/2} \prod_{k=1}^{N} \sqrt{\det(R_k)}} e^{-\frac{1}{2} \sum_{k=1}^{N} (y_k - h_k(x))^T R_k^{-1} (y_k - h_k(x))}$$

$$= \frac{1}{(2\pi)^{Nn_y/2} \prod_{k=1}^{N} \sqrt{\det(R_k)}} e^{-\frac{1}{2} V^{NWLS}(x)}.$$

The Gaussian ML estimate is then

$$\hat{x}^{ML} = \arg\max_x p(\mathbf{y}|x) = \arg\min_x -2\log\bigl(p(\mathbf{y}|x)\bigr) = \arg\min_x V^{NWLS}(x).$$

Maximizing the Gaussian likelihood is thus equivalent to minimizing the NWLS cost function. That is, the NWLS and ML estimates coincide in the Gaussian case. This important fact is in accordance with the linear case in (2.29).

- In case the noise covariance $R_k = R_k(x)$ is also parametrized with the parameters x, one has to be a bit careful. The Gaussian ML estimate then also contains the following log det term

$$\hat{x}^{GML} = \arg\min_x \left(V^{NWLS}(x) + \frac{1}{2} \sum_{k=1}^{N} \log\bigl(\det\bigl(R_k(x)\bigr)\bigr) \right).$$

This extra term is in many cases essential to avoid degenerate solutions.

Next, the NLS algorithm is illustrated on an example.

─── **Example 3.2: NLS for a pair of TOA sensors** ───

Consider a simple *time of arrival* (*TOA*) network with two sensors as depicted in Figure 3.2(a). The network is created, illustrated and simulated below, and a two-dimensional likelihood function is over-laid:

```
th=[0.4 0.1 0.6 0.1]; x0=[0.5 0.5];   % Positions
s=exsensor('toa',2);                   % TOA sensor model
s.th=th; s.x0=x0;                      % Change defaults
s.pe=0.001*eye(2);                     % Noise variance
plot(s), hold on                       % Plot network
y=simulate(s,1);                       % Generate observations
lh2(s,y,[0:0.02:1],[0:0.02:1]);        % Likelihood function plot
```

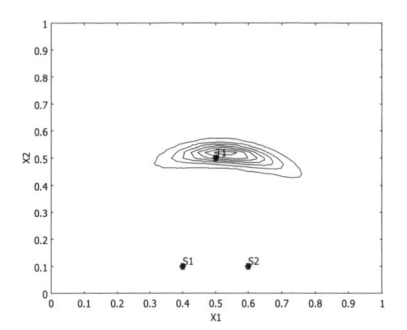

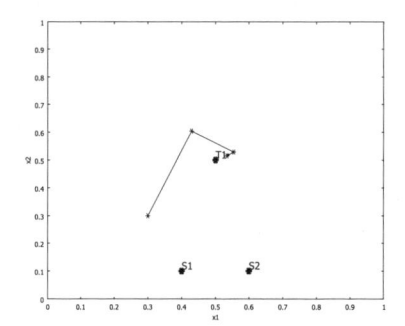

(a) The likelihood function assuming Gaussian measurement noise.

(b) The NLS estimate, where each iteration in the numerical Gauss-Newton search is marked.

Figure 3.2: *Illustrations of a network with one target and two range (TOA) sensors.*

Figure 3.2(a) illustrates how the fused likelihood from two circular likelihoods from each sensor forms a banana shape.

```
s0=s; s0.x0=[0.3;0.3];              % Prior model for estimation
[xhat,shat,res]=ml(s0,y);           % ML calls NLS
shat                                % Display estimated signal model
SIGMOD object: TOA (calibrated from data)
        / sqrt((x(1,:)-th(1)).^2+(x(2,:)-th(2)).^2) \
    y = \ sqrt((x(1,:)-th(3)).^2+(x(2,:)-th(4)).^2) / + e
    x0' =  [0.54      0.52] + N(0,[3.2e-013,2.3e-013;2.3e-013,1.7e-013])
    th' =  [0.4       0.1       0.6       0.1]

xplot2(xhat,'conf',90)              % Estimate and covariance plot
plot(res.TH(1,:),res.TH(2,:),'*-')  % Estimate for each iteration
```

Figure 3.2(b) shows how the Gauss-Newton method quite quickly converges from the initial guess to the most likely position given the measurement. Note that the ML estimator cannot find the true position from just one observation.

3.3 Linearizing the Measurement Equation

The basic idea in this section is to replace the nonlinear function $\mathbf{y} = \mathbf{h}(x)$ with one of the following approximations from Section A.3:

TT1 First order Taylor transformation leading to Gauss' approximation formula.

TT2 Second order Taylor expansion, which compensates the mean and covariance with the quadratic second order term.

3.3 Linearizing the Measurement Equation

UT The unscented transformation as described in Julier and Uhlmann (2004). Both versions in Table A.7 are considered.

MCT The Monte Carlo transformation approach, which in the limit should compute correct moments.

These methods will be briefly reviewed below, and the WLS framework is then applied to the approximate models.

3.3.1 Linearized Model

A first order Taylor expansion of a scalar-valued measurement equation as a function of a vector-valued parameter x around a point $\bar{x}$ is given by

$$y_k = h_k(x) + e_k \qquad (3.14a)$$

$$= h_k(\bar{x}) + h'_k(\bar{x})(x-\bar{x}) + \frac{1}{2}\underbrace{(x-\bar{x})^T h''_k(\xi)(x-\bar{x})}_{r_{k,2}(\xi)} + e_k. \qquad (3.14b)$$

Here, the rest term $r_{k,2}(\xi)$ is evaluated at a point ξ in the neighborhood of $\bar{x}$ to get equality.

It is implicitly assumed in (3.14) that the measurement y_k is scalar. For a vector-valued observation, the second derivative of $h(x)$ becomes a three-dimensional tensor. To avoid further notation and complex book-keeping, each component of the observation can be treated independently. This leads to the following component-wise Taylor expansion

$$y_k^i = h_k^i(x) + e_k^i \qquad (3.15a)$$

$$= h_k^i(\bar{x}) + (h_k^i)'(\bar{x})(x-\bar{x}) + \frac{1}{2}\underbrace{(x-\bar{x})^T (h_k^i)''(\xi)(x-\bar{x})}_{r_{k,2}^i(\xi)} + e_k^i. \qquad (3.15b)$$

---**Example 3.3: Ranging sensor: TT1**---

For the sensor model (3.2), we get (using $\frac{d \arctan(z)}{dz} = \frac{1}{1+z^2}$ and letting $r = \sqrt{x_1^2 + x_2^2}$ denote the distance)

$$\frac{dh(x)}{dx} = \begin{pmatrix} \frac{dr(x)}{dx_1} & \frac{dr(x)}{dx_2} \\ \frac{d\varphi(x)}{dx_1} & \frac{d\varphi(x)}{dx_2} \end{pmatrix} = \begin{pmatrix} \frac{x_1}{\sqrt{x_1^2+x_2^2}} & \frac{x_2}{\sqrt{x_1^2+x_2^2}} \\ \frac{-x_2/x_1^2}{1+x_2^2/x_1^2} & \frac{1/x_1}{1+x_2^2/x_1^2} \end{pmatrix} \qquad (3.16a)$$

$$= \begin{pmatrix} \frac{x_1}{\sqrt{x_1^2+x_2^2}} & \frac{x_2}{\sqrt{x_1^2+x_2^2}} \\ \frac{-x_2}{x_1^2+x_2^2} & \frac{x_1}{x_1^2+x_2^2} \end{pmatrix} = \begin{pmatrix} \frac{x_1}{r} & \frac{x_2}{r} \\ \frac{-x_2}{r^2} & \frac{x_1}{r^2} \end{pmatrix}. \qquad (3.16b)$$

We also get

$$\frac{d^2 h^1(x)}{dx^2} = \frac{1}{r^{3/2}} \begin{pmatrix} r + 2x_1^2 & 2x_2 x_1 \\ 2x_2 x_1 & r + 2x_2^2 \end{pmatrix}, \qquad (3.17a)$$

$$\frac{d^2 h^2(x)}{dx^2} = \frac{1}{r^4} \begin{pmatrix} 2x_1 x_2 & -r^2 - 2x_1 x_2 \\ -r^2 + 2x_1 x_2 & -2x_1 x_2 \end{pmatrix}. \qquad (3.17b)$$

Thus, the rest term vanishes faster than the linear term with distance r.

The TT1 approach just neglects the rest term $r_{k,2}(\xi)$, and applies LS or WLS estimation formulas to the re-formulated model $\bar{y}_k = H_k x + e_k$, where

$$\bar{y}_k = y_k - h_k(\bar{x}) + h_k'(\bar{x})\bar{x}, \qquad (3.18a)$$

$$H_k = h_k'(\bar{x}) = \left.\frac{dh_k(x)}{dx}\right|_{x=\bar{x}}. \qquad (3.18b)$$

The WLS estimate is then given by

$$\hat{x}^{WLS} = \left(\sum_{k=1}^{N} H_k^T R_k^{-1} H_k\right)^{-1} \sum_{k=1}^{N} H_k^T R_k^{-1} \bar{y}_k. \qquad (3.19)$$

The example below illustrates the method.

Example 3.4: Ranging sensors: WLS on linearized model

The setup is the same as in the preceding examples. This time 10 observations from each sensor are generated.

```
N=10;
R1=ndist(100*sqrt(2),5);   R2=ndist(100*sqrt(2),5);   % Range distribution
Phi1=ndist(pi/4,0.1);      Phi2=ndist(3*pi/4,0.1);    % Bearing distribution
p1=[0;0];                  p2=[200;0];                % Position of sensor
Y1=[R1;Phi1];              Y2=[R2;Phi2];              % Random variable
y1=rand(Y1,N);             y2=rand(Y2,N);             % 10 samples

% Vectorize measurements (rand returns (N,nx) matrix)
y1vec=y1';
y1vec=y1vec(:);
y2vec=y2';
y2vec=y2vec(:);
```

The function $h(y)$ and its gradient are defined as inline objects, and the linearization point is chosen as $\bar{x} = (100, 90)^T$, which is the (biased) estimate from the Monte Carlo approach.

```
% Define h(x) and J(x)
hstr='[sqrt((x(1,:)-p(1)).^2+(x(2,:)-p(2)).^2);
      atan2((x(2,:)-p(2)),(x(1,:)-p(1)))]';
h=inline(hstr,'x','p');
dhstr11='[(x(1,:)-p(1))./sqrt((x(1,:)-p(1)).^2+(x(2,:)-p(2)).^2),';
dhstr12=' (x(2,:)-p(2))./sqrt((x(1,:)-p(1)).^2+(x(2,:)-p(2)).^2);';
```

3.3 Linearizing the Measurement Equation

```
dhstr21=' -(x(2,:)-p(2))./((x(1,:)-p(1)).^2+(x(2,:)-p(2)).^2),';
dhstr22=' (x(1,:)-p(1))./((x(1,:)-p(1)).^2+(x(2,:)-p(2)).^2)]';
dh=inline([dhstr11 dhstr12 dhstr21 dhstr22],'x','p');
% Linearization point
xbar=[100;90];
```

The approximate linear model (3.18) and its corresponding WLS estimate in (3.19) are computed:

```
% Approximate linear model
y1bar=y1vec-repmat(h(xbar,p1)-dh(xbar,p1)*xbar,N,1);
H1=repmat(dh(xbar,p1),N,1);
y2bar=y2vec-repmat(h(xbar,p2)-dh(xbar,p2)*xbar,N,1);
H2=repmat(dh(xbar,p2),N,1);
ybar=[y1bar;y2bar];
H=[H1;H2];
% WLS estimate
xhat=[H1;H2]\[y1bar;y2bar]
xhat =
   100.3372
   101.1532
```

The bias has significantly decreased compared to the linearization point.

3.3.2 Iterated Refinement

A natural idea is to continue linearization around the new, hopefully better, estimate. That is, applying (3.18) and (3.19) in an iterative manner. This is exactly the main idea in one version of the nonlinear least squares algorithm in Section 3.2. Start with some arbitary but decent initial guess $\bar{x}^i$, and iterate

$$\bar{x}^{i+1} = \left(\sum_{k=1}^{N} h'_k(\bar{x}^i)^T R_k^{-1} h'_k(\bar{x}^i)\right)^{-1} \sum_{k=1}^{N} h'_k(\bar{x}^i)^T R_k^{-1} \left(y_k - h_k(\bar{x}^i) + h'_k(\bar{x}^i)\bar{x}^i\right), \quad (3.20)$$

until convergence or sufficient accuracy is obtained. One should be aware that such iterative schemes can diverge due to bad initialization.

The example below illustrates how the estimate successively improves over the iterations.

---- **Example 3.5: Ranging sensors: WLS on linearized model** ----

Using the same data as in Example 3.4, the Taylor linearization model (3.18) and its corresponding LS estimate in (3.19) are computed iteratively below, until the estimate does not change significantly.

```
xbar=[0;0];
xhat=[1;1];
iter=0;
while norm(xhat-xbar)>1e-4
    iter=iter+1;
    xhat=xbar;
```

```
    y1bar=y1vec-repmat(h(xbar,p1)-dh(xbar,p1)*xbar,N,1);
    H1=repmat(dh(xbar,p1),N,1);
    y2bar=y2vec-repmat(h(xbar,p2)-dh(xbar,p2)*xbar,N,1);
    H2=repmat(dh(xbar,p2),N,1);
    ybar=[y1bar;y2bar];
    H=[H1;H2];
    xbar=H\ybar;
end
xhat =
   99.5302
  100.4109
iter =
17
```

After 17 iterations the returned estimate is very accurate.

3.3.3 Bias Compensation

Consider again the Taylor expansion (3.14) with a rest term, now in a Bayesian setting where x is a stochastic variable with covariance P. The WLS estimate in (3.19) then gets a bias from the neglected rest term. For a scalar measurement, using an arbitrary linearization point $\bar{x}$, we get

$$\mathrm{E}(\hat{x}^{WLS}) = x^o +$$

$$\frac{1}{2}\left(\sum_{k=1}^{N} H_k^T R_k^{-1} H_k\right)^{-1} \sum_{k=1}^{N} H_k^T R_k^{-1} \mathrm{E}\left((x-\bar{x})^T h_k''(\xi)(x-\bar{x})\right). \quad (3.21)$$

The expectation can be simplified using the trace rule $\mathrm{tr}(AB) = \mathrm{tr}(BA)$,

$$\mathrm{E}\left((x-\bar{x})^T h_k''(\xi)(x-\bar{x})\right) \qquad (3.22a)$$
$$= \mathrm{E}\,\mathrm{tr}\left((x-\bar{x})(x-\bar{x})^T h_k''(\xi)\right) \qquad (3.22b)$$
$$= \sum_{k=1}^{N} \mathrm{tr}\left(\left[P + (\mathrm{E}(x)-\bar{x})(\mathrm{E}(x)-\bar{x})^T\right] h_k''(\xi)\right). \qquad (3.22c)$$

Here, the vector version of the well-known formula

$$\mathrm{E}\left((x-\bar{x})^2\right) = \mathrm{E}\left((x - \mathrm{E}(x) + \mathrm{E}(x) - \bar{x})^2\right)$$
$$= \mathrm{E}\left((x-\mathrm{E}(x))^2\right) + (\mathrm{E}(x)-\bar{x})^2) = \mathrm{Var}(x) + \left(\mathrm{E}(x)-\bar{x}\right)^2$$

is used. This bias cannot be computed exactly, since the rest term is evaluated in some ξ close to $\bar{x}$ but otherwise it is unknown. Still, a *bias compensated LS* solution is possible if the following conditions are satisfied:

- The second order Taylor term is approximately constant in the neighborhood of $\bar{x}$. The size of this neighborhood should be compared to the uncertainty reflected in P. Note that for models that can be expressed as quadratic forms $h_k(x) = H_k x + x^T G_k x$, the second order rest term is constant, so $h_k''(\xi) = G_k$.

- The linearization point is chosen as the mean of the prior $\bar{x} = \mathrm{E}(x)$, so the term $\mathrm{E}\big((\mathrm{E}(x) - \bar{x})^2\big)$ disappears.

In these cases, the bias of the measurement is known and can be compensated for. The bias compensated WLS estimate is thus

$$\hat{x}^{WLS} = \left(\sum_{k=1}^{N} H_k^T R_k^{-1} H_k\right)^{-1} \sum_{k=1}^{N} H_k^T R_k^{-1} \left(\bar{y}_k - \frac{1}{2}\mathrm{tr}\left(Ph_k''(\xi)\right)\right). \quad (3.23)$$

3.3.4 Variance and Bias Compensation

The bias compensation in the previous section takes care of the bias contribution, but does not compensate for the extra variance contribution from the stochastic (in the Bayesian setting) rest term. To include also this, more assumptions are needed to get an analytic expression. For a Gaussian prior distribution with zero mean, the *fourth order moment* can be computed using the following formula:

$$\mathrm{E}(x_1 x_2 x_3 x_4) = \mathrm{E}(x_1 x_2)\mathrm{E}(x_3 x_4) + \mathrm{E}(x_1 x_3)\mathrm{E}(x_2 x_4) + \mathrm{E}(x_1 x_4)\mathrm{E}(x_2 x_3). \quad (3.24)$$

For the case $x \in \mathcal{N}(0, \sigma^2)$, we have $\mathrm{E}(x^4) = 3\sigma^4$ (see Table A.1), $\mathrm{Var}(x^2) = \mathrm{E}(x^4) - \big(\mathrm{E}(x^2)\big)^2 = 2\sigma^4$, and

$$\mathrm{Cov}(r_{k,2}(\xi)) = \big(h_k''(\xi)\big)^2 2\sigma^4. \quad (3.25)$$

Section A.3 derives the formula $\mathrm{tr}\big(h_k''(\xi) P h_k''(\xi) P\big)$ in the multivariate case $x \sim \mathcal{N}(0, P)$. The result is the bias and variance compensated WLS estimate

$$\hat{x}^{WLS} = \left(\sum_{k=1}^{N} \frac{H_k^T H_k}{R_k + \frac{1}{2}\mathrm{tr}\big(h_k''(\xi) P h_k''(\xi) P\big)}\right)^{-1} \sum_{k=1}^{N} \frac{H_k^T \big(\bar{y}_k - \frac{1}{2}\mathrm{tr}\big(Ph_k''(\xi)\big)\big)}{R_k + \frac{1}{2}\mathrm{tr}\big(h_k''(\xi) P h_k''(\xi) P\big)}. \quad (3.26)$$

3.4 Inversion of the Measurement Equation

The ideal inverse mapping would be

$$\mathbf{y} = \mathbf{h}(x) + \mathbf{e} \leftrightarrow x = \mathbf{h}^{-1}(\mathbf{y} - \mathbf{e}) \stackrel{\triangle}{=} \mathbf{g}(\mathbf{y} - \mathbf{e}). \quad (3.27)$$

In the special case of additive noise $\mathbf{x} = \mathbf{g}(\mathbf{y}) - \mathbf{e}$, the methods in Chapter 2 could be applied directly. However, such a mapping from a high-dimensional measurement vector to a low-dimensional parameter is usually not explicit. What one can hope for is that each sensor observation can be inverted as

$$y_k = h_k(x) + e_k \leftrightarrow x = h_k^{-1}(y_k - e_k) \triangleq g_k(y_k - e_k). \tag{3.28}$$

Compare with the range and bearing sensor in (3.3). Each range and bearing can be inverted to a position rather straightforwardly, but if a whole set of range–bearing pairs is available, it is not easy to see how the estimation problem can be circumvented by a function inversion. In the sequel, we will assume that the observation model is given by the mapping

$$x = g_k(y_k - e_k) \approx g_k(y_k) + \bar{e}_k, \tag{3.29}$$

where the latter expression approximates the noise contribution with an additive term. Given this approximation, the mean and covariance of x can then be computed from each sensor observation, the fusion formula (2.23) provides a general tool for sensor fusion. More explicitely, (2.23) here becomes

$$P_k = \mathrm{Cov}(\bar{e}_k), \tag{3.30a}$$

$$P = \left(\sum_{k=1}^{N} P_k^{-1}\right)^{-1}, \tag{3.30b}$$

$$\hat{x} = P \sum_{k=1}^{N} P_k^{-1} g_k(y_k). \tag{3.30c}$$

Note that the fusion formula (3.30) is optimal in the minimum variance (MV) sense only when $\bar{e}_k$ happens to be Gaussian, otherwise it provides the best linear unbiased estimator (BLUE). Now when the goal with the inverse mapping is explained, the sensor index k will be dropped, and the focus is on how to construct the inverse mapping of one relation $y = h(x) + e$.

The following alternatives for how to compute the mean $\mathrm{E}(x) = \mathrm{E}(g(y))$ and covariance $\mathrm{Cov}(x) = \mathrm{Cov}(g(y))$ will be covered:

- Analytic approximations.

- Numeric approximations based on the general techniques in Appendix A.3 of nonlinear transformations $z = g(x)$.

    - Monte Carlo transformation (MCT), where a large number of random numbers propagated through the inverse mapping $x = g(y)$, and the mean and covariance are estimated from these samples.
    - Approximations based on first (TT1) and second (TT2) order Taylor expansion of $x = g(y)$.
    - Approximations based on the *unscented transform* (*UT*) of $x = g(y)$, where $2n_x + 1$ so called sigma points are propagated through the mapping $x = g(y)$, and the mean and covariance are estimated from these samples.

3.4.1 Analytic Approaches

Rather straightforward calculations on the example (3.3) using trigonometric relations show that the covariance matrix $\text{Cov}(x)$ in (3.3) can be neatly approximated as

$$\text{Cov}(x) = \frac{\sigma_r^2 - r^2 \sigma_\varphi^2}{2} \begin{pmatrix} b + \cos(2\varphi) & \sin(2\varphi) \\ \sin(2\varphi) & b - \cos(2\varphi) \end{pmatrix} \quad (3.31\text{a})$$

$$b = \frac{\sigma_r^2 + r^2 \sigma_\varphi^2}{\sigma_r^2 - r^2 \sigma_\varphi^2}. \quad (3.31\text{b})$$

This approximation is accurate for $r\sigma_\varphi^2/\sigma_r < 0.4$ and $\sigma_\varphi < 0.4$, as analyzed in Li and Bar-Shalom (1993), which is normally the case in radar applications. However, it does not hold in Example 3.1 where $r\sigma_\varphi^2/\sigma_r = 100\sqrt{2} \cdot 0.1/\sqrt{5} \approx 6.3$.

This approach can be tried on a case by case basis.

3.4.2 Monte Carlo Sampling

The Monte Carlo approach is quite generally applicable. The idea is to generate a large number of samples $y^{(i)}$ from the distribution of y, and propagate these through $g(y)$ to get the samples $x^{(i)}$. These samples might be real observations, or a true value with random samples of the noise e added. The mean and covariance are then approximated using standard formulas

$$\mu_x = \frac{1}{N} \sum_{i=1}^{N} x^{(i)}, \quad (3.32\text{a})$$

$$P_x = \frac{1}{N-1} \sum_{i=1}^{N} \left(x^{(i)} - \mu_x\right)\left(x^{(i)} - \mu_x\right)^T. \quad (3.32\text{b})$$

The normalization with $N-1$ in the covariance formula gives an unbiased estimate, while normalizing with N is also common.

---Example 3.6: Ranging sensors: MCT on inverse model---

A large number of samples from the distribution of y are generated and mapped to samples of x as shown in Figure 3.1. The approximations in (3.32) are then applied to each sensor observation, and the results are fused using (3.30).

```
V1=[R1;Phi1]
N([141;0.785],[5,0;0,0.1])
Nhat1=estimate(ndist,xhat1)
N([96.4;94],[950,-855;-855,948])
Nhat2=estimate(ndist,xhat2)
N([107;96.6],[1.06e+003,915;915,961])
xhat=fusion(Nhat1,Nhat2)
```

```
N([100;90.7],[91.3,0.372;0.372,86.9])
plot2(Nhat1,Nhat2,xhat,'legend','')
ans =
   352.0118   353.0114   356.0114   357.0114   358.0114   359.0114
```

Figure 3.3 shows the resulting Gaussian distribution before and after fusion. Note that the result gets a bias of about 10 meters in x_2, and a standard deviation of about $9.9 \approx \sqrt{98}$ meters.

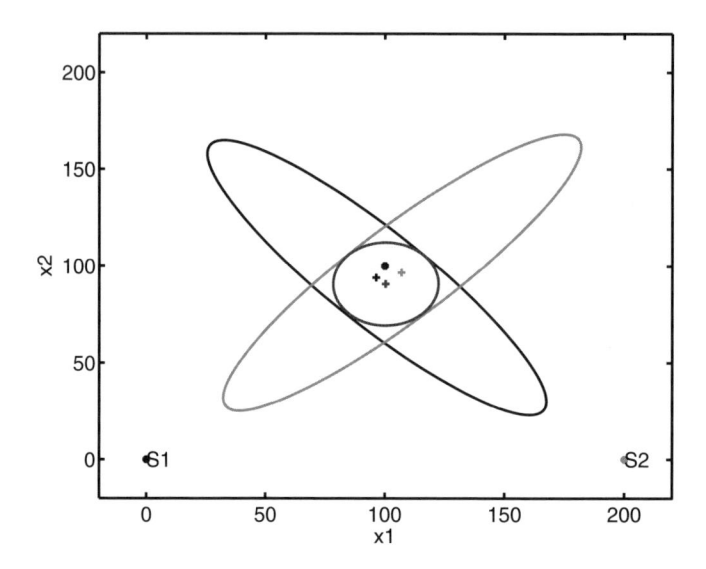

Figure 3.3: *The larger ellipsoids with center points in the plus sign represent two Gaussian distributions fitted to the Monte Carlo samples in Figure 3.1 for sensor 1 and 2, respectively. The smaller circle represent the result of applying the sensor fusion formula (2.23) to these Gaussian distributions.*

3.4.3 Taylor Expansion and Gauss Approximation Formula

Using the Taylor expansion around the observed y gives

$$x = g(y-e) \approx g(y) - g'(y)e \tag{3.33}$$

3.4 Inversion of the Measurement Equation

From this linear approximation, we get what is sometimes referred to as *Gauss' approximation formula*. Here, it gives

$$\mu_x = \mathrm{E}(x) \approx g(y), \tag{3.34a}$$

$$P_x = \mathrm{Cov}(x) \approx g'(y) R \big(g'(y)\big)^T, \tag{3.34b}$$

where $R = \mathrm{Cov}(e)$.

---**Example 3.7: Ranging sensors: TT1 on inverse model**---

Consider the radar problem in Example 3.1. The inverse model $x = g(y - e)$ in (3.3) is

$$x_1 = g_1(y) = r\sin(\varphi),$$
$$x_2 = g_2(y) = r\cos(\varphi).$$

The gradient is given by

$$g'(y) = \begin{pmatrix} \sin(\varphi) & r\cos(\varphi) \\ \cos(\varphi) & -r\sin(\varphi) \end{pmatrix}$$

Using the geometry of the problem, the resulting mean and covariance from (3.34) are computed below.

```
y1=[R1;Phi1];   y1.xlabel={'R','Phi'};
y2=[R2;Phi2];   y2.xlabel={'R','Phi'};
hinv=inline('[p(1)+x(1,:).*cos(x(2,:));
    p(2)+x(1,:).*sin(x(2,:))]','x','p');
Nhat1=tt1eval(y1,hinv,p1)
N([100;100],[1e+003,-997;-997,1e+003])
Nhat2=tt1eval(y2,hinv,p2)
N([100;100],[1e+003,998;998,1e+003])
xhat=fusion(Nhat1,Nhat2)
N([100;100],[4.99,-2.18e-008;-2.18e-008,4.99])
plot2(Nhat1,Nhat2,xhat,'legend','')
ans =
   538.0044   539.0039   542.0039   543.0039   544.0039   545.0039
```

Figure 3.4 shows the Gaussian ellipsoids, whose main axis basically follows the tangent of the transformation at the true position. Note that the this method indicates a too optimistic standard deviation of $2.2 \approx \sqrt{5}$ meters, since this is significantly smaller than the more accurate method in Figure 3.3.

3.4.4 The Unscented Transform

The *unscented transform* (*UT*) is a general method to transform a (Gaussian) distribution through nonlinear mappings. The goal is to approximate the distribution for $x = g(y - e)$ when $e \sim \mathcal{N}(0, R)$ as a Gaussian distribution $x \sim \mathcal{N}(\mu_x, P_x)$. As a summary, UT applied to this inverse sensor model works as follows:

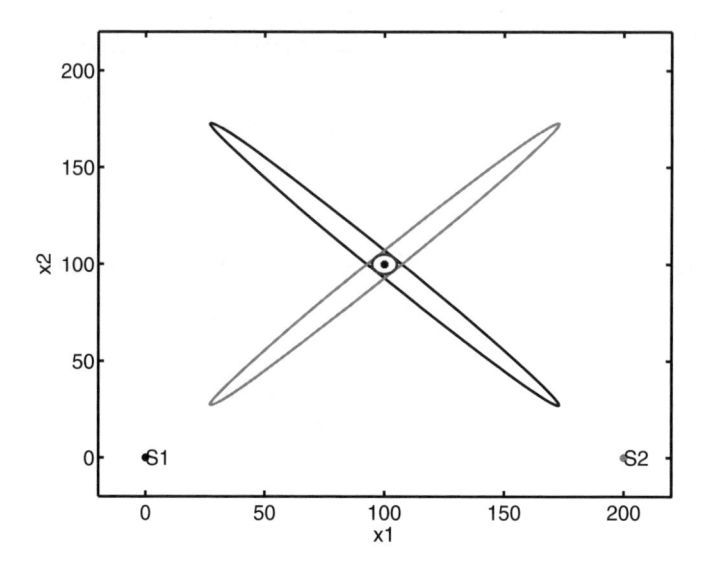

Figure 3.4: Resulting Gaussian approximations similar to Figure 3.3 but using a first order Taylor expansion.

1. The so called *sigma points* $e^{(i)}$ are computed. These are the mean and symmetric deviations around the mean computed from the covariance matrix R. For $e \in \mathcal{N}(0, \sigma^2)$, the three sigma points are

$$e^{(0)} = 0,$$
$$e^{(1)} = \sqrt{1+\lambda}\sigma,$$
$$e^{(-1)} = -\sqrt{1+\lambda}\sigma.$$

Here, λ is a design parameter.

2. The sigma points are mapped to $x^{(i)} = g(y - e^{(i)})$.

3. The mean and covariance are fitted to the mapped sigma points using a variant of (3.32) with a different set of weights. Basically, the mean has a different weight than the points at the ellipsoid, and there is no constraint that the weights are in the interval $[0, 1]$ as long as they sum to 1. For the case $e \in \mathcal{N}(0, \sigma^2)$ above, we get

$$\mu_x = \frac{\lambda x^{(0)}}{1+\lambda} + \frac{x^{(1)} + x^{(-1)}}{2(1+\lambda)}, \tag{3.35a}$$

$$P_x = \frac{\lambda(x^{(0)} - \mu_x)^2}{1+\lambda} + \frac{(x^{(1)} - \mu_x)^2 + (x^{(-1)} - \mu_x)^2}{2(1+\lambda)}. \tag{3.35b}$$

3.4 Inversion of the Measurement Equation 63

For more details, see Appendix A.3.

There are many claims and design rules for how to weigh together the approximation in the literature. Here, we apply the unscented transform to our standard example, which shows that it performs as well as the Monte Carlo approach, but with a much smaller computational effort.

─── **Example 3.8: Ranging sensors: UT on inverse model** ───

The sigma points of each sensor observation PDF are generated and illustrated in Figure 3.5. These are then mapped to samples of x and (3.32) is applied as shown in Figure 3.6. The approximations in (3.32) are then applied to each sensor PDF transformation, and the results are fused using (3.30).

The method `uteval` below returns the Gaussian approximation and also the sigma points before and after the transformation.

```
[Nhat1,S1,fS1]=uteval(y1,hinv,'std',[],p1)
N([95.1;95.1],[954,-854;-854,954])
S1 =
   141.4214   145.2943   141.4214   137.5484   141.4214
     0.7854     0.7854     1.3331     0.7854     0.2377
fS1 =
   100.0000   102.7386    33.2968    97.2614   137.4457
   100.0000   102.7386   137.4457    97.2614    33.2968
[Nhat2,S2,fS2]=uteval(y2,hinv,'std',[],p2)
N([105;95.1],[954,854;854,954])
S2 =
   141.4214   145.2943   141.4214   137.5484   141.4214
     2.3562     2.3562     2.9039     2.3562     1.8085
fS2 =
   100.0000    97.2614    62.5543   102.7386   166.7032
   100.0000   102.7386    33.2968    97.2614   137.4457
xhat=fusion(Nhat1,Nhat2)
N([100;90.8],[94.9,1.56e-013;-1.56e-013,94.9])
plot2(y1,y2,'legend','')
ans =
   724.0044   725.0039   728.0039   729.0039
hold on
plot(S1(1,:),S1(2,:),'bx',S2(1,:),S2(2,:),'g*','linewidth',3)
```

Note that the result gets a bias of about 10 meters in x_2, and a standard deviation of about $9.7 \approx \sqrt{94.9}$ meters. Figure 3.4 shows the distribution of the range and bearing observations from the two sensors, together with the sigma points.

```
plot2(Nhat1,Nhat2,xhat,'legend','')
ans =
   905.0044   906.0039   909.0039   910.0039   911.0039   912.0039
hold on
plot(100,100,'*','linewidth',3),
plot(fS1(1,:),fS1(2,:),'bx',fS2(1,:),fS2(2,:),'g*','linewidth',2)
```

Figure 3.6 shows the distribution returned by the unscented transformation and the corresponding transformed sigma points. The result from fusion is also illustrated.

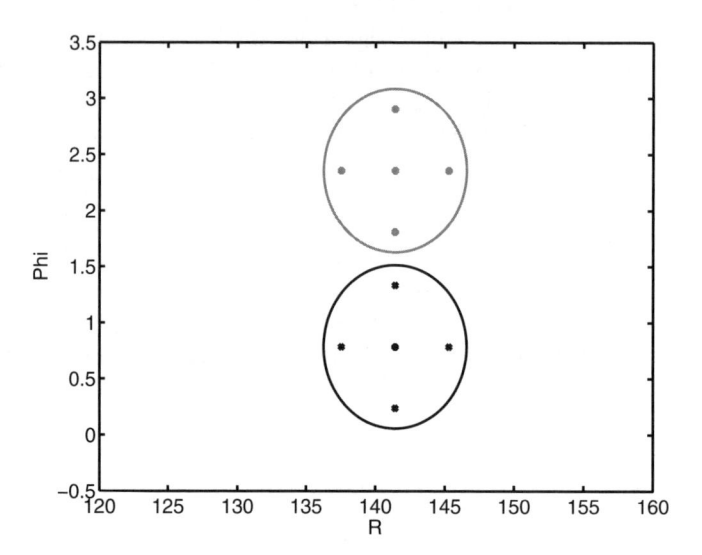

Figure 3.5: *The measurement distribution and the sigma points for the case in Figure 3.1.*

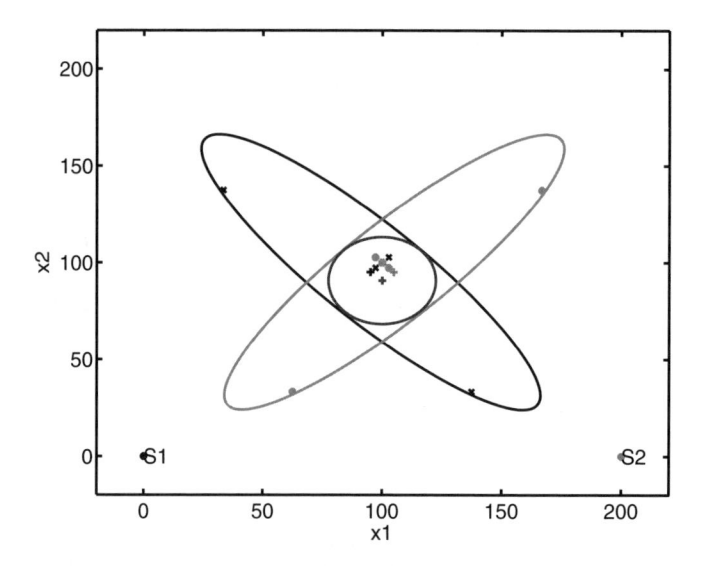

Figure 3.6: *The sigma points in Figure 3.5 transformed with $h^{-1}(y)$ and their Gaussian approximation. The fused estimate is also included.*

3.5 A General Approximation Strategy

The approaches in Section 3.4 all required that the inverse $h^{-1}(y)$ exists. This section will outline an important alternative utilizing an indirect approach, using the same transformation tools.

The strategy here is based on the the following result. If the joint distribution for the parameter and measurement is Gaussian and known,

$$\begin{pmatrix} x \\ y \end{pmatrix} \sim \mathcal{N}\left(\begin{pmatrix} \bar{x} \\ \bar{y} \end{pmatrix}, \begin{pmatrix} P^{xx} & P^{xy} \\ P^{yx} & P^{yy} \end{pmatrix} \right) \tag{3.36}$$

then the minimum variance unbiased (MVU) estimator of x is given by

$$\hat{x} = \bar{x} + P^{xy}\left(P^{yy}\right)^{-1}(y - \bar{y}), \tag{3.37a}$$

$$\mathrm{Cov}(\hat{x}) = P^{xx} - P^{xy}\left(P^{yy}\right)^{-1}P^{yx}. \tag{3.37b}$$

This formula is studied in more detail in Lemma 7.1, where a proof is given. In the non-Gaussian case, the result still holds in the sense of best linear unbiased estimate BLUE. This relation will be used to derive the Kalman filter in Chapter 7 and to derive the Riccati-free versions of EKF and UKF in Section 8.2.

Now, consider the composite prior vector

$$u = \begin{pmatrix} x \\ e \end{pmatrix} \in \mathcal{N}\left(\begin{pmatrix} x \\ 0 \end{pmatrix}, \begin{pmatrix} P & 0 \\ 0 & R \end{pmatrix} \right) \tag{3.38a}$$

The transformation approximation (UT, MC, TT1, TT2) gives the required posterior distribution using the mapping $z = g(u)$

$$z = \begin{pmatrix} x \\ y \end{pmatrix} = \begin{pmatrix} x \\ h(x) + e \end{pmatrix} \sim \mathcal{N}\left(\begin{pmatrix} \bar{x} \\ \bar{y} \end{pmatrix}, \begin{pmatrix} P^{xx} & P^{xy} \\ P^{yx} & P^{yy} \end{pmatrix} \right) \tag{3.38b}$$

In summary, TT1, TT2, UT and MCT can be applied to the mapping $z = g(u)$ to obtain the required approximation (3.38b), whose block matrices are plugged into the formula (3.37).

It can be shown quite easily, that the TT1 approach is equivalent to the direct Taylor expansion LS approach leading to (3.19).

When there are many independent sensor observations y_k, the approach can be applied repeatedly, each time using a new (improved) prior of x. Note that the ordering of y_k plays some role here, since iterative approximations are involved.

3.6 Conditionally Linear Models

Consider a structured nonlinear model

$$y_k = h_k^n(x^n) + h_k^l(x^n)x^l + e_k, \quad \mathrm{Cov}(e_k) = R_k(x^n), \tag{3.39}$$

which is conditionally linear in x^l, given that x^n is known. The fact that one part of the parameter vector enters the observation model linearly contains important structural information, and this can be used in two different ways:

- The *separable least squares* principle, that could also be called the *generalized least squares* method in parallel to the generalized likelihood method. Here, the LS estimate of the linear part is plugged in into the model.

- The Bayesian marginalization approach, where a prior distribution of the unknown parameters entering linearly in the model is used to eliminate x^l from the posterior distribution $p(x^n, x^l | y_{1:N})$.

3.6.1 Separable Least Squares

Consider the WLS estimate of (3.39),

$$\widehat{(x^n, x^l)} = \arg\min_{x^n, x^l} V^{WLS}(x^n, x^l). \tag{3.40}$$

For each value of x^n, the model (3.39) is linear in x^l and the WLS estimate of x^l can be computed as a function of x^n using the following variant of the standard WLS estimate:

$$\hat{x}^l(x^n) = \left(\sum_{k=1}^{N} h_k^{l,T}(x^n) R_k^{-1}(x^n) h_k^l(x^n)\right)^{-1} \sum_{k=1}^{N} h_k^{l,T}(x^n) R_k^{-1}(x^n) \big(y_k - h_k^n(x^n)\big). \tag{3.41a}$$

This conditional WLS estimate is unbiased $\mathrm{E}\big(\hat{x}^l(x^{n,0})\big) = x^{l,0}$ with covariance

$$P^l(x^n) = \mathrm{Cov}\big(\hat{x}^l(x^n)\big) = \left(\sum_{k=1}^{N} h_k^{l,T}(x^n) R_k^{-1}(x^n) h_k^l(x^n)\right)^{-1}. \tag{3.41b}$$

The NLS problem can then be rewritten in the following equivalent formulation:

$$\widehat{x^n} = \arg\min_{x^n} V^{WLS}\big(x^n, \hat{x}^l(x^n)\big). \tag{3.41c}$$

This approach will be referred to as *separable least squares* (SLS).

To analyze the properties of SLS, we will rewrite the loss function (3.41c) to highlight the noise structure. First, we use the batch formulation of the conditionally linear model (3.39), $\mathbf{y} = \mathbf{h}^n + \mathbf{h}^l x^l + \mathbf{e}$, surpressing the dependence of x^n in the sequel to simplify notation. The estimate can then be written

$$\hat{x}^l = \big(\mathbf{h}^{l,T} \mathbf{R}^{-1} \mathbf{h}^l\big)^{-1} \mathbf{h}^{l,T} \mathbf{R}^{-1}(\mathbf{y} - \mathbf{h}^n) = x^{l,0} + \big(\mathbf{h}^{l,T} \mathbf{R}^{-1} \mathbf{h}^l\big)^{-1} \mathbf{h}^{l,T} \mathbf{R}^{-1} \mathbf{e}, \tag{3.41d}$$

where the latter equality follows if $x^{l,0}$ is the value of x^l in the true system. The residual in the loss function $V^{WLS}(x^n, \hat{x}^l)$ can thus be rewritten as

$$\mathbf{y} - \mathbf{h}^n - \mathbf{h}^l \hat{x}^l = \mathbf{e} - \mathbf{h}^l \left(\mathbf{h}^{l,T} \mathbf{R}^{-1} \mathbf{h}^l\right)^{-1} \mathbf{h}^{l,T} \mathbf{R}^{-1} \mathbf{e}. \quad (3.41e)$$

It is straightforward to show that the covariance of this residual is $\mathbf{R} - \mathbf{h}^l \left(\mathbf{h}^{l,T} \mathbf{R}^{-1} \mathbf{h}^l\right)^{-1} \mathbf{h}^{l,T}$. The SLS estimate defined in (3.41c) can thus be expressed as

$$\widehat{x^n} = \arg\min_{x^n} \|\mathbf{y} - \mathbf{h}^n - \mathbf{h}^l \hat{x}^l\|_{\mathbf{R} - \mathbf{h}^l \left(\mathbf{h}^{l,T} \mathbf{R}^{-1} \mathbf{h}^l\right)^{-1} \mathbf{h}^{l,T}} \quad (3.41f)$$

Note that the matrix norm can be rewritten as $\mathbf{R} - \mathbf{h}^l P^l \mathbf{h}^{l,T}$. The derivations of (3.41f) are similar to the calculations in Section 2.2 for the standard WLS, compare for instance with (2.14) which shows the least squares loss function evaluated in the parameter estimate.

Which one of (3.40) and (3.41) is the best to use?

- The low-dimensional minimization in (3.41) is to prefer when numerical Monte Carlo methods are used for the minimization (compare with the marginalized particle filter later on). If x^n is one or two dimensional, the marginalized NLS solution can be evaluated over a grid quite efficiently.

- The original formulation (3.40) might have fewer local minima and for that reason be preferable to (3.41) in some cases, while in most cases (3.41) should be the standard choice when possible.

--- **Example 3.9: Gaussian signal with unknown mean and variance** ---

Consider the Gaussian observations $y_k = \mu + e_k$ where $e_k \in \mathcal{N}(0, P)$ and $x = (\mu, P)^T$. The model is conditionally linear in $x^l = \mu$ and nonlinear in $x^n = P$. Then, the ML estimator can be separated as follows:

$$\widehat{(\mu, P)} = \arg\max_{\mu, P} (2\pi P)^{-N/2} e^{-\frac{1}{2P} \sum_{k=1}^{N}(y_k - \mu)^2},$$

$$\hat{\mu}(P) = \frac{1}{N} \sum_k y_k,$$

$$\hat{P} = \arg\max_P (2\pi P)^{-N/2} e^{-\frac{1}{2P} \sum_{k=1}^{N}(y_k - \hat{\mu})^2}.$$

In this case, $\hat{\mu}(P)$ is independent of P. Note that there is an explicit solution in this simple case ($\hat{P} = \frac{1}{N} \sum_{k=1}^{N}(y_k - \hat{\mu})^2$).

3.6.2 Marginalized Nonlinear Least Squares

The marginalization approach in this section is identical to the theory in Section C.7.1. The basic idea is that under a Gaussian noise assumption, the conditional distribution of x^l is another Gaussian. Using a Gaussian prior, $x^l \sim \mathcal{N}(0, P^0)$, assuming zero mean for simplicity, gives

$$p(x^l | x^n, \mathbf{y}) = \mathcal{N}\big(\hat{x}^l(x^n), \big((P^l(x^n))^{-1} + (P^0)^{-1}\big)^{-1}\big). \quad (3.43\text{a})$$

The *marginalized NLS* (*MNLS*) estimate is then given as a prior compensated version of (3.41f),

$$\widehat{x^n} = \arg\min_{x^n} \|\mathbf{y} - \mathbf{h}^n - \mathbf{h}^l \hat{x}^l\|_{\mathbf{R} - \mathbf{h}^l\big((P^l(x^n))^{-1} + (P^0)^{-1}\big)^{-1} \mathbf{h}^{l,T}} \quad (3.43\text{b})$$

Example 3.9 can be extended to a MNLS approach as shown in Section E.5. The generalization to the filtering case is treated in detail in Section 9.8.

3.7 Implicit Measurement Equation

The nonlinear model has so far had additive noise and an explicit measurement. Both these assumptions can be dropped by considering a more general implicit measurement model

$$h_k(y_k, x, e_k) = 0, \quad \text{Cov}(e_k) = R_k. \quad (3.44)$$

One can still use the Taylor expansion approach with some obvious modifications. A first order Taylor approximation yields

$$0 = h_k(y_k, x, e_k) \approx h_k(y_k, \bar{x}, 0) + \bar{H}_k^x(x - \bar{x}) + \bar{H}_k^e e_k, \quad (3.45\text{a})$$

where

$$\bar{H}_k^x = \left.\frac{\partial h_k(y_k, x, e)}{\partial x}\right|_{x=\bar{x}, e=0}, \quad (3.45\text{b})$$

$$\bar{H}_k^e = \left.\frac{\partial h_k(y_k, x, e)}{\partial e}\right|_{x=\bar{x}, e=0}, \quad (3.45\text{c})$$

This can be recast into a linear model

$$\bar{y}_k = \bar{H}_k^x x + \bar{e}_k, \quad (3.45\text{d})$$

where

$$\bar{y}_k = -h_k(y_k, \bar{x}, 0) + \bar{H}_k^x \bar{x}, \quad (3.45\text{e})$$

$$\bar{R}_k = \text{Cov}(\bar{e}_k) = \bar{H}_k^e R_k \bar{H}_k^{e,T}. \quad (3.45\text{f})$$

3.7 Implicit Measurement Equation

If the observation is implicit but the noise additive, or if each equation can be solved for e_k,

$$h(y_k, x) = e_k, \qquad (3.46)$$

there are two alternatives:

- The maximum likelihood framework works as usual, leading to nonlinear optimization problems.

- The NLS algorithm works with minor modifications. The residual and gradient are here defined as

$$\varepsilon_k = h_k(y_k, x), \qquad (3.47\text{a})$$

$$J_k(x) = \frac{\partial h_k^T(y_k, x)}{\partial x}, \qquad (3.47\text{b})$$

respectively.

3.8 Summary

3.8.1 Theory

This chapter described methods to estimate the parameters x in the nonlinear model

$$y_k = h_k(x) + e_k, \quad \text{Cov}(e_k) = R_k, \quad k = 1, \ldots, N,$$
$$\mathbf{y} = \mathbf{h}(x) + \mathbf{e}, \quad \text{Cov}(\mathbf{e}) = \mathbf{R}.$$

The NLS solution minimizes

$$\hat{x}^{NLS} = \arg\min_x V^{NLS}(x) = \arg\min_x \frac{1}{2}\sum_{k=1}^{N}(y_k - h_k(x))^T R_k^{-1}(y_k - h_k(x))$$

using numeric iterative algorithms. ML corresponds to NLS for Gaussian noise, since $-\log(p(y|x)) = c + V^{NLS}(x)$. In the case of parameter dependent noise covariance $R_k(x)$, the ML estimate is given by

$$\hat{x}^{ML} = \arg\min_x \left[V^{NLS}(x) + \frac{1}{2}\sum_k \log\det(R_k(x))\right].$$

The principle of linearization is dominating in applications, where the idea is to replace $y_k = h_k(x) + e_k$ with $\bar{y}_k = y_k - h_k(\bar{x}) + h'_k(\bar{x})\bar{x} = h'_k(\bar{x})x + e$ and apply the WLS method to this linear model. The result depends on the linearization point $\bar{x}$ but also on the neglected second order rest term. There are two remedies here that can be tried separately or combined:

- Iterate the linearization process. This corresponds to one numerical solution to the NLS problem.

- Compensate for the bias and variance of the rest term. The bias is subtracted from the measurement y_k and the variance is added to R_k.

If the model contains a linear sub-structure $h(x) = h^n(x^n) + h^l(x^n)x^l$, the part x^l of the parameter vector that appears linearly in the sensor model can be eliminated with WLS, which gives $\hat{x}^l(x^n)$. The numerical search is then performed in a smaller parameter space by minimizing

$$\widehat{x^n} = \arg\min_{x^n} \sum_k \|y_k - h^n(x^n) - h^l(x^n)\hat{x}^l(x^n)\|^2_{R_k - h^l(x^n)\text{Cov}(\hat{x}^l(x^n))h^{l,T}(x^n)}.$$

A prior on x^l can also be taken into account, leading to a marginalization NLS principle studied in detail in Section 9.8 in a filter framework.

Other alternatives are based on nonlinear transform approximations that approximates the nonlinear mapping $z = g(u)$ of a Gaussian variable $u \sim \mathcal{N}(\hat{u}, P_u)$ with $z \sim \mathcal{N}(\hat{z}, P_z)$. Here, the TT1, TT2, UT or MCT approach, as defined in Appendix A.3, can be applied to the following formulations:

3.8 Summary

- The *direct approach*, where $x = \mathbf{h}^{-1}(\mathbf{y} - \mathbf{e})$ is approximated. The sensor fusion formula can be used to fuse $\hat{x}_k$ from different sensors y_k. An approximate Gaussian distribution of x can then be computed directly. Using a first order Taylor expansion, the standard (W)LS method can be used. Using a second order Taylor expansion, the standard (W)LS method can be compensated with mean and covariance of the second order term.

- The *indirect approach*, where the distribution of $\mathbf{y} = \mathbf{h}(x)$ is approximated using a prior of $x \sim \mathcal{N}(\hat{x}, P^{xx})$: The trick is to consider the mapping

$$u = \begin{pmatrix} x \\ e \end{pmatrix} \in \mathcal{N}\left(\begin{pmatrix} \bar{x} \\ 0 \end{pmatrix}, \begin{pmatrix} P^{xx} & 0 \\ 0 & R \end{pmatrix} \right)$$

$$z = \begin{pmatrix} x \\ y \end{pmatrix} = \begin{pmatrix} x \\ h(x, e) \end{pmatrix} \sim \mathcal{N}\left(\begin{pmatrix} \bar{x} \\ \bar{y} \end{pmatrix}, \begin{pmatrix} P^{xx} & P^{xy} \\ P^{yx} & P^{yy} \end{pmatrix} \right)$$

and then apply

$$\hat{x} = \bar{x} + P^{xy}(P^{yy})^{-1}(y - \bar{y}),$$
$$\text{Cov}(\hat{x}) = P^{xx} - P^{xy}(P^{yy})^{-1}P^{yx}.$$

This trick will be essential in Chapter 8.

3.8.2 Software

The central objects for this chapter are summarized below:

`sensormod` defines a *sensor model*

$$y(t) = h(t, x, u(t); \theta) + e(t), e(t) \sim p_e(e)$$

Here the nonlinear observation model is defined, where $u(t)$ is not used here and $p_e(e)$ is one object in the `pdfclass` family (see `list(pdfclass)`). The constructor has the syntax `sm=sensormod(h,[nx 0 ny nth])`, where `h` can be either a function handle, an inline function or a string with a function name. In all cases, the sensor model must be able to evaluate `h(t,x,u,th)` and returning a vector `y` of the specified dimension.

`sig` defines a signal with fields `y,t,u,x` and representation of uncertainty in terms of covariances or Monte Carlo samples.

The most important methods for the `sensormod` class that extends the linear case are

- Constructor `s=sensormod(h,[nx nu ny nth]); s.x0=x0; x.px0=P0;`, where `h` is a string or inline object in the variables `t,x,u,th`.

- For sensor networks, s.th should contain the sensor locations and x the target location(s). Here, plot(s) is useful.
- Pre-defined examples are available in exsensor.
- Simulation y=simulate(s,t), returns a SIG object y from a sensor model object s and time vector t.
- LS estimation xhat=ls(s,y), returns a SIG object xhat from a sensor model object s and signal object y. The method is based on a first order Taylor expansion with numeric gradient H_k. Here, x_k =xhat.x(k,:)' and P_k =xhat.P(k,:,:).
- WLS estimation xhat=wls(s,y), returns a SIG object xhat from a sensor model object s and signal object y. WLS works as LS but compensates for the covariance of e_k.
- NLS estimation xhat=estimate(s,y), returns a SIG object xhat from a sensor model object s and signal object y. For Gaussian noise, NLS and ML coincide. The Gauss-Newton method is default, where numeric gradients are used to compute H_k.
- CRLB xcrlb=crlb(s), returns a SIG object xcrlb with the same state as given in xcrlb.x=s.x0 and covariance xcrlb.Px as the (parametric) CRLB. crlb computes the CRLB for a specific x_0, while crlb1 and crlb2 evaluates a 1D and 2D grid, respectively.
- Likelihood computations of $p_e(\mathbf{y} - \mathbf{h}(x))$.
    - [lh,px,px0,x1]=lh1(s,y,x1,ind) computes the likelihood function $p_e(y - h(x))$ as lh over a gridded 1D state space specified in x1. When the parameter space is larger than one, the ind option specifies which index to vary. Without output arguments, a plot is generated.
    - [lh,px,px0,x1,x2]=lh2(s,y,x1,x2,ind) computes the likelihood function $p_e(y - h(x))$ as lh over a gridded 2D state space specified in x1,x2. When the parameter space is larger than two, the ind option specifies which two indices to vary. Without output arguments, a contour plot is generated.

There are also four transformation methods in the ndist object: tt1eval, tt2eval, uteval, mceval. These work in the same way, for instance the call z=tt1eval(x,h) computes the Gaussian approximation (3.34) of the nonlinear transformation $z = h(x)$, where h is an inline function and z, x are ndist objects.

4

Sensor Networks

Basically, localization in sensor networks is an application of sensor fusion theory in Chapters 3 and 2 to range, range difference and angle measurements. This chapter is first a summary of the theory in the preceeding chapters applied to the sensor network application, and second it provides some dedicated least squares solutions for this problem.

The basic sensor model for sensor network applications can be written

$$\mathbf{y} = \mathbf{h}(x; \mathbf{p}) + \mathbf{e}, \tag{4.1}$$

where $x = (x_1, x_2)^T$ denotes the (two-dimensional) position of a target, and $\mathbf{p} = (p_1^T, p_2^T, \ldots, p_N^T)^T$ contains the locations $p_k = (p_{k,1}, p_{k,2})^T$ of each sensor. This chapter discusses the following issues:

- Section 4.1 describes the background of the different types of observation models $\mathbf{h}(x; \mathbf{p})$.

- Section 4.2 overviews different ways to estimate the position $\hat{x}(\mathbf{p})$ when the sensor locations $\mathbf{p}$ are known. Different estimation criteria are overviewed in Section 4.2.1, numerical search procedures are briefly described in Section 4.2.2 and fundamental bounds are given in Section 4.2.3.

- Section 4.3 applies the NLS algorithm, and presents some SLS formulations.

- Section 4.4 describes some tricks that involve nonlinear transformations g of the observations and target location in order to recast the problem into a linear model of the kind $g^y(\mathbf{y}) = Hg^x(x) + g^e(\mathbf{e})$, where WLS can be used to estimate $g^x(x)$.

- Section 4.5 describes briefly some extended estimation problems. Section 4.5.1 describes the related problem of estimating the sensor locations $\hat{\mathbf{p}}(x)$ for a given target location(s). Section 4.5.2 outlines the general problem of simultanously estimating both target and sensor locations $\widehat{(x, \mathbf{p})}$. This problem is also the subject of Chapter 11. Section 4.5.3 briefly describes the idea of extending the target parameters with shape and size of the target rather than just its position as a point object.

4.1 Typical Observation Models

This section describes generic observation models that are common in sensor network applications. Detailed models are provided in Section 14.4. We start with a high level characterization of a *sensor network* (*SN*) using a rather generic vocabulary, and then proceed with more detailed sensor models.

The SN consists in general of a set of nodes, where each node can be seen as either a sensor or a target. A node can in this context be characterized with the following properties:

- The node either emits (*target*) or receives (*sensor*) energy at some frequency in some media, or both. Examples include radio waves propagated with the speed of light, acoustic energy propagated with the speed of sound, seismic waves propagated in the earth or sonic waves propagated in water.

- The node can be stationary or moving.

- The node can have known or unknown position at each time.

- As a consequence of the above, the reciprocal problems of localization of ones own or other nodes position are treated in the same way. The same applies to navigation and tracking for time-varying node positions.

- One emitting and one receiving node can be anything between perfectly synchronized and completely unsynchronized. The same applies to two emitting nodes or two receiving nodes.

- The communication between two nodes (two emitters or two receivers or one of each) can be wired or wireless, and the bandwidth can be low or high.

- The emitted wave may propagate directly to the receiver (*line of sight*, *LOS*) or only through reflections (*nonline of sight*, *NLOS*). In both cases, the signal might be subject to *multipath* effects.

4.1 Typical Observation Models

The difference of a *wireless sensor network* (*WSN*) and a sensor network in general is the assumption that at least two nodes exchange information using a wireless link and thus are subject to bandwidth limitations and nontrivial synchronization issues.

Depending on the nodes' synchronization and communication capabilities, three different types of observations can be distinguished:

- **Waveform observations.** A highly capable sensor node is able to operate on the signal waveform, and this observation can be shared with other nodes if bandwidth allows. This is only meaningful when these nodes are synchronized at an accuracy comparable to the inverse waveform frequency. Sensors very close to each other (in the order of half a wavelength) can form a sensor array and correlate the phase of the emitted signal to get direction of arrival estimates. For sensors further apart, there will be an integer problem in the ambiguity of the number of periods, that may be resolved by merging other information.

- **Timing observations.** If a known or easily distinguished signature is embedded in the signal, the sensor can correlate the signal with the signature to accurately estimate time of arrival. The timing estimation accuracy depends on the signature as well as the sensor capability. This is meaningful only if the sensor is synchronized either with the emitting node or another receiving node. In the former case, an absolute distance can be computed (*time of arrival* (*TOA*)), and in the latter case a relative distance (*time difference of arrival* (*TDOA*)) follows. Geometrically, these two cases constrain the position to be on a circle or on a hyberbolic function, respectively.

- **Power observations.** Another possibility is that the sensor estimates the received signal power, or *received signal strength* (*RSS*). In essence, this means integrating the received signal power within a certain frequency band during an integration interval to estimate the received signal energy during the time interval. If the emitted power is known, RSS provides coarse range information. Otherwise, two or more sensors can compare their RSS observations to eliminate the unknown emitted power.

These three types are further discussed and exemplified in subsequent subsections.

Figure 4.1 shows an example of a WSN, where each sensor node consists of a microphone, a geophone and a magnetometer. The sensor nodes are stationary with known position. The sensor nodes have a fairly low bandwidth in their wireless communication links, and they are synchronized to an accuracy of a few milliseconds.

The target node, on the other hand, is a moving vehicle with unknown position and on its way it is emitting acoustic and seismic waves as well as

Figure 4.1: *The sensor nodes (+) in a WSN overlaid on a aerial photo, with a plot of the emitting node's trajectory (o) where one particular position is highlighted (x).*

affecting the magnetic field. Depending on the type of vehicle, the range of these waves differs between some ten meters to some hundred meters.

4.2 Target Localization

Table 4.1 summarizes the basic categories of measurements from Section 14.4 in the standard form $h(x, p_k)$. The measurements are RSS (Okumura-Hata model), TOA, TDOA and DOA, respectively. The set of available measurements is summarized as $\mathbf{y} = \mathbf{h}(x) + \mathbf{e}$, and this section discusses how x can be estimated from this relation. First, an example scenario is presented.

⎯⎯ **Example 4.1: TOA and TDOA measurements** ⎯⎯

Consider the scenario in Figure 4.2, where four receivers are placed in the positions $(-1, 0)$, $(1, 0)$, $(0, -1)$ and $(0, 1)$, respectively. Each receiver measures the arrival time of a transmitted signal from an unknown position, using accurate and synchronized clocks. If the transmitter is also synchronized, the signal propagation time can be computed, which leads to a TOA mea-

4.2 Target Localization

Table 4.1: *Analytical expressions related to some measurements and gradients, where $r_i = \|x - p_i\|$ is used.*

Method	$h(x, p_i)$	$\partial h(x, p_i)/\partial x_1$	$\partial h(x, p_i)/\partial x_2$
RSS:	$P^0 + 10\beta \log_{10} r_i$	$\frac{10\alpha}{\log 10} \frac{x_1 - p_{i,1}}{r_i^2}$	$\frac{10\alpha}{\log 10} \frac{x_2 - p_{i,2}}{r_i^2}$
TOA:	r_i	$\frac{x_1 - p_{i,1}}{r_i}$	$\frac{x_2 - p_{i,2}}{r_i}$
TDOA:	$r_i - r_j$	$\frac{x_1 - p_{i,1}}{r_i} - \frac{x_1 - p_{j,1}}{r_j}$	$\frac{x_2 - p_{i,2}}{r_i} - \frac{x_2 - p_{j,2}}{r_j}$
DOA:	$\alpha_i + \text{arctan2}\left(x_1 - p_{i,1}, x_2 - p_{i,2}\right)$	$\frac{-(x_1 - p_{i,1})}{r_i^2}$	$\frac{x_2 - p_{i,2}}{r_i^2}$

surement. Propagation time corresponds to a distance, which leads to the distance circles around each receiver in Figure 4.2(a).

If the transmitter is unsynchronized, each pair of receivers can compute a time-difference of arrival TDOA. Four receivers can compute six such hyperbolic functions, which intersect in one unique point, see Figure 4.2(b).

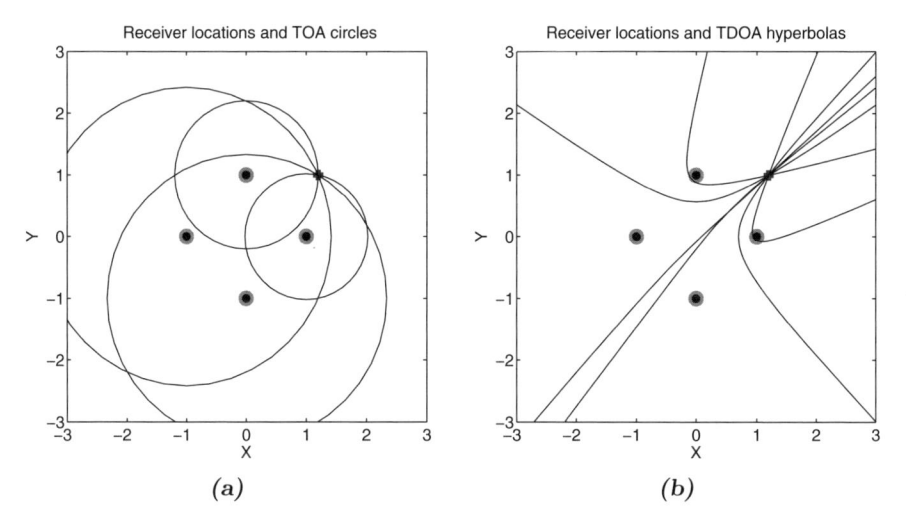

Figure 4.2: *Example scenario with four receivers placed in a square, and there is one transmitter at (1.2,1). With TOA measurements, each receiver measurement constrains the transmitter position to a circle, while with TDOA measurements, each pair of receiver measurements constrain the transmitter position to a hyperbola.*

4.2.1 Estimation Criteria

The general positioning problem is to find the position that minimizes a given norm of the difference of actual measurements and the model. Using the minimizing argument notation for a general loss function $V(x)$, we have

$$\hat{x} = \arg\min_x V(x) = \arg\min_x \|\mathbf{y} - \mathbf{h}(x)\|. \tag{4.2}$$

The most common choices of norms from Chapter 3 are discussed below, and summarized in Table 4.2. Table 4.2 starts with the nonlinear least squares

Table 4.2: Optimization criteria $V(x)$ for estimating position x from uncertain measurements $\mathbf{y} = \mathbf{h}(x) + \mathbf{e}$ using (4.2).

NLS	$V^{NLS}(x) = \|\mathbf{y} - \mathbf{h}(x)\|^2 = (\mathbf{y} - \mathbf{h}(x))^T(\mathbf{y} - \mathbf{h}(x))$
WNLS	$V^{WNLS}(x) = (\mathbf{y} - \mathbf{h}(x))^T \mathbf{R}^{-1}(x)(\mathbf{y} - \mathbf{h}(x))$
ML	$V^{ML}(x) = \log\, p_\mathbf{e}(\mathbf{y} - \mathbf{h}(x))$
GML	$V^{GML}(x) = (\mathbf{y} - \mathbf{h}(x))^T \mathbf{R}^{-1}(x)(\mathbf{y} - \mathbf{h}(x)) + \log\det \mathbf{R}(x)$

criterion. In a stochastic setting where the measurements are subject to a stochastic unknown error $\mathbf{e}$, optimizing the 2-norm is the best thing to do if the errors are independent identically distributed (i.i.d.) Gaussian variables, that is $p_\mathbf{e}(\mathbf{e}) = \mathcal{N}(\mathbf{e}; 0, \sigma_e^2 I)$. If there is a spatial correlation, or if the measurements are of different quality over time or for different sensors, improvements are possible. In these cases, the weighted NLS is to prefer. Further, with a given error probability distribution $p_\mathbf{e}(\cdot)$, the Maximum Likelihood (ML) approach asymptotically provides an efficient estimator. In the special case with a Gaussian error distribution with position dependent covariance $p_\mathbf{e}(\mathbf{e}) = \mathcal{N}(\mathbf{e}; 0, \mathbf{R}(x))$, the GML criterion is similar to the WNLS criterion, but with the term $\log\det \mathbf{R}(x)$. This term prevents the selection of positions with large uncertainty (large $\mathbf{R}(x)$), which could be the case with WNLS. In case $\mathbf{R}(x) = \mathbf{R}$ does not depend on x, ML coincides with WNLS.

---**Example 4.2: TOA and TDOA measurements**---

Consider the scenario in Example 4.1. The NLS and Gaussian likelihood loss functions from Table 4.2 are illustrated in Figure 4.3. The level curves are scaled by σ_e, so the level curve marked with one corresponds to σ_e scale units in standard deviation.

4.2.2 Numerical Algorithms

In the general case, there is no closed form solution to (4.2), and a numerical search method is needed similar to Section 3.2.3. As in any estimation

4.2 Target Localization

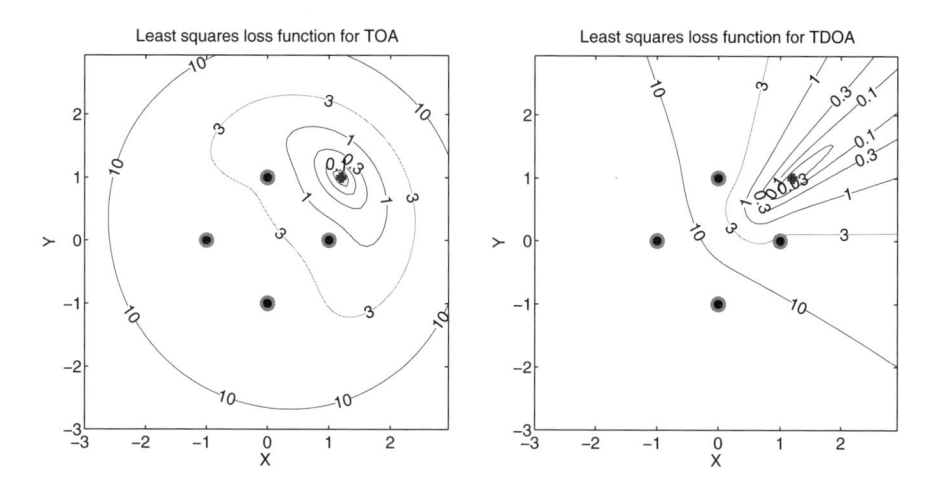

Figure 4.3: *Nonlinear least squares (NLS) loss functions, or the negative log likelihood for Gaussian error distribution, for TOA and TDOA measurements, respectively, for the sensor network in Figure 4.2.*

algorithm, the classical choice is between a gradient and Gauss-Newton algorithm, see the classic textbook Dennis Jr. and Schnabel (1983). The basic forms are given in Table 4.3. These local search algorithms generally require good initialization, otherwise the risk is to reach a local minimum in the loss function $V(x)$. Today, simulation based optimization techniques may provide an alternative. For further illustrations on computing the ML estimate for TDOA measurements, see Urruela and Riba (2004).

Table 4.3: *Estimation algorithms applicable to optimization criteria in Table 4.2. Here, $R = I$ for NLS and ML, $\mathbf{H}(x) = \nabla_x \mathbf{h}(x)$ for NLS and WNLS and $\mathbf{H}(x) = \nabla_x \log p_{\mathbf{e}}(\mathbf{y} - \mathbf{h}(x))$ for ML, and α_k is a step-size.*

Steepest descent	$\hat{x}_k = \hat{x}_{k-1} + \alpha_k \mathbf{H}^T(\hat{x}_{k-1}) R^{-1}(y - \mathbf{h}(\hat{x}_{k-1}))$
Gauss-Newton	$\hat{x}_k = \hat{x}_{k-1} + \alpha_k \left(\mathbf{H}^T(\hat{x}_{k-1}) R^{-1} \mathbf{H}(\hat{x}_{k-1})\right)^{-1}$ $\mathbf{H}^T(\hat{x}_{k-1}) R^{-1}(y - \mathbf{h}(\hat{x}_{k-1}))$

―― **Example 4.3: TOA and TDOA measurements** ――

Consider the scenario in Example 4.1. Figure 4.4 shows convergence trajectories for the stochastic gradient (steepest descent) and Gauss-Newton methods, respectively.

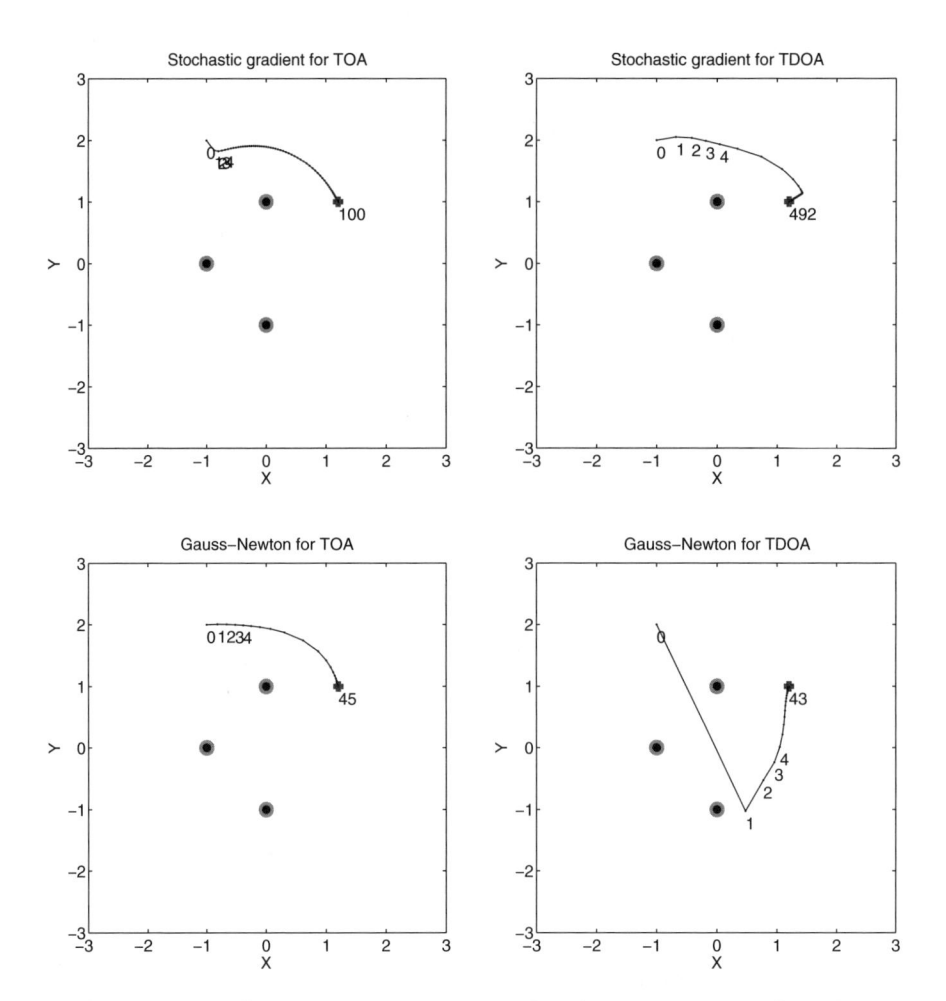

Figure 4.4: *Convergence trajectories for the sensor network in Figure 4.2. First column: The stochastic gradient (steepest descent) and Gauss-Newton algorithms, respectively, for TOA measurements. Second column: the stochastic gradient and Gauss-Newton algorithms, respectively, for TDOA measurements. The numbers indicate iteration index k in Table 4.3. Compare to the nonlinear least squares loss functions in Figure 4.3.*

4.2.3 Fundamental Performance Bounds

The *Fisher Information Matrix (FIM)* provides a fundamental estimation limit for unbiased estimators referred to as the *Cramér-Rao lower bound (CRLB)*, see Kay (1993) and Appendix C. This bound has been analyzed thoroughly in the literature, primarily for DOA, TOA and TDOA, but also for RSS and with specific attention to the impact from non-line-of-sight.

The 2×2 Fisher Information Matrix $\mathcal{I}(x)$ is defined as

$$\mathcal{I}(x) = \mathrm{E}\left(\nabla_x^T \log p_\mathbf{e}(\mathbf{y} - \mathbf{h}(x)) \nabla_x \log p_\mathbf{e}(\mathbf{y} - \mathbf{h}(x))\right) \quad (4.3\mathrm{a})$$

$$\nabla_x \log p_\mathbf{e}(\mathbf{y} - \mathbf{h}(x)) = \left(\frac{\partial \log p_\mathbf{e}(\mathbf{y} - \mathbf{h}(x))}{\partial x_1} \quad \frac{\partial \log p_\mathbf{e}(\mathbf{y} - \mathbf{h}(x))}{\partial x_2}\right) \quad (4.3\mathrm{b})$$

where p is the two-dimensional position vector and $p_\mathbf{e}(\mathbf{y} - \mathbf{h}(x))$ the likelihood given the error distribution.

In case of Gaussian measurement errors $p_\mathbf{e}(e) = \mathcal{N}(e; 0, \mathbf{R}(x))$, the FIM equals

$$\mathcal{I}(x) = \mathbf{H}^T(x) \mathbf{R}(x)^{-1} \mathbf{H}(x), \quad (4.4\mathrm{a})$$

$$\mathbf{H}(x) = \nabla_x \mathbf{h}(x). \quad (4.4\mathrm{b})$$

In the general case, numerical methods are needed to evaluate the CRLB. The larger the gradient $\mathbf{H}(x^o)$, or the smaller the measurement error, the more information is provided from the measurement, and the smaller potential estimation error.

Information is additive, so if two measurements are independent, the corresponding information matrices can be added. This is easily seen for instance from (4.4a) for $\mathbf{H}^T = (\mathbf{H}_1^T, \mathbf{H}_2^T)$ and $\mathbf{R}$ being block diagonal, in which case we can write $\mathcal{I} = \mathcal{I}_1 + \mathcal{I}_2$. Plausible approximative scalar information measures are the trace of the FIM and the smallest eigenvalue of FIM

$$\mathcal{I}_{\mathrm{tr}}(x) \triangleq \mathrm{tr}\,\mathcal{I}(x), \quad \mathcal{I}_{\min}(x) \triangleq \min \mathrm{eig}\,\mathcal{I}(x). \quad (4.5)$$

The former information measure is additive as FIM itself, while the latter is an under-estimation of the information useful when reasoning about whether the available information is sufficient or not.

The Cramer-Rao Lower Bound is given by

$$\mathrm{Cov}(\hat{x}) = \mathrm{E}(x^o - \hat{x})(x^o - \hat{x})^T \geq \mathcal{I}^{-1}(x^o), \quad (4.6)$$

where x^o denotes the true position. The CRLB holds for any unbiased estimate $\hat{x}$ of $\hat{x}$, in particular the ones based on minimizing the criteria in the previous sub-section. The lower bound may not be an attainable bound. It is known that asymptotically, the ML estimate is $\hat{x} \sim \mathcal{N}(x^o, \mathcal{I}^{-1}(x^o))$ and thus reaches this bound, but this may not hold for finite amount of data.

The right hand side of (4.6) gives an idea of how suitable a given sensor configuration is for positioning. It can also be used for *system design*, e.g. where to put the base stations or what information to include in the protocol standard. However, it should always be kept in mind though that this lower bound is quite optimistic and relies on many assumptions.

In practice, the root mean square error (RMSE) is perhaps of more importance. This can be interpreted as the achieved position error in meters. The CRLB implies the following bound:

$$\text{RMSE} = \sqrt{\text{E}[(x_1^o - \hat{x}_1)^2 + (x_2^o - \hat{x}_2)^2]} = \sqrt{\text{tr Cov}(\hat{x})} \geq \sqrt{\text{tr}\,\mathcal{I}^{-1}(x^o)} \quad (4.7)$$

If RMSE performance requirements are specified, one can try to include more and more measurements in the design until (4.7) indicates that the amount of information is enough.

── **Example 4.4: TOA and DOA measurements** ──

Consider a two sensor scenario similar to Example 4.1, but with only two sensors. The complete code for the TOA simulation is shown below.

```
M=2; N=1;
th=[0.4 0.1 0.6 0.1];
x0=[0.5 0.5];
stoa=exsensor('toa',M,N);
stoa.th=th;
stoa.x0=x0;
stoa.pe=0.005*eye(M);
y=simulate(stoa,1)
SIG object with continuous time stochastic state space data (no input)
    Sizes:        N = 1,   ny = 2,  nx = 2
    MC is set to: 30
    #MC samples:  0
stoa
SIGMOD object: TOA
            /  sqrt((x(1,:)-th(1)).^2+(x(2,:)-th(2)).^2) \
    y  = \  sqrt((x(1,:)-th(3)).^2+(x(2,:)-th(4)).^2) / + e
    x0'  = [0.5         0.5]
    th'  = [0.4         0.1         0.6         0.1]

    States:   x1        x2
    Outputs:  y1        y2
y.y
ans =
       0.4685      0.4319
plot(stoa)
hold on
lh2(stoa,y);
hold off
axis([0 1 0 1])
```

Figure 4.5 shows the NLS loss functions corresponding to this scenario. For this particular scenario, the nonlinear equations have no local minima, so both a gradient and a Newton-Raphson algorithm should converge from any initial point.

The CRLB for this scenario is computed and plotted below, resulting in Figure 4.6

4.2 Target Localization

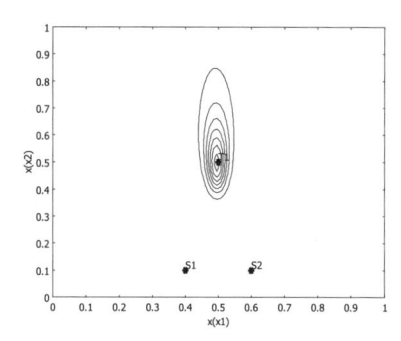

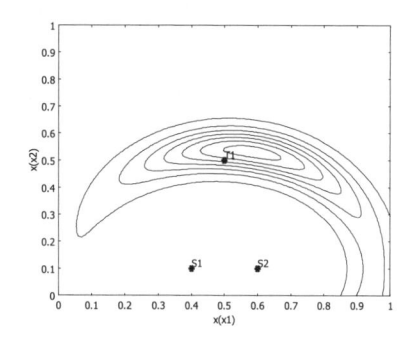

Figure 4.5: *Nonlinear least squares (NLS) loss functions, or the negative log likelihood for Gaussian error distribution, for TOA and DOA measurements, respectively.*

```
plot(stoa)
hold on
crlb(stoa);
hold off
```

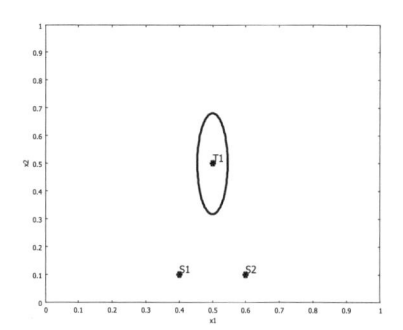

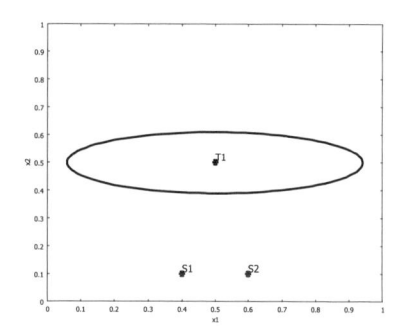

Figure 4.6: *CRLB for TOA and DOA measurements, respectively.*

The DOA illustrations in the right plots of Figures 4.5 and 4.6 are obtained with almost the same code, just change 'toa' to 'doa' in the call to sensormod.

4.3 NLS and SLS Solutions

This section applies the *nonlinear least squares* (*NLS*) method straightforwardly to the different observations *RSS*, *TOA*, *TDOA* and *DOA*. The *separable least squares* (*SLS*) principle is utilized whenever possible. That is, the parameter vector is split into two parts, and the NLS estimate is computed in two steps, $\widehat{(x^n, x^l)} = \arg\min_{x^n, x^l} V^{NLS}(x^n, x^l)$, starting with the explicit solution for the linear part of the parameter vector.

4.3.1 RSS Measurements

Consider the logarithmic observation model for power measurements with additive noise $\text{Var}(e_k) = \sigma_e^2$,

$$y_k = P^0 - \beta \log(\|x - p_k\|) + e_k, \quad k = 1, 2, \ldots, N. \tag{4.8}$$

The model is linear in both nuisance parameters $x^l = (P^0, \beta)^T$, where P_0 is the *emitted power* and β is the *path loss constant*. Thus, they can be eliminated using the SLS principle, see Section 3.6.1, leading to the LS estimate

$$\hat{x}^l(x^n) = \left(\sum_{k=1}^{N} H_k^T(x^n) H_k(x^n)\right)^{-1} \left(\sum_{k=1}^{N} H_k^T(x^n) y_k\right), \tag{4.9a}$$

$$P^l(x^n) = \sigma_e^2 \left(\sum_{k=1}^{N} H_k^T(x^n) H_k(x^n)\right)^{-1}, \tag{4.9b}$$

$$H_k(x^n) = (1, \log(\|x^n - p_k\|)). \tag{4.9c}$$

The measurement model using the separation principle is then

$$y_k = H_k(x^n)\hat{x}^l(x^n) + e_k, \tag{4.9d}$$

and the resulting NLS loss function is

$$V^{NLS}(x^n) = \sum_{k=1}^{N} \left(y_k - H_k(x^n)\hat{x}^l(x^n)\right)^T \left(\sigma_e^2 I + H_k(x^n) P^l(x^n) H_k^T(x^n)\right)^{-1}$$
$$\left(y_k - H_k(x^n)\hat{x}^l(x^n)\right). \tag{4.9e}$$

This can be minimized using numerical gradient algorithms or global grid based methods, where the loss function is evaluated over a position grid x^n.

4.3.2 TOA Measurements

Consider a TOA observation with additive noise $\text{Var}(e_k) = \sigma_e^2$,

$$r_k = \|x - p_k\| + e_k, \quad k = 1, 2, \ldots, N. \tag{4.10}$$

4.3 NLS and SLS Solutions

The straightforward NLS formulation of the position estimate is

$$\hat{x} = \arg\min_x \underbrace{\sum_{k=1}^{N} \frac{(r_k - \|x - p_k\|)^2}{\sigma_e^2}}_{V^{NLS}(x)}. \tag{4.11}$$

There is no explicit solution to this minimization problem, since the sensor model $\|x - p_k\|$ involves a square root. However, all the local algorithms in Table 4.3 can be applied, as well as global searches based on evaluating the cost function $V^{NLS}(x)$ over a grid on x. The latter approach is clearly quite attractive for positions in two dimensions.

The standard example of TOA measurements is when time of flight t_k is measured with a time error e_k. Let the speed of the media be c. Then, the TOA observation model becomes

$$t_k = \frac{1}{c}\|x - p_k\| + e_k, \quad k = 1, 2, \ldots, N. \tag{4.12}$$

For acoustic or underwater applications, the speed c might be uncertain. The SLS approach is then to include its inverse in the linear part, $x^l = 1/c$, and proceed similarly to (4.9).

4.3.3 TDOA Measurements using SLS

Consider a TDOA observation with additive noise $\text{Var}(e_k) = \sigma_e^2$,

$$r_k = \|x - p_k\| + r_0 + e_k, \quad k = 1, 2, \ldots, N. \tag{4.13}$$

This is linear in the unknown range r_0, so $x^l = r_0$. For time-of-flight measurements, r_0 corresponds to the clock error in the transmitter, while the receivers are assumed synchronized. The standard approach is to build pairwise differences as will be described below. Note that we stick to the standard notation in literature of TDOA, even though no time differences are formed. The most straightforward solution is to apply SLS, which is perhaps not sufficiently acknowledged in literature. The algorithm is very similar to above, where the range offset is first estimated with (let $x^n = x$ denote position)

$$\hat{x}^l(x^n) = \left(\sum_{k=1}^{N} H_k^T(x^n) H_k(x^n)\right)^{-1} \left(\sum_{k=1}^{N} H_k^T(x^n)(y_k - \|x^n - p_k\|)\right), \tag{4.14}$$

$$P^l(x^n) = \sigma_e^2 \left(\sum_{k=1}^{N} H_k^T(x^n) H_k(x^n)\right)^{-1}, \tag{4.15}$$

$$H_k(x^n) = 1. \tag{4.16}$$

This can be simplified to

$$\hat{x}^l(x^n) = \hat{r}_0(x^n) = \frac{1}{N}\sum_{k=1}^{N}\left(y_k - \|x^n - p_k\|\right), \qquad (4.17)$$

$$P^l(x^n) = \frac{\sigma_e^2}{N}, \qquad (4.18)$$

and the resulting NLS loss function becomes (compare to (3.41f))

$$V^{NLS}(x^n) = \sum_{k=1}^{N} \frac{(y_k - \|x^n - p_k\| - \hat{r}_0(x^n))^2}{\sigma_e^2 - \sigma_e^2/N}. \qquad (4.19)$$

This is again a two-dimensional optimization problem, where a global grid search can be applied efficiently.

4.3.4 TDOA Measurements using Pairwise Differences

There is a simple trick to get rid of the offset r_0 in (4.13), and it is to compare the observations pairwise,

$$r_{ij} = r_i - r_j, \quad 1 \le i < j \le N. \qquad (4.20)$$

Here, N is the number of receivers and (i,j) is an enumeration of all the K pairs of receivers, where

$$K = \binom{N}{2}. \qquad (4.21)$$

Below, we will show that each r_{ij} corresponds to positions x along a hyperbola. We let $i = 1$ and $j = 2$ here for simplicity. To simplify the derivation, we change the orientation of the coordinate system to a local one (X_1, X_2) where the receivers are both located at the X_1-axis at $p_1 = (D/2, 0)^T$ and $p_2 = (-D/2, 0)^T$, respectively. See Figure 4.7(a) for an illustration, where $D = 1$.

The range differences can then be expressed as

$$r_2 = \sqrt{X_2^2 + (X_1 + D/2)^2}, \qquad (4.22a)$$

$$r_1 = \sqrt{X_2^2 + (X_1 - D/2)^2}, \qquad (4.22b)$$

$$r_{12} = r_2 - r_1 = h(x, D) \qquad (4.22c)$$

$$= \sqrt{X_2^2 + (X_1 + D/2)^2} - \sqrt{X_2^2 + (X_1 - D/2)^2}. \qquad (4.22d)$$

After some simplifications, this equation can be rewritten in a more compact form as

$$\frac{X_1^2}{a} - \frac{X_2^2}{b} = \frac{X_1^2}{r_{12}^2/4} - \frac{X_2^2}{D^2/4 - r_{12}^2/4} = 1. \qquad (4.23)$$

4.3 NLS and SLS Solutions

This equation is on the standard form for a *hyperbolic function*. Figure 4.7 illustrates the hyperbolic function in the local coordinate system X.

The solution to this equation has asymptotes (let $|X_1| \to \infty$ and $|X_2| \to \infty$) along the lines

$$X_2 = \pm\sqrt{\frac{b}{a}}X_1 = \pm\sqrt{\frac{D^2/4 - r_{12}^2/4}{r_{12}^2/4}}X_1 = \pm X_1\sqrt{\left(\frac{D}{r_{12}}\right)^2 - 1}. \quad (4.24)$$

This defines the *direction of arrival (DOA)*, or *direction of arrival (DOA)*, φ, for far-away transmitters as

$$\varphi = \pm\arctan\left(\sqrt{\left(\frac{D}{r_{12}}\right)^2 - 1}\right). \quad (4.25)$$

In sensor arrays, *coherent detection* is used to directly estimate the time delays implicitely. There is a rather rich literature on the theory of sensor array signal processing, see for instance Krim and Viberg (1996), based on a far-field assumption (only the asymptotes of the hyperbolic functions below are considered).

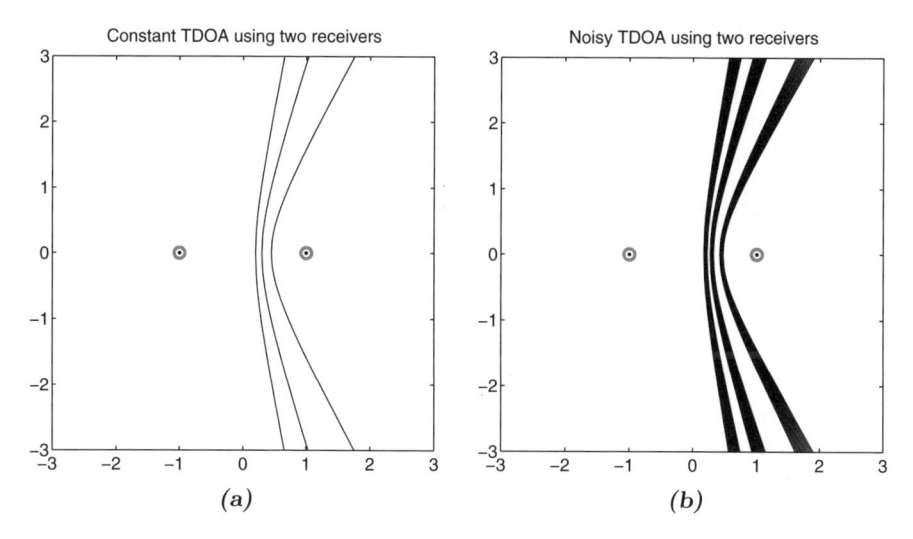

Figure 4.7: The hyperbolic function representing constant TDOA for three examples of TDOA observations ($r_{12} = -0.4$, -0.6 and -0.9 scale units, respectively). The left plot is for the noise-free case, and the right plot for the noisy case. The asymptotes define the DOA ±78°, ±73° and ±63°, respectively.

For general receiver position p_i and p_j, we simply translate the hyperbolic function (4.23) in local coordinates X to global coordinates x using

$$\begin{pmatrix} x_1 \\ x_2 \end{pmatrix} = \frac{1}{2} \begin{pmatrix} p_{i,1} + p_{j,1} \\ p_{i,2} + p_{j,2} \end{pmatrix} + \begin{pmatrix} \cos(\alpha) & -\sin(\alpha) \\ \sin(\alpha) & \cos(\alpha) \end{pmatrix} \begin{pmatrix} X_1 \\ X_2 \end{pmatrix}. \quad (4.26)$$

The hyperbolic function in global coordinates is thus given by $r_{ij} = h(X_{ij}, D_{ij})$ in (4.22). The general observation model with noise is thus

$$r_{ij} = \sqrt{X_{ij,2}^2 + (X_{ij,1} + D_{ij}/2)^2} - \sqrt{X_{ij,2}^2 + (X_{ij,1} - D_{ij}/2)^2} + e_i - e_j, \quad (4.27a)$$

$$D_{ij} = \sqrt{(p_{i,2} - p_{j,2})^2 + (p_{i,1} - p_{j,1})^2} \quad (4.27b)$$

$$\begin{pmatrix} X_{ij,1} \\ X_{ij,2} \end{pmatrix} = \begin{pmatrix} \cos(\alpha_{ij}) & \sin(\alpha_{ij}) \\ -\sin(\alpha_{ij}) & \cos(\alpha_{ij}) \end{pmatrix} \begin{pmatrix} x_1 - \frac{p_{i,1}+p_{j,1}}{2} \\ x_2 - \frac{p_{i,1}+p_{j,1}}{2} \end{pmatrix} \quad (4.27c)$$

$$\alpha_{ij} = \arctan2\left(p_{i,1} - p_{j,1}, p_{i,2} - p_{j,2}\right) \quad (4.27d)$$

We have now a functional form suitable for representing the measurement uncertainty in TDOA. The result is an uncertain hyperbolic area rather than a line. See Figure 4.7(b) for an illustration.

4.3.5 DOA Measurements

For the basic DOA observation

$$\varphi_k = \arctan2\left(x_1 - p_{k,1}, x_2 - p_{k,2}\right) + e_k, \quad k = 1, 2, \ldots, N, \quad (4.28)$$

the NLS estimate is defined by

$$\hat{x} = \arg\min_{x} \underbrace{\sum_{k=1}^{N} \left(\varphi_k - \arctan2\left(x_1 - p_{k,1}, x_2 - p_{k,2}\right)\right)}_{V^{NLS}(x)}. \quad (4.29)$$

Just as in the case of TOA observations, the estimate can be found by a grid or a gradient search.

Consider a sensor measuring relative angle subject to an unknown offset a and scale error b,

$$\varphi_k = a + b\arctan2(x_1 - p_{k,1}, x_2 - p_{k,2}) + e_k. \quad (4.30)$$

This is for instance the case for the marine radar in Figure 14.1. This case is a parallel to the power measurements with $x^l = (a, b)^T$, and the same methodology as in (4.9) applies.

4.4 Dedicated Least Squares Solutions

There are a couple of algorithms in literature with a quite nice common theme. The idea is to transform the observations such that the problem becomes linear and WLS algorithms can be applied. Notationally, the transformed observation model $\mathbf{y} = \mathbf{h}(x) + \mathbf{e}$ becomes

$$g^y(\mathbf{y}) = \mathbf{H}g^x(x) + g^e(\mathbf{e}), \qquad (4.31)$$

where g^* denotes a nonlinear function. The target position might appear as a transformed parameter vector $g^x(x)$, but this is invertible so an estimate $\widehat{g^x(x)}$ can be converted to $\hat{x}$. Also the noise may be transformed to $\bar{\mathbf{e}} = g^e(\mathbf{e})$. That is, in a stochastic framework, these approaches are quite *ad-hoc*, but their implementation is quite easy. The approaches are

- **Reference Sensor Trilateration.** Range observations r_k are squared and the squared range of a reference sensor, say 1, is substracted, yielding $r_k^2 - r_1^2$ as measurements. These differences of squares are linear in the target position $(x_1, x_2)^T$. The noise covariance matrix becomes non-diagonal.

- **Range Parameter Trilateration.** Range observations r_k are squared to r_k^2. These squares are linear in the (over-parametrized) parameter vector $(x_1, x_2, x_1^2 + x_2^2)^T$.

- **Triangulation.** DOA measurements φ_k are transformed by the tangent function to $\tan(\varphi_k)$. The problem is to handle the stochastic uncertainty in a proper way.

4.4.1 TOA Measurements

Consider the TOA measurement (4.10). This is a nonlinear function of the parameter x since the norm operator requires a root finding algorithm. We here consider the *squared distance measurements* r_k^2 as the observation. The basic idea is to obtain the squared norm $\|x - p_k\|^2$. This trick enables some nice explicit solutions. The assumption for using this trick without leaving a WLS framework is to assume that the noise enters the squared distance with zero mean. Otherwise, the mean has to be subtracted from each square below.

Reference Sensor Trilateration

The derivations in this section are based on the presentation in Sayed et al. (2005). First, assume without loss of generality that the first sensor is located at the origin, $p_1 = 0$. The first two measurements can thus be written

$$r_1^2 = \|x\|^2 = x_1^2 + x_2^2 + e_1, \qquad (4.32a)$$
$$r_2^2 = \|x - p_2\|^2 = (x_1 - p_{2,1})^2 + (x_2 - p_{2,2})^2 + e_2. \qquad (4.32b)$$

By expanding the squares and subtracting the first equation from the second (to get rid of the quadratic terms), we get

$$r_2^2 - r_1^2 = -2p_{2,1}x_1 - 2p_{2,2}x_2 + p_{2,1}^2 + p_{2,2}^2 + e_2 - e_1. \qquad (4.33)$$

This is a linear relation, and by continuing in the same way for the other sensors, a linear model is obtained

$$\mathbf{y} = \mathbf{H}x + \mathbf{e}, \qquad (4.34a)$$

$$\mathbf{y} = \begin{pmatrix} r_2^2 - r_1^2 - \|p_2\|^2 \\ r_3^2 - r_1^2 - \|p_3\|^2 \\ \vdots \\ r_N^2 - r_1^2 - \|p_N\|^2 \end{pmatrix}, \quad \mathbf{H} = \begin{pmatrix} -2p_{2,1} & -2p_{2,2} \\ -2p_{3,1} & -2p_{3,2} \\ \vdots & \vdots \\ -2p_{N,1} & -2p_{N,2} \end{pmatrix}, \quad \mathbf{e} = \begin{pmatrix} e_2 - e_1 \\ e_3 - e_1 \\ \vdots \\ e_N - e_1 \end{pmatrix}.$$

$$(4.34b)$$

Note that a possible constant bias in r_k^2 (for instance, stemming from range noise) cancels out when taking the difference $e_k - e_1$. The covariance matrix of the noise has elements

$$[\bar{\mathbf{R}}]_{ij} = \begin{cases} R_{i+1} + R_1, & i = j, \\ R_1, & i \neq j. \end{cases} \qquad (4.35)$$

The WLS estimate is then given by

$$\hat{x} = \left(\mathbf{H}^T \mathbf{R}^{-1} \mathbf{H}\right)^{-1} \mathbf{H}^T \mathbf{R}^{-1} \mathbf{y} \qquad (4.36)$$

The choice of reference sensor introduces a subtle problem. Taking the one with the smallest error is one natural choice. The covariance matrix $\bar{R}$ will then be the most diagonal dominant of all choices, and the computationally cheap LS solution can be shown to give the smallest covariance. The WLS estimate should give the same result anyhow, but it comes at the price of requiring inversion of the $2(N-1) \times 2(N-1)$ matrix $\mathbf{R}$.

Range Parameter Trilateration

The idea here is to use the parameter vector $x = (x_1, x_2, R^2)^T$, where $R^2 = \|x\|^2 = x^T x = x_1^2 + x_2^2$ is the squared range to the target. The range squares can then for an arbitrary origin be written

$$r_k^2 = \|x - p_k\|^2 + e_k = (x_1 - p_{k,1})^2 + (x_2 - p_{k,2})^2 + e_k \qquad (4.37a)$$

$$= -2x_1 p_{k,1} - 2x_2 p_{k,2} + \|p_k\|^2 + R^2 + e_k. \qquad (4.37b)$$

This can be seen as a linear model

$$y_k = H_k x + e_k, \qquad (4.38a)$$

$$y_k = r_k^2 - \|p_k\|^2, \qquad (4.38b)$$

$$H_k = (-2p_{k,1}, -2p_{k,2}, 1). \qquad (4.38c)$$

4.4 Dedicated Least Squares Solutions

The extended parameter vector is now estimated with the same WLS formulas as in (4.36):

$$\hat{x} = \left(\mathbf{H}^T\mathbf{R}^{-1}\mathbf{H}\right)^{-1}\mathbf{H}^T\mathbf{R}^{-1}\mathbf{y}. \tag{4.39}$$

There are two ways to proceed:

- Use only the first two components of x that correspond to the sought position.
- Try to use the range information, for instance by using any of the transform methods described in Chapter A.3.

The next section describes an SLS alternative.

Range Parameter SLS approach

The set of linear equations (4.38) can be rewritten in the following way using the dummy parameter $R^2 = \|x\|^2 = x^T x = x_1^2 + x_2^2$:

$$\mathbf{y} = \mathbf{H}^x x + \mathbf{H}^r R^2 + \mathbf{e}, \tag{4.40}$$

$$\mathbf{y} = \begin{pmatrix} r_{1,1}^2 - \|p_1\|^2 \\ r_{2,1}^2 - \|p_2\|^2 \\ \vdots \\ r_{N,1}^2 - \|p_N\|^2 \end{pmatrix}, \quad \mathbf{H}^x = \begin{pmatrix} -2p_{1,1} & -2p_{1,2} \\ -2p_{2,1} & -2p_{2,2} \\ \vdots \\ -2p_{N,1} & -2p_{N,2} \end{pmatrix}, \quad \mathbf{H}^r = \begin{pmatrix} 1 \\ 1 \\ \vdots \\ 1 \end{pmatrix}.$$

In the same fashion as the separable least squares algorithm, the LS solution of x as a function of R^2 is

$$\hat{x}(R^2) = \left(H^{x,T}\mathbf{R}^{-1}H^x\right)^{-1} H^{x,T}\mathbf{R}^{-1}\left(\mathbf{y} - H^r R^2\right). \tag{4.41}$$

We now just have to find the scalar R^2 such that

$$R^2 = \|\hat{x}(R^2)\|^2 = \hat{x}^T(R^2)\hat{x}(R^2). \tag{4.42}$$

This gives a second order equation for R^2, which is easily solvable for its positive root, which is then plugged into the expression (4.41).

4.4.2 DOA Measurements

Fusion of *direction of arrival* measurements is the most straightforward case from an LS perspective, since the measurements fit the linear model directly. However, additive noise in the observation model (4.28) becomes highly non-linear in the transformed model. Each sensor measures an angle φ_k, which can be written

$$\varphi_k = \arctan2(x_1 - p_{k,1}, x_2 - p_{k,2}), \tag{4.43a}$$

$$(x_1 - p_{k,1})\tan(\varphi_k) = x_2 - p_{k,2}. \tag{4.43b}$$

A linear model for all observations is then

$$\mathbf{y} = \mathbf{H}x + \mathbf{e}, \tag{4.44a}$$

$$\mathbf{y} = \begin{pmatrix} p_{1,1} \tan(\varphi_1) - p_{1,2} \\ p_{2,1} \tan(\varphi_2) - p_{2,2} \\ \vdots \\ p_{N,1} \tan(\varphi_N) - p_{N,2} \end{pmatrix}, \quad \mathbf{H} = \begin{pmatrix} \tan(\varphi_1) & -1 \\ \tan(\varphi_2) & -1 \\ \vdots & \vdots \\ \tan(\varphi_N) & -1 \end{pmatrix}. \tag{4.44b}$$

The noise in φ_k can be converted to an additive measurement noise using a Taylor expansion of $\tan(\varphi_k + e_k)$. However, this gives an approximate model, and even then it is not a standard WLS one. The problem here is that also the matrix $\mathbf{H}$ becomes uncertain, which gives rise to a so called *errors in variables* (*EIV*) problem in statistics, where the *total least squares* (*TLS*) algorithm can be used.

In summary, the transformation trick here is useful for manual solution of the system of equations, just as seamen have always drawn straight lines on maps to find the solution to triangulation problems. The most efficient solution in a statistical sense should be the NLS approach (4.29).

4.4.3 Least Squares Fusion of Hybrid Measurements

It should be obvious from the linear models formulated above for range, range difference and angle measurements, that any combination of such measurements can be formulated as a linear model by stacking (4.34), (4.40) and (4.44) on top of each other.

4.5 Extended Estimation Problems

So far the sensor network application has been restricted to estimate the target location x. Here, we briefly introduce the dual problem of sensor network calibration, simultaneous sensor and target localization and extended target models.

4.5.1 Sensor Localization Calibration

We have so far assumed the sensor location to be known. In a rapidly deployed network, the sensor positions may be partially or completely unknown. To calibrate the network, a collaborative target with known position can be used. That is, we consider the sensor observations $\mathbf{y} = \mathbf{h}(x; \mathbf{p}) + \mathbf{e}$ as a function of $\mathbf{p}$. Now, the roles of x and a sensor location p_k are completely reciprocal in sensor network measurements, since all measurements described here are a function of $x - p_k$. We can thus focus on estimating the location of each sensor position p_k separately. To get redundancy, the target should be moved to different positions $x_1, x_2, \ldots, x_M$. Let $\mathbf{x} = (x_1^T, x_2^T, \ldots, x_M^T)$. The observations from sensor k are thus of the form $\mathbf{y}_k = \mathbf{h}(\mathbf{x}; p_k) + \mathbf{e}_k$. The calibration problem is thus recast to the same form as the target localization problem, and the methods in Section 4.2 can be applied.

4.5.2 Simultaneous Localization and Mapping

A problem that falls into the more general class of *simultaneous localization and mapping (SLAM)* problems is when both sensor and target locations are unknown. For this problem to make sense, there has to be several sensors and targets at different locations, so their relative difference can be determined by comparing the set of observations to the targets.

To define a first problem, suppose there are N sensors and either M targets or one target observed at M different times. Assume that each sensor observes relative positions

$$y_{ij} = x_j - p_i + e_i, \quad i = 1, 2, \ldots, N, \quad j = 1, 2, \ldots, M. \quad (4.45a)$$

Defining a joint parameter vector x, a linear model can be defined as

$$x = \begin{pmatrix} x_1 \\ x_2 \\ \vdots \\ x_M \\ p_1 \\ p_2 \\ \vdots \\ p_N \end{pmatrix}, \mathbf{y} = \begin{pmatrix} y_{11} \\ y_{12} \\ \vdots \\ y_{1M} \\ y_{21} \\ \vdots \\ y_{NM} \end{pmatrix}, \mathbf{H} = \begin{pmatrix} I_2 & 0 & 0 & 0 & -I_2 & 0 & 0 & 0 \\ I_2 & 0 & 0 & 0 & 0 & -I_2 & 0 & 0 \\ I_2 & 0 & 0 & 0 & 0 & 0 & -I_2 & 0 \\ I_2 & 0 & 0 & 0 & 0 & 0 & 0 & -I_2 \\ 0 & I_2 & 0 & 0 & -I_2 & 0 & 0 & 0 \\ \vdots & & & & & & & \vdots \\ 0 & 0 & 0 & I_2 & 0 & 0 & 0 & -I_2 \end{pmatrix}.$$

(4.45b)

Note that this model is not identifiable, since $\mathbf{H}$ is rank deficient. This can be seen from the fact that there are two right eigenvectors to $\mathbf{H}$ with just ones that have eigenvalues zero. The least squares estimate delivered by the backslash operator is the one with minimum norm, that is all locations are maximally centralized. To get rid of this ambiguity, one sensor location can be fixed to the origin.

―― **Example 4.5: SLAM principle** ――

First, the scenario in Figure 4.8 is setup.

```
M=4; % Number of targets
N=4; % Number of sensors
c='bgrmkcy';
p=rand(2,N);
x=rand(2,M);
for i=1:N
    plot(p(1,i),p(2,i),['o',c(1+mod(i-1,7))],'linewidth',2)
    hold on
```

```
text(p(1,i),p(2,i),['S',num2str(i)],'fontsize',16)
for j=1:M
  if i==1
    plot(x(1,j),x(2,j),['*',c(1+mod(j-1,7))],'linewidth',2)
    text(x(1,j),x(2,j),['T',num2str(j)],'fontsize',16)
  end
  Y{i,j}=ndist(x(:,j)-p(:,i),0.0000001*eye(2));
  y(i,j,:)=rand(Y{i,j},1);
end
end
```

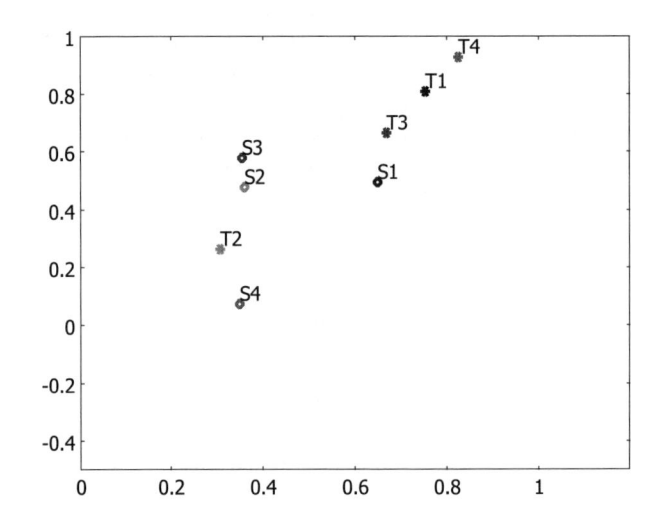

Figure 4.8: Four sensors and four targets, all with unknown position.

Note that the orientation of each sensor is assumed to be known, it is just the position that differs. The observations are illustrated as in Figure 4.9 as follows.

```
for i=1:N
  subplot(2,2,i)
  plot2(Y{i,1:4},'legend',[])
  title(['Observations from sensor ',num2str(i)])
  for j=1:M
    hold on
    plot(y(i,j,1),y(i,j,2),'.','linewidth',2)
    text(y(i,j,1),y(i,j,2),['O',num2str(i),num2str(j)],'fontsize',16)
  end
end
```

Next, the joint estimation of target and sensor locations are computed.

```
H=[kron(kron(eye(M),ones(N,1)),eye(2)) kron(ones(N,1),eye(2*M))];
Y=permute(y,[3 1 2]);
Y=Y(:);
rank(H)
ans =
14
H=[H;eye(2) zeros(2,14)];
```

4.5 Extended Estimation Problems

```
Y=[Y;zeros(2,1)];
Xhat=H\Y;
norm(Y-H*Xhat)
ans =
0.1401
```

The last line shows that the estimated position is consistent with the observations. The size of H is 16 while the rank is only 14. That is expected, since the absolute position of the network can never be estimated. For that reason, an extra dummy measurement is added, that says that the first sensor is located in the origin, to get a well-defined solution.

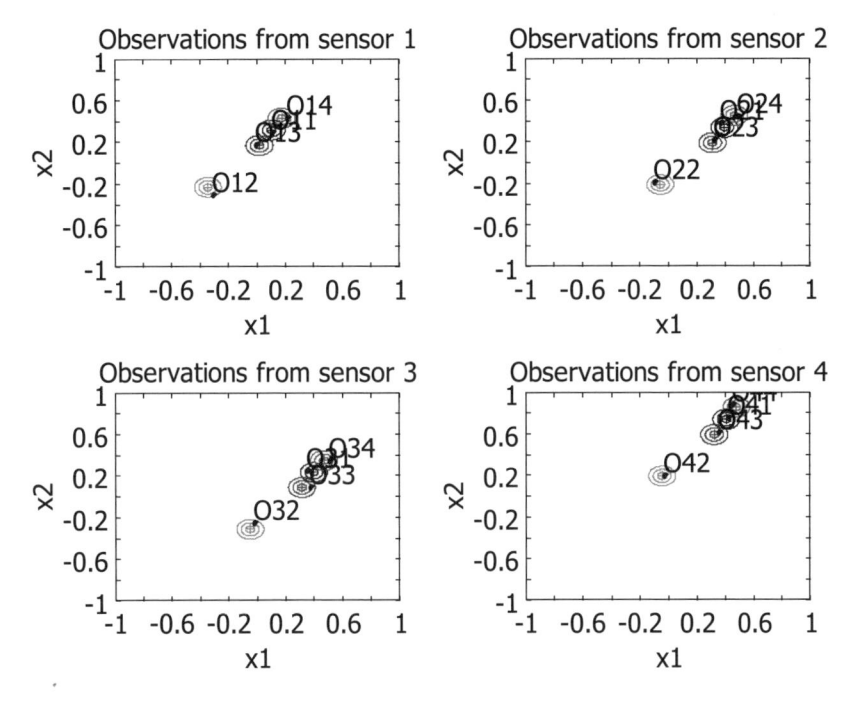

Figure 4.9: Observations $O_{ij} = x_j - p_i$, as seen from each sensor.

In a more realistic problem formulation, the sensor orientation α_i is an unknown parameter too. The linear model (4.45) then becomes nonlinear in α_i

$$y_{ij} = R(\alpha_i)(x_j - p_i + e_i), \quad i = 1, 2, \ldots, N, \quad j = 1, 2, \ldots, M, \quad (4.46a)$$

$$R(\alpha_i) = \begin{pmatrix} \cos(\alpha_i) & \sin(\alpha_i) \\ -\sin(\alpha_i) & \cos(\alpha_i) \end{pmatrix}. \quad (4.46b)$$

Since the model is nonlinear in the N angles α_i, the SLS approach here comes as a natural approach, besides nonlinear optimization.

4.5.3 Extended Target Models

So far, the target has been represented by its coordinates as a point. This is a good assumption when the observation uncertainty is much larger than the target size. For large objects or good sensors, this assumption does not hold. The problem is that the observations get a larger spread in space than the sensor itself promises. This can be taken care of by introducing a model of the target's shape, size and orientation. Common shapes include rectangles and ellipsoids, and the parameters to estimate include the position, size and orientation. The following example is typical for detection, localization and classification of ground vehicles.

---**Example 4.6: Extended target model**---

Consider a sensor network, where four sensors with good range and bearing accuracy deliver multiple measurement from a rectangular target as depicted in Figure 4.10. Using a model with $\mathbf{x} = (x_1, x_2, L, W, \psi)^T$, the position (x_1, x_2),

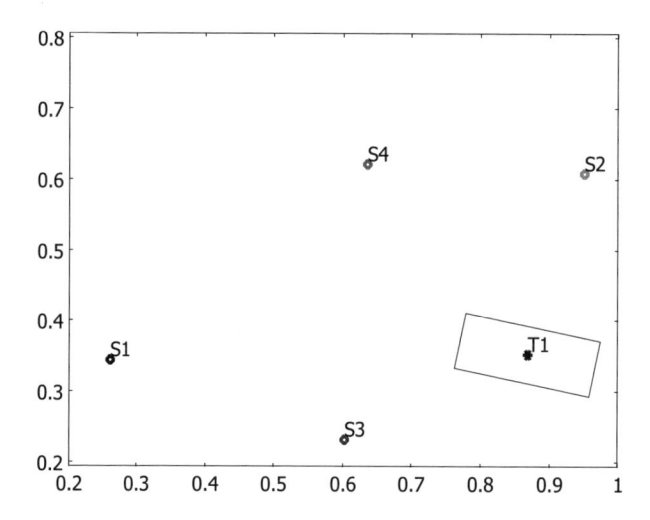

Figure 4.10: *The target is known to be rectangular with a size considerably larger than the observation uncertainty.*

length (L), width (W) and orientation ψ can be estimated with dedicated NLS algorithms.

4.6 Summary

4.6.1 Theory

This chapter described methods to estimate the parameters x in the nonlinear model

$$y_k = h_k(x - p_k) + e_k, \qquad \text{Cov}(e_k) = R_k, \quad k = 1, \ldots, N,$$
$$\mathbf{y} = \mathbf{h}(x - \mathbf{p}) + \mathbf{e}, \qquad \text{Cov}(\mathbf{e}) = \mathbf{R}.$$

The basic network measurements can be summarized as:

- TOA $r_k = \|x - p_k\| + e_k$,
- TDOA $r_k = \|x - p_k\| + e_k$,
- DOA $\varphi_k = \arctan2\left(x_1 - p_{k,1}, x_2 - p_{k,2}\right) + e_k$ and
- RSS $y_k = P^0 - \beta \log\left(\|x - p_k\|\right)$.

These can be combined to form a hybrid measurement $\mathbf{y} = \mathbf{h}(x - \mathbf{p}) + \mathbf{e}$, and the NLS algorithm can be applied. The separable least squares principle is another useful idea to obtain a low-dimensional problem where numerical or grid based search can be applied efficiently. One example concerns RSS measurement, where emitted power P^0 and path loss constant β are parameters entering linearly.

There is also a bag of tricks to get least squares problems where the thematic idea is to transform the measurements (and possibly the parameters if necessary) to

$$g^y(\mathbf{y}) = \mathbf{H} g^x(x) + g^e(\mathbf{e}). \tag{4.47}$$

The approaches are

- TOA range parameter trilateration: r_k^2 is linear in $(x_1, x_2, x_1^2 + x_2^2)^T$.
- TOA reference sensor trilateration: $r_k^2 - r_1^2$ is linear in x.
- DOA triangulation approach: x is an affine function in $\tan(\varphi_k)$.

Sensor network calibration using a target with known position forms a dual problem, where the NLS methodology can be applied to the usual network measurements considering the sensor location p_k as unknown. The NLS framework also enables more parameters to be estimated jointly with x, for instance the sensor locations or size and shape parameters for the target.

4.6.2 Software

A sensor network is conveniently represented by a `sensormod` object

$$y_k = h(k, x, u_k; \theta) + e_k, e_k \sim p_e(e),$$

where u_k is not used, and

- x contains the M unknown target locations.
- θ contains the N (partially) known sensor locations.
- h describes the measurement relation between sensors and targets (RSS, TOA, TDOA, DOA).
- e_k models the measurement noise, where $p_e(e)$ is one object in the `pdfclass` family (see `list(pdfclass)`).

Assume a 2D sensor network where each sensor delivers one measurement, and assume x is a $2 \times M$ matrix with target (prior) locations, p is a $2 \times N$ matrix with sensor locations. The SENSORMOD object is created by

```
nn=[2*M 0 ns 2*N]    %Structure parameters [nx nu ny nth]
s=sensormod(h,nn);
s.x0=x(:); x.px0=Px;
s.th=p(:); x.P=Pp;
```

Standard structures of SN are created by

```
s=exsensor(type,N,M)
```

where `type` is 'RSS', 'TOA', 'TDOA' or 'DOA'. The sensor and target positions are randomized over the unit square, but can be changed as above.

Use `plot(s)` to plot the network, and `y=simulate(s,t)` for simulating measurements (`t=1` can be used).

For estimating network parameters, the following methods are available:

- LS estimation `xhat=ls(s,y)`, returns a SIG object `xhat` from a sensor model object `s` and signal object `y`. The method is based on a first order Taylor expansion with numeric gradient H_k.

- WLS estimation `xhat=wls(s,y)`, returns a SIG object `xhat` from a sensor model object `s` and signal object `y`. WLS works as LS but compensates for the covariance of e_k.

- NLS estimation `shat=estimate(s,y)`, returns a calibrated SENSORMOD object `shat` from a nominal model `s` using a signal object `y`. The Gauss-Newton method is default, where numeric gradients are used for H_k.

4.6 Summary

- CRLB xcrlb=crlb(s), returns a SIG object xcrlb with the same state as given in xcrlb.x=s.x0 and covariance xcrlb.Px as the (parametric) CRLB. xcrlb=crlb1(s) and xcrlb=crlb2(s) plots the trace or determinant of the CRLB over a 1D and 2D grid, respectively, of the state(s). xcrlb=fim(s) returns the Fisher information matrix.

- Likelihood approach. Compute a one-dimensional/two-dimensional likelihood function by
  ```
  [lh,px,px0]=lh1(s,y,x1,ind)
  [lh,px,px0]=lh2(s,y,x1,x2,ind)
  ```
  The likelihood function is evaluated over a gridded 1D/2D state space, where ind specifies which index/indexes to vary. Without output arguments, a (contour) plot is returned.

- Calibration shat=calibrate(s,y), returns a SENSORMOD object shat where the sensor locations are calibrated using the NLS algorithm. The target locations are here assumed given in y.x with uncertainy y.Px.

5

Detection and Classification Problems

Statistical approaches to estimation are based on the likelihood $p(\mathbf{y}|x)$, or posterior $p(x|\mathbf{y})$, see Appendix A. Similarly, statistical approaches to detection problems concern the comparison of likelihoods $p(\mathbf{y}|H_i)$ of the observations given a hypothesis H_i. We will here in parallel apply the theory in Appendix D to some sensor network problems of discrete nature:

- Section 5.1: Detect whether a target is present or not.

- Section 5.2: Decide which target, from a list of known targets, an observation stems from.

- Section 5.3: Associate a number of observations to a number of targets.

The linear model $\mathbf{y} = \mathbf{H}x + \mathbf{e}$ and nonlinear model $\mathbf{y} = \mathbf{h}(x) + \mathbf{e}$ are treated in turn for each problem.

5.1 Detection

The detection problem in sensor networks is to detect whether a target is present or not. An observation that does not correspond to a target is usually referred to as *clutter*. The null hypothesis H_0 is that there is no target and the alternate hypothesis H_1 can involve a model. The general hypotheses can be stated as

$$H_0 : \mathbf{y} = \mathbf{e}^0, \qquad \mathbf{e}^0 \sim p^0(\mathbf{e}^0), \qquad (5.1a)$$

$$H_1 : \mathbf{y} = \mathbf{h}(x) + \mathbf{e}^1, \qquad \mathbf{e}^1 \sim p^1(\mathbf{e}^1). \qquad (5.1b)$$

One typical application is a radar that gives random detections (*clutter*) or range and bearing to a target. Under H_0, the clutter $\mathbf{e}^0$ can be modeled with a uniform distribution or an exponential distribution for instance. Under H_1, the range error might be uniformly distributed and the bearing error Gaussian.

The underlying assumption to get explicit expressions in this section is that the observation noise is Gaussian $\mathbf{e} \sim \mathcal{N}(0, \mathbf{R})$. However, the problem setup is general, and extensions to general distributions are straightforward. The background theory is described in Appendix D.

5.1.1 No Model

The simplest test is to assume nothing about the target, and just check if the sensor observation is just noise or not,

$$H_0 : \mathbf{y} = \mathbf{e}, \tag{5.2a}$$
$$H_1 : \mathbf{y} \neq \mathbf{e}. \tag{5.2b}$$

A Gaussian assumption $\mathbf{e} \sim \mathcal{N}(0, \mathbf{R})$ implies that the *test statistic*

$$T(\mathbf{y}) = \mathbf{y}^T \mathbf{R}^{-1} \mathbf{y} \sim \chi^2_{Nn_y} \tag{5.3}$$

is χ^2 distributed under H_0. The number of freedoms Nn_y is given by the dimension of the vector $\mathbf{y}$, assuming N sensors, where each y_k is of dimension n_y. The hypothesis test is to compute $T(\mathbf{y})$ from the available observations $\mathbf{y}$, and then compare this test statistic to a predefined threshold h. If it exceeds this value, the H_0 hypothesis is rejected.

The detection threshold h is chosen such that $\mathrm{P}(T(\mathbf{y}) > h | H_0) = P_{FA}$. The advantages of having a model in the alternate hypothesis are:

- The threshold h is taken from a $\chi^2_{n_x}$ distribution rather than $\chi^2_{Nn_y}$, which implies that the threshold will be smaller and thus the detection rate P_D increases.

- One can compute the theoretical detection rate $P_D(x)$ as a function of the true parameters x for a given false alarm rate P_{FA}. One can also plot the so called *receiver operating characteristic (ROC)* curve with $P_D(x)$ and P_{FA} as a function of the threshold h for a given x.

5.1.2 Linear model

Using the linear model $\mathbf{y} = \mathbf{H}x + \mathbf{e}$, where $\mathbf{e} \sim p_{\mathbf{e}}(\cdot)$, the detection test is

$$H_0 : x = 0, \tag{5.4a}$$
$$H_1 : x \neq 0. \tag{5.4b}$$

A Suboptimal Approach

A suboptimal approach is to again use the test statistic $T(\mathbf{y}) = \mathbf{y}^T \mathbf{R}^{-1} \mathbf{y} \sim \chi^2_{Nn_y}$. Under H_1 and the given model, $T(\mathbf{y})$ is distributed according to the non-central χ^2 distribution $\chi^2_{Nn_y}(x^T \mathbf{H}^T R^{-1} \mathbf{H} x)$. From this distribution, the theoretical $P_D(x)$ can be computed for each value of x, and the ROC curve can be plotted for any value of x. See Example 5.1 for an illustration. This approach, however, does not take the structure of the model into account. One effectively loses $Nn_y - n_x$ degrees of freedom compared to the approach below.

The GLR Test

The *generalized likelihood ratio test (GLRT)* becomes

$$T(\mathbf{y}) = 2\log \frac{\max_x p_{\mathbf{e}}(\mathbf{y} - \mathbf{H}x)}{p_{\mathbf{e}}(\mathbf{y})} = 2\log \frac{p_{\mathbf{e}}(\mathbf{y} - \mathbf{H}\hat{x}^{ML})}{p_{\mathbf{e}}(\mathbf{y})}$$

where the ML estimate is defined as

$$\hat{x}^{ML} = \arg\max_x p_{\mathbf{e}}(\mathbf{y} - \mathbf{H}x),$$

For Gaussian noise, the ML estimate is given by the WLS estimate, and we have

$$\begin{aligned}
T(\mathbf{y}) &= V^{WLS}(0) - \min_x V^{WLS}(x) = V^{WLS}(0) - V^{WLS}(\hat{x}^{WLS}) \\
&= \mathbf{y}^T \mathbf{R}^{-1} \mathbf{y} - (\mathbf{y} - \mathbf{H}\hat{x}^{WLS})^T \mathbf{R}^{-1} (\mathbf{y} - \mathbf{H}\hat{x}^{WLS}) \\
&= 2\mathbf{y}^T \mathbf{R}^{-T} \mathbf{H}(\mathbf{H}^T \mathbf{R}^{-1} \mathbf{H})^{-1} \mathbf{H}^T \mathbf{R}^{-1} \mathbf{y} \\
&\quad - \mathbf{y}^T \mathbf{R}^{-T} \mathbf{H}(\mathbf{H}^T \mathbf{R}^{-1} \mathbf{H})^{-1} (\mathbf{H}^T \mathbf{R}^{-1} \mathbf{H})(\mathbf{H}^T \mathbf{R}^{-1} \mathbf{H})^{-1} \mathbf{H}^T \mathbf{R}^{-1} \mathbf{y} \\
&= \mathbf{y}^T \mathbf{R}^{-T} \mathbf{H}(\mathbf{H}^T \mathbf{R}^{-1} \mathbf{H})^{-1} \mathbf{H}^T \mathbf{R}^{-1} \mathbf{y} \\
&= \mathbf{y}^T \mathbf{R}^{-T/2} \underbrace{\left(\mathbf{R}^{-T/2} \mathbf{H}(\mathbf{H}^T \mathbf{R}^{-1} \mathbf{H})^{-1} \mathbf{H}^T \mathbf{R}^{-1/2} \right)}_{\Pi} \mathbf{R}^{-1/2} \mathbf{y}
\end{aligned}$$

We have here defined the projection matrix

$$\Pi = \mathbf{R}^{-T/2} \mathbf{H}(\mathbf{H}^T \mathbf{R}^{-1} \mathbf{H})^{-1} \mathbf{H} \mathbf{R}^{-1/2}, \tag{5.5}$$

which obeys the relation $\Pi\Pi = \Pi$. Note that this projection matrix Π has the property that n_x eigenvalues are one and the remaining ones are zero. We also define the *residual* as the normalized measurement,

$$r = \mathbf{R}^{-1/2} \mathbf{y}, \tag{5.6}$$

which gets the covariance matrix equal to the identity matrix. Assuming a true parameter in the model $\mathbf{y} = \mathbf{H}x^o + \mathbf{e}$, we get for the two hypotheses

$$T(\mathbf{y}) \sim \begin{cases} \chi^2_{n_x} & \text{under } H_0, \\ \chi^2_{n_x}(\lambda) & \text{under } H_1, \end{cases} \tag{5.7a}$$

$$\lambda = (x^o)^T \mathcal{I}(x^o) x^o, \tag{5.7b}$$

$$\mathcal{I}(x^o) = \mathbf{H}^T \mathbf{R}^{-1} \mathbf{H}. \tag{5.7c}$$

The number n_x of degrees of freedom in the $\chi^2_{n_x}$ distribution comes from the projection matrix Π that projects the residual to an n_x dimensional subspace. For a given $\mathbf{H}$ this can be used to compute the theoretical threshold h for a certain false alarm rate P_{FA} using the χ^2 distribution. For each x^o, the detection probability P_D can be computed from the noncentral χ^2 distribution, and the complete ROC (see Chapter D) curve $(P_{FA}(h), P_D(h))$ can be plotted.

In the two-dimensional sensor network case, the probability for false alarm can be computed explicitly using a change of variable (Cartesian to polar) as

$$P_{FA} = \int_{r^T r > h} \frac{1}{2\pi} e^{-\frac{r^T r}{2}} dr = \int_0^{2\pi} \int_h^\infty \frac{x}{2\pi} e^{-\frac{x^2}{2}} dx d\phi = e^{-\frac{h^2}{2}}.$$

which means that the threshold design is to choose P_{FA} and then letting $h = \sqrt{-2\log(P_{FA})}$.

──── **Example 5.1: Two sensors with good range resolution** ────

Let us revisit Example 2.1, where the observation model is

$$y_1 = x + e_1, \quad \text{Cov}(e_1) = R_1 \tag{5.8a}$$

$$y_2 = x + e_2, \quad \text{Cov}(e_2) = R_2 \tag{5.8b}$$

$$\mathbf{y} = \mathbf{H}x + \mathbf{e}, \quad \text{Cov}(\mathbf{e}) = \mathbf{R} \tag{5.8c}$$

$$\mathbf{H} = \begin{pmatrix} I \\ I \end{pmatrix} \tag{5.8d}$$

$$\mathbf{R} = \begin{pmatrix} R_1 & 0 \\ 0 & R_2 \end{pmatrix} \tag{5.8e}$$

The projection matrix (5.5) is

$$\Pi = \begin{pmatrix} 22.8 & 22.2 & 5.0 & 0.0 \\ 22.2 & 22.8 & 0.0 & 5.0 \\ 5.0 & 0.0 & 22.8 & -22.2 \\ -0.0 & 5.0 & -22.2 & 22.8 \end{pmatrix}$$

The threshold corresponding to a false alarm rate of $1 - 0.999$ is computed by

```
h=erfinv(chi2dist(2),0.999)
h =
10.1405
```

5.1 Detection

We add the following lines to generate N Monte Carlo samples from each sensor

```
N=1000; % Number of samples
y1=rand(X1,N);
y2=rand(X2,N);
plot(y1(1:10,1),y1(1:10,2),'.b')
plot(y2(1:10,1),y2(1:10,2),'.g')
```

and compute a histogram of the test statistic for each pair of observations

```
for i=1:N
   T(i)=[y1(i,:) y2(i,:)]*Pi*[y1(i,:) y2(i,:)]';
end
hist(T,50)
```

Figure 5.1 shows that the noncentral χ^2 distribution looks approximately Gaussian shaped, and that there is no risk to miss this detection.

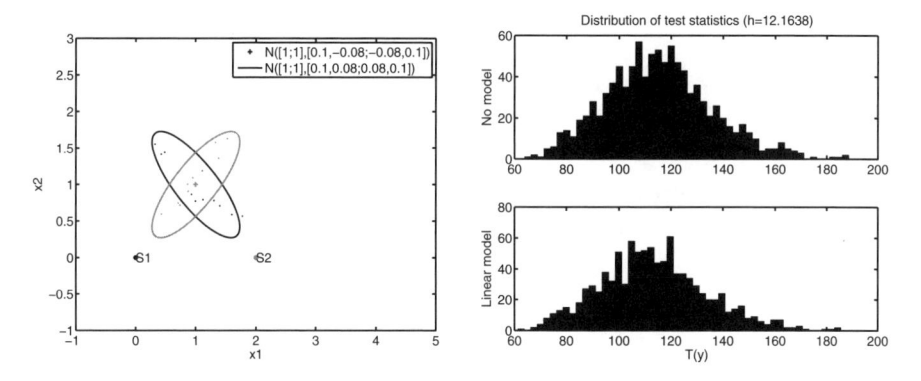

Figure 5.1: *Two sensors S1 and S2 measure a target with good range resolution but poor angle resolution. The histogram shows the test statistics in (5.3) and (5.7), respectively.*

5.1.3 Nonlinear Model

Using the nonlinear model $\mathbf{y} = \mathbf{h}(x) + \mathbf{e}$, where $\mathbf{e} \sim p_\mathbf{e}(\cdot)$, the detection test is

$$H_0 : \mathbf{h}(x) = 0, \tag{5.9a}$$
$$H_1 : \mathbf{h}(x) \neq 0. \tag{5.9b}$$

Note that $\mathbf{h}(0) \neq 0$ generally. Applying the no model approach, without using the structure in $h(x)$, gives the following test statistic:

$$T(\mathbf{y}) = \mathbf{y}^T \mathbf{R}^{-1} \mathbf{y} \sim \begin{cases} \chi^2_{Nn_y} & \text{under } H_0, \\ \chi^2_{Nn_y}(\mathbf{h}^T(x)\mathbf{R}^{-1}\mathbf{h}(x)) & \text{under } H_1, \end{cases} \quad (5.10)$$

Now, just as in the linear case, the structure contains a lot of information, and the *generalized likelihood ratio test (GLRT)* is the tool to use, see Appendix D.4.1.

The GLRT test statistic is here

$$T(\mathbf{y}) = 2 \log \left(\frac{\max_x p_\mathbf{e}(\mathbf{y} - \mathbf{h}(x))}{p_\mathbf{e}(\mathbf{y})} \right) = 2 \log \left(\frac{p_\mathbf{e}(\mathbf{y} - \mathbf{h}(\hat{x}^{ML}))}{p_\mathbf{e}(\mathbf{y})} \right), \quad (5.11)$$

where the ML estimate is defined as

$$\hat{x}^{ML} = \arg\max_x p_\mathbf{e}(\mathbf{y} - \mathbf{h}(x)). \quad (5.12)$$

Applying the asymptotic relation in (D.6), we get the asymptotic extension of the result (5.7) for the linear case

$$T(\mathbf{y}) \sim \begin{cases} \chi^2_{n_x} & \text{under } H_0, \\ \chi^2_{n_x}(\lambda) & \text{under } H_1, \end{cases} \quad \frac{n_y}{n_x} \to \infty \quad (5.13a)$$

$$\lambda = (x^o)^T \mathcal{I}(x^o) x^o \quad (5.13b)$$

where $\mathcal{I}(x^o)$ is the Fisher information matrix.

The difference of using no model and the model is thus essentially that the number of degrees of freedom has decreased from Nn_y to n_x. The non-central parameter is also changed. The following example illustrates the gain in performance of using the model for the case $n_x = 2$, $n_y = 1$ and $N = 5$.

───── **Example 5.2: TOA network** ─────

A random TOA network with five sensors is first generated, and illustrated in Figure 5.2. With a model of the background noise, the no model test can easily be checked. The code below returns no detection, and the level for this decision. The TOA model is nonlinear in the state, and thus linearization is required. The smaller detection threshold implies that the target is now detected with a high confidence level.

```
ny=5; nx=2;
s=exsensor('toa',ny,1);     % Default network
s.pe=0.2*eye(ny);           % Set noise level
plot(s)
y=simulate(s);
T=y.y*inv(cov(s.pe))*y.y';  % Test statistic
[b,level]=detect(chi2dist(ny),y.y*inv(cov(s.pe))*y.y')
b =
```

5.1 Detection

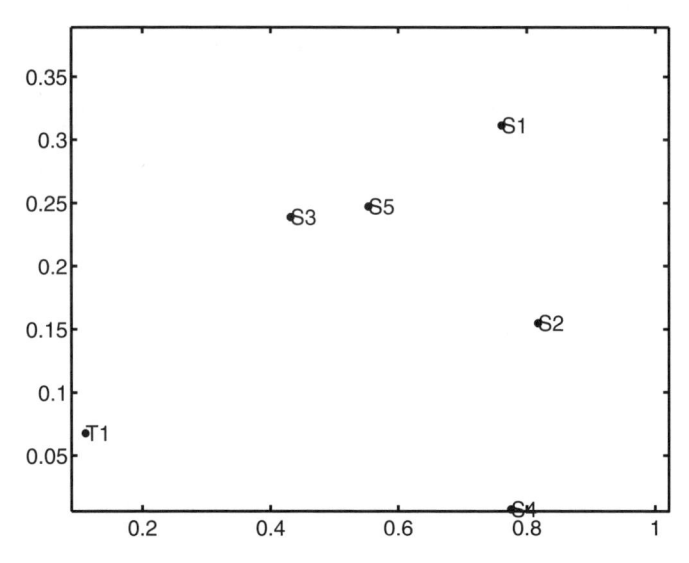

Figure 5.2: TOA network in Example 5.2.

```
         0
level =
    0.9626
[b,level]=detect(s,y)
b =
         1
level =
    0.9985
```

The distributions of the two test statistics assuming no model and model $\mathbf{h}(x)$, respectively, are computed and plotted in the code below.

```
s0=s;
s0.pe=zeros(ny);     % Dummy model without noise
y0=simulate(s0);     % Simulate h(x) without noise
hx=y0.y';            % Noise-free data
lambda=hx'*inv(cov(s.pe))*hx;  % Noncentral chi2 parameter
subplot(2,1,1)
plot(chi2dist(ny),ncchi2dist(ny,lambda))
hold on, plot([T T],[0 0.2],'k*--')  % Mark T(y)
axis([0 30 0 0.2])
subplot(2,1,2)
plot(chi2dist(nx),ncchi2dist(nx,lambda))
hold on, plot([T T],[0 0.2],'k*--')  % Mark T(y)
axis([0 30 0 0.2])
```

Figure 5.3 shows that the statistic $T(\mathbf{y})$ indeed is in the tail of the χ_5^2 distribution, but apparently from the test it is not sufficiently large to give a detection on the 1% level. Now, the χ_2^2 distribution decays much quicker, so here $T(\mathbf{y})$ is large enough. The distribution in the alternate hypothesis indicates the detection level.

The practical way to illustrate how the level of the test depends on the

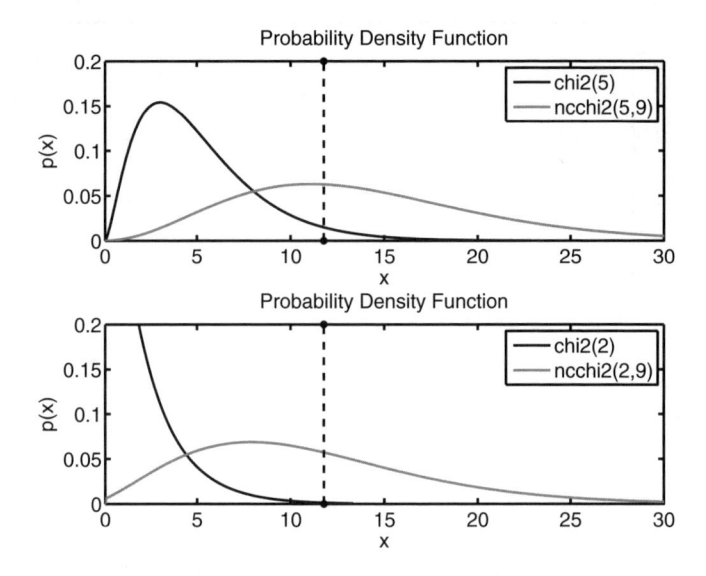

Figure 5.3: The distribution of the test statistic for the TOA network in Figure 5.2 under H_0 and $H_1(x^o)$, respectively. The observed $T(\mathbf{y})$ is marked with a vertical dashed line. (a) No model with $T(\mathbf{y})$ as in (5.10). (b) Model with $T(\mathbf{y})$ as in (5.11).

chosen false alarm rate is to plot the ROC curve. The ROC curves in Figure 5.4 is generated with the code below.

```
roc(s);
hold on
roc(chi2dist(2),lambda);
hold off
```

Note that only the network needs to be specified, since P_D is a function of P_{FA} and the true target position x^o. Note that there is one ROC curve for each value of x^o.

The detection probability as a function of target position, $P_D(x)$, is an interesting property of a network deployment. The code below generates Figure 5.5.

```
pdplot2(s,[],0:0.05:1,0:0.05:1);
```

5.1 Detection

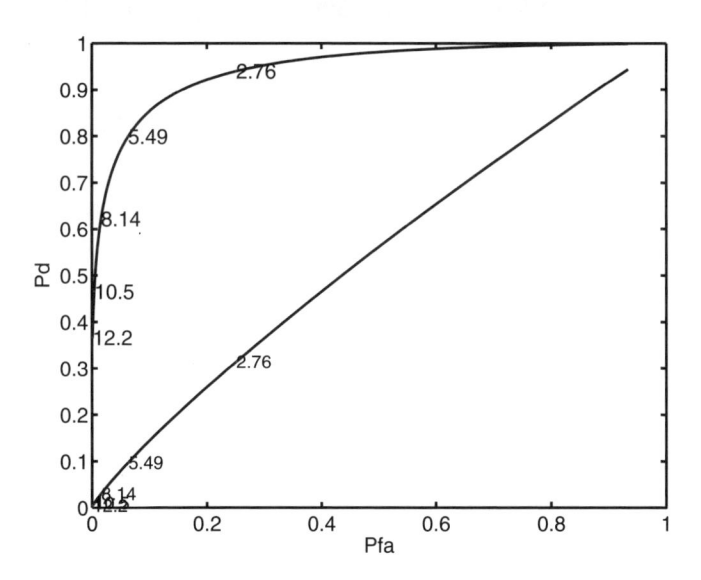

Figure 5.4: ROC curve $P_D(P_{FA})$ in the TOA network in Figure 5.2 for the cases of no model and model, respectively. The structure in the model allows a much stronger test (the upper curve with higher P_D for each P_{FA}).

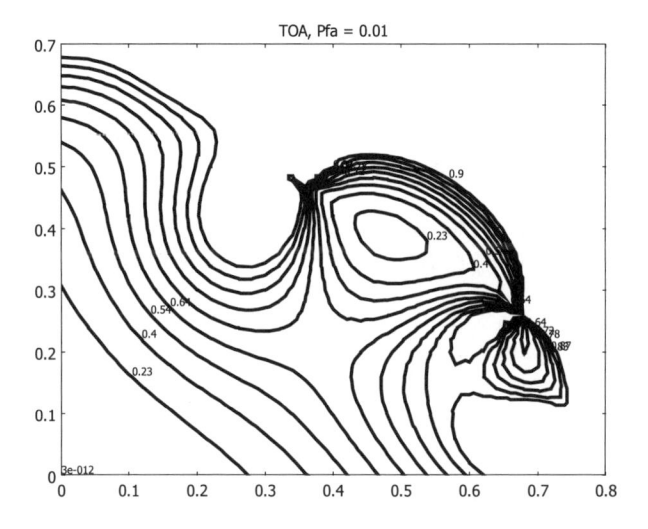

Figure 5.5: $P_D(x)$ for $P_{FA} = 0.01$ in the TOA network in Figure 5.2.

5.1.4 Nonlinear Model with Different Noise Distributions

Next, the general hypothesis test (5.1) is considered,

$$H_0 : \mathbf{y} = \mathbf{e}^0, \qquad\qquad \mathbf{e}^0 \sim p_\mathbf{e}^0(\mathbf{e}^0), \qquad (5.14a)$$
$$H_1 : \mathbf{y} = \mathbf{h}(x) + \mathbf{e}^1, \qquad\qquad \mathbf{e}^1 \sim p_\mathbf{e}^1(\mathbf{e}^1). \qquad (5.14b)$$

The straightforward generalization of (5.11) gives the GLRT test statistic

$$T(\mathbf{y}) = 2\log\left(\frac{\max_x p_\mathbf{e}^1(\mathbf{y}-\mathbf{h}(x))}{p_\mathbf{e}^0(\mathbf{y})}\right) = 2\log\left(\frac{p_\mathbf{e}^1(\mathbf{y}-\mathbf{h}(\hat{x}^{ML}))}{p_\mathbf{e}^0(\mathbf{y})}\right), \qquad (5.15)$$

where the ML estimate is defined as

$$\hat{x}^{ML} = \arg\max_x p_\mathbf{e}^1(\mathbf{y}-\mathbf{h}(x)). \qquad (5.16)$$

5.1.5 Local Detection

So far, we have assumed a centralized detection algorithm which has access to all observations in the network. This may imply too much signalling, and a local detection approach may be prefered for rare events. The idea is that each sensor applies a detection test, and only if it indicates an event, it transmits a message to a central decision unit.

Consider for that reason a local hypothesis test at each sensor:

$$H_{0,k} : h_k(x) = 0, \qquad (5.17a)$$
$$H_{1,k} : h_k(x) \neq 0. \qquad (5.17b)$$

The test statistic (5.10) gives now

$$T(y_k) = y_k^T R_k^{-1} y_k \sim \begin{cases} \chi_{n_y}^2 & \text{under } H_{0,k}, \\ \chi_{n_y}^2(h_k^T(x) R_k^{-1} h_k(x)) & \text{under } H_{1,k}, \end{cases} \qquad (5.18)$$

This test can be applied also in the case $n_y < n_x$, in particular it can be used for scalar observations $n_y = 1$. If $n_y > n_x$, one can still apply the model-based GLRT test as in (5.11).

Suppose each sensor only communicates its observation if its test statistics exceeds a threshold corresponding to a local false alarm rate $P_{FA,\text{local}}$. There are many principles for how the local decision can be made. Suppose the rule is to give an alarm if there is at least one detection in the network. The total false alarm rate is then

$$P_{FA} = 1 - \left(1 - P_{FA,\text{local}}\right)^N. \qquad (5.19)$$

For instance, to get a false alarm rate of 0.01 in a network with $N = 5$ sensors, the local false alarm rate has to be designed to 0.002.

5.2 Classification

The detection rate $P_{D,k}(x)$ will depend on both the sensor k and the parameter x. The total detection rate is

$$P_D(x) = 1 - \prod_{k=1}^{N} \left(1 - P_{D,k}(x)\right). \tag{5.20}$$

Note that it suffices that one $P_{D,k}(x)$ is close to one to get a total $P_D(x)$ close to one.

---Example 5.3: TOA network---

The code below computes the individual test statistics $T(y_k)$, the non-central parameter $h_k^T(x)h_k(x)$, and generates Figure 5.6.

```
hk=hx.^2.*diag(inv(cov(s.pe)))
hk =
    0.0131
    0.2268
   11.1119
    3.4106
    4.9507
Tk=y.y'.^2.*diag(inv(cov(s.pe)))
Tk =
    0.0678
    0.4121
   11.8875
    0.2479
    3.9312
for k=1:ny;
   H1k{k}=ncchi2dist(1,hk(k));
end
plot(H1k{:},chi2dist(1))
hold on, plot([Tk'; Tk'],repmat([0; 0.2],1,ny),'*--'), % Mark T(y)
axis([0 30 0 0.2])
```

5.2 Classification

Classification extends detection to decide between several competing alternate hypotheses H_i. Again, three slightly different cases of increasing difficulty are discussed in turn.

5.2.1 Direct Observation

If the parameter is directly observed, the alternate hypotheses are

$$H_i : \mathbf{y} = x_i + \mathbf{e}, \quad i = 1, 2, \ldots, M. \tag{5.21}$$

Define the classification index $c \in [1, M]$ which picks out the best hypothesis H_c. The simplest solution is the *nearest neighbor* (NN) solution,

$$\hat{c}^{NN} = \arg\min_i \|\mathbf{y} - x_i\|. \tag{5.22}$$

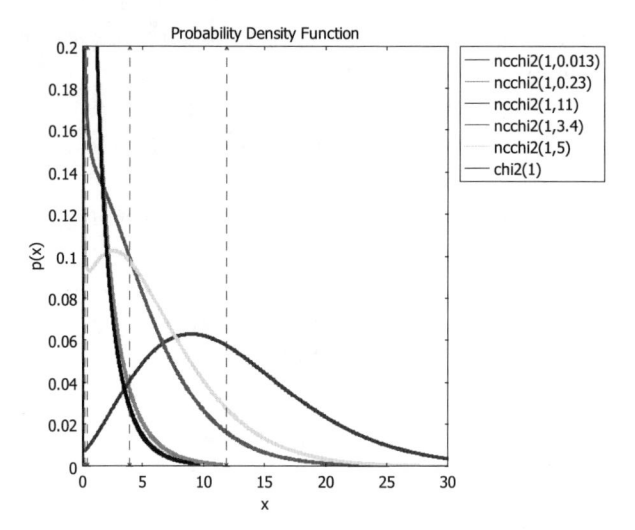

Figure 5.6: *The distribution of the $N = 5$ local test statistics for the TOA network in Figure 5.2 under H_0 and $H_1(x^o)$, respectively. The observed $T(y_k)$ are marked with vertical dashed lines.*

The ML solution is

$$\hat{c}^{ML} = \arg\min_i (\mathbf{y} - x_i)^T \mathbf{R}^{-1}(\mathbf{y} - x_i). \tag{5.23}$$

The NN approach is the simplest one and geometrically appealing. When the noise covariance matrix is not a scaled identity matrix, the ML approach might perform much better.

──── **Example 5.4: Classification of target identity** ────

Figure 5.7(a) shows the distribution and ten samples of the measurements from sensor 1 to target 1, which in this classification framework is assumed to have known position. The problem is that there are four possible targets, and the problem is to choose which one the measurement stems from.

First, a setup with one sensor and four targets is randomized.

```
p1=[0;0];
x=[1 1 0.5 0.5;1 1.5 1 1.5];
X=ndist(x(:,1),0.1*[1 -0.8;-0.8 1]);
plot2(X,'legend',[])
N=1000; % Number of samples
y=rand(X,N);
plot(y(1:10,1),y(1:10,2),'.b')
```

The transformed measurements and target locations are computed next.

```
P=sqrtcov(inv(cov(X)));
ybar=y*P;
Xbar=P'*X
```

5.2 Classification

```
xbar=P'*x
plot2(Xbar,'legend',[])
hold on
plot(ybar(1:10,1),ybar(1:10,2),'.b')
```

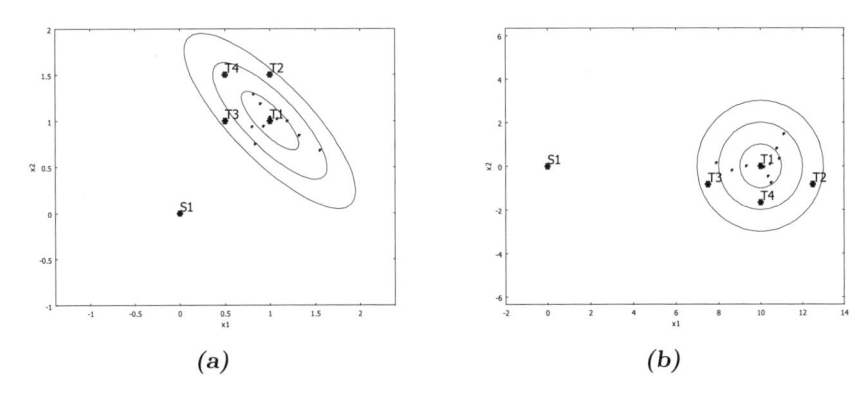

Figure 5.7: One sensor S1 measures the position of one out of four known targets T1–T4. (a) original observations $\mathbf{y}$ and (b) normalized observations $\bar{x}$.

Figure 5.7(b) shows a normalized measurement distribution with ten samples, where the transformation is chosen such that the covariance matrix becomes a unit matrix. This transformation is the key for the analysis.

5.2.2 Linear Model

The alternate hypotheses are for a linear model

$$H_i: \mathbf{y} = \mathbf{H}x_i + \mathbf{e}, \quad i = 1, 2, \ldots, M.$$

The NN solution is

$$\hat{c}^{NN} = \arg\min_i \|\hat{x}^{WLS} - x_i\|, \tag{5.24a}$$

$$\hat{x}^{WLS} = \left(\mathbf{H}^T \mathbf{R}^{-1} \mathbf{H}\right)^{-1} \mathbf{H}^T \mathbf{R}^{-1} \mathbf{y}. \tag{5.24b}$$

The direct ML solution is

$$\hat{c}^{ML} = \arg\min_i (\mathbf{y} - \mathbf{H}x_i)^T \mathbf{R}^{-1} (\mathbf{y} - \mathbf{H}x_i). \tag{5.25}$$

Using the concept of *equivalent measurement*, the ML solution can also be expressed as

$$\hat{c}^{ML} = \arg\min_i (\hat{x}^{WLS} - x_i)^T P_i^{-1} (\hat{x}^{WLS} - x_i), \tag{5.26a}$$

$$P = \left(\mathbf{H}^T \mathbf{R}^{-1} \mathbf{H}\right)^{-1}. \tag{5.26b}$$

That is, the equivalent measurement $(\hat{x}^{WLS}|H_i) \sim \mathcal{N}(x_i, P_i)$ projects the problem to the direct observation case of the previous section. The advantages with this projection are that the vectors $\hat{x}^{WLS} - x_i$ are of smaller dimension, and that it is easier to get an intuitive feeling for the problem in the parameter space.

5.2.3 Nonlinear Model

The alternate hypotheses are for a nonlinear model

$$H_i: \quad \mathbf{y} = \mathbf{h}(x_i) + \mathbf{e}, \quad i = 1, 2, \ldots, M. \tag{5.27}$$

The NN solution can be approximated by using the ML parameter estimate $\hat{x}^{ML}$ and covariance P instead of the WLS in the previous section,

$$\hat{c}^{NN} = \arg\min_i \|\hat{x}^{ML} - x_i\|. \tag{5.28}$$

Similarly, the ML solution can be approximated with

$$\hat{c}^{ML} = \arg\min_i \left(\hat{x}^{ML} - x_i\right)^T P_i^{-1} \left(\hat{x}^{ML} - x_i\right). \tag{5.29}$$

5.2.4 Different Parameter Spaces

One can distinguish the following spaces for classification.

- In the *physical parameter space*, classification is defined by nonlinear lines (manifolds).

- In the *normalized parameter space*, the classification is defined by straight lines.

- There exists a *structured parameter space* if $M \leq n_x$, where a nonoptimal classification is performed by just checking which component of the parameter is the largest one.

We formalize these latter two spaces as follows.

Definition 5.1 (Normalized parameter space)
The normalized parameter space is defined by a transformation T such that $\text{Cov}(\tilde{\hat{x}}) = \text{Cov}(T\hat{x}) = I$. For the linear model (2.1) and the WLS estimate (2.7), the transformation is given by $T = (\mathbf{H}^T \mathbf{R}^{-1} \mathbf{H})^{-1/2}$.

5.2 Classification

Definition 5.2 (Structured parameter space)
The structured parameter space is defined by the fact that all hypotheses have $\bar{x}_i$ as corners on a unit cube. For instance, $\bar{x}_i = e_i$ can be the i'th unit vector. Let S be a matrix with the selected corners on the unit cube The transformation is given by $T = S(x_1, x_2, \ldots x_M)^{-\dagger}$. The necessary requirement for the pseudo-inverse $\dagger$ to exist is that there are at least as many parameters as hypotheses, $n_x \geq M$.

Common choices of S include $S = I$ and $S = \mathbf{1}\mathbf{1}^T - I$.

5.2.5 Classification Algorithm

The ML approach to classification is summarized in Algorithm 5.1.

Algorithm 5.1 ML Classification

Assume a direct observation $\mathbf{y} = \mathbf{H}x + \mathbf{e}$ ($\mathbf{H} = I$), a linear model $\mathbf{y} = \mathbf{H}x + \mathbf{e}$ or a nonlinear model $\mathbf{y} = \mathbf{h}(x) + \mathbf{e}$ ($\mathbf{H} = d\mathbf{h}(x)/dx$).

1. Compute the parameter estimate

$$\hat{x} = \left(\mathbf{H}^T \mathbf{R}^{-1} \mathbf{H}\right)^{-1} \mathbf{H}^T \mathbf{R}^{-1} \mathbf{y}. \tag{5.30a}$$

2. Compute the normalized parameter estimate as

$$\hat{\bar{x}} = \left(\mathbf{H}^T \mathbf{R}^{-1} \mathbf{H}\right)^{-1/2} \hat{x}. \tag{5.30b}$$

3. Compute the transformed hypotheses

$$\bar{x}_i = \left(\mathbf{H}^T \mathbf{R}^{-1} \mathbf{H}\right)^{-1/2} x_i. \tag{5.30c}$$

4. Classify $\hat{c}$ as

$$\hat{c} = \arg\min_i \|\hat{\bar{x}} - \bar{x}_i\|_2^2 \tag{5.30d}$$

5.2.6 Classification Error

We can compare classification to the demodulation problem in digital communication. In modulation theory, using an additive Gaussian error assumption, it is straightforward to compute the risk for incorrect symbol detection. We will here extend these expressions from regular 2D (complex plane) patterns to general vectors in $\mathcal{R}^{n_x}$.

The risk of incorrect classification can be computed exactly in the case of only two alternate hypotheses as follows. It relies on the symmetric distribution of $\hat{\bar{x}}$, where the decision region becomes a straight line. The first step is a change of coordinates to one where one axis is perpendicular to the decision plane. Because of the normalization, the Jacobian of this transformation equals one. The second step is to marginalize all dimensions except the one perpendicular to the decision plane. All these marginals integrate to one. The third step is to evaluate the Gaussian error function. Here we use the definition

$$\mathrm{erfc}(x) = 2\int_x^\infty \frac{1}{\sqrt{2\pi}} e^{-x^2/2} dx$$

The result in $\mathcal{R}^2$ (cf. Figure 5.7(b)) can be written

$$\mathrm{P}(i|j) = \frac{1}{2}\mathrm{erfc}\left(\frac{\|\bar{x}_j - \bar{x}_i\|}{2}\right)$$

We can now define the matrix P with classification probabilities as

$$P^{(i,j)} = \mathrm{P}(\hat{c} = i|x = x_j), i \neq j$$
$$P^{(j,j)} = 1 - \sum_{i \neq j} P^{(i,j)}. \tag{5.31}$$

The missed detection probabilities are computed in a similar way as

$$P(\hat{c} = 0|x = x_j) = \frac{1}{2}\mathrm{erfc}\left(\frac{\|\bar{x}_j\|}{2}\right) \tag{5.32a}$$
$$P^{(0,0)} = 1 - \sum_j P^{(0,j)} < P_{FA}. \tag{5.32b}$$

Algorithm 5.2 ML classification error analysis

Consider the case $H_i : \mathbf{y} = \mathbf{H}x_i + \mathbf{e}$.

1. Compute the normalized estimates as

$$\bar{x}_i = \left(\mathbf{H}^T\mathbf{R}^{-1}\mathbf{H}\right)^{-1/2} x_i. \tag{5.33}$$

2. The classification matrix is computed by

$$\mathrm{P}(i|j) = \frac{1}{2}\mathrm{erfc}\left(\frac{\|\bar{x}_j - \bar{x}_i\|}{2}\right)$$

The convention $\bar{x}_0 = x_0 = 0$ can be used to also include the probability of false alarms and missed detections.

For more than two alternate hypotheses, this expression is an approximation but, as in modulation theory, generally quite a good one. The approximation becomes worse when there are several conflicting hypotheses, which means that there are three or more hypotheses vectors in about the same direction.

Furthermore, in the classification we should allow the null hypothesis (0), where $x = 0$, to decrease the false alarm rate by neglecting residual vectors, though having large amplitude, being far from the known fault vectors. Consider for instance a normalized estimate $\bar{x} = (-1, 0)^T$ in Figure 5.7(b). This would most likely be caused by noise, not one of the parameters.

5.2.7 Classification with Range Uncertainty

We here modify and extend the classification theory in the previous section to the case where range r is uncertain. The measurement can then be seen a scaled version of the target position,

$$H_0 : \mathbf{y} = \mathbf{e}, \tag{5.34a}$$
$$H_i : \mathbf{y} = rx_i + \mathbf{e}, \quad i = 1, 2, \ldots, M. \tag{5.34b}$$

Here, r can be seen as a nuisance parameter. This case occurs for all bearings only sensors, including camera views where the depth is unknown.

Basically, we remain in the normalized parameter space but as a consequence of the unknown scaling the normalized distances are replaced with angles. The last steps in Algorithms 5.1 and 5.2, respectively, are modified as follows:

Algorithm 5.3 ML Classification with Range Uncertainty

1. Compute the normalized parameter estimate as

$$\hat{\bar{x}} = \left(\mathbf{H}^T \mathbf{R}^{-1} \mathbf{H}\right)^{-1/2} \hat{x}. \tag{5.35a}$$

2. Compute the transformed hypotheses

$$\bar{x}_i = \left(\mathbf{H}^T \mathbf{R}^{-1} \mathbf{H}\right)^{-1/2} x_i. \tag{5.35b}$$

3. Diagnose fault $\hat{c}$ as

$$\hat{c} = \arg\min_i \left| \frac{\hat{\bar{x}}}{\|\hat{\bar{x}}\|} - \frac{\bar{x}_i}{\|\bar{x}_i\|} \right|_2^2 = \arg\min_i \angle(\hat{\bar{x}}, \bar{x}_i) \tag{5.35c}$$

where $\angle(a, b)$ denotes the angle between the vectors a and b.

Quite interestingly, the error analysis in Algorithm 5.2 can be be modified as follows.

Algorithm 5.4 ML Classification Matrix with Range Uncertainty

1. Compute the normalized faults as in Algorithm 5.3, step 2.
2. The diagnosis matrix is computed by

$$P(i|j,r) = \text{erfc}\left(r\left\|\bar{x}_j - \frac{(\bar{x}_j, \bar{x}_j + \bar{x}_i)}{(\bar{x}_j + \bar{x}_i, \bar{x}_j + \bar{x}_i)}(\bar{x}_j + \bar{x}_i)\right\|\right)$$

5.3 Association

Consider the hypotheses

$$H_{ij}: \quad \mathbf{y}_j = x_i + \mathbf{e}_j, \quad i = 1, 2, \ldots, M, j = 1, 2, \ldots, n_y. \tag{5.36}$$

The problem is to find the pairs $(j, i) = (j, c_j)$ that associates a target number c_j to each observation j. Conversely, the inverse mapping c_i^{-1} specifies the pairs $(j, i) = (c_i^{-1}, i)$ where each target is associated with one measurement. The convention is that $c_i = 0$ means that there is no target associated with observation $\mathbf{y}_i$, and similarly for $c_i^{-1} = 0$. With this convention, the cases $M > n_y$ and $M < n_y$, respectively, are covered.

Let $D_{ij} = (\mathbf{y}_j - x_i)^T \mathbf{R}_j^{-1}(\mathbf{y}_j - x_i)$ be the normalized distances for each hypothesis. The overall optimal (ML) solution is given by

$$c = (c_1, \ldots, c_M) = \arg\min_c \sum_j D_{c_j j}. \tag{5.37}$$

The straightforward solution makes an enumeration of all $(\min(ny, M))!$ combinations, and evaluates $\sum_j D_{c_j j}$ for each of them. A much more efficient solution to this optimization problem is provided by the *auction algorithm*, also known as the *Hungarian algorithm* or *Munkre's algorithm*. The implementation has the syntax `[c,cinv]=auction(D)`.

―――― Example 5.5: Target association ――――

Consider a sensor network scenario where one sensor gets nine observations, each one originating from one target with known position. It is the observation uncertainty that makes the association problem nontrivial. Figure 5.8(a) shows one random realization. It is not obvious how to associate the observations. In fact, optimization using the auction algorithm gives that the first pair of observations–targets is mixed up.

First, the scenario is setup.

```
M=9; % Number of targets
N=1; % Number of sensors
c='bgrmkcy';
```

5.3 Association

```
p=rand(2,N);
x=rand(2,M);
for i=1:N
   plot(p(1,i),p(2,i),['o',c(1+mod(i-1,7))],'linewidth',2)
   hold on
   text(p(1,i),p(2,i),['S',num2str(i)],'fontsize',16)
   for j=1:M
     if i==1
        plot(x(1,j),x(2,j),['*',c(1+mod(j-1,7))],'linewidth',2)
        text(x(1,j),x(2,j),['T',num2str(j)],'fontsize',16)
     end
     X{i,j}=ndist(x(:,j),0.01*eye(2));
     y(i,j,:)=rand(X{i,j},1);
     plot(y(i,j,1),y(i,j,2),'.','linewidth',2)
     text(y(i,j,1),y(i,j,2),['O',num2str(i),num2str(j)],'fontsize',16)
   end
end
```

Then, all distances from observation j_1 to target j_2 are computed, and the auction algorithm is applied to find the smallest sum of row elements.

```
for j1=1:M
   for j2=1:M
      D(j1,j2)=norm(squeeze(y(1,j1,:))-x(:,j2));
   end
end
auction(D)
ans =
     2    1    3    4    5    6    7    8    9
```

Since the observations were already sorted in the same order as the targets, the perfect solution would be 1:9. However, in this realization, two nearby targets (1 and 2) are mixed up. Using this association vector, the association can be graphically illustrated as in Figure 5.8(b).

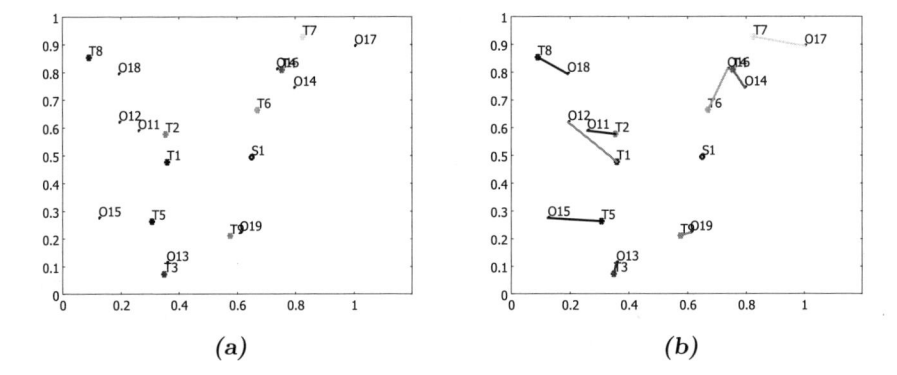

Figure 5.8: *Association of nine observations from one sensor. There are nine targets, so there are no false or missing observations, the only question is which observation that belongs to which target. (a) Sensor location S_1, target locations T_i and observations O_{1j}. (b) Association pairs (i,j) from the auction algorithm (where target 1 and 2 are mixed up).*

5.4 Summary

5.4.1 Theory

The discrete problem areas in this chapter relate to the following sensor network problems:

- Detection, test

$$H_0 : \mathbf{y} = \mathbf{e},$$
$$H_1 : \mathbf{y} \neq \mathbf{e}, \quad \text{or}$$
$$\mathbf{y} = \mathbf{H}x + \mathbf{e}, \quad \text{or}$$
$$\mathbf{y} = \mathbf{h}(x) + \mathbf{e}.$$

The first alternative gives the test $T(\mathbf{y}) = \mathbf{y}^T \mathbf{R}^{-1} \mathbf{y} \underset{H_0}{\overset{H_1}{\gtrless}} h$, where h is given by a $\chi^2_{n_y}$ distribution, while the threshold in the latter two cases is taken from $\chi^2_{n_x}$ distribution and gives a smaller threshold when $n_x < n_y$. As a consequence, the model gives higher P_D for a given P_{FA}.

The probability of detection $P_D(x^0)$ as a function of the parameter x^0 is computed from the noncentral χ^2 distribution

$$T(\mathbf{y}) \sim \chi^2_{n_x}\left((x^0)^T \mathbf{H}^T \mathbf{R}^{-1} \mathbf{H} x^0\right),$$

where $\mathbf{H}$ denotes the gradient of $\mathbf{H} = \mathbf{h}'(x^0)$ for the nonlinear model.

- Classification, test

$$H_i : \mathbf{y} = x_i + \mathbf{e}, \quad \text{or}$$
$$\mathbf{y} = \mathbf{H}x_i + \mathbf{e}, \quad \text{or}$$
$$\mathbf{y} = \mathbf{h}(x_i) + \mathbf{e}, \quad i = 1, 2, \ldots, M.$$

In the first case, the nearest neighbor (NN) solution is to choose the classification index $c \in [1, M]$ as $\hat{c} = \arg\min_i \|\mathbf{y} - x_i\|$, and the ML solution is $\hat{c} = \arg\min_i (\mathbf{y} - x_i)^T \mathbf{R}^{-1} (\mathbf{y} - x_i)$. For the other models, replace $\mathbf{y}$ with $\hat{x}^{ML}$ and $\mathbf{R}$ with P. The classification probabilities $P(\hat{c} = i | c = j)$ can be computed approximately.

- Association, test

$$H_{ij} : \mathbf{y}_j = x_i + \mathbf{e}_j, \quad \text{or}$$
$$\mathbf{y}_j = \mathbf{H}x_i + \mathbf{e}_j, \quad \text{or}$$
$$\mathbf{y}_j = \mathbf{h}(x_i) + \mathbf{e}_j, \quad i = 1, 2, \ldots, M, j = 1, 2, \ldots, n_y.$$

Let $D_{ij} = (\mathbf{y}_j - x_i)^T \mathbf{R}_j^{-1} (\mathbf{y}_j - x_i)$ be the normalized distances for each hypothesis. The auction algorithm computes the overall optimal solution $c = (c_1, \ldots, c_M) = \arg\min_c \sum_j D_{c_j j} = \arg\max_c \sum_j -D_{c_j j}$ such that $c_i \neq c_j$ for all $i \neq j$.

5.4.2 Software

The following methods of the object `sensormod` relate to the theory in this chapter:

- `[b,level]=detect(chi2(ny),T)` for detection without having an alternate model in H_1. Here, b is a binary alarm, and `level` is the significance level of $T(\mathbf{y})$.

- `[b,level]=detect(s,y)` for detection with an alternate model defined as the signal model s in H_1. Here, b is a binary alarm, and `level` is the significance level of $T(\mathbf{y})$.

- `[pd,pfa]=roc(s)` plots the ROC curve for the sensor network s. The vector of threshold values and the true parameter x_0 can be provided as `roc(s,x0,h)`, otherwise the x_0 specified in s is used and a default vector h is generated. This can be compared to the ROC curve when no model is used, which is given by the $\chi^2_{n_y}$ distribution as `[pd,pfa]=roc(chi2dist(ny))`.

- `pdplot2(s0,s1,x1min:x1step:x1max,x2min:x2step:x2max,pfa,ind)` plots $P_D(x_1, x_2)$ as a contour plot for the model s0 under H_0 and s1 under H_1 as a function of `x(ind(1))` and `x(ind(2))`.

The m-file `[c,cinv]=auction(D)` solves the auction problem.

Part II
Fusion in the Dynamic Case

6

Filter Theory

Filtering can be seen as an extension of estimation to a non-stationary parameter x_k governed by a dynamic model. The estimation problem in sensor network, that was used as an illustration in the previous chapters, becomes a filtering problem when the target is moving according a motion model.

The most general nonlinear *state space model* considered here is

$$x_{k+1} = f(x_k, u_k, v_k), \tag{6.1a}$$
$$y_k = h(x_k, u_k, e_k). \tag{6.1b}$$

Here, u_k is a known (control) input to the system, which will in the sequel often be omitted. An alternate slightly more general model is expressed in conditional densities for state transition and observation,

$$p(x_{k+1}|x_k), \tag{6.2a}$$
$$p(y_k|x_k). \tag{6.2b}$$

The special case of a nonlinear model with additive noise is sometimes useful for intuition and to get more concrete results,

$$x_{k+1} = f(x_k, u_k) + v_k, \tag{6.3a}$$
$$y_k = h(x_k, u_k) + e_k. \tag{6.3b}$$

Finally, the linear model will be one important special case of particular interest,

$$x_{k+1} = A_k x_k + B_{u,k} u_k + B_{v,k} v_k, \tag{6.4a}$$
$$y_k = C_k x_k + D_k u_k + e_k. \tag{6.4b}$$

The optimal solution given by Bayes' law is well-known, see Jazwinsky (1970), as will be surveyed in Section 6.3. In the case of a linear model, the Kalman filter theory applies, see Kailath et al. (2000), which is briefly discussed in this chapter and more thoroughly in Chapter 7. In the nonlinear case, several numerical approximations have appeared in the past, for instance in Kramer and Sorenson (1988) and Alspach and Sorenson (1972), see Section 6.4.1, though quite few applications of these have been reported. It was first with the invention of the particle filter in Gordon et al. (1993), which presented a general working solution with a sound theoretical basis, that the signal processing community started to apply approximate nonlinear filtering to real-world problems. This is the topic of Chapter 9.

6.1 Introduction

6.1.1 The Nonlinear Filtering Problem

A *dynamic system* can in general terms be characterized by a state space model with a hidden state from which partial information is obtained by observations. For the applications in mind, the state vector may include position, velocity and acceleration of a moving platform, and the observations may come from either internal on-board sensors (the *navigation problem*) measuring inertial motion or absolute position relative some landmarks, or from external sensors (the *tracking problem*) measuring for instance range and bearing to the target.

The *nonlinear filtering* problem is to make inference on the state from the observations. In the Bayesian framework, this is done by computing or approximating the posterior distribution for the state vector given all available observations at that time. For the applications in mind, this means that the position of the platform is represented with a conditional probability density function given the observations.

6.1.2 Brief History

Classical approaches to Bayesian nonlinear filtering described in literature include the following algorithms:

- The *Kalman filter* (KF) Kailath et al. (2000); Kalman (1960) computes the posterior distribution exactly for linear Gaussian systems by updating finite dimensional statistics recursively.

- For nonlinear non-Gaussian models, the KF algorithm can be applied to a linearized model with Gaussian noise with the same first and second order moments. This approach is commonly referred to as the *extended Kalman filter* (EKF). This may work well, but without any guarantees, for mildly nonlinear systems where the true posterior is unimodal (just one peak) and essentially symmetric.

- The *unscented Kalman filter* (*UKF*) Julier et al. (1995) propagates a number of points in the state space from which a Gaussian distribution is fit at each time step. UKF is known to accomodate also the quadratic term in nonlinear models, and is often more accurate than EKF. The *divided difference filter* (*DFF*) Norgaard et al. (2000) and the *quadrature Kalman filter* (*QKF*) Arasaratnam et al. (2007) are two other variants of this principle. Again, the applicability of these filters is limited to unimodal posterior distributions.

- *Gaussian sum Kalman filters* (*GS-KF*) (*Kalman filter banks*) as proposed in Alspach and Sorenson (1972), represent the posterior with a Gaussian mixture distribution, and filters in this class can handle multimodal posteriors. The idea can be extended to Kalman filter approximations as the GS-QKF in Arasaratnam et al. (2007).

- The point mass filter (PMF) surveyed in Kramer and Sorenson (1988) grids the state space and computes the posterior over this grid recursively. PMF applies to any nonlinear and non-Gaussian model and is able to represent any posterior distribution. The main limiting factor is the curse of dimensionality of the grid size in higher state dimensions, and that the algorithm itself is of quadratic complexity in the grid size.

It should be stressed that both EKF and UKF approximate the model and propagates Gaussian distributions as the posterior, while the PMF uses the original model and approximates the posterior over a grid. The *particle filter* (*PF*) also provides a numerical approximation to the nonlinear filtering problem similar to the PMF, but uses an adaptive stochastic grid that automatically selects relevant grid points in the state space, and in contrast to the PMF the standard PF has linear complexity in the number of grid points.

6.1.3 Basic Ideas

The task of nonlinear filtering can be split into two parts: representation of the filtering probability density function and propagation of this density during the time and measurement update stages. Figure 6.1 illustrates different representations of the filtering density for a two-dimensional example (similar to the example used in Section 9.8.6). The *extended Kalman filter* (EKF) can be interpreted as using one Gaussian distribution for representation and the propagation is performed according to a linearized model. The *Gaussian sum filter* extends the EKF to be able to represent multi-modal distributions, still with an approximate propagation.

(a) True PDF

(b) Gaussian approximation

(c) Gaussian mixture approximation

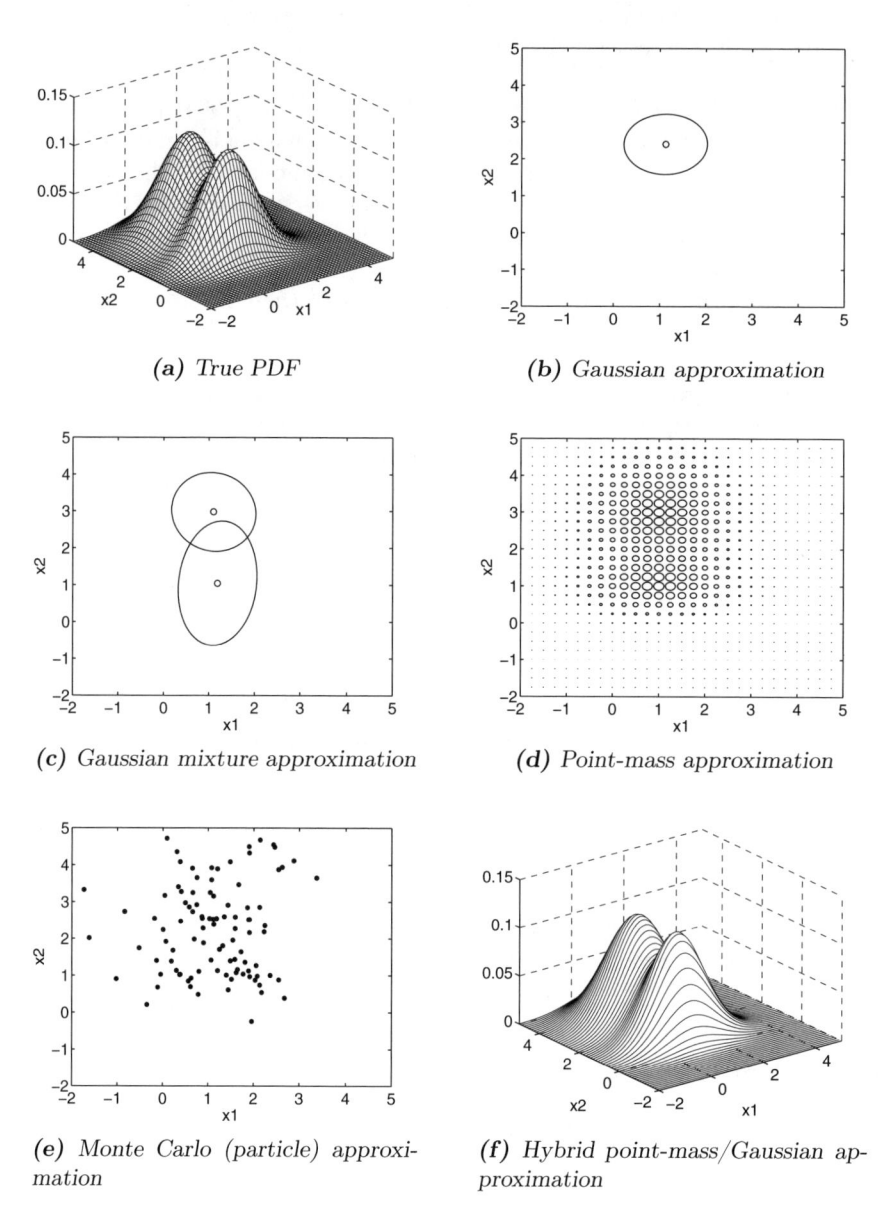

(d) Point-mass approximation

(e) Monte Carlo (particle) approximation

(f) Hybrid point-mass/Gaussian approximation

Figure 6.1: True probability density function and different approximate representations.

Figure 6.1(d)–(f) illustrates numerical approaches where the exact nonlinear relations present in the model are used for propagation. The *point-mass filter* (grid-based approximation) employs a regular grid, where the grid weight is proportional to the posterior. The *particle filter* (PF), represents the posterior by a stochastic grid in form of a set of samples, where all particles (samples) have the same weight. Finally, the *marginalized particle filter* (MPF) uses a stochastic grid for some of the states, and Gaussian distributions for the rest. That is, the MPF can be interpreted as a particle representation for a subspace of the state space, where each particle has an associated Gaussian distribution for the remaining state dimensions, Figure 6.1(f). It will be demonstrated in Section 9.8 that an exact nonlinear propagation is still possible if there is a linear sub-structure in the model.

6.2 The Fusion–Diffusion Approach to Filtering

The somewhat unconventional approach to filtering presented in this section applies to the model (6.3). The approach is inspired by the work in Levy et al. (1996). It reuses two modules already developed in the previous chapters. Essentially, all known nonlinear filters will be covered using the following four high-level primitives:

- *Fusion* to combine information.

- *Diffusion* to add uncertainty.

- *Estimation* for extracting information from an observation.

- *Transformation* to keep track of information in nonlinear transformations.

Fusion and diffusion act as a kind of *transform pair* in the uncertainty–information domains, and are in many ways complementary. The formal transform relation becomes clear in Section 6.3, where diffusion is given by convolution and fusion is given by multiplication of the posterior density, and the other way around for the Fourier transform of the posterior (similar to the definition of the *characteristic function*).

The fusion formula as given in (2.23) is here given in a Bayesian framework as a definition.

Definition 6.1 (Bayesian Fusion)
Given two posterior densities for x, $p(x|y_1)$ and $p(x|y_2)$, where y_1 and y_2 are independent. The joint posterior is then

$$p(x|y_1, y_2) = \frac{p(x|y_1)p(x|y_2)}{p(x)}. \tag{6.5}$$

The information matrix and information state are fused as

$$\mathcal{I} = \mathcal{I}_1 + \mathcal{I}_2 - \mathcal{I}_0, \tag{6.6a}$$
$$\iota = \iota_1 + \iota_2 - \iota_0. \tag{6.6b}$$

Note that the prior $p(x)$ and $\iota_0, \mathcal{I}_0$ may in a Bayesian context be used when forming both estimates, so to avoid double-counting the prior it has to be removed. For the information form, this is done quite intuitively. For the PDF expression (6.5), it follows from Bayes law as

$$p(x|y_1, y_2) = p(y_1, y_2|x)\frac{p(x)}{p(y_1, y_2)} \tag{6.7a}$$
$$= \frac{p(x)p(y_1|x)p(y_2|x)}{p(y_1)p(y_2)} \tag{6.7b}$$
$$= \frac{p(x|y_1)p(x|y_2)}{p(x)}. \tag{6.7c}$$

In case of two Gaussian posteriors $x \sim \mathcal{N}(\hat{x}_i, P_i)$, the standard fusion formula (2.23) applies. Note that this form allows for singular information matrices, and the estimate is unambiguous when the sum $\mathcal{I}$ is non-singular. The complementary operation of diffusion is defined next.

Definition 6.2 (Diffusion)
Given a posterior $p(x|y)$ and a pdf $p_w(\cdot)$. The sum operation $z = x + w$ denotes a diffusion of the posterior of x, and the posterior of z is given by

$$p(z|y) = p(x|y) \star p_w(w) = \int_{-\infty}^{\infty} p(v - x|y)p_w(v)dv, \tag{6.8}$$

where $\star$ denotes convolution. The mean and covariance of these fusion operations are given by the diffusion formula

$$\hat{z} = \hat{x} + \mathrm{E}[w], \tag{6.9a}$$
$$\mathrm{Cov}[\hat{z}] = \mathrm{Cov}[x] + \mathrm{Cov}[w]. \tag{6.9b}$$

That is, diffusion is additive in the covariance domain and fusion is additive in the information domain (where information is the (pseudo-)inverse of covariance). Algorithm 6.1 gives a general algorithm for nonlinear filtering. The first two steps in Algorithm 6.1 will be referred to as the measurement update, and the last two steps the time update. Remarks:

- The algorithm will in most cases include approximations, and this is one of the main topics of the coming chapters.

- The fusion and diffusion formulas are exact for the mean and covariance. It is only estimation and transformation that are approximate generally.

6.3 The Classical Approach to Nonlinear Filtering 131

Algorithm 6.1 Filtering as fusion–diffusion recursion

Given a prior mean $\hat{x}_{1|0}$ and covariance $P_{1|0}$. The posterior filtering mean $\hat{x}_{k|k}$ with covariance $P_{k|k}$ and prediction $\hat{x}_{k+1|k}$ with covariance $P_{k+1|k}$ for the nonlinear model with additive noise (6.3) are given by the recursion:

1. Estimate x from y_k to get $p(x_k|y_k)$ represented by $\hat{x}_k$ and P_k (single index is now used for an estimate of just one measurement similar to the static case).

2. Apply the fusion formula (6.6) to $\hat{x}_k$ and the prior $\hat{x}_{k|k-1}$ to get $\hat{x}_{k|k}$, $P_{k|k}$.

3. Apply a transformation approximation to the temporary variable $z = f(x_k, u_k)$ to get $\mathrm{E}(z)$ and $\mathrm{Cov}(z)$. The methods in Section 3.4 apply, including analytical approaches in Section 3.4.1 the Monte Carlo approach in Section 3.4.2, Taylor expansion in Section 3.4.3 and the unscented transformation in Section 3.4.4.

4. Apply the diffusion formula (6.9) to the relation $x_{k+1} = z + v_k$ to get $\hat{x}_{k+1|k}$, $P_{k+1|k}$ or the complete prediction posterior $p(x_{k+1}|y_{1:k})$.

- However, when f and h are both linear as in model (6.4), both transformation and estimation using WLS (2.7) are optimal in the first two moments (unbiased and minimum variance). The mean and covariance are then computed analytically.

6.3 The Classical Approach to Nonlinear Filtering

We here focus on the most general model in (6.2). The goal is to compute the filtering density $p(x_k|y_{1:k})$ and prediction density $p(x_{k+1}|y_{1:k})$ using only $p(x_{k+1}|x_k)$ and $p(y_k|x_k)$ as provided by the model (6.2).

6.3.1 The General Bayesian Solution

Applying Bayes' rule of the form

$$p(A|B, C) = \frac{p(B|A, C)p(A|C)}{p(B|C)} \tag{6.10}$$

to the posterior distribution of the state vector, identifying $A = x_k$, $(B,C) = y_{1:k}$, $B = y_k$ and $C = y_{1:k-1}$, yields

$$p(x_k|y_{1:k}) = \frac{p(y_k|x_k, y_{1:k-1})p(x_k|y_{1:k-1})}{p(y_k|y_{1:k-1})} = \frac{p(y_k|x_k)p(x_k|y_{1:k-1})}{p(y_k|y_{1:k-1})}. \quad (6.11)$$

In the last equality, the conditioning of $y_{1:k-1}$ can be removed if the state x_k is known, which follows from the Markov property of a state space model. The expression (6.11) is referred to as the *measurement update* in the Bayesian recursion. The denominator can be expressed through the law of total probability, that is, by marginalization.

Bayes' rule $p(A,B|C) = p(A|B,C)p(B|C)$ gives the *time update*

$$p(x_{k+1}, x_k|y_{1:k}) = p(x_{k+1}|x_k, y_{1:k}) \, p(x_k|y_{1:k}) = p(x_{k+1}|x_k) \, p(x_k|y_{1:k}).$$

Again, the Markov property was used in the last equality. Marginalization of x_k by integrating both sides with respect to x_k yields

$$p(x_{k+1}|y_{1:k}) = \int_{\mathbb{R}^n} p(x_{k+1}|x_k)p(x_k|y_{1:k}) \, dx_k. \quad (6.12)$$

This relation is known as the *Chapman-Kolmogorov* equation, see Jazwinsky (1970). After (6.12) has been evaluated, the time index can be increased and the effect of a new measurement incorporated as in (6.11). To summarize, the Bayesian recursion is stated as a theorem.

Theorem 6.3
The recursive Bayesian solution for the model (6.2) is given by

$$p(x_k|y_{1:k}) = \frac{p(y_k|x_k)p(x_k|y_{1:k-1})}{p(y_k|y_{1:k-1})} \quad (6.13a)$$

$$p(x_{k+1}|y_{1:k}) = \int_{\mathbb{R}^n} p(x_{k+1}|x_k)p(x_k|y_{1:k}) \, dx_k \quad (6.13b)$$

where $k = 1, 2, \ldots, N$ *and*

$$p(y_k|y_{1:k-1}) = \int_{\mathbb{R}^n} p(y_k|x_k)p(x_k|y_{1:k-1}) \, dx_k \quad (6.13c)$$

The recursion is initiated by $p(x_1|y_0) = p(x_0)$.

The *minimum variance* (MV) and *maximum a posteriori* (MAP) estimates can be extracted from the posterior as

$$x_{k|k}^{\text{MV}} = \int_{\mathbb{R}^n} x_k p(x_k|y_{1:k}) \, dx_k,$$

$$x_{k|k}^{\text{MAP}} = \arg\max_{x_k} p(x_k|y_{1:k}),$$

6.3 The Classical Approach to Nonlinear Filtering

respectively. The estimation error covariance is computed by

$$P_{k|k} = \int_{\mathbb{R}^n} (x_k - x_{k|k}^{\text{MV}})(x_k - x_{k|k}^{\text{MV}})^T p(x_k|y_{1:k})\, dx_k.$$

6.3.2 Existence Conditions for Finite-Dimensional Filters

To understand the problems involved in finding cases where Theorem 6.3 can be applied, we consider the more structured estimation problem in (6.3). In this case, the Bayesian solution is given by

$$p(x_k|y_{1:k}) = \frac{1}{\alpha} p_{e_k}(y_k - h(x_k)|x_k) p(x_k|y_{1:k-1}), \tag{6.14a}$$

$$p(x_{k+1}|y_{1:k}) = \int_{\mathbb{R}^{n_x}} p_{v_k}(x_{k+1} - f(x_k)|x_k) p(x_k|y_{1:k})\, dx_k, \tag{6.14b}$$

$$\alpha = \int_{\mathbb{R}^{n_x}} p_{e_k}(y_k - h(x_k)|x_k) p(x_k|y_{1:k-1})\, dx_k. \tag{6.14c}$$

The question is, what requirements are necessary on f, h, p_v and p_e to get a finite dimensional posterior density after each step in the recursion? The main steps are:

- The nonlinear transformation $f(x_k)$. There are a few nonlinear functions $f(x)$ and distributions for x that admit a parametric distribution for the transformed variable. For instance, $p(x) = \mathcal{N}(0,1)$ and $f(x) = x^2$ gives a $\chi^2(1)$ distribution, $p(x) = \mathcal{N}(0, I_2)$ and $z = x_1^2 + x_1^2$ gives a $\chi^2(2) = \exp(1)$ distribution and so on. However, this works only once, not recursively.

- The addition of $f(x_k)$ and v_k in the time update. This gives a finite dimensional density only in a few cases, including the examples where both are Gaussian or both are χ^2 distributed.

- The inference of x_k from y_k done in the measurement update.

For the inference problem in the last item, we first define another version of the fusion operation as an alternative to Definition 6.1.

Definition 6.4 (Fusion as Inference)
Given the possibly dependent information in y_1 and y_2, respectively. The conditional density of x given both pieces y_1, y_2 of information is by Bayes' law

$$p(x|y_1, y_2) = \frac{p(y_2|x, y_1)}{p(y_2|y_1)} p(x|y_1). \tag{6.15}$$

This inference is called fusion of y_1 and y_2, where the roles of y_1 and y_2 above are interchangable.

Table 6.1: *A few examples of feasible dynamic models for a posterior distribution. See Table A.6 for a more comprehensive list.*

Posterior $p(x_k\|y_{1:k})$	Dynamic model	Prediction $p(x_{k+1}\|y_{1:k})$
$\mathcal{N}(\hat{x}_{k\|k}, P_{k\|k})$	$x_{k+1} = Ax_k + v_k$	$\mathcal{N}(A\hat{x}_{k\|k}, AP_{k\|k}A^T + Q)$
$\text{GAMMA}(a_{k\|k}, b_{k\|k})$	$x_{k+1} = Ax_k + \text{GAMMA}(q, Ab_{k\|k})$	$\text{GAMMA}(a_{k\|k} + q, Ab_{k\|k})$

The question here is when the posterior distribution preserves its functional form after the fusion step. This is a well-known problem in statistics, where the term *conjugate priors* is used, see Raiffa and Schlaifer (1961). Table A.6 summarizes some known cases. It should be pointed out that actually all members of the exponential family has a conjugate prior.

Tables A.6 and 6.1 can be used to contruct some special cases of the model (6.1) where Theorem 6.3 has an explicit finite-dimensional solution, as in the following two examples.

- If both $f(x) = Fx$ and $h(x) = Hx$ are linear functions, the prior is Gamma distributed, there is no process noise and the measurement noise has an exponential distribution, then the posterior will be Gamma distributed.

- If both $f(x) = F_0 + F_1 x$ and $h(x) = H_0 + H_1 x$ are affine functions, the prior is Pareto distributed, there is no process noise and the measurement noise has a uniform Pareto distribution, then the posterior will be Pareto distributed.

However, such restrictive models are hardly useful in practice. One toy example is given below as an illustration.

Example 6.1: Gamma prior and exponential noise

Consider a scalar state x and let the model be

$$x_{k+1} = Fx_k, \tag{6.16a}$$
$$y_k \sim \text{EXP}(x_k), \tag{6.16b}$$
$$x_1 \sim \text{GAMMA}(a_{1|0}, b_{1|0}). \tag{6.16c}$$

The model is setup and simulated below.

```
lh=expdist;
x0=0.4;
x(1)=x0;
F=0.9;
for k=1:5
    y(k)=rand(expdist(1/x(k)),1);
    x(k+1)=F*x(k);
end
```

The conjugate prior to the exponential likelihood is the Gamma distribution. The gamma distribution is also preserved under multiplication. The

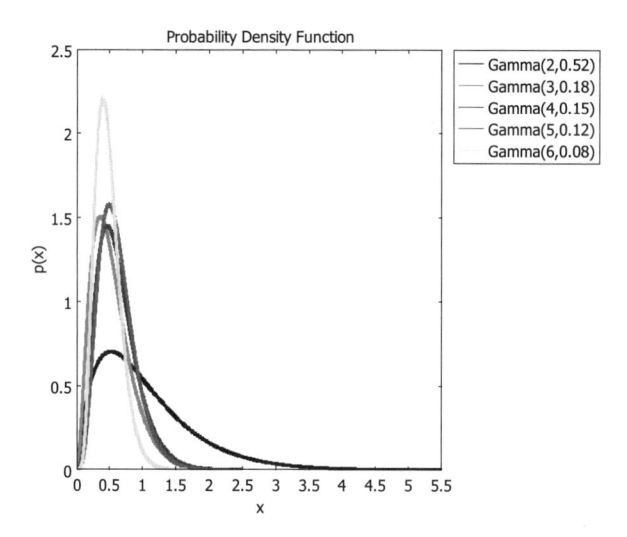

6.3 The Classical Approach to Nonlinear Filtering

Figure 6.2: *Posterior distributions for the parameter x_k in the exponential likelihood model as a function of time k.*

following code implements the exact Bayesian nonlinear filter, initialized with $a_{1|0} = 1$ and $b_{1|0} = 1$:

```
p{1}=gammadist(1,1);
for k=1:5
   p{k}=posterior(p{k},lh,y(k));   % Measurement update
   p{k+1}=F*p{k};                  % Time update
end
plot(p{1:5})
```

The result is illustrated in Figure 6.2.

The next sub-section lists the most important cases in applications.

6.3.3 Cases with Analytic Solutions

The first two cases are the only known finite dimensional algorithms for nonlinear filtering. The last two cases have analytical solutions to the Bayesian recursions Theorem 6.3, but requires an exponentially increasing memory and computational power.

Case 1: Linear Gaussian Model

If both f and h are linear as in model (6.4) and both e_k and v_k are Gaussian and the prior $p(x_1|y_0)$ is Gaussian, then Algorithm 6.1 computes the posterior Gaussian distribution exactly (given that the model is correct). This is true because the Gaussian distribution is preserved for the basic operations in Algorithm 6.1 for a linear model (6.4) with Gaussian noise:

1. Estimation: the WLS estimate is Gaussian distributed if the noise is Gaussian. The estimate is given by

$$\hat{x}_{k|y_k} = (C^T R^{-1} C)^{-1} C^T R^{-1} y_k, \tag{6.17a}$$

$$\mathcal{I}_{k|y_k} = C^T R^{-1} C. \tag{6.17b}$$

2. Fusion: fusing two Gaussian distributions is again Gaussian,

$$\mathcal{I}_{k|k} = \mathcal{I}_{k|k-1} + \mathcal{I}_{k|y_k}, \tag{6.17c}$$

$$\hat{x}_{k|k} = \mathcal{I}_{k|k}^{-1} \big(\mathcal{I}_{k|k-1} \hat{x}_{k|k-1} + \mathcal{I}_{k|y_k} \hat{x}_{k|y_k} \big) \tag{6.17d}$$

3. Transformation: Linear transformation $z = Ax_k$ preserves the Gaussian distribution,

$$\hat{z} = A\hat{x}_{k|k}, \tag{6.17e}$$

$$P_z = AP_{k|k} A^T. \tag{6.17f}$$

4. Diffusion: the sum $z + v_k$ of two Gaussian variables is Gaussian,

$$\hat{x}_{k+1|k} = \hat{z}, \tag{6.17g}$$

$$P_{k+1|k} = P_z + Q_k. \tag{6.17h}$$

In fact, this is one possible implementation of the Kalman filter. To avoid the conversion between information and covariance in this algorithm, the standard formulation utilizes the *matrix inversion lemma* to eliminate the information matrix in the measurement update. We here just state the Kalman filter equations for recursively computing the mean and covariance of the Gaussian posterior without proof, just to give a hint of the involved calculations:

$$\hat{x}_{k+1|k} = A_k \hat{x}_{k|k} + B_{u,k} u_k \tag{6.18a}$$

$$P_{k+1|k} = A_k P_{k|k} A_k^T + B_{v,k} Q_k B_{v,k}^T \tag{6.18b}$$

$$\hat{x}_{k|k} = \hat{x}_{k|k-1} + P_{k|k-1} C_k^T (C_k P_{k|k-1} C_k^T + R_k)^{-1} (y_k - C_k \hat{x}_{k|k-1} - D_k u_k) \tag{6.18c}$$

$$P_{k|k} = P_{k|k-1} - P_{k|k-1} C_k^T (C_k P_{k|k-1} C_k^T + R_k)^{-1} C_k P_{k|k-1}. \tag{6.18d}$$

Derivations and properties of the KF are given in Chapter 7. There is also a kind of dual formulation, called the *information filter*, see Section 7.1.4, where the matrix inversion lemma is used to eliminate the covariance in the time update.

Case 2: Hidden Markov Model (HMM)

The other case is hidden Markov models (HMM),

$$x_{k+1} = Ax_k. \tag{6.19a}$$

Though there are formal similarities with (6.4), there are fundamental differences. In the basic HMM formulation, there is only one unknown discrete state $\xi_k \in \{1, 2, \ldots, n_x\}$. The state vector x_k contains the probability of the Markov variable ξ being j, so

$$x_k^{(j)} = P(\xi_k = j), \quad j = 1, 2, \ldots, n_x. \tag{6.19b}$$

The posterior filter distribution is denoted

$$\pi_{k|k}^{(j)} = P(\xi_k = j | y_{1:k}), \quad j = 1, 2, \ldots, n_x. \tag{6.19c}$$

At each time instant, an observation is made of the Markov variable, with

$$P(y_k = i | \xi_k = j) = C^{(i,j)}, \quad i = 1, 2, \ldots, n_y, \tag{6.19d}$$

$$P(y_k = i) = \sum_{j=1}^{n_x} C^{(i,j)} P(\xi_k = j) = \sum_{j=1}^{n_x} C^{(i,j)} x_k^{(j)}, \quad i = 1, 2, \ldots, n_y. \tag{6.19e}$$

Applying the time update (6.13b) in Theorem 6.3 to the HMM (6.19), we get

$$\pi_{k|k-1}^{(i)} = \sum_{j=1}^{n_x} p(\xi_k = i | \xi_{k-1} = j) p(\xi_{k-1} = j | y_{k-1}) \tag{6.20a}$$

$$= \sum_{j=1}^{n_x} A^{(i,j)} \pi_{k-1|k-1}(j), \quad i = 1, 2 \ldots, n_x. \tag{6.20b}$$

Similarly, the measurement update (6.13a) gives

$$\pi_{k|k}^{(i)} = \frac{p(\xi_k = i | y_{k-1}) p(y_k | \xi_k = i)}{p(y_k | y_{k-1})} \tag{6.21a}$$

$$= \frac{\pi_{k|k-1}^{(i)} C^{(y_k, i)}}{\sum_{j=1}^{n_x} \pi_{k|k-1}^{(j)} C^{(y_k, j)}}, \quad i = 1, 2 \ldots, n_x. \tag{6.21b}$$

Eliminating the time update, the filter resulting from Theorem 6.3 for computing the *a posteriori* probabilities can be expressed as the recursion

$$\pi_{k|k}^{(i)} = p(\xi_k = i | y_{1:k}) = \frac{\sum_{j=1}^{n_x} \pi_{k-1|k-1}^{(j)} A^{(i,j)} C^{(y_k, i)}}{\sum_{l=1}^{n_y} \sum_{j=1}^{n_x} \pi_{k-1|k-1}^{(j)} A^{(l,j)} C^{(y_k, l)}}. \tag{6.21c}$$

The MAP estimate is $\hat{\xi}_k = \arg\max_i \pi_{k|k}^{(i)}$.

Case 3: Discrete Prior and Process Noise

Assume the following discrete *multinomial distribution* for the prior and process noise in (6.1):

1. The prior takes on one of n_0 different values $x_0^{(i)}$ with probability $P(x_0 = x_0^{(i)}) = \pi_0^{(i)}$.

2. Each process noise v_k takes on one of n_w different values $v_k^{(i)}$ with probability $P(v_k = v_k^{(i)}) = \pi_v^{(i)}$.

3. The measurement noise p_{e_k} can be arbitrary.

Then it is rather obvious from Theorem 6.3 that the posterior is a combinatorial enumeration of all possible sequences, and the posterior for each sequence is given by the following discrete distribution:

$$N_0 = n_0, \tag{6.22a}$$

$$N_{k+1} = N_k n_v, \tag{6.22b}$$

$$x_{k+1}^{(i)} = f(x_k^{(m)}, v_k^{(n)}), \quad m = 1, \ldots, N_k, \ n = 1, \ldots, n_v, \ i = 1, \ldots, N_{k+1}, \tag{6.22c}$$

$$\pi_{k+1|k}^{(i)} = \pi_{k|k}^{(m)} \pi_v^{(n)}, \quad m = 1, \ldots, N_k, \ n = 1, \ldots, n_v, \ i = 1, \ldots, N_{k+1}, \tag{6.22d}$$

$$\pi_{k|k}^{(i)} \propto \pi_{k|k-1}^{(i)} p_{e_k}(y_k - x_k^{(i)}), \quad i = 1, \ldots, N_{k+1}. \tag{6.22e}$$

The $\propto$ symbol indicates that a normalization is needed in the last equation. The main implementation problem is the book keeping of the index sequence, while the limiting factor is that the number of indices grows exponentially.

Case 4: Gaussian Mixtures and Mode Dependent Linear Model

As a practical extension of the discrete case in the previous section, the prior, process noise and measurement noise can all be Gaussian mixtures with n_0, n_v and n_e modes, respectively. Also, the dynamics might have n_m discrete modes.

$$x_{k+1} = A(\xi_k)x_k + v_k(\xi_k), \tag{6.23a}$$

$$y_k = C(\xi_k)x_k + e_k(\xi_k), \tag{6.23b}$$

$$v_k(\xi_k) \sim \mathcal{N}(\mu_v(\xi_k), Q(\xi_k)), \tag{6.23c}$$

$$e_k(\xi_k) \sim \mathcal{N}(\mu_e(\xi_k), R(\xi_k)), \tag{6.23d}$$

$$x_0(\xi_k) \sim \mathcal{N}(\mu_{x_0}(\xi_k), P_{x_0}(\xi_k)). \tag{6.23e}$$

The discrete variable ξ_k contains all $n_v n_e n_m$ permutations of modes at time k. The important point is that conditioned on the sequence $\xi_{1:k}$, the model (6.23) is linear and Gaussian. That is, the posterior $p(x_k|y_{1:k},\xi_{1:k})$ is Gaussian. The sought posterior is then a *Gaussian mixture*

$$p(x_k|y_{1:k}) = \sum_{\xi_{1:k}} \frac{p(\xi_{1:k}|y_{1:k})}{\sum_{\xi_{1:k}} p(\xi_{1:k}|y_{1:k})} p(x_k|y_{1:k},\xi_{1:k}) \qquad (6.24a)$$

$$= \sum_{\xi_{1:k}} \frac{p(\xi_{1:k}|y_{1:k})}{\sum_{\xi_{1:k}} p(\xi_{1:k}|y_{1:k})} \mathcal{N}\big(\hat{x}_{k|k}(\xi_{1:k}), P_{k|k}(\xi_{1:k})\big), \qquad (6.24b)$$

$$p(\xi_{1:k}|y_{1:k}) = \frac{p(y_{1:k}|\xi_{1:k})p(\xi_{1:k})}{p(y_{1:k})}. \qquad (6.24c)$$

The problem is again the exponential increase in the number of modes, which in this formulation is $n_0(n_v n_e n_m)^k$. However, there are many good ways to recursively prune or merge this growing tree of modes to get a finite dimensional approximation. This has turned out to be a quite powerful tool in many applications. This approach is more thoroughly discussed in Chapter 10, where different pruning and merging strategies are surveyed.

6.3.4 General Approximations

There are two conceptually different ways to obtain approximate nonlinear filters:

1. Approximate the model to a case where an optimal algorithm exists.

2. Approximate the optimal nonlinear filter for the original model.

The classical approach to nonlinear filtering uses the first approach, and the *extended Kalman filter (EKF)* has been used in practice almost as long as the Kalman filter. The EKF applies a Taylor expansion to the nonlinear functions f and h in (6.1). The EKF will be treated in more detail in Chapter 7. Modern approaches applies the second approach. The particle filter in Chapter 9 and the Kalman filter bank in Chapter 10 are two examples on numerical approximation of the analytical posterior distribution.

Rather interestingly, one can define an *extended HMM filter* to fill up the gap in the reasoning above. The HMM admits a finite dimensional solution, so all that is needed is to approximate (6.1) with an HMM. This involves approximating the continuous state space with a fixed grid. The transition probability $A^{(j,i)}$ for going from one state $x_k^{(i)}$ to another $x_{k+1}^{(j)}$ can be computed from the model (6.1). Similarly, the observation likelihood gives the entries in C_k. This is actually one example of a grid method, as described in the next section.

6.4 Grid Based Methods

Numerical approximations to the nonlinear filtering problem was an active research area in the early 1970's, see Kramer and Sorenson (1988), Jazwinsky (1970), Alspach and Sorenson (1972) and Anderson and Moore (1979). The main ideas at that time are summarized below.

- Numerical approximation of the integrals. Replace the infinite integrals with Riemann sums over finite intervals.

- Discretization of the state space. Limit the range of values of x_k from $\mathbb{R}^{n_x}$ to a finite number of values.

- Probability region equivalent. Divide the state space into regions (cubes, splines) and express the probability of being in each region. Use this probability as a weight on each region.

- Piecewise constant approximation of the posterior. Numerically approximate the posterior as a sum of weighted and shifted indicator functions.

- Nyquist approach. Assume that the posterior is band-limited, i.e., has an upper bound on the frequency of spatial variation. Sample the function spatially and update these samples.

All these approaches basically end up with the same algorithm, the *point mass filter* (*PMF*).

6.4.1 The Point-Mass Filter

We here make the derivation of the PMF based on the first item above, where the basic idea is to replace the integrals in Theorem 6.3 with Riemann approximations,

$$\int_{\mathbb{R}^{n_x}} f(x_k)\, dx_k \approx \Delta \sum_{j=1}^{N} f(x_k^{(j)}), \qquad (6.25)$$

where $\Delta = \prod_{i=1}^{n_x} \delta_i$ is the volume of each grid point, and δ_i is the (uniform) grid resolution in each dimension $x_k^{(i)}$. The number N of grid points is the product of the grid resolution in each dimension. Except for a fixed memory size, this is the main limitation for high-dimensional state vectors, where the number of grid points quickly increases.

On the positive side, the PMF is simple to implement, there is no risk for filter divergence, it works excellently at least when $n_x \leq 2$, and it applies a global search, so there is no risk to get stuck in a local minimum.

On the other hand, a grid search is quite inefficient in higher dimensions. Further, the grid should be adaptive to allow for a non-informative prior and

6.4 Grid Based Methods

a gradually improved posterior. In the case of multi-modal distributions, the grid might need to be too large, and the use of sparse matrices can be used. The PMF is further discussed in Section 9.2.3.

---**Example 6.2: PMF and range measurements**---

First, a nonlinear model object is defined:

```
NL object
x(t+1) = [x(1)+u(1); x(2)+u(2)]; + N([0;0],[3,0;0,3])
    y = norm(x) + N(0,1)
   x0' = [1e+002     1e+002]
   th' = []

   States: x1        x2
   Outputs: r
   Inputs: v         psi
```

This model is first used to simulate data, then the same model is used in the PMF. To initialize the PMF, a grid is created for the 2D state vector, and the grid is moved to the prior location $(100, 100)$. The process noise PDF is evaluated in the grid points, and the prior weights defined.

```
% Grid
[X1,X2]=meshgrid(-20:1:20,-20:1:20);
pv = exp(-(X1.^2+X2.^2)/2/1);
% Prior
p0=20;
P0=p0*eye(2);
lh= exp(-(X1.^2+X2.^2)/2/p0);
X1=X1+100;
X2=X2+100;
```

The PMF recursion is now only a few lines of code:

```
for k=1:10
   % Measurement update
   lh=lh.* exp(-(y.y(1)-sqrt(X1.^2+X2.^2)).^2/2/var(m.pe));
   lh=lh/sum(sum(lh));
   xhat(k,:)=[sum(sum(lh.*X1)) sum(sum(lh.*X2))];
   yhat(k,:)=m.h(k,xhat(k,:).',y.u(k,:).',[]);
   contour(X1,X2,lh)
   % Time update
   X1=X1+u.y(k,1);
   X2=X2+u.y(k,2);
   lh=conv2(lh,pv,'same');
   contour(X1,X2,lh)
end
```

The PMF is available as the method `nl.pmf`. The contour plots with some additional labels and a title are illustrated in Figure 6.3.

6.4.2 A Frequency Domain Variant

With a frequency domain approach, the convolution can be performed using the efficient *FFT* algorithm with $N \log(N)$ complexity. Consider first a scalar state vector $n_x = 1$, and assume that the prior, process noise and measurement

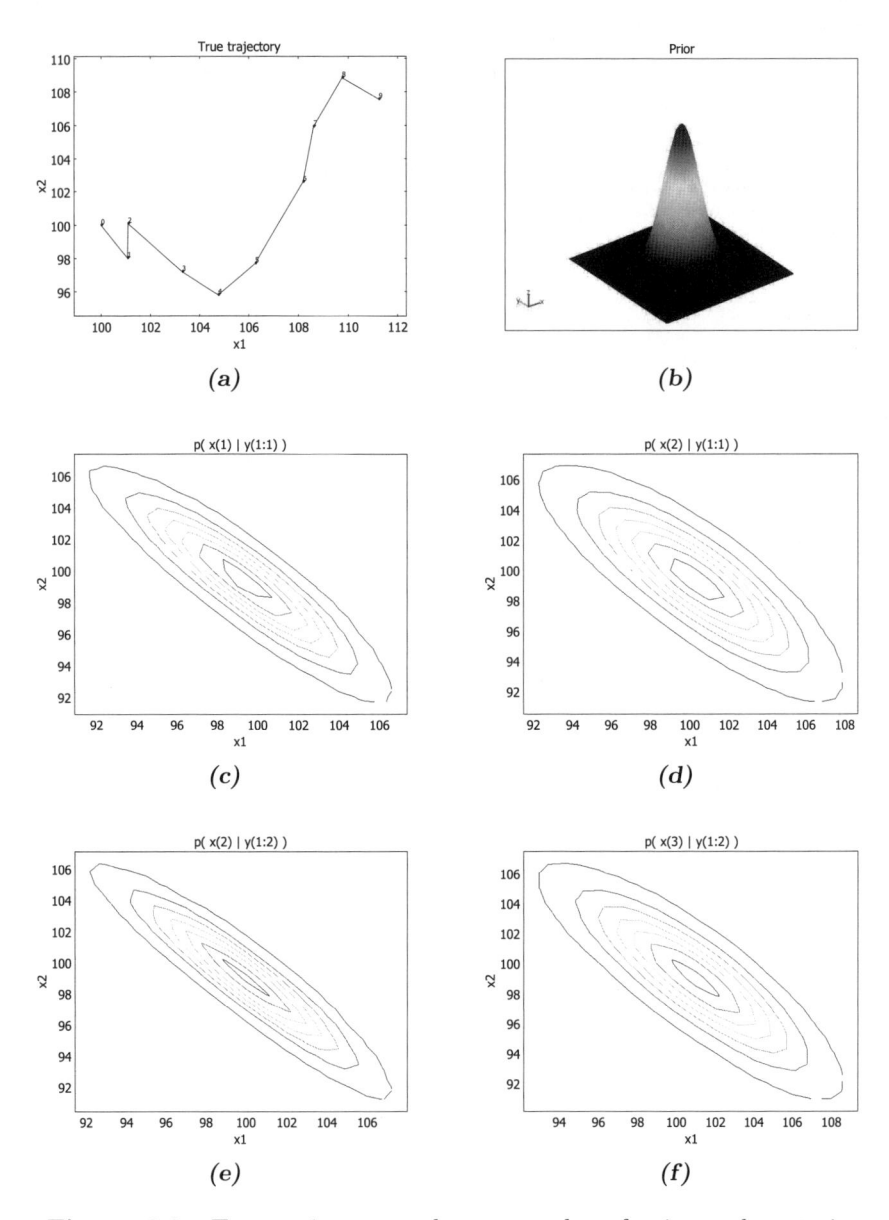

Figure 6.3: True trajectory and contour plot of prior and posterior distributions.

noise PDF's are bandlimited in the frequency domain. Under this assumption, these distributions can be sampled and exactly reconstructed again. Let the posterior and measurement noise distributions at some time be sampled sufficiently fast with respect to their respective bandwidth. The measurement update in Theorem 6.3 can then be performed using straightforward elementwise multiplication. The convolution in the time update of Theorem 6.3 can be implemented as multiplication of the discrete Fourier transform using a well-known transform property. The main open question is whether this recursion keeps the posterior band-limited. The algorithm outlined above for $n_x = 1$ can be extended to higher state dimensions under certain conditions.

6.5 Nonlinear Filtering Bounds

Consider again the general state-space model (6.1) and the more general formulation (6.2)

$$\begin{cases} x_{k+1} = f(x_k, v_k) \\ y_k = h(x_k, e_k) \end{cases} \longleftrightarrow \begin{cases} p(x_{k+1}|x_k) \\ p(y_k|x_k) \end{cases}. \quad (6.26)$$

The *Cramér-Rao lower bound* (*CRLB*) provides a lower bound on the covariance matrix for any unbiased estimator $\hat{x}_{k|k}$,

$$\operatorname{Cov}(\hat{x}_{k|k}) \geq P_{k|k}^{\text{CRLB}}. \quad (6.27)$$

This is the filtering CRLB. There are similar bounds for predictions, for instance $\hat{x}_{k+1|k}$, and smoothing.

There are two important instances of the CRLB:

- The *parametric CRLB*: this holds for a specific trajectory only, so the CRLB is a function of this trajectory which could be expressed as $P_{k|k}^{\text{ParCRLB}}(x_{1:k})$. This is the simplest case, leading to a Riccati equation of the same functional form as the covariance matrix update in the Kalman filter.

- The *posterior CRLB*: this is the parametric CRLB averaged over all possible trajectories $P_{k|k}^{\text{PostCRLB}} = \int P_{k|k}^{\text{ParCRLB}}(x_{1:k}) p(x_{1:k}) dx_{1:k}$.

The posterior CRLB assumes that the state trajectory is a stochastic process. If the trajectory is assumed to be a parameter to be estimated, the *parametric CRLB* applies instead. The parametric CRLB is easier to compute, but on the other hand generally depends on the state trajectory for which it is evaluated. We start with the simpler parametric CRLB and then continue with the posterior CRLB.

6.5.1 Parametric CRLB

Gaussian Case

The recursion for the parametric CRLB is studied in Tichavsky et al. (1998), Bergman (1999) and N. Bergman and Gordon (2001). It is algorithmically identical to the Kalman filter Riccati equation:

$$P_{k+1|k}^{\text{ParCRLB}} = F_k P_{k|k}^{\text{ParCRLB}} F_k^T + G_k Q_k G_k \tag{6.28a}$$
$$P_{k+1|k+1}^{\text{ParCRLB}} = P_{k+1|k}^{\text{ParCRLB}} - P_{k+1|k}^{\text{ParCRLB}} H_k^T (H_k P_{k+1|k}^{\text{ParCRLB}} H_k^T + R_k)^{-1} H_k P_{k+1|k}^{\text{ParCRLB}} \tag{6.28b}$$

with

$$F_k^T = \frac{\partial}{\partial x_k} f^T(x_k, v_k),$$

$$G_k^T = \frac{\partial}{\partial v_k} f^T(x_k, v_k),$$

$$H_k^T R_k^{-1} H_k = \mathrm{E}\Big(-\frac{\partial^2}{\partial x_k^2} \log p(y_k|x_k)\Big),$$

$$Q_k^{-1} = \mathrm{E}\Big(-\frac{d^2}{dv_k^2} \log p_v\Big),$$

where all expressions are evaluated for the true trajectory. The factorization $H_k^T R_k^{-1} H_k$ is ambigious, but should be done so that R_k^{-1} exists. One way to obtain the factorization is to use a thin SVD to get $\mathrm{E}\big(-\frac{d^2}{dx_k^2} \log p(y_k|x_k)\big) = UDU^T$ and let $R_k^{-1} = D$ and $H_k = U$. For the model (6.3), the natural option is to choose $H_k = \frac{\partial}{\partial x_k} h^T(x_k)$ and $R = \text{Cov}(e_k)$.

Non-Gaussian Case

The parametric CRLB given by recursion (6.28a) must be modified for non-Gaussian noise. As demonstrated in Hendeby (2005), this modification can quite elegantly be expressed in terms of *intrinsic accuracy*, see (2.47). Similarly to (2.48), a *relative accuracy* of the noise distribution can be defined, which scales the covariance matrices in the CRLB recursion (6.28). The fundamental result in Hendeby (2005) is that (6.28) gives the parametric CRLB, if the covariances Q_k, R_k and P_0 are replaced with the inverse intrinsic accuracies $\mathcal{I}_{v_k}^{-1}$, $\mathcal{I}_{e_k}^{-1}$ and $\mathcal{I}_{x_0}^{-1}$, respectively.

The standard Riccati form becomes

$$P_{k|k}^{\text{ParCRLB}} = P_{k|k-1}^{\text{ParCRLB}} \tag{6.29}$$
$$- P_{k|k-1}^{\text{ParCRLB}} H_k^T (\mathcal{I}_{e_k}^{-1} + H_k P_{k|k-1}^{\text{ParCRLB}} H_k^T)^{-1} H_k P_{k|k-1}^{\text{ParCRLB}},$$
$$P_{k|k-1}^{\text{ParCRLB}} = F_k P_{k-1|k-1}^{\text{ParCRLB}} F_k^T + \mathcal{I}_{v_k}^{-1}. \tag{6.30}$$

6.5 Nonlinear Filtering Bounds

An information filter version of this is

$$P_{k|k}^{-1} = \mathcal{I}_{v_{k-1}} + H_k^T \mathcal{I}_{e_k} H_k - \mathcal{I}_{v_{k-1}} F_k (P_{k-1|k-1}^{-1} + F_k^T \mathcal{I}_{v_{k-1}} F_k)^{-1} F_k^T \mathcal{I}_{v_{k-1}}. \quad (6.31)$$

The computation of these expressions is greatly simplified in the case of scalar noise processes, or when these are vectors with independent identically distributed variables.

Example 6.3

The *relative accuracy* (RA) is computed using MC simulations for two χ^2 distributions below.

```
X{1}=ndist(0,1);      X{1}.MC=1e5;  xra(1)=ra(X{1});
X{2}=chi2dist(4);     X{2}.MC=1e4;  xra(2)=ra(X{2});
X{3}=chi2dist(10);    X{3}.MC=1e4;  xra(3)=ra(X{3});
plot(X{:})
title(['Relative Accuracy:',mat2str(xra,2)])
```

Similarly, the RA for three Gaussian mixtures are computed below.

```
Y{1}=gmdist([0.9 0.1],[0 0],[1 10]);  Y{1}.MC=1e4; yra(1)=ra(Y{1});
Y{2}=gmdist([0.85 0.075 0.075],[0 -2.5 2.5],[0.065 0.065 0.065]);
     Y{2}.MC=1e4; yra(2)=ra(Y{2});
Y{3}=gmdist([0.75 0.25],[0 12],[9 36]);  Y{3}.MC=1e4; yra(3)=ra(Y{3});
plot(Y{:},'legend','off')
title(['Relative Accuracy:',mat2str(yra,2)])
set(gca,'Xlim',[-10 10])
```

Figure 6.4 illustrates the result and gives the RA's in the plot titles.

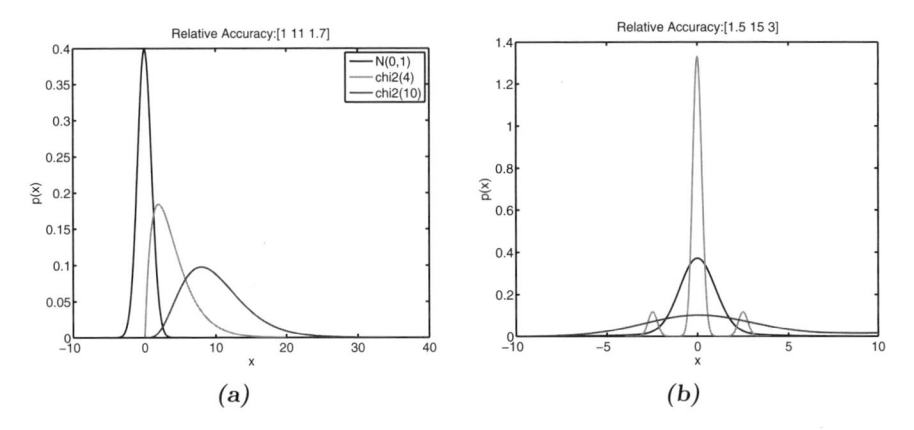

Figure 6.4: *Relative accuracy (RA) for some selected distributions. Basically, 1/RA scales the covariances P_0, Q and R in the CRLB Riccati equation to smaller values. In this sense, a Gaussian noise is worst case, since RA> 1 for all other distributions.*

Stationary CRLB

The covariance matrices of a stationary filter are defined by

$$\bar{P}_p = \lim_{k \to \infty} P_{k|k-1}^{\mathrm{ParCRLB}}, \qquad (6.32\mathrm{a})$$

$$\bar{P}_f = \lim_{k \to \infty} P_{k|k}^{\mathrm{ParCRLB}}. \qquad (6.32\mathrm{b})$$

The CRLB bound will thus converge to a stationary solution of (6.28), whose solution is defined by

$$\bar{P}_p = F\bar{P}_f F^T + GQG \qquad (6.33\mathrm{a})$$

$$\bar{P}_f = \bar{P}_p - \bar{P}_p H^T (H\bar{P}_p H^T + R)^{-1} H \bar{P}_p, \qquad (6.33\mathrm{b})$$

whenever such stationary values $\bar{P}_f$ and $\bar{P}_p$ exist. A necessary condition is that the system $(F_k, G_k, H_k, Q_k, R_k) = (F, G, H, Q, R)$ is time-invariant.

One can eliminate for instance the filter form $\bar{P}_f$ and get the following nonlinear system of equations

$$\bar{P}_p = F\big(\bar{P}_p - \bar{P}_p H^T (H\bar{P}_p H^T + R)^{-1} H \bar{P}_p\big) F^T + GQG. \qquad (6.33\mathrm{c})$$

There is in general no closed form solution for stationary Riccati equations. However, one such is given in the next section.

Explicit Solution for Random Walk Dynamics

Assume a linear model without dynamics,

$$x_{k+1} = x_k + v_k, \qquad (6.34\mathrm{a})$$

$$y_k = Hx_k + e_k. \qquad (6.34\mathrm{b})$$

Let the information in each observation be defined as usual as $\mathcal{I}$, which for Gaussian noise can be computed as $\mathcal{I} = H^T R^{-1} H$. The time update on standard form in the case $F = I$ and measurement update on information form are given by

$$\bar{P}_p = \bar{P}_f + Q, \qquad (6.35\mathrm{a})$$

$$\bar{P}_f^{-1} = \bar{P}_p^{-1} + \mathcal{I}, \qquad (6.35\mathrm{b})$$

respectively. The stationary filtering CRLB expressed in the information form is thus

$$\bar{P}_f^{-1} = (\bar{P}_f + Q)^{-1} + \mathcal{I}. \qquad (6.36\mathrm{a})$$

This equation can be rewritten by multiplying from right with $(\bar{P}_f + Q)$ and from left with $\bar{P}_f$, which gives after removing two identical matrices $\bar{P}_f$,

$$\bar{P}_f \mathcal{I} \bar{P}_f + Q\mathcal{I}\bar{P}_f = Q \qquad (6.36\mathrm{b})$$

6.5 Nonlinear Filtering Bounds

Completing the squares gives

$$\left(\bar{P}_f \mathcal{I}^{1/2} + \frac{1}{2}Q\mathcal{I}^{1/2}\right)\left(\mathcal{I}^{1/2}\bar{P}_f + \frac{1}{2}\mathcal{I}^{1/2}Q\right) = Q + \frac{1}{4}Q\mathcal{I}Q \quad (6.36c)$$

$$\Rightarrow \left(\mathcal{I}^{1/2}\bar{P}_f \mathcal{I}^{1/2} + \frac{1}{2}\mathcal{I}^{1/2}Q\mathcal{I}^{1/2}\right)\left(\mathcal{I}^{1/2}\bar{P}_f \mathcal{I}^{1/2} + \frac{1}{2}\mathcal{I}^{1/2}Q\mathcal{I}^{1/2}\right) =$$

$$\mathcal{I}^{1/2}\left(Q + \frac{1}{4}Q\mathcal{I}Q\right)\mathcal{I}^{1/2} \quad (6.36d)$$

$$\Rightarrow \bar{P}_f = -\frac{1}{2}Q + \mathcal{I}^{-1/2}\left(\mathcal{I}^{1/2}(Q + \frac{1}{4}Q\mathcal{I}Q)\mathcal{I}^{1/2}\right)^{1/2}\mathcal{I}^{-1/2} \quad (6.36e)$$

All involved matrices are symmetric, so it follows from (6.36b) that $Q\mathcal{I}\bar{P}_f$ is also symmetric, a fact used in the factorization (6.36c). The left and right multiplication with $\mathcal{I}^{1/2}$ in (6.36d) is done to get a symmetric solution for $\bar{P}_f$. We have found one symmetric positive semi-definite solution, and we know from the general Riccati theory that the solution is unique, hence we are done.

The result (6.36e) is an explicit expression for how the filtering lower bound depends on the state noise covariance and measurement information. For instance, one can use it to analyze how different sensor deployments affect the measurement information in order to optimize the lower bound with respect to sensor locations in a sensor network. For non-Gaussian noise, Q and R can be replaced by the inverse intrinsic accuracy of the respective noise, and the result still holds.

6.5.2 Posterior CRLB

The *posterior* filtering CRLB for the general nonlinear model (6.2) is given by the recursion, see (Bergman, 1999, Theorem 4.5),

$$P_{k+1|k+1}^{-1} = \tilde{Q}_k + \tilde{R}_k - \tilde{S}_k^T (P_{k|k}^{-1} + \tilde{V}_k)^{-1}\tilde{S}_k, \quad (6.37)$$

where

$$\tilde{Q}_k = \mathrm{E}\Big(-\frac{d^2}{dx_{k+1}^2}\log p(x_{k+1}|x_k)\Big),$$

$$\tilde{R}_k = \mathrm{E}\Big(-\frac{d^2}{dx_k^2}\log p(y_k|x_k)\Big),$$

$$\tilde{S}_k = \mathrm{E}\Big(-\frac{d^2}{dx_{k+1}x_k}\log p(x_{k+1}|x_k)\Big),$$

$$\tilde{V}_k = \mathrm{E}\Big(-\frac{d^2}{dx_k^2}\log p(x_{k+1}|x_k)\Big)$$

and the iteration is initiated with

$$P_0^{-1} = \mathrm{E}\Big(-\frac{d^2}{dx_0^2}\log p(x_0)\Big).$$

The expectation is taken with respect to x_k, x_{k+1}, and y_k where applicable. The main recursion resembles the Riccati equation in the Kalman filter, but the definitions of $\tilde{V}_k$, $\tilde{R}_k$, $\tilde{S}_k$, $\tilde{Q}_k$, and P_0^{-1} might look repelling. To understand what they say, consider Gaussian distributions in the additive model (6.3)

$$p(x_{k+1}|x_k) = p_{v_k}(x_{k+1} - f(x_k)) = \mathcal{N}(x_{k+1}; f(x_k), Q), \tag{6.38a}$$
$$p(y_k|x_k) = p_{e_k}(y_k - h(x_k)) = \mathcal{N}(y_k; h(x_k), R). \tag{6.38b}$$

The different gradients are readily computed using the chain rule and the PDF for Gaussian distributions as

$$\tilde{Q}_k = -\mathrm{E}\left(\frac{d^2}{dx_{k+1}^2}\log p(x_{k+1}|x_k)\right) = Q^{-1}, \tag{6.39a}$$

$$\tilde{R}_k = -\mathrm{E}\left(\frac{d^2}{dx_k^2}\log p(y_k|x_k)\right) = -\mathrm{E}\left(\left(\frac{dh^T(x_k)}{dx_k}\right)R^{-1}\left(\frac{dh^T(x_k)}{dx_k}\right)^T\right), \tag{6.39b}$$

$$\tilde{S}_k = -\mathrm{E}\left(\frac{d^2}{dx_{k+1}dx_k}\log p(x_{k+1}|x_k)\right) = -\mathrm{E}\left(\left(\frac{df^T(x_k)}{dx_k}\right)Q^{-1}\right), \tag{6.39c}$$

$$\tilde{V}_k = -\mathrm{E}\left(\frac{d^2}{dx_k^2}\log p(x_{k+1}|x_k)\right) = -\mathrm{E}\left(\left(\frac{df^T(x_k)}{dx_k}\right)Q^{-1}\left(\frac{df^T(x_k)}{dx_k}\right)^T\right). \tag{6.39d}$$

In summary, the posterior CRLB for the model with additive noise (6.3) is given by the recursion (6.37) with the matrices defined in (6.39).

6.6 Summary

6.6.1 Theory

In a Bayesian framework, filtering consists of applying the following high-level primitives recursively:

1. *Estimation* that provides $p(x_k|y_k)$ as discussed in Chapters 2 and 3. Here, the observation y_k might give only partial information about x_k, or in other words the information matrix is singular.

2. *Fusion*, where the estimation information $p(x_k|y_k)$ is merged with the prior information $p(x_k|y_{1:k-1})$ to obtain $p(x_k|y_{1:k})$.

3. *Transformation*, to propagate information through the dynamics $z = f(x_k, u_k)$. This gives $p(z|y_{1:k})$.

4. *Diffusion*, for adding uncertainty from the process noise. This gives $p(x_{k+1}|y_{1:k})$.

Estimation and fusion can be elegantly solved by finite-dimension filters for all observation models in the exponential family, if the prior is taken from its conjugate prior. The problem is that in most cases, the freedom in the dynamic model is severely limited. The most explored exceptions are the linear Gaussian models and the hidden Markov model, where finite dimensional filter exists.

In a functional form, the Bayesian recursion is

$$p(x_k|y_{1:k}) = \frac{p(y_k|x_k)p(x_k|y_{1:k-1})}{p(y_k|y_{1:k-1})},$$

$$p(y_k|y_{1:k-1}) = \int_{\mathbb{R}^n} p(y_k|x_k)p(x_k|y_{1:k-1})\,dx_k,$$

$$p(x_{k+1}|y_{1:k}) = \int_{\mathbb{R}^n} p(x_{k+1}|x_k)p(x_k|y_{1:k})\,dx_k.$$

The first equation is the measurement update (estimation and fusion), the second one a normalization, and the third one the time update (transformation and diffusion). From the posterior distribution $p(x_k|y_{1:k})$, one can extract the conditional mean $\hat{x}_{k|k}$ as the state estimate, and its covariance $P_{k|k}$ as a measure of certainty.

The point mass filter (PMF) provides an easy to implement numerical approximation that is useful when $n_x \leq 2$. The PMF computes the measurement update exactly for each grid point $x_k^{(i)}$, and approximates the integral in the time update with a Riemann sum.

Independently of how the state is estimated, the covariance matrix of the estimate is bounded by the Cramér Rao lower bound (CRLB):

$$\operatorname{Cov}(\hat{x}_{k|k}) \geq P_{k|k}^{\mathrm{CRLB}}.$$

The most useful version of CRLB is computed recursively by a Riccati equation which has the same functional form as the Kalman filter,

$$P_{k+1|k}^{\text{ParCRLB}} = F_k P_{k|k}^{\text{ParCRLB}} F_k^T + G_k Q_k G_k,$$
$$P_{k+1|k+1}^{\text{ParCRLB}} = P_{k+1|k}^{\text{ParCRLB}} - P_{k+1|k}^{\text{ParCRLB}} H_k^T (H_k P_{k+1|k}^{\text{ParCRLB}} H_k^T + R_k)^{-1} H_k P_{k+1|k}^{\text{ParCRLB}}.$$

- In the nonlinear case, F_k, G_k and H_k above denote the gradients of $f(x_k, v_k)$ and $h(x_k)$, respectively.

- In the non-Gaussian case, the covariances Q_k, R_k and P_0 are replaced with the inverse intrinsic accuracies $\mathcal{I}_{v_k}^{-1}$, $\mathcal{I}_{e_k}^{-1}$ and $\mathcal{I}_{x_0}^{-1}$, respectively.

- The parametric CRLB is a function of the true state trajectory $x_{1:k}$.

- The posterior CRLB is the parametric CRLB averaged over all possible trajectories $P_{k|k}^{\text{PostCRLB}} = \mathrm{E}\bigl(P_{k|k}^{\text{ParCRLB}}\bigr)$. The expecation makes its computation quite complex in general.

- In the linear Gaussian case, the parametric and posterior bounds coincide.

- The CRLB bound is attainable in the linear Gaussian case, and the Kalman filter provides the optimal estimate.

6.6.2 Software

For nonlinear filtering, the NL object is central. The underlying model is in continuous time

$$\dot{x}(t) = f(t, x(t), u(t); \theta) + v(t),$$
$$y(t_k) = h(t_k, x(t_k), u(t_k); \theta) + e_k,$$

and in discrete time

$$x_{k+1} = f(k, x_k, u_k; \theta) + v_k,$$
$$y_k = h(t_k, x_k, u_k; \theta) + e_k.$$

The contructor is `m=nl(f,h,nn)`:

- `f` and `h` are each either a string that fits the inline object, an inline object, or a string that defines an m-file. For the cases of inline and m-file, the syntax must be `h(t,x,u,th)`, even if not all arguments are used. Also, the variable names have to be `t,x,u,th`.

- `nn=[nx, nu, ny, nth]` is a vector defining the model orders.

6.6 Summary

- The fields `pv,pe,px0` can be set to arbitrary distributions in the `pdfclass`. They can also be specified as a covariance, for instance `m.pe=2*eye(2)`, in which case a Gaussian distribution is default.

- The fields `th,P` denote the parameters in the model and their covariance matrix, respectively.

- Other fields of interest: `x0, fs, xlabel, ylabel, name`.

The constructor checks that the dimensions are appropriate and that the functions `f,h` are feasible and consistent with the specified dimensions in `nn`. `exnl` contains a number of examples of nonlinear models.

All algorithms become methods of the NL model object:

- `y=simulate(m,u)` or `y=simulate(m,t)` simulates the system.

- `nl2ss(m,x)` returns a linear state space model, which is linearized around the specified state.

- `m=ss2nl(m)` converts a state space model object SS to an NL object.

- `xhat=pmf(x,y,x1grid,x2grid)` implements the point-mass filter for $n_x \leq 2$.

- `xcrlb=crlb(x,y)` computes the parametric CRLB for the trajectory in `y.x`.

There are many more methods implementing different nonlinear filter approximations that will be presented later on.

Typical plots:

- `xplot(y,xhat,xcrlb,'conf',90)` compares the true states to the estimated ones and the CRLB, where the latter ones correspond to a 90% confidence interval. A trailing `ind` argument can be used to select the state indices to include in the plot (`ind=1` is default).

- `xplot2(y,xhat,xcrlb,'conf',90)` plots the first two states against each other for the true states, the estimated ones and the CRLB, where the latter ones are complemented with ellipsoids corresponding to 90% confidence regions. A trailing `ind` argument can be used to select the two state indices (`ind=[1 2]` is default).

7
The Kalman Filter

The *Kalman filter* (*KF*) estimates the states in a linear state space model

$$x_{k+1} = F_k x_k + G_{u,k} u_k + G_{v,k} v_k, \qquad \text{Cov}(v_k) = Q_k, \qquad (7.1a)$$
$$y_k = H_k x_k + D_k u_k + e_k, \qquad \text{Cov}(e_k) = R_k, \qquad (7.1b)$$
$$\text{E}(x_0) = \hat{x}_{1|0}, \qquad (7.1c)$$
$$\text{Cov}(x_0) = P_{1|0}. \qquad (7.1d)$$

Without any observations, (7.1a) implies that the initial state (7.1c) and covariance (7.1d) are propagated as

$$\hat{x}_{k|0} = F_k \hat{x}_{k-1|0}, \qquad (7.2a)$$
$$P_{k|0} = F_k P_{k-1|0} F_k^T + G_k Q_k G_k^T. \qquad (7.2b)$$

The double time index $k|m$ should be read *at time k given measurements up to time m*.

The Kalman filter finds the best possible linear filter, where the observation y_k is the input to the filter,

$$\hat{x}_{k+1|k} = M_k \hat{x}_{k|k-1} + L_k y_k, \qquad (7.3a)$$
$$P_{k+1|k} = M_k P_{k|k-1} M_k^T + L_k R_k L_k^T. \qquad (7.3b)$$

The different derivations optimize, in one way or another, the matrices M_k and L_k such that the estimate is unbiased and the covariance as small as possible. The chapter is organized as follows:

- The main algorithm is given in Section 7.1 and two derivations are provided in Sections 7.1.2 and 7.1.3, respectively. The first one is based on direct optimization of M and L above, while the second one utilizes a powerful lemma on conditional expectations.

- Practical issues are described in Section 7.2.

- Computational issues are described in Section 7.3.

- Section 7.4 gives three versions of the Kalman smoother.

- Section 7.5 derives the numerically attractive square root form of the KF.

- Section 7.6 gives some detection algorithms to decide if the measurements are inliers or outliers, if the KF has diverged and if two distributed filters are consistent with each other.

- Section 7.7 provides some examples.

7.1 Kalman Filter Algorithms

7.1.1 Standard Form

The standard algebraic form of the Kalman filter is given in Algorithm 7.1.

Algorithm 7.1 The Kalman filter

For the linear model (7.1), the best linear unbiased filter is given by the following recursions initialized with $\hat{x}_{1|0} = \mathrm{E}(x_0)$ and $P_{1|0} = \mathrm{Cov}(x_0)$:

1. **Measurement update.**

$$\hat{x}_{k|k} = \hat{x}_{k|k-1} + P_{k|k-1}H_k^T(H_k P_{k|k-1}H_k^T + R_k)^{-1}(y_k - H_k\hat{x}_{k|k-1} - D_k u_k)$$
$$P_{k|k} = P_{k|k-1} - P_{k|k-1}H_k^T(H_k P_{k|k-1}H_k^T + R_k)^{-1}H_k P_{k|k-1}.$$

2. **Time update.**

$$\hat{x}_{k+1|k} = F_k \hat{x}_{k|k} + G_{u,k} u_k,$$
$$P_{k+1|k} = F_k P_{k|k} F_k^T + G_{v,k} Q_k G_{v,k}^T.$$

To structure the computations, one can define the innovation

$$\varepsilon_k = y_k - H_k \hat{x}_{k|k-1} - D_k u_k, \qquad (7.4)$$

the innovation covariance

$$S_k = H_k P_{k|k-1} H_k^T + R_k, \qquad (7.5)$$

and Kalman gain

$$K_k = P_{k|k-1} H_k^T (H_k P_{k|k-1} H_k^T + R_k)^{-1}. \qquad (7.6)$$

With these definitions, the measurement update can be written

$$\hat{x}_{k|k} = \hat{x}_{k|k-1} + K_k \varepsilon_k, \qquad (7.7a)$$
$$P_{k|k} = P_{k|k-1} - K_k H_k P_{k|k-1} = P_{k|k-1} - K_k S_k K_k^T. \qquad (7.7b)$$

These versions avoid computing the same quantity twice. A further trick is to compute a local variable $H_k P_{k|k-1}$, since this product occurs several times.

7.1.2 Best Linear Unbiased Filter

The derivation in this section directly addresses the unknowns matrices in the general linear filter (7.3). For simplicity, the time index is dropped from F, G, H in the derivations, and the input u_k is also omitted.

Error Model

Suppose the observations up to time $k-1$ have been used to form the estimate $\hat{x}_{k-1|k-1}$ of the state x_{k-1}. The prediction error, or innovation, is defined by

$$\varepsilon_k = y_k - H\hat{x}_{k|k-1}.$$

A general linear recursive filter (7.3) gives a state prediction error defined as

$$\tilde{x}_{k+1|k} = x_{k+1} - \hat{x}_{k+1|k}, \qquad (7.8)$$

and in the same way the state filtering error is defined as $\tilde{x}_{k|k} = x_k - \hat{x}_{k|k}$.

The state prediction error using the model (7.1) and general linear recursive filter (7.3) is given by

$$\tilde{x}_{k+1|k} = x_{k+1} - \hat{x}_{k+1|k} \qquad (7.9a)$$
$$= Fx_k + Gv_k - \big(M_k \hat{x}_{k|k-1} + L_k y_k\big) \qquad (7.9b)$$
$$= Fx_k - M_k \hat{x}_{k|k-1} + Gv_k - L_k(Hx_k + e_k). \qquad (7.9c)$$

Next, the matrices M_k and L_k will be determined.

Unbiased Linear Filter

It is immediate from (7.9) that the prediction error is unbiased if $M_k = F - L_k H$. The unbiased linear filter in (7.3) can be rearranged to

$$\hat{x}_{k+1|k} = F\hat{x}_{k|k-1} + L_k(y_k - H\hat{x}_{k|k-1}), \tag{7.10}$$

and we have

$$\tilde{x}_{k+1|k} = (F - L_k H)\tilde{x}_{k|k-1} + Gv_k - L_k e_k. \tag{7.11}$$

Minimum Variance Filter

This section derives the minimum variance unbiased filter by induction. First, the filter is initiated with $P_{1|0} = P_0$. The derivation is done inductively, so the model can be time-varying without increasing the complexity in the derivation.

Suppose that all observations up to time $k-1$ have been optimally processed, and we now want to find out the optimal gain L_k in (7.10).

The recursive form of the covariance matrix follows from (7.11) as

$$P_{k+1|k} = (F_k - L_k H_k)P_{k|k-1}(F_k - L_k H_k)^T + G_k Q_k G_k^T + L_k R_k L_k^T. \tag{7.12}$$

Completing the squares, using the more compact notation $P = P_{k|k-1}$, and temporary dropping the time index, give

$$P_{k+1|k} = FPF^T - LHPH^T L^T + LHPF^T + FPH^T L^T + GQG^T - LRL^T$$
$$= \left(L - FPH^T(HPH^T + R)^{-1}\right)\left(HPH^T + R\right)\left(L - FPH^T(HPH^T + R)^{-1}\right)^T$$
$$+ FPF^T - FPH^T(HPH^T + R)^{-1}HPF^T + GQG^T$$
$$\geq FPF^T - FPH^T(HPH^T + R)^{-1}HPF^T + GQG^T.$$

with equality if and only if (since R is positive definite)

$$L_k = F_k P_{k|k-1} H_k^T (H_k P_{k|k-1} H_k^T + R_k)^{-1}. \tag{7.13}$$

With this choice of L_k, the minimal state prediction error covariance becomes

$$P_{k+1|k} = F_k \left(P_{k|k-1} - P_{k|k-1} H_k^T (H_k P_{k|k-1} H_k^T + R_k)^{-1} H_k P_{k|k-1}\right) F_k^T + G_k Q_k G_k^T. \tag{7.14}$$

This completes the induction proof of the filter gain and error covariance.

Stationary Kalman Filter

Let the gain $L_k = L$ be time-invariant. The error recursion (7.11) can be expanded to

$$\tilde{x}_{k+1|k} = (F - LH)^k(x_0 - \hat{x}_{1|0}) + \sum_{m=1}^{k}(F - LH)^{k-m}(Gv_m - Le_m).$$

This expression illustrates the following:

- The influence of the initial error $(x_0 - \hat{x}_{1|0})$ decays to zero if L is chosen such that the eigenvalues to $F - LH$ are inside the unit circle, in which case $\|F - LH\|^k \to 0$, $k \to \infty$. This is the *stability constraint*.

- The choice of L is a compromize between minimizing the transient error and suppression of process and measurement noise.

The latter requirement can be expressed as minimizing the error covariance matrix

$$P_{k+1|k} = \text{Cov}(\tilde{x}_{k+1|k}) = (F - LH)^k P_{1|0}(F - LH)^T)^k + \qquad (7.15)$$
$$\sum_{m=1}^{k}(F - LH)^{k-m}(GQG^T + LRL^T)((F - LH)^T)^{k-m}.$$

The stochastic uncertainties represented with the covariance matrices $P_{1|0}$, Q and R, respectively, thus correspond to one term each. Some special cases in the design of L include:

1. *Observer.* If there is no noise ($Q = 0$, $R = 0$), then only the transient is to be minimized. It is known from control theory that the eigenvalues to $F - LH$ can be chosen arbitrarily if the model is observable. For instance, a dead beat observer is obtained if all eigenvalues are put in the origin, in which case the transient vanishes after at most n_x samples.

2. *The stationary Kalman filter* focuses on balancing the noise terms, neglecting the transient (stability assures that the transient has vanished in a stationarity).

3. *The time-varying Kalman filter* minimizes the sum of all three error sources, as derived in the next section. This approach is the focus here.

7.1.3 Conditional Expectation

The next derivation is based on the following versatile result from statistics (which has already been used in Section 3.5).

Lemma 7.1 (Conditional Gaussian Distributions.)
Suppose the vectors X and Y are simultaneously Gaussian distributed

$$\begin{pmatrix} X \\ Y \end{pmatrix} \sim \mathcal{N}\left(\begin{pmatrix} \mu_x \\ \mu_y \end{pmatrix}, \begin{pmatrix} P_{xx} & P_{xy} \\ P_{xy} & P_{yy} \end{pmatrix}\right) = \mathcal{N}\left(\begin{pmatrix} \mu_x \\ \mu_y \end{pmatrix}, P\right). \quad (7.16)$$

Then, the conditional distribution for X, given the observed $Y = y$, is Gaussian distributed:

$$(X|Y=y) \sim \mathcal{N}(\mu_x + P_{xy}P_{yy}^{-1}(y-\mu_y), P_{xx} - P_{xy}P_{yy}^{-1}P_{yx}). \quad (7.17)$$

Proof: Start with the easily checked formula

$$\begin{pmatrix} I & -P_{xy}P_{yy}^{-1} \\ 0 & I \end{pmatrix} \underbrace{\begin{pmatrix} P_{xx} & P_{xy} \\ P_{xy} & P_{yy} \end{pmatrix}}_{P} \begin{pmatrix} I & 0 \\ -P_{yy}^{-1}P_{yx} & I \end{pmatrix} = \begin{pmatrix} P_{xx} - P_{xy}P_{yy}^{-1}P_{yx} & 0 \\ 0 & P_{yy} \end{pmatrix},$$
(7.18)

and Bayes' rule

$$p_{X|Y}(x,y) = \frac{p_{X,Y}(x,y)}{p_Y(y)}$$

$$= \frac{1}{(2\pi)^{n_x/2}} \frac{\det(P_{yy})^{1/2}}{\det(P)^{1/2}} \frac{\exp\left(-\frac{1}{2}\begin{pmatrix} x-\mu_x \\ y-\mu_y \end{pmatrix}^T P^{-1} \begin{pmatrix} x-\mu_x \\ y-\mu_y \end{pmatrix}\right)}{\exp\left(-\frac{1}{2}(y-\mu_y)^T P_{yy}^{-1}(y-\mu_y)\right)}.$$

From (7.18) we get

$$\det(P) = \det(P_{xx} - P_{xy}P_{yy}^{-1}P_{yx})\det(P_{yy}),$$

and the ratio of determinants can be simplified. We note that the new Gaussian distribution must have $P_{xx} - P_{xy}P_{yy}^{-1}P_{yx}$ as covariance matrix.

Next, we simplify the exponent using (7.18),

$$\begin{pmatrix} x-\mu_x \\ y-\mu_y \end{pmatrix}^T P^{-1} \begin{pmatrix} x-\mu_x \\ y-\mu_y \end{pmatrix}$$

$$= \begin{pmatrix} x-\mu_x \\ y-\mu_y \end{pmatrix}^T \begin{pmatrix} I & 0 \\ -P_{yy}^{-1}P_{yx} & I \end{pmatrix} \begin{pmatrix} (P_{xx}-P_{xy}P_{yy}^{-1}P_{yx})^{-1} & 0 \\ 0 & (P_{yy})^{-1} \end{pmatrix}$$
$$\cdot \begin{pmatrix} I & -P_{xy}P_{yy}^{-1} \\ 0 & I \end{pmatrix} \begin{pmatrix} x-\mu_x \\ y-\mu_y \end{pmatrix}$$

$$= \begin{pmatrix} x-\hat{x} \\ y-\mu_y \end{pmatrix}^T \begin{pmatrix} (P_{xx}-P_{xy}P_{yy}^{-1}P_{yx})^{-1} & 0 \\ 0 & (P_{yy})^{-1} \end{pmatrix} \begin{pmatrix} x-\hat{x} \\ y-\mu_y \end{pmatrix}$$

$$= (x-\hat{x})^T (P_{xx}-P_{xy}P_{yy}^{-1}P_{yx})^{-1}(x-\hat{x}) + (y-\mu_y)^T(P_{yy})^{-1}(y-\mu_y),$$

7.1 Kalman Filter Algorithms

where
$$\hat{x} = \mu_x + P_{xy}P_{yy}^{-1}(y - \mu_y).$$

From this, we can conclude that
$$p_{X|Y}(x,y) = \frac{1}{(2\pi)^{n_x/2}\det(P_{xx} - P_{xy}P_{yy}^{-1}P_{yx})^{1/2}}$$
$$\cdot \exp\left(-\frac{1}{2}(x-\hat{x})^T(P_{xx} - P_{xy}P_{yy}^{-1}P_{yx})^{-1}(x-\hat{x})\right),$$

which is a Gaussian distribution with mean and covariance as given in (7.17). □

An inductive derivation of the Kalman filter can now be performed. Assume again time-invariant model (7.1) and neglect the input term $G_u u_k$.

1. Let $x_0 \sim \mathcal{N}(\hat{x}_{0|0}, P_{0|0})$.

2. Suppose that x_k, conditioned on the observations of $y_{1:k}$ up to time k, is $N(\hat{x}_{k|k}, P_{k|k})$

3. **Time update.**
$$\hat{x}_{k+1|k} = E[x_{k+1}|y_{1:k}] = E[Fx_k + G_v v_k|y_{1:k}] = F\hat{x}_{k|k}$$
$$P_{k+1|k} = E[(x_{k+1} - \hat{x}_{k+1|k})(x_{k+1} - \hat{x}_{k+1|k})^T|y_{1:k}]$$
$$= E[(F(x_k - \hat{x}_{k|k}) + G_v v_k)(F(x_k - \hat{x}_{k|k}) + G_v v_k)^T|y_{1:k}]$$
$$= FP_{k|k}F^T + G_v Q G_v^T.$$

Measurement update. We have derived the conditional distribution for X, given Y, as
$$E(X|Y) = \mu_x + P_{xy}P_{yy}^{-1}(Y - \mu_\gamma)$$
$$\text{Cov}(X|Y) = P_{xx} - P_{xy}P_{yy}^{-1}P_{yx}.$$

Identify the stochastic variable X with $(x_k|y_{1:k-1}) \sim \mathcal{N}(\hat{x}_{k|k-1}, P_{k|k-1})$ and Y with $(y_k|y_{1:k-1}) \sim \mathcal{N}(H\hat{x}_{k|k-1}, HP_{k|k-1}H^T + R)$. We have
$$P_{xy} = E[(x_k - E[x_k])(y_k - E[y_k])^T|y_{1:k-1}]$$
$$= E((x_k - \hat{x}_{k|k-1})(Hx_k + G_v v_k - H\hat{x}_{k|k-1})^T)$$
$$= P_{k|k-1}H^T = P_{yx}^T.$$

This gives
$$\hat{x}_{k|k} = \hat{x}_{k|k-1} + P_{k|k-1}H^T(HP_{k|k-1}H^T + R)^{-1}(y_k - H\hat{x}_{k|k-1})$$
$$P_{k|k} = P_{k|k-1} - P_{k|k-1}H^T(HP_{k|k-1}H^T + R)^{-1}HP_{k|k-1}.$$

Induction implies that x_k, given $y_{1:k}$, is normally distributed.

7.1.4 The Information Filter

The information form will be used frequently in the sequel:

- To understand the relation between centralized and de-centralized filtering.

- As a feasible implementation alternative in sensor fusion problems where $n_y > n_x$.

- As a tool in several derivations.

When deriving sequential version of the least squares solution in Appendix E, we started with formulas like

$$\iota_k = \iota_{k-1} + H_k^T R_k^{-1} y_k$$
$$\mathcal{I}_k = \mathcal{I}_{k-1} + H_k^T R_k^{-1} H_k$$
$$\hat{x}_k = \mathcal{I}_k^{-1} \iota_k.$$

After noticing that the matrix inversion requires many operations, we used the matrix inversion lemma to derive Kalman filter like equations, where no matrix inversion is needed (for scalar y).

Here we will go the other way around. We have a Kalman filter, and we are attracted by the simplicity of the measurement update. Define $\mathcal{I}_{k|k} = P_{k|k}^{-1}$ and $\iota_{k|k} = \mathcal{I}_{k|k} \hat{x}_{k|k}$. The matrix inversion lemma can then be used to rewrite the time update

$$\hat{x}_{k+1|k} = F_k \hat{x}_{k|k} + G_{u,k} u_k \tag{7.19a}$$
$$P_{k+1|k} = F_k P_{k|k} F_k^T + G_{v,k} Q G_{v,k}^T, \tag{7.19b}$$

in terms of information state and information matrix, and Algorithm 7.2 follows. Note that the measurement update here is trivial, while the main computational burden comes from the time update. This is in contrast to the Kalman filter. The proof is left as an exercise.

The time update has been divided into two operations: one concerning the linear transformation with F_k and one originating in the two input terms $G_{u,k} u_k + G_{b,k} v_k$. Note the structural similarity of the input contribution to the time update and the measurement in the standard Kalman filter.

Note the following:

- $H_k^T R_k^{-1} H_k$ is the information in a new measurement, and $H_k^T R_k^{-1} y_k$ is a sufficient statistic for updating the estimate. The interpretation is that $\mathcal{I}_{k|k}$ contains the information contained in all past measurements. The time update implies a forgetting of information.

7.1 Kalman Filter Algorithms

Algorithm 7.2 The information filter

For the linear model (7.1), the best linear unbiased filter is given by the following information recursions initialized with $\mathcal{I}_{1|0} = \left(\mathrm{Cov}(x_0)\right)^{-1}$ and $\iota_{1|0} = \mathcal{I}_{1|0}\mathrm{E}(x_0)$:

1. **Measurement update:**

$$\iota_{k|k} = \iota_{k|k-1} + H_k^T R_k^{-1} y_k + \mathcal{I}_{k|k} D_{u,k} u_k,$$
$$\mathcal{I}_{k|k} = \mathcal{I}_{k|k-1} + H_k^T R_k^{-1} H_k.$$

2. **Time update, dynamics:**

$$\bar{\iota}_{k+1|k} = F_k^{-T} \iota_{k|k} + \mathcal{I}_{k+1|k} G_{u,k} u_k,$$
$$\bar{\mathcal{I}}_{k+1|k} = F_k^{-T} \mathcal{I}_{k|k} F_k^{-1}.$$

3. **Time update, inputs:**

$$\iota_{k+1|k} = \bar{\iota}_{k+1|k} - \bar{\mathcal{I}}_{k+1|k} G_{v,k} (G_{v,k}^T \bar{\mathcal{I}}_{k+1|k} G_{v,k} + Q_k^{-1})^{-1} G_{v,k}^T \bar{\iota}_{k+1|k}$$
$$\quad + \bar{\mathcal{I}}_{k+1|k} G_{u,k} u_k$$
$$\mathcal{I}_{k+1|k} = \bar{\mathcal{I}}_{k+1|k} - \bar{\mathcal{I}}_{k+1|k} G_{v,k} (G_{v,k}^T \bar{\mathcal{I}}_{k+1|k} G_{v,k} + Q_k^{-1})^{-1} G_{v,k}^T \bar{\mathcal{I}}_{k+1|k}.$$

- Note that this is one occasion where it is important to factorize the state noise covariance, so that Q_k is non-singular. In most other applications, the term $G_{v,k} v_k$ can be replaced with $\bar{v}_k$ with (singular) covariance matrix $\bar{Q}_k = G_{v,k} Q_k G_{v,k}^T$.

There is an alternative form with just one equation for the time update as given in Anderson and Moore (1979). To give this, introduce two auxiliary matrices to simplify notation

$$M_k = F_k^{-T} \mathcal{I}_{k|k} F_k^{-1} \tag{7.20a}$$
$$N_k = M_k G_{v,k} (G_{v,k}^T M_k G_{v,k} + Q_k^{-1})^{-1} \tag{7.20b}$$

The information filter can then be implemented as

$$\iota_{k+1|k} = (I - N_k G_{v,k}^T) F_k^{-T} \iota_{k|k} + P_{k+1|k}^{-1} G_{u,k} u_k \tag{7.20c}$$
$$\mathcal{I}_{k+1|k} = (I - M_k G_{v,k} (G_{v,k}^T M_k G_{v,k} + Q_k^{-1})^{-1} G_{v,k}^T) M_k \tag{7.20d}$$
$$\quad = (I - N_k G_{v,k}^T) M_k \tag{7.20e}$$
$$\iota_{k|k} = \iota_{k|k-1} + H_k^T R_k^{-1} y_k + \mathcal{I}_{k|k} D_{u,k} u_k \tag{7.20f}$$
$$\mathcal{I}_{k|k} = \mathcal{I}_{k|k-1} + H_k^T R_k^{-1} H_k. \tag{7.20g}$$

The main advantages of the information filter are:

- Very vague prior knowledge of the state can now be expressed as $\mathcal{I}_{1|0} = P_{1|0}^{-1} = 0$.

- The Kalman filter requires an $n_y \times n_y$ matrix to be inverted ($n_y = \dim y$), while the information filter requires an $n_v \times n_v$ matrix to be inverted. In tracking and navigation applications, the number of state noise inputs may be quite small (2 or 3, say), while the number of outputs can be large.

- There is an iterated time update implementation, further reducing the complexity. If the state noise can be split up into independent contributions as $G_{v,k} v_k = \sum_{i=1}^{M} G_{v,k}^{(i)} v_k^{(i)}$, where $Q = \operatorname{diag}\left(Q^{(11)}, \ldots, Q^{(MM)}\right)$ then the input part of the time update can be applied to each term. The advantage is that only the sub-blocks $Q^{(ii)}$ needs to be inverted. See also Anderson and Moore (1979) (pp. 146–147). A similar algorithm for the case that the covariance R_k is block diagonal is derived in Section 7.3.2.

7.2 Practical Issues

7.2.1 Observability

We have snapshot observability if H_k has rank n_x, in which case the state can be estimated as

$$(\hat{x}_k|y_k) = (H_k^T R_k^{-1} H_k)^{-1} H_k^T R_k^{-1} y_k \sim \mathcal{N}(x_k, (H_k^T R_k^{-1} H_k)^{-1}). \qquad (7.21)$$

In the filtering case, the quite conservative full rank condition on H_k can be relaxed substantially. The classical definition of observability is to check the rank of the observability matrix

$$\mathcal{O}_{n_x} = \begin{pmatrix} H \\ HF \\ HF^2 \\ \vdots \\ HF^{n_x-1} \end{pmatrix} \qquad \mathcal{O}_{k,n_x} = \begin{pmatrix} H_{k-n_x+1} \\ H_{k-n_x+2} F_{k-n_x+1} \\ H_{k-n_x+3} F_{k-n_x+2} F_{k-n_x+1} \\ \vdots \\ H_k F_{k-1} \ldots F_{k-n_x+1} \end{pmatrix}. \qquad (7.22)$$

Note that $\mathcal{O}_{n_x}$ is deterministic and depends on the model only.

In the time-varying case, the time horizon that needs to be checked can be longer than n_x, so investigating the rank of larger observability matrices $\mathcal{O}_n$ for $n > n_x$ makes sense.

7.2 Practical Issues

For stochastic time-varying systems, the observability Grammian is more useful. It is defined as

$$\mathcal{O}_N = \sum_{k=0}^{N} \Phi_{k,0}^T H_k^T R_k^{-1} H_k \Phi_{k,0}, \qquad (7.23)$$

where $\Phi_{k,0}$ is the state transition matrix from time 0 to time k. This is a weighted version of $\mathcal{O}^T \mathcal{O}$ if $n_x = N$, which thus has the same rank as $\mathcal{O}$. If this has full rank, then the initial state x_0 can be estimated, and thus all future states predicted. An eigenvector decomposition of $\mathcal{O}$ reveals which linear combinations of the state vector are more difficult to estimate. The observability Grammian does not take the process noise into account. The Kalman filter covariance P_k is here the best overall indicator of observability, where the eigenvector decomposition reveals the accuracy in the state estimate.

7.2.2 Sliding Window Kalman Filter

If $\mathcal{O}_{n_x}$ (time invariant case) or $\mathcal{O}_{k,n_x}$ (time-varying case) has full rank, then the state vector can be computed algebraically from noise free data, using the relation

$$\hat{x}^{LS}_{k-n_x+1|k-n_x+1:k} = (\mathcal{O}_{n_x}^T \mathcal{O}_{n_x})^{-1} \mathcal{O}_{n_x}^T \begin{pmatrix} y_{k-n_x+1} \\ y_{k-n_x+2} \\ y_{k-n_x+3} \\ \vdots \\ y_k \end{pmatrix} \qquad (7.24)$$

This is the *least squares* (*LS*) estimate of the state. The *weighted least squares* (*WLS*) estimate is obtained by a scaling with the covariance matrix,

$$\hat{x}^{WLS}_{k-n_x+1|k-n_x+1:k} = (\mathcal{O}_{n_x}^T R^{-1} \mathcal{O}_{n_x})^{-1} \mathcal{O}_{n_x}^T R^{-1/2} \begin{pmatrix} y_{k-n_x+1} \\ y_{k-n_x+2} \\ y_{k-n_x+3} \\ \vdots \\ y_k \end{pmatrix} \qquad (7.25)$$

Since WLS just as KF is BLUE and this is unique, the estimate $\hat{x}^{WLS}_{k-n_x+1|k-n_x+1:k}$ can be termed the Kalman filter estimate of the initial state over a sliding window.

7.2.3 Cross Correlated Noise

If there is a correlation between the noise processes $\mathrm{E}(v_k e_k^T) = M_k$, so that

$$\mathrm{E}\left(\begin{pmatrix} v_k \\ e_k \end{pmatrix} (v_l,\ e_l) \right) = \begin{pmatrix} Q_k & M_k \\ M_k^T & R_k \end{pmatrix} \delta_{k-l},$$

one trick to get back on track is to replace the state noise with

$$\bar{v}_k = v_k - M_k R_k^{-1} e_k \tag{7.26a}$$
$$= v_k - M_k R_k^{-1}(y_k - H_k x_k). \tag{7.26b}$$

Equation (7.26a) gives

$$\mathrm{E}\left(\begin{pmatrix}\bar{v}_k\\e_k\end{pmatrix}(\bar{v}_k,\ e_k)\right) = \begin{pmatrix}Q_k - M_k R_k^{-1} M_k^T & 0\\ 0 & R_k\end{pmatrix}.$$

Substitution with (7.26b) implies the equivalent signal model

$$x_{k+1} = (F_k - G_{v,k} M_k R_k^{-1} H_k) x_k + G_{u,k} u_k + G_{v,k} \bar{v}_k + G_{v,k} M_k R_k^{-1} y_k \tag{7.27a}$$
$$y_k = H_k x_k + e_k. \tag{7.27b}$$

The Kalman filter applies to this model. The last output dependent term in the state equation should be interpreted as an extra input.

7.2.4 Bias Error

A *bias error* due to modeling errors can be analyzed by distinguishing the true system and the design model, denoted by super-indices o and d, respectively. We here summarize one suggested approach in (Anderson and Moore, 1979). The state space model for the true system and the Kalman filter for the design model is

$$\begin{pmatrix}x_{k+1}^o\\ \hat{x}_{k+1|k}^d\end{pmatrix} = \begin{pmatrix}F_k^o & 0\\ F_k^d K_k^d H_k^o & F_k^d - F_k^d K_k^d H_k^d\end{pmatrix}\begin{pmatrix}x_k^o\\ \hat{x}_{k|k-1}^d\end{pmatrix} \tag{7.28}$$
$$+ \begin{pmatrix}G_{v,k}^o & 0\\ 0 & F_k^d K_k^d\end{pmatrix}\begin{pmatrix}v_k\\ e_k\end{pmatrix} + \begin{pmatrix}G_{u,k}^o & 0\\ 0 & G_{u,k}^d\end{pmatrix}\begin{pmatrix}u_k^0\\ u_k^d\end{pmatrix}.$$

Let $\bar{x}$ denote the augmented state vector with covariance matrix $\bar{P}$. This is a state space model (without measurements), so the covariance matrix time update can be applied,

$$\bar{P}_{k+1} = \begin{pmatrix}F_k^o & 0\\ F_k^d K_k^d H_k^o & F_k^d - F_k^d K_k^d H_k^d\end{pmatrix}\bar{P}_k\begin{pmatrix}F_k^o & 0\\ F_k^d K_k^d H_k^o & F_k^d - F_k^d K_k^d H_k^d\end{pmatrix}^T$$
$$+ \begin{pmatrix}G_{v,k}^o & 0\\ 0 & F_k^d K_k^d\end{pmatrix}\begin{pmatrix}Q_k^o & 0\\ 0 & R_k^o\end{pmatrix}\begin{pmatrix}G_{v,k}^o & 0\\ 0 & F_k^d K_k^d\end{pmatrix}^T. \tag{7.29}$$

The total error (including both bias and variance) can be expressed as

$$x_k^o - \hat{x}_{k|k-1}^d = (I,\ -I)\bar{x}_k.$$

7.2 Practical Issues

and the error correlation matrix (note that P is no covariance matrix when there is a bias error) is

$$P^p_{k|k-1} = (I, \ -I) \bar{P}_{k|k-1} \begin{pmatrix} I \\ -I \end{pmatrix}.$$

This matrix is a measure of the *performance* (hence the superscript p) of the mean square error. This should not be confused with the optimal performance P^o and the measure that comes out from the Kalman filter P^d. Note that $P^p > P^o$, and the difference can be interpreted as the bias. Note also that although the Kalman filter is stable, P^p may not even be bounded.

The main use of (7.28) is a simple test procedure for evaluating the influence of parameter variations. This kind of *sensitivity analysis* should always be performed before implementation. That is, vary each single partially unknown parameter, and compare P^p to P^o. Sensitivity analysis can here be interpreted as a Taylor expansion

$$P^p \approx P^o + \frac{dP^o}{d\theta} \Delta \theta.$$

The more sensitive P^o is to a certain parameter, the more important it is to model it correctly. The sensitivity aspects of filters in general are discussed further in Section 15.1.

7.2.5 Out of Sequence Updates

Normally, the measurements y_{t_k}, $t_1 \leq t_2 \leq t_3 \ldots$ are received in the natural order $k = 1, 2, 3, \ldots$. Out of sequence measurement updates are needed when the measurements arrive scrambled in time. One cause for this action is delayed measurements. There is sometimes a preprocessing time-delay (for instance when using sophisticated computer vision algorithms). Further, the time delays can be different for different sensors, leading to the problem of losing the natural order $k = 1, 2, 3, \ldots$ of the measurements and the corresponding Kalman filter updates.

One can distinguish three cases:

- It is known at time t_k that a measurement will be available and what the measurement equation H_{t_k} is, but the numeric value of the measurement is delayed. This case is typical for remote sensors which are scheduled to perform certain measurements at predefined times, but the communication causes a delay.

- The measurement time t_k is known, but the relation H_{t_k} and y_{t_k} are both delayed. This is the case when computer vision algorithms compute for feature positions in an image frame, and then converts these into measurement relations.

- At some time, the information t_k, $y(t_k)$ and $h(x_{t_k})$ reaches the filter, all being delayed. This is a typical problem in distributed asynchronous sensor networks with communication delays.

The following sub-sections will outline algorithms for these three cases of increasing difficulty. The derivations will be based on the linear property of the state update. First, to simplify notation, let the nonuniformly sampled linear state space model be

$$x(t_{k+1}) = F(t_{k+1})x(t_k) + v(t_k), \qquad (7.30a)$$
$$y(t_k) = H(t_k)x(t_k) + e(t_k). \qquad (7.30b)$$

Using the index k to denote the recursion counter in the Kalman filter, we obtain the usual model,

$$x_{k+1} = F_k x_k + v_k, \qquad (7.31a)$$
$$y_k = H_k x_k + e_k. \qquad (7.31b)$$

That is, this standard model implicitly covers both cases of uniform and nonuniform sampled measurements. The correction procedure below applies in both cases.

References on this subject include Zhang et al. (2005), Bar-Shalom and Li (1993) and Challa et al. (2003).

Known t_k and H_k but Delayed y_k

We will here outline a procedure for post-updating the state estimate for the first case, when there is a relatively simple solution.

The state update recursion can be expanded to the following form,

$$\hat{x}_{k|k} = F_k \hat{x}_{k-1|k-1} + K_k(y_k - H_k F_k \hat{x}_{k-1|k-1}) \qquad (7.32a)$$
$$= (I - K_k H_k) F_k \hat{x}_{k-1|k-1} + K_k y_k \qquad (7.32b)$$
$$= \sum_{m=0}^{k} \prod_{l=m+1}^{k} (I - K_l H_l) F_l K_m y_m + \prod_{l=1}^{k} (I - K_l H_l) F_l \hat{x}_{0|0}. \qquad (7.32c)$$

From this, the solution is rather obvious. If y_n is not available at time t_n, for some n with $t_n < t$, then enter a dummy value and compensate for the correct value afterwards. For instance, $y_n = 0$ is one option, in which case the term $\prod_{l=n+1}^{k}(I - K_l H_l) F_l K_n y_n$ has to be added to the state estimate when the measurement arrives. A better dummy value is to use the expected value, that is, $y_n = H_n \hat{x}_{n|n-1}$, which corresponds to just skipping the measurement update. Now, the innovation term

$$\prod_{l=n+1}^{k} (I - K_l H_l) F_l K_n (y_n - H_n \hat{x}_{n|n-1}) \qquad (7.33)$$

has to be added when the measurement arrives. Note that this corresponds to doing nothing in the state measurement update at time t_n. This solution is clearly to prefer, when (i) the application requires the state estimate before y_n is available, or (ii) when the model is nonlinear and the EKF linearizes the dynamics around the current state estimate.

Remarks:

- The matrix $\prod_{l=n+1}^{k}(I - K_l H_l) F_l K_n$ can be computed recursively with a minor extension of the basic Kalman filter equations.

- The covariance matrix lives its own life, and is not affected by the measurement delay. However, it does not reflect the covariance of the state estimate until the correction has been done.

- This result, including the points above, relies on the assumption that the time instant t_n and measurement relation H_n are both known in real-time.

The second item might be a limiting factor in applications. If this is a serious problem, the solution in the next sub-section must be used.

Known t_k but Delayed H_k and y_k

The Bayesian solution leads to an augmented state space model, where the state vector is augmented with the state x_n at time t_n. In this way, the correlation between x_n and the current state vector is kept track of automatically by the Kalman filter. This solution bears a lot in common with *fixed-point smoothing*, see Section 7.4. The algorithm is outlined below.

1. At time t_{n-1}, perform a time update to time t_n.

2. Augment the state vector to get the dynamic model

$$\bar{x}_k = \begin{pmatrix} x_k \\ x_n \end{pmatrix}, \tag{7.34a}$$

$$\bar{x}_{k+1} = \begin{pmatrix} F_k & 0 \\ 0 & I \end{pmatrix} \bar{x}_k + \begin{pmatrix} v_k \\ 0 \end{pmatrix} \tag{7.34b}$$

initialized with

$$\hat{\bar{x}}_{n|n-1} = \begin{pmatrix} \hat{x}_{n|n-1} \\ \hat{x}_{n|n-1} \end{pmatrix}, \tag{7.34c}$$

$$\bar{P}_{n|n-1} = \begin{pmatrix} P_{n|n-1} & P_{n|n-1} \\ P_{n|n-1} & P_{n|n-1} \end{pmatrix}. \tag{7.34d}$$

3. Continue with time update to time t_{n+1} followed by the usual recursive measurement and time updates.

4. When the measurement relation arrives, use the measurement relation

$$y_n = \begin{pmatrix} 0 & H_n \end{pmatrix} \bar{x}_k + e_n \qquad (7.34e)$$

for a measurement update.

5. The augmented state can now be deleted.

Delayed t_k, H_k and y_k

This case is naturally even harder than the case above, since we do not know when to augment the state vector. A similar idea can be used, however, in the case of (i) an underlying uniform sampling interval and (ii) a maximum time delay D is known. The solution resembles *fixed-interval smoothing*, which in fact is a feature obtained for free here.

Before the filter is started, augment the state vector to get the dynamic model

$$\bar{x}_k = \begin{pmatrix} x_k \\ x_{k-1} \\ \vdots \\ x_{k-D+1} \end{pmatrix} \qquad (7.35a)$$

$$\bar{x}_{k+1} = \begin{pmatrix} F_k & 0 & \cdots & 0 \\ I & 0 & \cdots & 0 \\ \vdots & \ddots & & \vdots \\ 0 & \cdots & I & 0 \end{pmatrix} \bar{x}_k + \begin{pmatrix} v_k \\ 0 \\ \vdots \\ 0 \end{pmatrix} \qquad (7.35b)$$

When a measurement with delay $0 \le d < D$ arrives, use the measurement relation

$$y_{k-d} = H_{k-d} I_{d+1,:} \bar{x}_k + e_{k-d}. \qquad (7.35c)$$

In the general case, there might be no other solution than the non-realtime one based on re-computing both the Riccati and state update equations from time t_n.

7.3 Computational Aspects

First, some brief comments on how to improve the numerical properties are given before we examine major issues:

- In case the measurement noise covariance is *singular*, the filter may be numerically unstable, although the solution to the estimation problem is well defined.

7.3 Computational Aspects

- One solution to this is to replace the inverse of S_k in K_k with a pseudo-inverse.
- Another solution is regularization. That is to add a small identity matrix to R_k. This is a common suggestion, especially when the measurement is scalar.
- A singular R means that one or more linear combinations of the state vector can be exactly computed from one measurement. A third solution is to compute a reduced order filter, see for instance (Anderson and Moore, 1979). This is quite similar to the Luenberger observer, see (Kailath, 1980) and Luenberger (1966). The idea is to transform the state and observation vector to $\bar{x}, \bar{y}$ such that the exactly known part of $\bar{y}$ is a sub-vector of $\bar{x}$, which does not need to be estimated.

• Sometimes we have linearly dependent measurements, which give cause to a rank deficient H_k. Then there is the possibility to reduce the complexity of the filter, by replacing the observation vector with an *equivalent observation*, see Murdin (1998), obtained by solving the least squares problem $y_k = H_k x_k + e_k$. For instance, when there are more measurements than states we get the least squares (snapshot) estimate

$$\hat{x}_k = (H_k^T H_k)^{-1} H_k^T y_k = x_k + (H_k^T H_k)^{-1} H_k^T e_k \triangleq \bar{y}_k.$$

Then

$$\bar{y}_k = I x_k + \bar{e}_k, \quad \bar{e}_k \in \mathcal{N}(0, (H_k^T H_k)^{-1} H_k^T R_k H_k (H_k^T H_k)^{-1})$$

can be used as the observation model in the Kalman filter without loss of performance.

• *Outliers* should be removed from the filter. Section 7.6.1 describes an outlier detection algorithm. Alternatively, prior knowledge and data pre-processing can be used, or impulsive disturbances can be incorporated into the noise model, see Niedzwiecki and Cisowski (1996) and Settineri et al. (1996).

• All numerical problems are significantly decreased by using square root algorithms. Section 7.5 is devoted to this issue.

7.3.1 Computational Complexity Issues

Since the measurement vector is potentially very large in sensor fusion, the following implementation issues should be considered:

• A direct implementation of KF requires inversion of a $n_y \times n_y$ symmetric and positive definite matrix, which can be done with $\mathcal{O}(n_y^{2.69})$ complexity Thrun et al. (2005).

- A general solution is based on a first projection to a *sufficient statistics*, which may be called an *equivalent measurement*, that is based on the snapshot solution (7.21)

$$\bar{y}_k = (H^T R^{-1} H)^{-1} H^T R^{-1} y_k \sim \mathcal{N}(\underbrace{I}_{\bar{H}} x, \underbrace{(H^T R^{-1} H)^{-1}}_{\bar{R}}).$$

Note that the $n_y \times n_y$ covariance matrix R_k needs to be inverted.

- The information form of the KF in Algorithm 7.2 is the general recommendation when $n_y > n_x$, where the time update requires inversion of a $n_v \times n_v$ matrix.

- The iterated Kalman filter in the next section is the recommended solution when there are many independent sensors.

7.3.2 Iterated Kalman filter

If the measurements are independent, that is, R is (block-)diagonal, then the *iterated Kalman filter* applies. The basic algebraic result is that one can make one measurement update per independent measurement sequentially, without time updating in between. This gives in the most favourable case n_y one-dimensional measurement updates with only a scalar division and no matrix inverse.

One key advantage of the Kalman filter for sensor fusion is its flexibility. The measurement can come at any time and have any dimension. In particular, we may get two measurements with zero time difference, so the time update does not affect x or P. That is, we get two measurement updates after each other. This is an intuitive proof of the iterated Kalman filter.

A formal derivation can be based on the *information filter*, whose measurement update propagates information (inverse covariance) rather than uncertainty and takes a particularly simple form,

$$\mathcal{I}_{k|k} = \mathcal{I}_{k|k-1} + H_k^T R_k^{-1} H_k. \tag{7.36}$$

Let the number of independent observations be denoted M. Using a block diagonal structure of R, this immediately gives

$$\mathcal{I}_{k|k} = \mathcal{I}_{k|k-1} + \sum_{i=1}^{M} (H_k^i)^T (R_k^{ii})^{-1} H_k^i. \tag{7.37}$$

Specifically, it implies that the all-in-one measurement update gives exactly the same result as the iterated measurement update, where M iterations of the KF measurement update are performed at each time k. The iterations are initialized with

$$\hat{x}_k^0 = \hat{x}_{k|k-1}, \tag{7.38a}$$
$$P_k^0 = P_{k|k-1}, \tag{7.38b}$$

and for $i = 1, 2, \ldots, M$,

$$K_k^i = P_k^{i-1}(H_k^i)^T (H_k^i P_k^{i-1}(H_k^i)^T + R_k^{ii})^{-1}, \tag{7.38c}$$
$$P_{k|k}^i = P_k^{i-1} - K_k^i H_k^i P_k^{i-1}, \tag{7.38d}$$
$$\hat{x}_{k|k}^i = \hat{x}_k^{i-1} + K_k^i(y_k^i - H_k^i \hat{x}_k^{i-1}). \tag{7.38e}$$

The final step is

$$\hat{x}_{k|k} = \hat{x}_k^M, \tag{7.38f}$$
$$P_{k|k} = P_k^M, \tag{7.38g}$$

after which a time update follows.

This scheme can be proven to be algebraically equivalent to (6.18c). We also have the relation useful for the detection problems in Section 7.6,

$$(y_k - H_k \hat{x}_{k|k-1})^T (H_k P_{k|k-1} H_k^T + R_k)^{-1} (y_k - H_k \hat{x}_{k|k-1})$$
$$= \sum_{i=1}^M (y_k^i - H_k^i \hat{x}_k^{i-1})^T (H_k^i P_k^{i-1}(H_k^i)^T + R_k^{ii})^{-1} (y_k^i - H_k^i \hat{x}_k^{i-1}). \tag{7.39}$$

Note, that the order of measurement processing is no issue, the same result will always be obtained.

7.3.3 Distributed Versus Centralized Implementations

An important issue is whether the fusion should be made at a central computer or in a distributed fashion. *Central fusion* means that we have access to all measurements when deriving the optimal filter; see Figure 7.1. In contrast, in *decentralized fusion* a filter is applied to each measurement, and the global fusion process has access only to the estimates and their error covariances. Decentralized filtering has certain advantages in fault detection in that the different sub-systems can apply a 'voting' strategy for fault detection. An obvious disadvantage is increased signaling, since the state, and in particular its covariance matrix, can be of much larger dimension than the measurement vector. One might argue that the global processor can process the measurements in a decentralized fashion to get the advantages of both alternatives. However, many sensing systems have built-in filters.

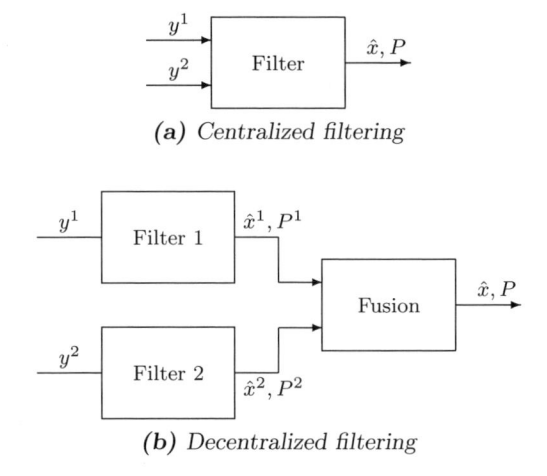

Figure 7.1: *The concepts of centralized and decentralized filtering.*

Centralized Fusion

Much of the beauty of the Kalman filter theory is the powerful state space model. A large set of measurements is simply collected into one measurement equation,

$$y_k = \begin{pmatrix} y_1 \\ y_2 \\ \vdots \\ y_m \end{pmatrix} = \begin{pmatrix} H_1 \\ H_2 \\ \vdots \\ H_m \end{pmatrix} x_k + \begin{pmatrix} e_1 \\ e_2 \\ \vdots \\ e_m \end{pmatrix}. \tag{7.40}$$

The only problem that might occur is that the sensors are working in different state coordinates.

The General Fusion Formula

If there are two independent state estimates $\hat{x}^1$ and $\hat{x}^2$, with covariance matrices P^1 and P^2 respectively, fusion of these pieces of information is straightforward using the sensor fusion formula (2.23):

$$\hat{x} = P\left((P^1)^{-1}\hat{x}^1 + (P^2)^{-1}\hat{x}^2\right) \tag{7.41a}$$

$$P = \left((P^1)^{-1} + (P^2)^{-1}\right)^{-1}. \tag{7.41b}$$

7.3 Computational Aspects

The sensor fusion formula can be formulated as a Kalman filter solution using a rather awkward but instructive state space model,

$$x_{k+1} = v_k, \qquad \mathrm{Cov}(v_k) = \infty I$$

$$\begin{pmatrix} \hat{x}^1 \\ \hat{x}^2 \end{pmatrix} = \begin{pmatrix} I \\ I \end{pmatrix} x_k + \begin{pmatrix} e^1 \\ e^2 \end{pmatrix}, \quad \mathrm{Cov}(e^i) = P^i.$$

To verify it, use the information filter. The infinite state variance gives $\mathcal{I}_{k+1|k} = 0$ for all k. The measurement update becomes

$$\mathcal{I}_{k|k} = 0 + H^T R^{-1} H = (P^1_{k|k})^{-1} + (P^2_{k|k})^{-1}.$$

Decentralized Fusion

Suppose we get estimates from two Kalman filters working with the same state vector. The total state space model for the *decentralized filters* is

$$\begin{pmatrix} x^1_{k+1} \\ x^2_{k+1} \end{pmatrix} = \begin{pmatrix} F_k & 0 \\ 0 & F_k \end{pmatrix} \begin{pmatrix} x^1_k \\ x^2_k \end{pmatrix} + \begin{pmatrix} G_k & 0 \\ 0 & G_k \end{pmatrix} \begin{pmatrix} v^1_k \\ v^2_k \end{pmatrix}, \quad \mathrm{Cov}(v^i_k) = Q$$

$$y_k = \begin{pmatrix} H^1_k & 0 \\ 0 & H^2_k \end{pmatrix} \begin{pmatrix} x^1_k \\ x^2_k \end{pmatrix} + \begin{pmatrix} e^1_k \\ e^2_k \end{pmatrix}.$$

The diagonal forms of $\bar{A}$, $\bar{B}$ and $\bar{C}$ imply:

1. The updates of the estimates can be done separately, and hence decentralized.

2. The state estimates will be independent.

Thus, because of independence the general *fusion formula* (7.41) applies. The state space model for *decentralized filters* and the *fusion filter* is now

$$\begin{pmatrix} x^1_{k+1} \\ x^2_{k+1} \\ x_{k+1} \end{pmatrix} = \begin{pmatrix} F_k & 0 & 0 \\ 0 & F_k & 0 \\ 0 & 0 & 0 \end{pmatrix} \begin{pmatrix} x^1_k \\ x^2_k \\ x_k \end{pmatrix} + \begin{pmatrix} G_k & 0 & 0 \\ 0 & G_k & 0 \\ 0 & 0 & I \end{pmatrix} \begin{pmatrix} v^1_k \\ v^2_k \\ v^3_k \end{pmatrix}$$

$$\begin{pmatrix} y^1_k \\ y^2_k \\ 0 \\ 0 \end{pmatrix} = \begin{pmatrix} H^1_k & 0 & 0 \\ 0 & H^2_k & 0 \\ I & 0 & -I \\ 0 & I & -I \end{pmatrix} \begin{pmatrix} x^1_k \\ x^2_k \\ x_k \end{pmatrix} + \begin{pmatrix} e^1_k \\ e^2_k \\ 0 \\ 0 \end{pmatrix}$$

$$\mathrm{Cov}(v^1_k) = Q, \; \mathrm{Cov}(v^2_k) = Q, \; \mathrm{Cov}(v^3_k) = \infty I.$$

That is, the Kalman filter applied to this state space model provides the fusioned state estimate as the third sub-vector of the augmented state vector.

The only problem in this formulation is that we have *lost information*. The decentralized filters do not use the fact that they are estimating the same system, which have only one state noise. It can be shown that the covariance matrix will become too small, compared to that which the Kalman filter provides for the model

$$x_{k+1} = F_k x_k + G_k v_k, \quad \text{Cov}(v_k) = Q \qquad (7.42a)$$

$$y_k = \begin{pmatrix} H_k^1 \\ H_k^2 \end{pmatrix} x_k + \begin{pmatrix} e_k^1 \\ e_k^2 \end{pmatrix}. \qquad (7.42b)$$

To obtain the optimal state estimate, given the measurements and the state space model, we need to invert the Kalman filter and recover the raw *information in the measurements*. One way to do this is to ask all decentralized filters for the Kalman filter transfer functions $\hat{x}_{k|k} = H_k(q)y_k$, and then apply inverse filtering. It should be noted that the inverse Kalman filter is likely to be high-pass, and thus amplifying numerical errors. Perhaps a better alternative is as follows. As indicated, it is the information in the measurements, not the measurements themselves, that are needed. Sufficient statistics are provided by the *information filter*; see Section 7.1.4. We recapitulate the measurement update for convenience:

$$\iota_{k|k} = \iota_{k|k-1} + H_k^T R_k^{-1} y_k$$

$$\mathcal{I}_{k|k} = \mathcal{I}_{k|k-1} + H_k^T R_k^{-1} H_k,$$

where $\iota_{k|k} = \hat{P}_{k|k}^{-1} \hat{x}_{k|k}$. Since the measurement error covariance in (7.42b) is block diagonal, the time update can be split into two parts

$$\iota_{k|k} = \iota_{k|k-1} + (H_k^1)^T (R_k^1)^{-1} y_k^1 + (H_k^2)^T (R_k^2)^{-1} y_k^2,$$

$$\mathcal{I}_{k|k} = \mathcal{I}_{k|k-1} + (H_k^1)^T (R_k^1)^{-1} H_k^1 + (H_k^2)^T (R_k^2)^{-1} H_k^2.$$

Each Kalman filter, interpreted as information filters, has computed a measurement update

$$\iota_{k|k}^i = \iota_{k|k-1}^i + (H_k^i)^T (R_k^i)^{-1} y_k^i$$

$$(P_{k|k}^i)^{-1} = (P_{k|k-1}^i)^{-1} + (H_k^i)^T (R_k^i)^{-1} H_k^i.$$

By backward computations, we can now recover the information in each measurement, expressed below in the available Kalman filter quantities:

$$(H_k^i)^T (R_k^i)^{-1} y_k^i = \iota_{k|k}^i - \iota_{k|k-1}^i = (\hat{P}_{k|k}^i)^{-1} \hat{x}_{k|k}^i - (\hat{P}_{k|k-1}^i)^{-1} \hat{x}_{k|k-1}^i$$

$$(H_k^i)^T (R_k^i)^{-1} H_k^i = (P_{k|k}^i)^{-1} - (P_{k|k-1}^i)^{-1}.$$

The findings are summarized and generalized to several sensors in the algorithm below.

The price paid for the decentralized structure is heavy signaling and for the standard form also a lot of matrix inversions. Also for this approach, there

7.4 Smoothing

Algorithm 7.3 Decentralized filtering

Given the filter and prediction quantities from m Kalman filters working on the same state space model, the optimal sensor fusion is given by the usual time update and the following measurement update:

$$\hat{P}_{k|k}^{-1}\hat{x}_{k|k} = \hat{P}_{k|k-1}^{-1}\hat{x}_{k|k-1} + \sum_{i=1}^{m}\left((\hat{P}_{k|k}^i)^{-1}\hat{x}_{k|k}^i - (\hat{P}_{k|k-1}^i)^{-1}\hat{x}_{k|k-1}^i\right)$$

$$\hat{P}_{k|k}^{-1} = \hat{P}_{k|k-1}^{-1} + \sum_{i=1}^{m}\left((\hat{P}_{k|k}^i)^{-1} - (\hat{P}_{k|k-1}^i)^{-1}\right).$$

The information form of these equations is

$$\iota_{k|k} = \iota_{k|k-1} + \sum_{i=1}^{m}\left(\iota_{k|k}^i - \iota_{k|k-1}^i\right),$$

$$\mathcal{I}_{k|k} = \mathcal{I}_{k|k-1} + \sum_{i=1}^{m}\left(\mathcal{I}_{k|k}^i - \mathcal{I}_{k|k-1}^i\right).$$

---

might be numerical problems, since differences of terms of the same order are to be computed. These differences correspond to a high-pass filter.

In case of possible information loops, the safe fusion algorithm can be used to get a conservative covariance. See Algorithm 2.1 for the details.

7.4 Smoothing

7.4.1 Fixed-Lag Smoothing

Fixed-lag smoothing is to estimate the state vector x_k from measurements $y(s)$, $s \leq k + m$. If m is chosen in the order of two, say, time constants of the system, almost all of the information in the data is utilized. The trick to derive the optimal filter is to augment the state vector with delayed states,

$$x_k^i = x_{k-i}, \quad i = 1, \ldots, m+1. \tag{7.43}$$

The augmented state vector is

$$\bar{x}_k = \begin{bmatrix} x_k^0 \\ \vdots \\ x_k^{m+1} \end{bmatrix}. \tag{7.44}$$

For simplicity, we will drop the time indices on the state matrices in this subsection. The full signal model is

$$\bar{x}_{k+1} = \bar{F}\bar{x}_k + \bar{G}_v v_k \tag{7.45a}$$

$$y_k = \bar{H}\bar{x}_k + e_k, \tag{7.45b}$$

where

$$\bar{F} = \begin{pmatrix} F & 0 & 0 & \cdots & 0 & 0 \\ I & 0 & 0 & \cdots & 0 & 0 \\ 0 & I & 0 & \cdots & 0 & 0 \\ 0 & 0 & I & \cdots & 0 & 0 \\ \vdots & \vdots & \vdots & \ddots & \vdots & \vdots \\ 0 & 0 & 0 & \cdots & I & 0 \end{pmatrix} \qquad \bar{G}_{v,k} = \begin{pmatrix} G_{v,k} \\ 0 \\ \vdots \\ 0 \end{pmatrix} \tag{7.45c}$$

$$\bar{H} = \begin{pmatrix} H & 0 & \cdots & 0 \end{pmatrix}. \tag{7.45d}$$

Estimate $\hat{x}_{k-m|k}, \hat{x}_{k-m+1|k}, \ldots, \hat{x}_{k|k}$ by applying the Kalman filter on (7.45a)–(7.45b). The estimate of x_{k-m} given observations of $y_s, 0 \leq s \leq k$ is given by $\hat{x}_{k+1|k}^{m+1}$. This is recognized as the last sub-vector of the Kalman filter prediction $\hat{\bar{x}}_{k+1|k}$.

To simplify the equations somewhat, let us write out the Kalman predictor formulas

$$\hat{\bar{x}}_{k+1|k} = \bar{F}\hat{\bar{x}}_{k|k-1} + \bar{F}\bar{K}_k \left(y_k - \bar{H}^T \hat{\bar{x}}_{k|k-1}\right) \tag{7.46a}$$

$$\bar{K}_k = \bar{P}_{k|k-1}\bar{H}^T \left(\bar{H}\bar{P}_{k|k-1}\bar{H}^T + R\right)^{-1} \tag{7.46b}$$

$$\bar{P}_{k+1|k} = \bar{F}\bar{P}_{k|k-1}\bar{F}^T + \bar{G}_{v,k} Q \bar{G}_{v,k}^T$$
$$- \bar{F}\bar{P}_{k|k-1}\bar{H}^T \left(\bar{H}\bar{P}_{k|k-1}\bar{H}^T + R\right)^{-1} \bar{H}\bar{P}_{k|k-1}\bar{F}^T. \tag{7.46c}$$

Next, we aim at simplifying (7.46a)–(7.46c). It follows from (7.45d) that the innovation covariance can be simplified to its usual low dimensional form

$$\bar{H}\bar{P}_{k|k-1}\bar{H}^T + R = H P_{k|k-1} H^T + R. \tag{7.47}$$

Split $\bar{P}_{k|k-1}$ into $n_x \times n_x$ blocks

$$\bar{P}_{k|k-1} = \begin{pmatrix} \bar{P}_{k|k-1}^{0,0} & \bar{P}_{k|k-1}^{0,1} & \cdots & \bar{P}_{k|k-1}^{0,m+1} \\ \bar{P}_{k|k-1}^{1,0} & \bar{P}_{k|k-1}^{1,1} & \cdots & \bar{P}_{k|k-1}^{1,m+1} \\ \vdots & \vdots & \ddots & \vdots \\ \bar{P}_{k|k-1}^{m+1,0} & \bar{P}_{k|k-1}^{m+1,1} & \cdots & \bar{P}_{k|k-1}^{m+1,m+1} \end{pmatrix}, \tag{7.48}$$

which gives

$$\bar{K}_k = \begin{pmatrix} \bar{K}_k^0 \\ \vdots \\ \bar{K}_k^{m+1} \end{pmatrix} = \begin{pmatrix} \bar{P}_{k|k-1}^{0,0} \\ \bar{P}_{k|k-1}^{1,0} \\ \vdots \\ \bar{P}_{k|k-1}^{m+1,0} \end{pmatrix} H^T \left(H P_{k|k-1} H^T + R\right)^{-1}. \tag{7.49}$$

7.4 Smoothing

To compute the Kalman gain, we need to know $\bar{P}^{i,0}$, $i = 0, \ldots, m+1$. These are given by a substitution of (7.48) in the Riccati equation (7.46c)

$$\bar{P}^{i+1,0}_{k+1|k} = \bar{P}^{i,0}_{k|k-1}(F - FK_k H)^T, \quad i = 0, \ldots, m, \qquad (7.50)$$

where $K_k = \bar{K}^0_k$. That is, to compute the state update we only need to know the quantities from the standard Kalman predictor and the original state space matrices of size $n \times n$, and the covariance matrices updated in (7.50). To compute the diagonal matrices $P^{i,i}_{k|k-1}, i = 0, \ldots, m+1$, corresponding to the smoothed estimate's covariance matrix, a substitution of (7.48) in the Riccati equation (7.46c) gives

$$\bar{P}^{i+1,i+1}_{k+1|k} = \bar{P}^{i,i}_{k|k-1} - \bar{K}^i_k H \bar{P}^{0,i}_{k|k-1} \quad i = 0, \ldots, m. \qquad (7.51)$$

$\bar{P}^{0,0}_{k|k-1}$ is found from the Riccati equation for the original signal model.

7.4.2 Fixed-Point Smoothing

The *fixed-point smoothing* algorithm described here is based on a similar state augmentation trick as in fixed-interval smoothing. Suppose state x_n is critical, and we want to use future observations to improve its estimate. The procedure is as follows. At time n, augment the state vector to get the new dynamic model

$$\bar{x}_k = \begin{pmatrix} x_k \\ x_n \end{pmatrix}, \qquad (7.52a)$$

$$\bar{x}_{k+1} = \begin{pmatrix} F_k & 0 \\ 0 & I \end{pmatrix} \bar{x}_k + \begin{pmatrix} v_k \\ 0 \end{pmatrix}, \qquad (7.52b)$$

$$y_k = \begin{pmatrix} H_k & 0 \end{pmatrix} \bar{x}_k + e_k. \qquad (7.52c)$$

initialized with

$$\hat{\bar{x}}_{n|n-1} = \begin{pmatrix} \hat{x}_{n|n-1} \\ \hat{x}_{n|n-1} \end{pmatrix}, \qquad (7.52d)$$

$$\bar{P}_{n|n-1} = \begin{pmatrix} P_{n|n-1} & P_{n|n-1} \\ P_{n|n-1} & P_{n|n-1} \end{pmatrix}. \qquad (7.52e)$$

It is easily shown that the augmented model is observable if the original model is observable. The Kalman filter will automatically process the information optimally, so the state estimate $\hat{x}_{n|k}$ is delivered as the last part of $\hat{\bar{x}}_{k|k}$.

7.4.3 Fixed-Interval Smoothing

In this off-line filtering situation, we have access to all N measurements and want to find the best possible state estimate $\hat{x}_{k|N}$. One rather naive way is to use the fixed-lag smoother from the previous section, let $m = N$ and introduce N fictive measurements (e.g. zeros) with infinite variance. Then we apply the fixed-lag smoother and at time $k = 2N$ we have all smoothed estimates available. The reason for introducing information less measurements is solely to being able to run the Kalman filter until all useful information in the measurements have propagated down to the last sub-vector of the augmented state vector in (7.44).

A better way is to use so-called *forward-backward* filters. We will study two different approaches.

Rauch–Tung–Striebel Formulas

Algorithm 7.4 gives the *Rauch–Tung–Striebel formulas* for fixed-interval smoothing, which are given without proof. The notation $\hat{x}_{k|N}$ means as usual the estimate of x_k given measurements up to time N.

Algorithm 7.4 Fixed interval smoothing: Rauch–Tung–Striebel

Available observations are $y_k, k = 1, \ldots, N$.

1. Forward filter: Run the standard Kalman filter and store both the time and measurement updates, $\hat{x}_{k|k}, \hat{x}_{k|k-1}, P_{k|k}, P_{k|k-1}$.

2. Backward filter: Apply the following time recursion backwards in time:

$$\hat{x}_{k-1|N} = \hat{x}_{k-1|k-1} + P_{k-1|k-1} F_{k-1}^T P_{k|k-1}^{-1} \left(\hat{x}_{k|N} - \hat{x}_{k|k-1} \right). \quad (7.53a)$$

The covariance matrix of the estimation error $P_{k|N}$ is

$$\begin{aligned} P_{k-1|N} =& P_{k-1|k-1} \\ &+ P_{k-1|k-1} F_{k-1}^T P_{k|k-1}^{-1} \left(P_{k|N} - P_{k|k-1} \right) P_{k|k-1}^{-1} F_{k-1} P_{k-1|k-1}. \end{aligned} \quad (7.53b)$$

This algorithm is quite simple to implement, but it requires that all state and covariance matrices, both for the prediction and filter errors, are stored. That also means that we must explicitely apply both time and measurement updates.

Two-Filter Smoothing Formulas

We will here derive a *two-filter smoothing formula*.

7.4 Smoothing

Denote the conditional expectations of past and future data

$$(x_k|y_{1:k}, x_0) \sim \mathcal{N}(\hat{x}^F_{k|k}, P^F_{k|k})$$
$$(x_k|y_{k+1:N}) \sim \mathcal{N}(\hat{x}^B_{k|k+1}, P^B_{k|k+1}),$$

respectively. Here $\hat{x}^F_{k|k}$ and $P^F_{k|k}$ are the estimates from the Kalman filter (these will not be given explicitly). The index F is introduced to stress that the filter runs forwards in time, in contrast to the filter running backwards in time, yielding $\hat{x}^B_{k|k+1}$ and $P^B_{k|k+1}$, to be introduced. Quite logically, $\hat{x}^B_{k|k+1}$ is the estimate of x_k based on measurements $y_{k+1}, y_{k+2}, \ldots, y_N$, which is up to time $k+1$ backwards in time.

The smoothed estimate is, given these distributions, a standard fusion problem, whose solution is given in Section 7.3.3. The result is

$$(x_k|y_1^N, x_0) \sim \mathcal{N}(\hat{x}_{k|N}, P_{k|N})$$
$$P_{k|N} = \left((P^F_{k|k})^{-1} + (P^B_{k|k+1})^{-1}\right)^{-1}$$
$$\hat{x}_{k|N} = P_{k|N}\left((P^F_{k|k})^{-1}\hat{x}^F_{k|k} + (P^B_{k|k+1})^{-1}\hat{x}^B_{k|k+1}\right).$$

There is an elegant way to use the Kalman filter backwards on data to compute $\hat{x}^B_{k|k+1}$ and $P^B_{k|k+1}$. What is needed is a so-called backward model.

The following lemma gives the desired backwards Markovian model that is sample path equivalent to (7.1). The Markov property implies that the noise process v^B_k, and the final value of the state vector x_N, are independent. Not only are the first and second order statistics equal for these two models, but they are indeed sample path equivalent, since they both produce the same state and output vectors.

Lemma 7.2
Backward Model The following model is sample path equivalent to (7.1) and is its corresponding backward Markovian model

$$x_k = F^B_k x_{k+1} + v^B_{k+1}$$
$$y_k = H_k x_k + e_k. \tag{7.54a}$$

Here

$$F^B_k = \Pi_k F^T_k \Pi^{-1}_{k+1} = F^{-1}_k - F^{-1}_k Q_k \Pi^{-1}_{k+1} \tag{7.54b}$$
$$Q^B_k = \Pi_k - \Pi_k F^T_k \Pi^{-1}_{k+1} F_k \Pi_k \tag{7.54c}$$
$$= F^{-1}_k Q_k F^{-T}_k - F^{-1}_k Q_k \Pi^{-1}_{k+1} Q_k F^{-T}_k, \tag{7.54d}$$

where $\Pi_k = \mathrm{E}(x_k x_k^T)$ is the a priori covariance matrix of the state vector, computed recursively by $\Pi_{k+1} = F_k \Pi_k F^T_k + Q_k$. The last equalities of (7.54b) and (7.54d) can be used when F_k is invertible.

Proof: See Verghese and Kailath (1979). ■

Note that no inversion of the state noise covariance matrix is needed, so we do not have to deal with the factorized form here.

The Kalman filter applied to the model (7.54a) in reverse time provides the quantities sought:

$$\hat{x}^B_{k|k+1}, \quad P^B_{k|k+1}.$$

The backward model can be simplified by assuming no initial knowledge of the state, and thus $\Pi_0 = P_{1|0} = \infty I$, or more formally, $\Pi_0^{-1} = 0$. Then $\Pi_k^{-1} = 0$ for all k, and the latter expressions assuming invertible F_k give

$$F_k^B = F_k^{-1}$$
$$Q_k^B = F_k^{-1} Q_k F_k^{-T}.$$

These last two formulas are quite intuitive, and the result of a straightforward inversion of the state equation as $x_k = F^{-1} x_{k+1} - F^{-1} v_k$.

Algorithm 7.5 Fixed interval smoothing: forward-backward filtering

Consider the state space model

$$x_k = F_k^B x_{k+1} + v_{k+1}^B$$
$$y_k = H_k x_k + e_k.$$

1. Run the standard Kalman filter forwards in time and store the filter quantities $\hat{x}_{k|k}, P_{k|k}$.

2. Compute the backward model

$$F_k^B = \Pi_k F_k^T \Pi_{k+1}^{-1} = F_k^{-1} - F_k^{-1} Q_k \Pi_{k+1}^{-1}$$
$$Q_k^B = \Pi_k - \Pi_k F_k^T \Pi_{k+1}^{-1} F_k \Pi_k$$
$$= F_k^{-1} Q_k F_k^{-T} - F_k^{-1} Q_k \Pi_{k+1}^{-1} Q_k F_k^{-T}$$
$$\Pi_{k+1} = F_k \Pi_k F_k^T + Q_k,$$

   and run the standard Kalman filter backwards in time and store the 'predictions' $\hat{x}_{k|k+1}, P_{k|k+1}$.

3. Merge the information from the forward and backward filters

$$P_{k|N} = \left((P^F_{k|k})^{-1} + (P^B_{k|k+1})^{-1} \right)^{-1}$$
$$\hat{x}_{k|N} = P_{k|N} \left((P^F_{k|k})^{-1} \hat{x}^F_{k|k} + (P^B_{k|k+1})^{-1} \hat{x}^B_{k|k+1} \right).$$

7.5 Square Root Implementation

Square root algorithms are motivated by numerical problems when updating the covariance matrix P that often occur in practice:

- P is not *symmetric*. This is easily checked, and one remedy is to replace it by $0.5(P + P^T)$.

- P is not *positive definite*. This is not as easy to check as symmetry, and there is no good remedy. One computationally demanding solution might be to compute the SVD of $P = UDU^T$ after each update, and to replace negative values in the singular values in D with zero or a small positive number.

- Due to large differences in the scalings of the states, there might be numerical problems in representing P. Assume, for instance, that $n_x = 2$ and that the states are rescaled

$$\bar{x} = \begin{pmatrix} 10^{10} & 0 \\ 0 & 1 \end{pmatrix} x \Rightarrow \bar{P} = \begin{pmatrix} 10^{10} & 0 \\ 0 & 1 \end{pmatrix} P \begin{pmatrix} 10^{10} & 0 \\ 0 & 1 \end{pmatrix},$$

and there will almost surely be numerical difficulties in representing the covariance matrix, while there is no similar problem for the state. The solution is a thoughtful scaling of measurements and states from the beginning.

- Due to a numerically sensitive state space model (e.g an almost singular R matrix), there might be numerical problems in computing P.

The square root implementation resolves all these problems. We define the square root as any matrix $P^{1/2}$ of the same dimension as P, satisfying $P = P^{1/2}P^{T/2}$. The reason for the first requirement is to avoid solutions of the kind $\sqrt{1} = (1/\sqrt{2}, 1/\sqrt{2})$. Note that sqrtm in MATLAB™ defines a square root without transpose as $P = AA$. This definition is equivalent to the one used here, since P is symmetric.

The idea of updating a *square root* is very old, and fundamental contributions have been done in Bierman (1977), Park and Kailath (1995) and Potter (1963). The theory now seems to have reached a very mature and unified form, as described in Kailath et al. (1998).

7.5.1 Time and Measurement Updates

The idea is best described by studying the time update,

$$P_{k+1|k} = F_k P_{k|k} F_k^T + G_{v,k} Q_k G_{v,k}^T. \tag{7.55}$$

A first attempt of factorization

$$P_{k+1|k} = \begin{pmatrix} F_k P_{k|k}^{1/2} & G_{v,k} Q_k^{1/2} \end{pmatrix} \begin{pmatrix} F_k P_{k|k}^{1/2} & G_{v,k} Q_k^{1/2} \end{pmatrix}^T \quad (7.56)$$

fails to the condition that a square root must be quadratic. We can, however, apply a *QR factorization* to each factor in (7.56). A QR factorization is defined as

$$X = QR, \quad (7.57)$$

where Q is a unitary matrix such that $Q^T Q = I$ and R is a upper triangular matrix. The R and Q here should not be confused with R_k and Q_k in the state space model.

Example 7.1: QR factorization

The MATLAB™ function qr efficiently computes the factorization

$$\begin{pmatrix} 1 & 2 \\ 3 & 4 \\ 5 & 6 \end{pmatrix} = \begin{pmatrix} -0.1690 & 0.8971 & 0.4082 \\ -0.5071 & 0.2760 & -0.8165 \\ -0.8452 & -0.3450 & 0.4082 \end{pmatrix} \begin{pmatrix} -5.9161 & -7.4374 \\ 0 & 0.8281 \\ 0 & 0 \end{pmatrix}.$$

Applying QR factorization to the transpose of the first factor in (7.56) gives

$$\begin{pmatrix} F_k P_{k|k}^{1/2} & G_{v,k} Q_k^{1/2} \end{pmatrix} = R^T Q^T. \quad (7.58)$$

That is, the time update can be written

$$P_{k+1|k} = R^T Q^T Q R = R^T R. \quad (7.59)$$

It is clear that Q is here instrumental, and does not have to be saved after the factorization is completed. Here R consists of a quadratic part (actually triangular) and a part with only zeros. We can identify the square root as the first part of R,

$$R^T = \begin{pmatrix} P_{k+1|k}^{1/2} & 0 \end{pmatrix}. \quad (7.60)$$

To summarize, the time update is as follows.

The measurement update can be treated analogously. However, we will give a somewhat more complex form that also provides the Kalman gain and the innovation covariance matrix. Apply the QR factorization to the matrix

$$\begin{pmatrix} R_k^{1/2} & H_k P_{k|k-1}^{1/2} \\ 0 & P_{k|k-1}^{1/2} \end{pmatrix} = R^T Q^T, \quad (7.62)$$

where

$$R^T = \begin{pmatrix} X & 0 \\ Y & Z \end{pmatrix}. \quad (7.63)$$

7.5 Square Root Implementation

Algorithm 7.6 Square root algorithm, time update

1. Form the matrix in the left hand side of (7.58). This involves computing one new square root of Q_k, which can be done by the QR factorization (R is taken as the square root) or `sqrtm` in MATLAB^{TM}.

2. Apply the QR factorization in (7.58).

3. Identify the square root of $P_{k+1|k}$ as in (7.60).

More compactly, the relations are

$$\left(F_k P_{k|k}^{1/2} \quad G_{v,k} Q_k^{1/2} \right) = \left(P_{k+1|k}^{1/2} \quad 0 \right) Q^T. \tag{7.61}$$

The Kalman filter interpretations of the matrices X, Y, Z can be found by squaring the QR factorization

$$R^T Q^T Q R = R^T R = \begin{pmatrix} XX^T & XY^T \\ YX^T & YY^T + ZZ^T \end{pmatrix}$$

$$= \begin{pmatrix} \underbrace{R_k + H_k P_{k|k-1} H_k^T}_{S_k} & H_k P_{k|k-1} \\ P_{k|k-1} H_k^T & P_{k|k-1} \end{pmatrix}, \tag{7.64}$$

from which we can identify

$$XX^T = S_k$$
$$YX^T = P_{k|k-1} H_k^T \Rightarrow Y = P_{k|k-1} H_k^T S_k^{-1/2} = K_k S_k^{1/2}$$
$$ZZ^T = P_{k|k-1} - YY^T = P_{k|k-1} - P_{k|k-1} H_k^T X^{-T} X^{-1} H_k P_{k|k-1}$$
$$= P_{k|k-1} - P_{k|k-1} H_k^T S_k^{-1} H_k P_{k|k-1} = P_{k|k}.$$

This gives the algorithm below.

Note that the Y can be interpreted as the Kalman gain on a normalized innovation $S_k^{-1/2} \varepsilon_k \in \mathcal{N}(0, I)$. That is, we have to multiply either the gain Y or the innovation by the inverse of X.

Remarks:

- The only non-trivial part of these algorithms, is to come up with the matrix to be factorized in the first place. Then, in all cases here it is trivial to verify that it works.

- There are many ways to factorize a matrix to get a square root. The QR factorization is recommended here for several reasons. First, there are

Algorithm 7.7 Square root algorithm, measurement update

1. Form the matrix in the left-hand side of (7.62).
2. Apply the QR factorization in (7.62).
3. Identify R with (7.63), where

$$X = S_k^{1/2}$$
$$Y = K_k S_k^{1/2}$$
$$Z = P_{k|k}^{1/2}.$$

All in one equation:

$$\begin{pmatrix} R_k^{1/2} & H_k P_{k|k-1}^{1/2} \\ 0 & P_{k|k-1}^{1/2} \end{pmatrix} = \begin{pmatrix} S_k^{1/2} & 0 \\ K_k S_k^{1/2} & P_{k|k}^{1/2} \end{pmatrix} Q^T. \qquad (7.65)$$

many efficient implementations available, e.g. MATLAB™'s qr. Secondly, this gives a unique square root. Further, the triangular matrix R is simple to invert (this is needed in the state update).

7.5.2 Kalman Predictor

The predictor form, eliminating the measurement update, can be derived using the matrix (7.66) below. The derivation is done by squaring up in the same way as before, and identifying the blocks.

Algorithm 7.8 Square root Kalman predictor

Apply the QR factorization

$$\begin{pmatrix} R_k^{1/2} & H_k P_{k|k-1}^{1/2} & 0 \\ 0 & F_k P_{k|k-1}^{1/2} & G_{v,k} Q_k^{1/2} \end{pmatrix} = \begin{pmatrix} S_k^{1/2} & 0 & 0 \\ F_k K_k S_k^{1/2} & P_{k+1|k}^{1/2} & 0 \end{pmatrix} Q^T. \qquad (7.66)$$

The state update is computed by

$$\hat{x}_{k+1|k} = F_k \hat{x}_{k|k-1} + G_{u,k} u_k + F_k K_k S_k^{1/2} S_k^{-1/2} (y_k - H_k \hat{x}_{k|k-1}). \qquad (7.67)$$

We need to multiply the gain vector $K_k S_k^{1/2}$ in the lower left corner with the inverse of the upper left corner element $S_k^{1/2}$. Here the triangular structure

7.5 Square Root Implementation

can be utilized for matrix inversion. However, this matrix inversion can be avoided, by factorizing a larger matrix; see Kailath et al. (1998) for details.

7.5.3 Kalman Filter

Similarly, a square root algorithm for the Kalman filter is given below.

Algorithm 7.9 Square root Kalman filter

Apply the QR factorization

$$\begin{pmatrix} R_{k+1}^{1/2} & H_{k+1}F_kP_{k|k}^{1/2} & H_{k+1}G_{v,k}Q_k^{1/2} \\ 0 & F_kP_{k|k}^{1/2} & G_{v,k}Q_k^{1/2} \end{pmatrix} = \begin{pmatrix} S_{k+1}^{1/2} & 0 & 0 \\ K_{k+1}S_{k+1}^{1/2} & P_{k+1|k+1}^{1/2} & 0 \end{pmatrix} Q^T. \quad (7.68)$$

The state update is computed by

$$\hat{x}_{k|k} = F_{k-1}\hat{x}_{k-1|k-1} + K_k S_k^{1/2} S_k^{-1/2}(y_k - H_k F_{k-1}\hat{x}_{k-1|k-1}). \quad (7.69)$$

Example 7.2: DC motor: square root filtering

Consider the DC motor model

$$F = \begin{pmatrix} 1 & 0.3297 \\ 0 & 0.6703 \end{pmatrix}, \quad G = \begin{pmatrix} 0.0703 \\ 0.3297 \end{pmatrix}, \quad H = \begin{pmatrix} 1 & 0 \end{pmatrix}.$$

Assume that $Q_k = R_k = 1$ and we initially have

$$\hat{x}_{0|0} = \begin{pmatrix} 0 \\ 0 \end{pmatrix}$$

$$P_{0|0} = \begin{pmatrix} 1 & 0 \\ 0 & 1 \end{pmatrix}.$$

The left-hand side (LHS) required for the QR factorization in the filter formulation is given by

$$\text{LHS} = \begin{pmatrix} 1 & 1 & 0.3297 & 0.0703 \\ 0 & 1 & 0.3297 & 0.0703 \\ 0 & 0 & 0.6703 & 0.3297 \end{pmatrix}.$$

The sought factor is computed by [Q,R]=qr(LHS'), which gives

$$R^T = \begin{pmatrix} -1.4521 & 0 & 0 & 0 \\ -0.7635 & -0.7251 & 0 & 0 \\ -0.1522 & -0.1445 & -0.6444 & 0 \end{pmatrix}.$$

From this matrix, we identify the Kalman filter quantities

$$P_{1|1}^{1/2} = \begin{pmatrix} -0.7251 & 0 \\ -0.1445 & -0.6444 \end{pmatrix}$$

$$K_1 S_1^{1/2} = \begin{pmatrix} -0.7635 \\ -0.1522 \end{pmatrix}$$

$$S_1^{1/2} = -1.4521.$$

Note that the QR factorization in this example gives negative signs of the square roots of S_k and K_k. The usual Kalman filter quantities are recovered as

$$P_{1|1} = \begin{pmatrix} 0.3911 & 0.1015 \\ 0.1015 & 0.1891 \end{pmatrix}$$

$$K_1 = \begin{pmatrix} 0.3911 \\ 0.1015 \end{pmatrix}$$

$$S_1 = 1.6423.$$

7.6 Filter Monitoring

This section overviews a couple of detection problems related to monitoring the performance of the Kalman filter. The detection tests are similar to the ones in Section 5.1.

7.6.1 Outlier Detection and Rejection

Outliers are important to discard from the measurement update. This can be done using a test of the two hypotheses of inliers and outliers. A natural implementation of an *outlier rejection* algorithm is based on the normalized residual

$$\bar{\varepsilon}_k = (H_k^T P_{k|k-1} H_k + R_k)^{-1/2} (y_k - H_k \hat{x}_{k|k-1}) \sim \mathcal{N}(0, I), \tag{7.70}$$

which is Gaussian distributed in case of a linear Gaussian model and if all measurements are inliers. Otherwise, the central limit theorem can be used to motivate the approximate Gaussian distribution. Standard hypothesis tests can be used to reject the null hypothesis of inliers. The natural test is based on the test statistic

$$T(y_k) = (y_k - H_k \hat{x}_{k|k-1})^T (H_k^T P_{k|k-1} H_k + R_k)^{-1} (y_k - H_k \hat{x}_{k|k-1}) \sim \chi_{n_y}^2. \tag{7.71}$$

7.6 Filter Monitoring

In an iterated Kalman filter measurement update implementation, the outlier rejection should be performed measurement by measurement as:

$$T(y_k^i) = (y_k^i - H_k^i \hat{x}_{k|k-1})^T (H_k^i P_{k|k-1}(H_k^i)^T + R_k^i)^{-1}(y_k^i - H_k^i \hat{x}_{k|k-1}) \sim \chi_{n_y^i}^2. \quad (7.72)$$

A measurement is rejected as inlier if

$$T(y_k^i) > \chi_{\alpha, n_y^i}^2. \quad (7.73)$$

The result using iterated outlier rejection depends on the order of the features. The more correct information used already in the measurement update, the better decisions can be taken.

7.6.2 Divergence Monitoring

Divergence monitoring is related to outlier rejection, since it is usually based on monitoring similar whiteness statistics of the normalized residual in (7.70). A variance based statistic is obtained by summing up the squared normalized prediction errors over time to get

$$T(y_{1:N}) = \sum_{k=1}^{N} (y_k - H_k \hat{x}_{k|k-1})^T (H_k^T P_{k|k-1} H_k + R_k)^{-1}(y_k - H_k \hat{x}_{k|k-1}), \quad (7.74a)$$

$$T(y_{1:N}) \sim \chi_{Nn_y}^2. \quad (7.74b)$$

According to the central limit theorem, the test statistic is approximately Gaussian distributed,

$$\frac{T(y_{1:N})}{Nn_y} \sim \mathcal{N}\left(1, \frac{2}{Nn_y}\right). \quad (7.75)$$

The test then becomes

$$\left(\frac{T(y_{1:N})}{Nn_y} - 1\right)\sqrt{Nn_y/2} > \phi_\alpha, \quad (7.76)$$

where ϕ_α is the standard Gaussian error function at level α.

This test above is based on summing up all errors since the filter started. A better alternative is to use a sliding window over a fixed time horizon, for instance just the last set of measurements, or a fixed number of measurements. An exponential window may also be used, which for the iterated update becomes

$$T(y_{1:k}) = \lambda T(y_{1:k-1}) + \frac{1-\lambda}{n_y} \quad (7.77a)$$

$$\cdot (y_k - H_k \hat{x}_{k|k-1})^T (H_k P_{k|k-1}(H_k)^T + R_k)^{-1}(y_k - H_k \hat{x}_{k|k-1})$$

$$\sim \mathcal{N}(1, 2(1-\lambda)/(1+\lambda)), \quad (7.77b)$$

where $0 \ll \lambda < 1$ is the forgetting factor in the exponential filter.

Now, if this check is combined with the outlier rejection, we know that each term is upper bounded by $h = \chi^2_{\alpha, n^i_y}$. After total divergence, all prediction errors are more or less random numbers. The test statistic will in that case be significantly larger than 1 ($T \gg 1$), but implicitly bounded by the outlier threshold. Assume that $n^i_y = 1$, then uniformly distributed measurements over $[-h, h]$ give

$$\mathrm{E}(T(y_{1:k})) = \frac{4h^2}{12} \frac{2(1-\lambda)}{1+\lambda}. \qquad (7.78)$$

7.6.3 Distributed Kalman Filter

The distributed implementation of the Kalman filter described in Section 7.3.3 gives a possibility to compare any two filters. Assuming that both filters have the same state vector and independent observations, the following test statistic can be used,

$$T_{12} = \left(\hat{x}^1_{k|k} - \hat{x}^2_{k|k}\right)^T \left(P^1_{k|k} + P^2_{k|k}\right)^{-1} \left(\hat{x}^1_{k|k} - \hat{x}^2_{k|k}\right) \qquad (7.79)$$

7.6.4 Time-Interleaved Kalman Filters

One efficient solution to obtain robust filters that can withstand a high amount of outliers is based on time-interleaved Kalman filter banks. The basic condition is that the sample rate is higher than needed, so each Kalman filter processes the decimated measurement sequence $y_{m:M:kM}$ for $m = 1, 2, \ldots M$. The filters can then be compared for instance pair-wise using (7.79). Voting decides which filters to trust, and filters that have diverged can be restarted using the state and covariance from a healthy filter. This is a simple and powerful approach to *robust filtering*.

7.7 Examples

7.7.1 Target Tracking

For the target tracking example, a standard constant velocity model is used for two-dimensional tracking. It naturally extends the sensor network example in Chapter 2 to dynamically moving targets. Only one position sensor is used here. First, the model is created.

```
T=0.5;
A=[1 0 T 0; 0 1 0 T; 0 0 1 0; 0 0 0 1];
B=[T^2/2 0; 0 T^2/2; T 0; 0 T];
C=[1 0 0 0; 0 1 0 0];
R=0.01*eye(2);
m=ss(A,[],C,[],B*B',R,1/T);
m.xlabel={'X','Y','vX','vY'};
```

7.7 Examples

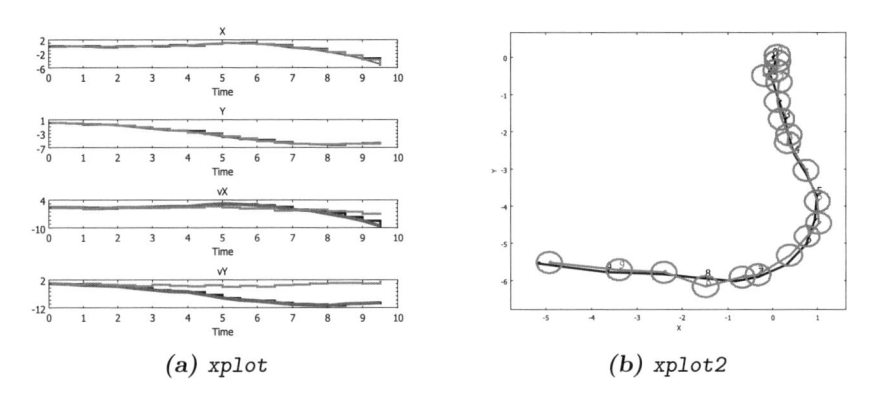

(a) xplot

(b) xplot2

Figure 7.2: *KF tracking example: comparison of two plotting options.*

```
m.ylabel={'X','Y'};
m.name='Constant velocity motion model';
```

A faster alternative is to use the model database. `m=exlti('cv2D')` provides the same model. Next, the model is first displayed and then simulated.

```
m
              / 1   0   0.5   0  \
              | 0   1   0    0.5 |
x[k+1]   =    | 0   0   1    0   | x[k] + v[k]
              \ 0   0   0    1   /

              /1  0  0  0\
y[k]     =    \0  1  0  0/  x[k] + e[k]


              /0.016    0      0.063    0  \
              |  0    0.016      0    0.063 |
Q = Cov(v) =  | 0.063   0      0.25     0   |
              \  0    0.063     0     0.25 /

              / 0.01    0  \
R = Cov(e) =  \  0    0.01 /
```

```
z=simulate(m,20);
SIG object with discrete time (fs = 2) stochastic state space data (no
    input)
    Name:         Simulation of Constant velocity motion model
    Sizes:        N = 20,  ny = 2,  nx = 4
```

Now, when both a model and a signal are generated, the Kalman filter can be applied. The estimated states with confidence bounds are compared to the true ones, see Figure 7.2(a).

```
xhat=kalman(m,z);
xplot(xhat,z,'conf',99)
```

Alternatively, one state can be plotted versus another state, which for tracking provides an intuitive picture of tracking performance, see Figure 7.2(b).

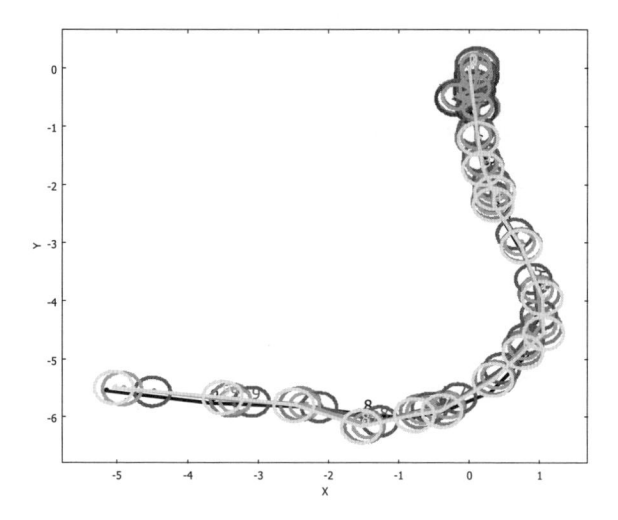

Figure 7.3: *KF tracking example: comparison of stationary KF, time-varying KF, square-root implementation, RTS smoother and sliding window KF.*

```
xplot2(z,xhat,'conf',99,[1 2])
```

Next, different implementations are compared. The algorithm switch provides stationary KF (1), time-varying KF (2), square-root implementation (3), RTS smoother (4) and sliding window KF (5).

```
xhat10=kalman(m,z,'alg',1,'pred',0);
xhat20=kalman(m,z,'alg',2,'pred',0);
xhat30=kalman(m,z,'alg',3,'pred',0);
xhat40=kalman(m,z,'alg',4,'pred',0);
xhat50=kalman(m,z,'alg',5,'pred',5);
xplot2(z,xhat10,xhat20,xhat30,xhat40,xhat50,'conf',99,[1 2])
```

Finally, the filter and predictor performances are compared.

```
xhat10=kalman(m,z,'alg',1,'pred',0);
xhat11=kalman(m,z,'alg',1,'pred',1);
xplot2(z,xhat10,xhat11,'conf',99,[1 2])
```

7.7 Examples

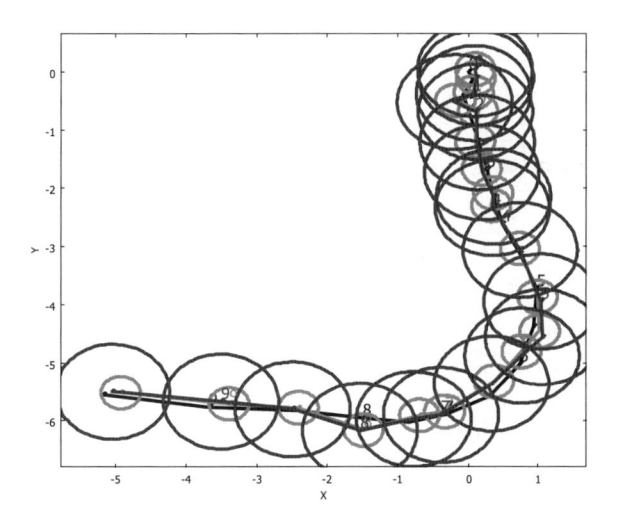

Figure 7.4: KF tracking example: comparison of filter and predictor.

7.8 Summary

7.8.1 Theory

This chapter described methods to estimate the states x_k in the linear model

$$x_{k+1} = F_k x_k + G_{u,k} u_k + G_{v,k} v_k, \quad \text{Cov}(v_k) = Q_k,$$
$$y_k = H_k x_k + D_k u_k + e_k, \quad \text{Cov}(e_k) = R_k.$$

The Kalman filter implements the following recursion:

1. **Initialization.** Set $\hat{x}_{1|0}$ and $P_{1|0}$.

2. **Measurement update.**

$$\hat{x}_{k|k} = \hat{x}_{k|k-1} + P_{k|k-1} H_k^T (H_k P_{k|k-1} H_k^T + R_k)^{-1} (y_k - H_k \hat{x}_{k|k-1} - D_k u_k)$$
$$P_{k|k} = P_{k|k-1} - P_{k|k-1} H_k^T (H_k P_{k|k-1} H_k^T + R_k)^{-1} H_k P_{k|k-1}.$$

3. **Time update.**

$$\hat{x}_{k+1|k} = F \hat{x}_{k|k} + G_{u,k} u_k,$$
$$P_{k+1|k} = F P_{k|k} F^T + G_{v,k} Q_k G_{v,k}^T.$$

The KF is the best linear unbiased estimate (BLUE) and the conditional expectation estimate. For Gaussian distributions on v_k, e_k and x_0, it is also the minimum variance (MV) and maximum likelihood (ML) estimate.

User aspects include:

- Implementation forms:
  - Stationary or time-varying KF.
  - Standard or information filter.
  - Standard or square root implementations to improve numerical accuracy.
  - Filter, prediction or smoothing form.

- Practical issues:
  - Centralized or distributed implementation in fusion problems.
  - Out of sequence problems, where the KF is updated with old information.
  - Excitation and observability aspects for how accurately the states can be estimated. These relate closely to the condition number of the covariance matrix $P_{k|k}$.

7.8 Summary

- Outlier detection and removal.
- Divergence monitoring by comparing the covariance of the innovations from the KF with their theorectical covariance provided by the model. This also relates to outlier monitoring.
- How to deal with cross-correlated noise and analyze model errors.

• Computational aspects.

- Structuring the computations to avoid computing the same thing twice.
- Assuring a positive definite symmetric covariance matrix.
- Choosing the standard or information form depending on n_x and n_y.
- The iterated KF where a measurement update is done for each sensor observation.

7.8.2 Software

The linear state space model object lss is central for this chapter. It has the constructor:

m=lss([nx nu nv ny]) for defining an empty black-box model. This is relevant for the rand and estimate methods for obtaining random and estimated models, respectively.

m=lss(F,G,H,D,Q,R,fs); to specify a stochastic state space model. This can be simulated with y=simulate(m,t); (if $G = D = [\,]$) and then the Kalman filter can be applied by xhat=kalman(m,y);

There are fields m.x0, m.px0 to set the initial conditions. The noise distributions can be changed from the default Gaussian to any available distribution, using the fields m.pv and m.pe.

There are methods for most conceivable operations on linear systems, see help lss for a complete list of methods.

The Kalman filter is called by

[xhat,V]=kalman(m,y,Property1,Value1,...)

where

• y is the data SIG object, m the LSS state space model, and xhat the estimated SIG object with xhat.x states and xhat.y output estimates with covariances as three-dimensional matrices xhat.Px and xhat.Py, respectively.

- V is the normalized sum of squared innovations, which should be a sequence of $\chi^2(n_y)$ variables when the model is correct.

- The tuning properties x0 and P0 overwrite the corresponding field in the LSS object, which is the default.

- The algorithm property alg is an integer 1–5, corresponding to the implementations

  1. Stationary KF (default).
  2. Time-varying KF.
  3. Square root filter.
  4. Fixed interval KF smoother Rauch-Tung-Striebel.
  5. Sliding window KF, delivering $\hat{x}_{k|k-L+1:k}$, where L is the length of the sliding window.

Prediction horizon m is the property called 'pred'. $m = 0$ for filter (default), $m = 1$ for one-step ahead predictor, and generally $m > 1$ (only options 1 and 2) gives $\hat{x}_{k+m|k}$ and $y_{k+m|k}$ for alg=1,2. In case alg=5, $m = L$ is the size of the sliding window.

Typical plots:

- plot(y,xhat) compares the measurement with the Kalman filter estimate $\hat{y} = F\hat{x}$.

- xplot(xhat) plots the states of the SIG objects in subplots. When there are many states, xplot(xhat,ind) can be used to only plot the states listed in ind. Note that several tunings of the KF can be compared directly with multiple arguments xplot(xhat1,xhat2).

- xplot2(xhat,ind) creates a two-dimensional state plot of xhat(ind(1)) versus xhat(ind(2)). Default ind=[1 2].

Typical tracking example:

```
m=exlti('ca2D');            % Constant acc motion model ex
y=simulate(m,10);           % Simulate 10 samples
xhat=kalman(m,y);           % Estimate state
xplot(y,xhat,'conf',90,[1 2]);   % Subplots of states 1,2 (X,Y)
xplot2(y,xhat,'conf',90,[1 2]);  % Plot states 1 vs 2 (X vs Y)
```

8

The Extended and Unscented Kalman Filters

This chapter compares various approaches for how to propagate a Gaussian approximate state distribution for a nonlinear system

$$x_{k+1} = f(x_k, u_k, v_k), \quad (8.1a)$$
$$y_k = h(x_k, u_k, e_k). \quad (8.1b)$$

Similar to the Kalman filter (KF) in the linear case, the nonlinear filter recursion is split into one measurement update providing $\hat{x}_{k|k}, P_{k|k}$ and a time update yielding $\hat{x}_{k+1|k}, P_{k+1|k}$.

The nonlinear filters in this chapter are in one way or another related to the Taylor expansion of a nonlinear function $z = g(x)$ around $\hat{x}$,

$$z = g(x) = g(\hat{x}) + g'(\hat{x})(x - \hat{x}) + \underbrace{\tfrac{1}{2}(x-\hat{x})^T g''(\xi)(x-\hat{x})}_{r(x;\hat{x},g''(\xi))}, \quad (8.2)$$

where $x \sim \mathcal{R}^{n_x}$ and (initially for notational convenience) $z \sim \mathcal{R}^1$. Here, g' denotes the Jacobian and g'' the Hessian of the function $g(x)$, and $\xi(x)$ is a point in the neighborhood of $\hat{x}$. Basically, as a preview,

- The *extended Kalman filter* (EKF) developed in Smith et al. (1962) and Schmidt (1966) is based on the first two terms in (8.2). This works fine as long as the rest term is small. Small here relates both to the state estimation error and the degree of nonlinearity of g. Basically, as a rule of thumb, the rest term is negligible if either the model is almost linear, or the SNR is high, in which case the estimation error can be considered sufficiently small.

- The second order compensated EKF, see for instance, Athans et al. (1968), Maybeck (1982) and Bar-Shalom et al. (2001), approximates the rest term $r(x; \hat{x}, g''(\xi))$ with $r(x; \hat{x}, g''(\hat{x}))$, and compensates for the mean and variance of this term.

- The *unscented Kalman filter* (UKF) first presented in Julier et al. (1995), see also the survey Julier and Uhlmann (2004), can be interpreted, as will be demonstrated, as implicitly estimating the first terms in the nonlinear transformation in (8.2).

There are several links and interpretations between UKF and EKF as will be pointed out later on.

The standard forms of the KF and EKF include a discrete-time algebraic Riccati equation (DARE) for propagating the state covariance, while the UKF in its proposed form is based on a different principle in linear estimation and has no explicit DARE. Further, UKF is based on function evaluations $g(x)$ only, so neither the Jacobian nor the Hessian are needed. This is another claimed advantage of UKF:

> Julier and Uhlmann (2004): "... [UKF] is not the same as using a central difference scheme to calculate the Jacobian."

This is indeed true, but we show that there is a duality in the implementation. UKF can be implemented with Riccati equations, and the EKF (and even the linear KF) can be implemented without Riccati equations.

The core tool in the analytic results is the underlying transform approximations of a nonlinear mapping $z = g(x)$, providing a Gaussian approximation $\mathcal{N}(\mu_z, P_z)$ of the stochastic variable z. It is often stated that the unscented transform (UT) gives the correct first and second order moments ($\mu_z = \mathrm{E}(z)$ and $P_z = \mathrm{Cov}(z)$):

> Julier and Uhlmann (2004): "Any set of sigma points that encodes the mean and covariance correctly, ..., calculates the projected mean and covariance correctly to the second order"

However, we show with a simple counter-example that this is not the case, even for a quadratic function of Gaussian variables. We also show analytically that the UT generally does not give the same elements in the covariance as the second order Taylor expansion, which should at least be optimal for quadratic functions of Gaussian variables. On the other hand, we show that the UT gives a good approximation of many common sensor models in tracking and navigation applications.

The outline is as follows. Section 8.1 discusses the classical implementation of EKF, and shows how the sigma points of the UT can be used to estimate derivatives such that the need for Jacobians and Hessians is eliminated. Section 8.2 gives a general version of the Riccati-free nonlinear filter, where the transform approximation can be chosen individually for the time and measurement update, respectively.

8.1 DARE-based Extended Kalman Filter

Here, detailed recursions are given for the *extended Kalman filter* (EKF) without and with second order compensation, respectively. The function $f(x, u, 0)$ is here more compactly written $f(x)$, and similarly $h(x) = h(x, u, 0)$.

8.1.1 EKF Algorithms

Using the transformation approximation TT1 and TT2, respectively, described in Section A.3, immediately gives the two Riccati-based EKF filters in Algorithm 8.1.

Algorithm 8.1 DARE-based EKF1 and EKF2

The EKF2, using the TT2 transformation, for the model (8.1) with additive noises v_k and e_k is given by the following recursions initialized with $\hat{x}_{1|0}$ and $P_{1|0}$:

$$S_k = R_k + h'(\hat{x}_{k|k-1})P_{k|k-1}(h'(\hat{x}_{k|k-1}))^T$$
$$+ \tfrac{1}{2}\left[\operatorname{tr}(h''_i(\hat{x}_{k|k-1})P_{k|k-1}h''_j(\hat{x}_{k|k-1})P_{k|k-1})\right]_{ij}$$

$$K_k = P_{k|k-1}(h'(\hat{x}_{k|k-1}))^T S_k^{-1} \tag{8.3a}$$

$$\varepsilon_k = y_k - h(\hat{x}_{k|k-1}) - \tfrac{1}{2}\left[\operatorname{tr}(h''_i P_{k|k-1})\right]_i \tag{8.3b}$$

$$\hat{x}_{k|k} = \hat{x}_{k|k-1} + K_k \varepsilon_k \tag{8.3c}$$

$$P_{k|k} = P_{k|k-1} \tag{8.3d}$$
$$\qquad - P_{k|k-1}(h'(\hat{x}_{k|k-1}))^T S_k^{-1} h'(\hat{x}_{k|k-1})P_{k|k-1}$$

$$\hat{x}_{k+1|k} = f(\hat{x}_{k|k}) + \tfrac{1}{2}\left[\operatorname{tr}(f''_i P_{k|k})\right]_i \tag{8.3e}$$

$$P_{k+1|k} = Q_k + f'(\hat{x}_{k|k})P_{k|k}(f'(\hat{x}_{k|k}))^T$$
$$+ \tfrac{1}{2}\left[\operatorname{tr}(f''_i(\hat{x}_{k|k})P_{k|k}f''_j(\hat{x}_{k|k})P_{k|k})\right]_{ij}. \tag{8.3f}$$

The EKF 1, using the TT1 transformation, is obtained by letting both Hessians f''_x and h''_x be zero.

The common EKF should work well when the bias and variance contribution of the second order Taylor term is negligible to the noise,

$$\frac{1}{4}\left[\operatorname{tr}\left(f_i''(\hat{x}_{k|k})P_{k|k}\right)^T \operatorname{tr}\left(f_j''(\hat{x}_{k|k})P_{k|k}\right)\right]_{ij}$$
$$+\frac{1}{2}\left[\operatorname{tr}(f_i''(\hat{x}_{k|k})P_{k|k}f_j''(\hat{x}_{k|k})P_{k|k})\right]_{ij} \ll Q_k, \quad (8.4a)$$
$$\frac{1}{4}\left[\operatorname{tr}\left(h_i''(\hat{x}_{k|k-1})P_{k|k-1}\right)^T \operatorname{tr}\left(h_j''(\hat{x}_{k|k-1})P_{k|k-1}\right)\right]_{ij}$$
$$+\frac{1}{2}\left[\operatorname{tr}(h_i''(\hat{x}_{k|k-1})P_{k|k-1}h_j''(\hat{x}_{k|k-1})P_{k|k-1})\right]_{ij} \ll R_k. \quad (8.4b)$$

These are conditions that can be monitored on-line, but with a large computational overhead, or analyzed off-line based on only the model and typical operating points.

There are some variations in how the linearization is performed for the EKF1 in Algorithm 8.1:

- The standard EKF linearizes around the current state estimate.

- The *linearized Kalman filter* linearizes around some reference trajectory. The classical application is in *inertial navigation systems* (*INS*), where the measurements from an *IMU* are dead-reckoned to form a reference trajectory. To estimate the IMU sensor offsets and avoid long-term drifts, additional support sensors as GPS are used in linearized KF.

- The *error state Kalman filter*, also known as the *complementary Kalman filter*, estimates the state error $\tilde{x}_k = x_k - \hat{x}_k$ with respect to some approximate or reference trajectory, see Brown and Hwang (1997) and Wagner and Wieneke (2003). The basic idea is that the error filter corrects the nominal trajectory $\hat{x}_k$ by adding an estimate of the error state $\hat{\tilde{x}}_k$. There are two versions referred to as feedforward or feedback configurations. In feedforward configuration, the reference trajectory is not affected by the Kalman filter estimate, while in the feedback mode the reference state is corrected.

It can be shown that the linearized Kalman filter is the same as the feedforward error state Kalman filter, and that the EKF is the same as the feedback error state Kalman filter. These relations are purely algebraic. There might be numerical or computations reasons to choose the error model. For instance, the error state Kalman filter can be run on a slower sampling rate, which is often used in navigation systems, where the IMU dead-reckoning provides the reference trajectory at a high sampling rate.

8.1.2 Numeric Gradients

The standard form of the EKF involves symbolic derivatives. However, numeric derivatives may be preferred in the following cases:

8.1 DARE-based Extended Kalman Filter

- The nonlinear function is too complex to be differentiated. For instance, it may involve a computer vision algorithm or a database look-up.
- The derivatives are too complex functions, requiring too much computer code, memory or computations to be evaluated.
- A user-friendly algorithm is desired, with as few user inputs as possible.

These can then be approximated numerically for instance by

$$\frac{\partial g(x)}{\partial x_i} \approx \frac{g(x + \Delta e_i) - g(x)}{\Delta}, \quad (8.5a)$$

$$\frac{\partial g(x)}{\partial x_i \partial x_j} \approx \frac{1}{\Delta^2}\big(g(x + \Delta e_i + \Delta e_j) - g(x + \Delta e_i)$$
$$- g(x + \Delta e_j) + g(x)\big). \quad (8.5b)$$

The number of function evaluations is $n_x + 1$ for the difference in (8.5a) ($2n_x$ for a central difference) and $n_x(n_x + 1)/2 + n_x + 1$ for difference in (8.5a) ($2n_x(n_x + 1) + 2n_x + 1$ for a central difference). This should be compared to the total complexity of EKF2, which is of order n_x^3.

These numerical approximations of the Jacobian and the Hessian can be used in (8.3). However, we next show an alternative using the sigma points, where these matrices never need to be formed. This algorithm is fundamentally different from other approaches in literature for derivative free (with derivative free is meant that neither analytical derivatives nor numerical approximations of these are required) implementation of the EKF, such as *DF-EKF* in Quine (2006).

Theorem 8.1 (Sigma-point based DARE EKF)
Consider the mapping $z = g(x)$. Given the transformed sigma-points $z^{(i)} = g(x^{(i)})$ in (A.19), the terms in Algorithm 8.1 involving Jacobians and Hessians can be approximated arbitrarily well as $\alpha \to 0$ with

$$g'_k(\hat{x})P \approx \sum_{i=1}^{n_x} \frac{z_k^{(i)} - z_k^{(-i)}}{2\alpha\sqrt{n_x \sigma_i}} u_i^T,$$

$$g'_k(\hat{x})P(g'_k(\hat{x}))^T \approx \sum_{i=1}^{n_x} \sigma_i \left(\frac{z_k^{(i)} - z_k^{(-i)}}{2\alpha\sqrt{n_x \sigma_i}}\right)\left(\frac{z_k^{(i)} - z_k^{(-i)}}{2\alpha\sqrt{n_x \sigma_i}}\right)^T,$$

$$\operatorname{tr}(g''_k(\hat{x})P) \approx \sum_{i=1}^{n_x} \sigma_i \frac{z_k^{(i)} - 2z^{(0)} + z_k^{(-i)}}{\alpha^2 \sigma_i n_x}.$$

Proof: Using the SVD

$$P = UDU^T = \sum_{i=1}^{n_x} \sigma_i u_i u_i^T, \quad (8.6)$$

the sigma points in (A.19) can be written, using $n_x + \lambda = \alpha^2 n_x$ when $\kappa = 0$,

$$x^0 = \hat{x}, \tag{8.7a}$$

$$x^{(\pm i)} = \hat{x} \pm \alpha\sqrt{n_x \sigma_i} u_i, \quad i = 1, 2, \ldots, n_x. \tag{8.7b}$$

The Taylor expansion (8.2) for the transformed sigma points can then be written

$$z_k^{(\pm i)} = g_k(x^{(\pm i)}) = g_k(\hat{x}) \pm \alpha\sqrt{n_x \sigma_i} g_k'(\hat{x}) u_i + \frac{\alpha^2 n_x \sigma_i}{2} u_i^T g_k''(\hat{x}) u_i. \tag{8.8}$$

Note that the second order rest term is accurate only in a small neighborhood of $\hat{x}$, so the sigma points should be chosen close to $\hat{x}$, which means that α should be small.

The first and second order terms in the Taylor expansion can now be resolved using the following linear combinations,

$$\frac{z_k^{(i)} - z_k^{(-i)}}{2\alpha\sqrt{n_x \sigma_i}} = g_k'(\hat{x}) u_i, \tag{8.9a}$$

$$\frac{z_k^{(i)} - 2z_k^{(0)} + z_k^{(-i)}}{\alpha^2 \sigma_i n_x} = u_i^T g_k''(\hat{x}) u_i. \tag{8.9b}$$

Finally, the following relations give the first and second moments in the Taylor expansion (using the rule $\text{tr}(AB) = \text{tr}(BA)$):

$$g_k'(\hat{x}) P = \sum_{i=1}^{n_x} g_k'(\hat{x}) \sigma_i u_i u_i^T \approx \sum_{i=1}^{n_x} \frac{z_k^{(i)} - z_k^{(-i)}}{2\alpha\sqrt{n_x \sigma_i}} u_i^T \tag{8.10a}$$

$$g_k'(\hat{x}) P (g_k'(\hat{x}))^T = g_k'(\hat{x}) \sum_{i=1}^{n_x} \sigma_i u_i u_i^T (g_k'(\hat{x}))^T \tag{8.10b}$$

$$= \sum_{i=1}^{n_x} \sigma_i g_k'(\hat{x}) u_i \big(g'(\hat{x}) u_i\big)^T$$

$$\approx \sum_{i=1}^{n_x} \sigma_i \left(\frac{z_k^{(i)} - z_k^{(-i)}}{2\alpha\sqrt{n_x \sigma_i}}\right) \left(\frac{z_k^{(i)} - z_k^{(-i)}}{2\alpha\sqrt{n_x \sigma_i}}\right)^T$$

$$\text{tr}(g_k''(\hat{x}) P) = \text{tr}\left(g_k''(\hat{x}) \sum_{i=1}^{n_x} \sigma_i u_i u_i^T\right) \tag{8.10c}$$

$$= \sum_{i=1}^{n_x} \sigma_i u_i^T g_k''(\hat{x}) u_i,$$

$$\approx \sum_{i=1}^{n_x} \sigma_i \frac{z_k^{(i)} - 2z^{(0)} + z_k^{(-i)}}{\alpha^2 \sigma_i n_x}$$

$$\mathrm{tr}(g_k''(\hat{x})Pg_l''(\hat{x})P) = \mathrm{tr}\left(g_k''(\hat{x})\sum_{i=1}^{n_x}\sigma_i u_i u_i^T g_l''(\hat{x})\sum_{j=1}^{n_x}\sigma_j u_j u_j^T\right)$$

$$= \sum_{i=1}^{n_x}\sum_{j=1}^{n_x}\sigma_i\sigma_j\left(u_j^T g_k''(\hat{x})u_i\right)\left(u_i^T g_l''(\hat{x})u_j\right). \quad (8.10\mathrm{d})$$

This concludes the proof. □

From Theorem 8.1, we make the following remarks on the EKF, assuming for simplicity additive noise processes:

- Equations (8.10a,b) give the gradient needed in the standard EKF (8.3). That is, h_x' can be substituted with one of these approximations on all occasions.

- Equation (8.10c) provides the mean corrections in the second order EKF (8.3a,e).

- Equation (8.10d) provides the covariance corrections in the second order EKF (8.3f,h), but *only for the diagonal elements*. This is due to that (8.9) only approximates the Hessian for the same u_i from left and right, while (8.10d) requires asymmetric terms. A remaining challenge is to extend the UT to include off-diagonal terms.

In summary, the transformed sigma points can be used to approximate the linear term and rest term in the Taylor expansion (8.2), without explicitly computing the Jacobian and Hessian of f and h. This is one sound motivation for propagating the sigma points through the nonlinearity.

8.2 Riccati-Free EKF and UKF

The Kalman filter equations are often obscured by the complexity of the Riccati equation. However, one key idea in the UKF is based on a result from optimal filtering, where UT but also TT1, TT2 and MCT in Section A.3 can be applied. The basic idea is to consider the nonlinear transformation

$$z = \begin{pmatrix} x \\ g(x,w) \end{pmatrix} \quad (8.11)$$

of the state x and a stochastic variable w, both assumed Gaussian distributed, using the prior

$$\bar{x} = \begin{pmatrix} x \\ w \end{pmatrix} \sim \mathcal{N}\left(\begin{pmatrix} \hat{x} \\ 0 \end{pmatrix}, \begin{pmatrix} P^x & 0 \\ 0 & P^w \end{pmatrix}\right). \quad (8.12)$$

The transformed variables can then be approximated with the following Gaussian distribution, using TT1, TT2, UT, or MCT,

$$z \sim \mathcal{N}\left(\begin{pmatrix} z^x \\ z^g \end{pmatrix}, \begin{pmatrix} P^{xx} & P^{xg} \\ P^{gx} & P^{gg} \end{pmatrix}\right). \quad (8.13)$$

Assuming an observation g^{obs} of the nonlinear relation $g(x, w)$, Lemma 7.1 gives

$$K = P^{xg}(P^{gg})^{-1}, \qquad (8.14a)$$
$$\hat{x} = z^x + K(g^{obs} - z^g). \qquad (8.14b)$$

The algorithm is thus to approximate the covariance matrix for $(x^T, g^T)^T$ numerically and compute the Kalman gain K from its block matrix decomposition. Algorithm 8.2 gives the general algorithm. Note that the process noise

---

Algorithm 8.2 Nonlinear Transformation-Based Filtering

The nonlinear transform-based filter for the model (8.1) is given by the following recursions initialized with $\hat{x}_{1|0}$ and $P_{1|0}$:

1. **Measurement update:** Let

$$\bar{x} = \begin{pmatrix} x_k \\ e_k \end{pmatrix} \sim \mathcal{N}\left(\begin{pmatrix} \hat{x}_{k|k-1} \\ 0 \end{pmatrix}, \begin{pmatrix} P_{k|k-1} & 0 \\ 0 & R_k \end{pmatrix} \right) \qquad (8.15a)$$

$$z = \begin{pmatrix} x_k \\ y_k \end{pmatrix} = \begin{pmatrix} x_k \\ h(x_k, u_k, e_k) \end{pmatrix} \qquad (8.15b)$$

The transformation approximation (UT, MC, TT1, TT2) gives

$$z \sim \mathcal{N}\left(\begin{pmatrix} \hat{x}_{k|k-1} \\ \hat{y}_{k|k-1} \end{pmatrix}, \begin{pmatrix} P^{xx}_{k|k-1} & P^{xy}_{k|k-1} \\ P^{yx}_{k|k-1} & P^{yy}_{k|k-1} \end{pmatrix} \right) \qquad (8.15c)$$

The measurement update is then

$$K_k = P^{xy}_{k|k-1}(P^{yy}_{k|k-1})^{-1}, \qquad (8.15d)$$
$$\hat{x}_{k|k} = \hat{x}_{k|k-1} + K_k(y_k - \hat{y}_{k|k-1}), \qquad (8.15e)$$
$$P^{xx}_{k|k} = P^{xx}_{k|k-1} - K_k P^{yy}_{k|k-1} K_k^T. \qquad (8.15f)$$

2. **Time update:** Let

$$\bar{x} = \begin{pmatrix} x_k \\ v_k \end{pmatrix} \sim \mathcal{N}\left(\begin{pmatrix} \hat{x}_{k|k} \\ 0 \end{pmatrix}, \begin{pmatrix} P_{k|k} & 0 \\ 0 & Q_k \end{pmatrix} \right) \qquad (8.15g)$$

$$z = x_{k+1} = f(x_k, u_k, v_k). \qquad (8.15h)$$

The transformation approximation (UT, MC, TT1, TT2) gives

$$z \sim \mathcal{N}(\hat{x}_{k+1|k}, P_{k+1|k}). \qquad (8.15i)$$

8.2 Riccati-Free EKF and UKF

does not need to be additive. The NLTF provides a framework for nonlinear filtering from which the following different combinations of transforms can be done:

- The EKF obtained using TT1 above is equivalent to the EKF in (8.3).

- The EKF version obtained using TT2 above should be equivalent to the second order compensated EKF in (8.3).

- One should be aware of that it is not advisable to start with a large initial covariance P_0 when using the UKF version of Algorithm 8.2, in contrast to the standard implementation in Algorithm 8.1.

- The Monte Carlo approach should potentially be the most accurate, given that a sufficient number of samples are used, since it asymptotically computes the correct first and second order moments.

- The UKF is obtained by using the UT (1, 2 or other variants) in both time and measurement update above.

- There is a freedom to mix transform approximations in the time and measurement update.

- If the observation model is linear, the usual Kalman filter measurement update should be performed. The same holds for a linear dynamic model.

Finally, it should be stressed that the algorithms in this chapter can only represent state uncertainty with Gaussian distributions, and are thus not able to handle multi-modal distributions. For such cases, the point-mass filter in Section 6.4.1, the Kalman filter bank in 10, or the particle filter in Chapter 9 must be used.

The actual performance for the 16 different combinations depends of course on the degree of nonlinearity in the system model. As a rule of thumb, the choice can be guided by studying the nonlinear mappings in the dynamic model and sensor model individually. For target tracking and navigation applications, it is often the nonlinear sensor model that gives the greatest filtering challenge as pointed out in Gustafsson et al. (2002).

Example 8.1: Bearings-Only Tracking

The next example exemplifies the common bearings only problem depicted in Figure 8.1. Here a situation where a target, known to be in an approximate location quantified by $\hat{x}_{0|0}$ and $P_{0|0}$, is being triangulated using bearings-only measurements. This can mathematically be described as

$$x_{k+1} = f(x_k) + v_k = x_k + v_k$$
$$y_k = h(x_k) + e_k = \arctan2\left(\mathsf{x}_k - \mathsf{x}_k^0, \mathsf{y}_k - \mathsf{y}_k^0\right) + e_k,$$

where the state $x = \begin{pmatrix} x & y \end{pmatrix}^T$ is the Cartesian position of the target, $w_k \equiv 0$ for clarity, and $\mathrm{Cov}(e_k) = R$. For this situation, the gradients needed to perform filtering using an EKF are

$$F = I, \qquad H = \frac{1}{(\mathsf{x} - \mathsf{x}^0)^2 + (\mathsf{y} - \mathsf{y}^0)^2} \begin{pmatrix} -(\mathsf{y} - \mathsf{y}^0) \\ \mathsf{x} - \mathsf{x}^0 \end{pmatrix}.$$

Note, the first order approximation of arctan2 is best for $\left|\frac{\mathsf{y}}{\mathsf{x}}\right| \gg 0$.

Now, assume

$$x_0 = \begin{pmatrix} 1.5 \\ 1.5 \end{pmatrix}, \qquad \hat{x}_{0|0} = \begin{pmatrix} 2 \\ 2 \end{pmatrix}, \qquad P_{0|0} = \begin{pmatrix} 1 & 0 \\ 0 & 1 \end{pmatrix},$$

and that the bearing to the target is measured first from the position $\mathcal{M}_1 = (0,0)$ and then from $\mathcal{M}_2 = (2,0)$, as depicted in Figure 8.1. Figure 8.2 depicts the estimates based on this new $\hat{x}_0$ and noise-free measurements for the different filters. Here, also the iterated EKF is shown.

The variance of the estimation error based on Monte Carlo simulations of the problem specified above, using the described filters, yield the result in Table 8.1. The table somewhat contradicts the previous results. One thing to observe is that the UKF outperforms the EKF. Hence, it seems that the conservative P matrix actually pays off.

Finally, note that the PF and the inferred distribution is almost identical. Worth noticing, though, is the substantially better estimates achieved with the PF compared to the other used filters. Hence, this is a situation when the PF pays off.

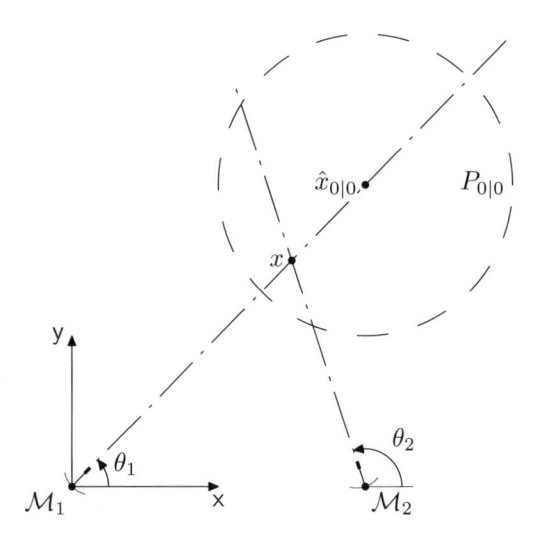

Figure 8.1: Bearings-only problem. Two measurements are used, one from $\mathcal{M}_1$ and one from $\mathcal{M}_2$.

Table 8.1: *Mean square error filter performance for 1 000 Monte Carlo simulations. True posterior is computed by a point-mass filter (PMF) with a dense grid, and the particle filter (PF) performance is given for comparison.*

Filter	None	$\mathcal{M}_1$	$\mathcal{M}_1 + \mathcal{M}_2$
True (PMF)	2.01	0.95	0.06
EKFI	2.01	1.33	1.23
EKFII	2.01	1.38	0.79
UKF2	2.01	1.33	1.00
CKF	2.01	1.45	1.12
PF	2.01	0.95	0.06

8.3 Target Tracking Examples

8.3.1 Linear Motion Model

The first model is linear to be able to compare to the Kalman filter that provides the optimal estimate. The example makes use of two different objects:

- Signal object where the state $x_{1:k}$ and observation $y_{1:k}$ sequences are

(a) Measurements from: $\mathcal{M}_1$ (b) Measurements from: $\mathcal{M}_1 + \mathcal{M}_2$

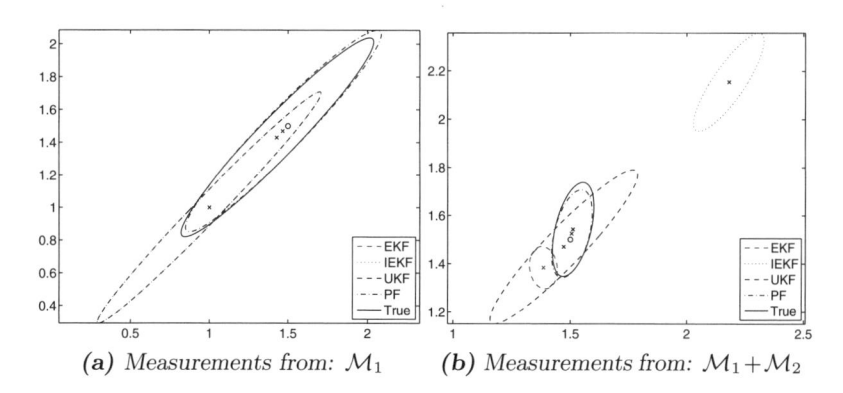

Figure 8.2: *Estimate with $\hat{x}_0 = (0,0)^T$, based on one and two measurements. (Estimates are denoted with × in the center of each covariance ellipse, and the true target position is denoted with ○.)*

stored, with their associated uncertainty (covariances P_k^x, P_k^y or particle representation).

- Model objects for linear and nonlinear models, with methods for simulation and filtering algorithms.

First, the model is loaded from an example database as a linear state space model. It is then converted to the general nonlinear model structure, which does not make use of the fact that the underlying model is linear.

```
mss=exlti('cv2d');
mnl=nl(mss);
```

Now, the following state trajectories are compared:

- The true state from the simulation.

- The Cramér-Rao lower bound (CRLB) computed from the nonlinear model.

- The Kalman filter (KF) estimate using the linear model.

- The extended Kalman filter (EKF1) using the nonlinear model.

- The second order extended Kalman filter (EKF2) using the nonlinear model.

- The unscented Kalman filter (UKF) using the nonlinear model.

For all except the first one, a confidence ellipsoid indicates the position estimation uncertainty.

```
y=simulate(mss,10);
xhat1=kalman(mss,y);
xhat2=ekf(mnl,y);
xhat3=ekf2(mnl,y);
xhat4=ukf(mnl,y);
xhat5=pf(mnl,y);
xcrlb=crlb(mnl,y);
xplot2(xcrlb,xhat5,xhat4,xhat3,xhat2,xhat1,'conf',90)
```

Figure 8.3 validates that all algorithms provide comparable estimates in accordance with the CRLB.

8.3 Target Tracking Examples

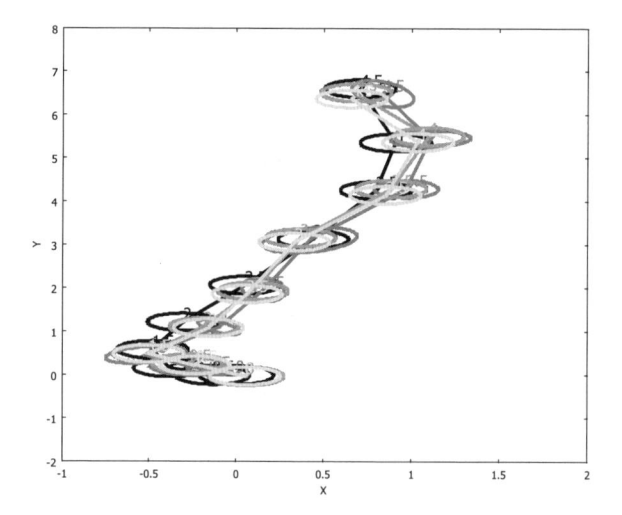

Figure 8.3: Simulated trajectory with its CRLB (darkest), and estimates from KF, EKF1, EKF2 and UKF, respectively.

8.4 Summary

8.4.1 Theory

The Kalman filter is characterized by its measurement and time updates, respectively,

$$\hat{x}_{k|k} = \hat{x}_{k|k-1} + K_k(y_k - H\hat{x}_{k|k-1}),$$
$$\hat{x}_{k+1|k} = F\hat{x}_{k|k}.$$

The following recursion extends this structure to nonlinear filtering,

$$\hat{x}_{k|k} = \hat{x}_{k|k-1} + K_k(y_k + \bar{u}_k^h - h(\hat{x}_{k|k-1})),$$
$$\hat{x}_{k+1|k} = f(\hat{x}_{k|k}) + \bar{u}_k^f,$$

Here, $\bar{u}_k^h$ and $\bar{u}_k^f$ are optional compensation terms for higher order effects. There are two approaches for how to compute the Kalman gain K_k:

- Riccati-based approaches, where the Kalman gain is computed as

$$K_k = P_{k|k-1}(h'_x(\hat{x}_{k|k-1}))^T \left(h'_x(\hat{x}_{k|k-1}) P_{k|k-1}(h'_x(\hat{x}_{k|k-1}))^T + R_k \right)^{-1},$$

where $P_{k|k-1}$ is updated with a Riccati equation. There are two options:

  - First order Taylor expansion, leading to the standard extended Kalman filter in (8.3).
  - Second order Taylor expansion, leading to the second order compensated extended Kalman filter in Algorithm 8.1, where the terms $\bar{u}_k^h$ and $\bar{u}_k^f$ are used.

  Initialization is fairly important. Using too large a $P_{1|0}$ would imply that the involved gradients are quite unreliable initially.

- Riccati-free approaches, where the Kalman gain is computed as

$$K_k = P_{k|k-1}^{xy} \left(P_{k|k-1}^{yy} \right)^{-1},$$

where $P_{k|k-1}^{xy}$ and $P_{k|k-1}^{yy}$ are obtained by a nonlinear transformation of the Gaussian vector $(x^T, e^T)^T$ to $(x^T, y^T)^T = (x^T, h^T(x,e))^T$. There are here four different alternatives each for the time and measurement update:

  - Unscented transformation, leading to the unscented Kalman filter if used in both time and measurement update.
  - First order Taylor transformation, leading to the extended Kalman filter in (8.3) when used in both time and measurement updates, but with a completely different implementation compared to above.

- Second order Taylor transformation, leading to the second order compensated extended Kalman filter in Algorithm 8.1 when used in both time and measurement updates, but with a completely different implementation compared to above.
- Monte Carlo transformation, leading to an asymptotically, in the number of samples, best possible time and measurement update, at least for each step at the time.

These four transformation can be combined arbitrarily. Note that if either the time or measurement model is linear, the standard Kalman filter update applies.

Initialization is even more important for this approach. The usual paradigm in Kalman filtering using $P_{1|0} \approx \infty I$ does not work, since the transformations will return unrealistic distributions.

Dithering, that is, increasing the noise covariances, is a sound tuning option to compensate for linearization and transformation errors, respectively.

The standard EKF is to prefer if one of the following holds:

- The model is almost linear, in which case the Hessian is small.

- The SNR is high, in which case the estimation error can be considered sufficiently small.

In both cases, the neglected rest term is small. For severely nonlinear models or low SNR, the second order EKF or UKF may improve performance.

8.4.2 Software

The following nl methods are related to the content of this chapter:

- [xhat,V]=ekf(m,y) implements the standard EKF algorithm in 8.3. Numerical gradients $f'_x(x)$ and $h'_x(x)$ are computed by numgrad.

- [xhat,V]=nltf(m,y) implements the general nonlinear filtering algorithm in Algorithm 8.2. The methods ndist.tt1eval, ndist.tt2eval, ndist.uteval and ndist.mceval, respectively, are called. The following methods call this general implementation:
    - [xhat,V]=ekf1(m,y) should give the same result as Algorithm 8.1, but using a completely different implementation.
    - [xhat,V]=ekf2(m,y) gives a version of the EKF in Algorithm 8.1.
    - [xhat,V]=ukf(m,y) gives the UKF.

9

The Particle Filter

The nonlinear filtering problem concerns state estimation in a nonlinear non-Gaussian model, which in a general form is given by

$$x_{k+1} = f(x_k, u_k, v_k), \qquad (9.1a)$$
$$y_k = h(x_k, u_k) + e_k. \qquad (9.1b)$$

The probability density functions (PDF) of the process noise v_k and the measurement noise e_k are arbitrary but assumed to be known. The *particle filter* (*PF*) provides an approximation to the posterior distribution $p(x_k|y_{1:k})$ of the state x_k conditioned on the set of measurements $y_{1:k}$. The approximation is based on a set of N samples $\{x^i\}_{i=1}^N$, referred to as particles, and their associated weights $\{w^i\}_{i=1}^N$. The resulting posterior approximation is then $p(x_k|y_{1:k}) \approx \sum_{i=1}^N w_k^i \delta(x_k - x_k^i)$, where $\delta(u)$ is Dirac's delta function. The PF recursively processes data using the following operations:

1. The *measurement update*: Modify the weights according to the likelihood function of the difference between observed and predicted measurement.

2. A *resampling* step: Take N new samples of the state from the existing set of particles.

3. The *time update*: Simulate a trajectory from one measurement time to the next using the dynamic model.

This material can be approached in different ways:

- Section 9.1 introduces the basic concepts, provides the standard example of the chapter, gives a historic background, and also summarizes the theory on nonlinear filtering from Chapter 6.

- One can have a code first approach, starting with Section 9.9 to get a complete simulation code for a concrete example. This section also provides some other examples using *Signals and Systems Lab*.

- An overview of the PF theory as found in Section 9.3.

- A summary of the MPF theory is provided in Section 9.8.

Section 16.3 provides an overview of a number of positioning applications of the PF and MPF, respectively, and conclusions from practice, which can also be read standalone. This chapter is an extended version of Gustafsson (2010), which is the preferred reference for this material.

9.1 Introduction

9.1.1 Brief History

The first traces of the PF dates back to the fifties Hammersley and Morton (1954); Rosenbluth and Rosenbluth (1956), and the control community made some attempts in the seventies Akashi and Kumamoto (1977); Handshin (1970). However, the PF era started with the seminal paper Gordon et al. (1993), and the independent developments in Isard and Blake (1998); Kitagawa (1996). Here, an important resampling step was introduced. The timing for proposing a general solution to the nonlinear filtering problem was perfect, in that the computer development enabled the use of computational complex algorithms to quite realistic problems. The research has since the paper Gordon et al. (1993) steadily intensified, see the article collection Doucet et al. (2001a), the surveys Arulampalam et al. (2002); Cappé et al. (2007); Djuric et al. (2003); Liu and Chen (1998), and the monograph Ristic et al. (2004). The particle filters may be a serious alternative for real-time applications classically approached by the (extended) Kalman filter. The more nonlinear model, or the more non-Gaussian noise, the more potential particle filters have, especially in applications where computational power is rather cheap and the sampling rate is moderate.

9.1.2 Positioning Applications

Positioning of moving platforms has been a *technical driver* for real-time applications of the particle filter (PF) in both the signal processing and the robotics communities. For this reason, we will spend some time to explain several such applications in detail, and to summarize the experiences of using the PF in

9.1 Introduction

practice. The applications concern positioning of underwater (UW) vessels, surface ships, cars, and aircraft using geographical information systems (GIS) containing a database with features of the surrounding. These applications provide conclusions from practice supporting the theoretical survey part.

The following example will be used for illustrating concepts throughout this chapter.

Example 9.1: 1D terrain navigation

We here consider a one-dimensional terrain navigation problem. Assume here that the speed u_k can be measured,

$$x_{k+1} = x_k + u_k + v_k, \qquad (9.2a)$$
$$y_k = h(x_k) + e_k, \qquad (9.2b)$$
$$v_k \sim \mathcal{N}(0, Q_k), \qquad (9.2c)$$
$$e_k \sim \mathcal{N}(0, R_k). \qquad (9.2d)$$

The Bayesian terrain navigation problem is to compute $p(x_k|y_{1:k}, u_{1:k})$ based on the observed terrain altitude $y_{1:k}$, the measured speed $u_{1:k}$ and the terrain elevation map (GIS) $h(x)$. The EKF approach would be to linearize $h(x) = h(\hat{x}_{k|k-1}) + h'(\hat{x}_{k|k-1})(x_k - \hat{x}_{k|k-1})$ and apply the Kalman filter recursions. This would work well if the terrain can be approximated well with a linear function in the whole uncertainty area of $\hat{x}_{k|k-1}$. Otherwise, the PF (or PMF) is the main alternative.

In the robotics community, the PF has been developed into one of the main algorithms (fastSLAM) Montemerlo et al. (2002) for solving the simultaneous localization and mapping (SLAM) problem Bailey and Durrant-Whyte (2006); Durrant-Whyte and Bailey (2006); Thrun et al. (2005). This can be seen as an extension to the aforementioned applications, where the features in the GIS are dynamically detected and updated on the fly.

The common denominator of these applications of the PF is the use of a low-dimensional state vector consisting of horizontal position and course (three dimensional pose). The PF performs quite well in a three dimensional state-space. In higher dimensions the curse of dimensionality quite soon makes the particle representation too sparse to be a meaningful representation of the posterior distribution. That is, the PF is *not* practically useful when extending the models to more realistic cases with

- motion in three dimensions (six-dimensional pose),
- more dynamic states (accelerations, unmeasured velocities, etc),
- or sensor biases and drifts.

A *technical enabler* for such applications is the marginalized particle filter (MPF), also referred to as the Rao-Blackwellized particle filter (RBPF). It allows for the use of high-dimensional state-space models as long as the (severe) nonlinearities only affect a small subset of the states. In this way, the structure of the model is utilized, so that the particle filter is used to solve the most difficult tasks, and the (extended) Kalman filter is used for the (almost) linear Gaussian states. The fastSLAM algorithm is in fact a version of the MPF, where hundreds or thousands of feature points in the state vector are updated using the (extended) Kalman filter. The need for the MPF in the list of applications will be motivated by examples and experience from practice.

The particle filter should be the nonlinear filtering algorithm that appeals to engineers the most, since it intimately addresses the system model. The filtering code is thus very similar to the simulation code that the engineer working with the application should already be quite familiar with.

9.2 Recapitulation of Nonlinear Filtering

This section summarizes the most essential information in Chapter 6.

9.2.1 Models and Notation

Applied nonlinear filtering is based on discrete time nonlinear state space models relating a hidden state x_k to the observations y_k:

$$x_{k+1} = f(x_k, v_k), \qquad v_k \sim p_{v_k}, \quad x_0 \sim p_{x_0}, \qquad (9.3\text{a})$$

$$y_k = h(x_k) + e_k, \qquad e_k \sim p_{e_k}. \qquad (9.3\text{b})$$

Here, v_k is a stochastic noise process specified by its known probability density function (PDF) p_{v_k}, which is compactly expressed as $v_k \sim p_{v_k}$. Similarly, e_k is an additive measurement noise also with known PDF p_{e_k}. The PDF of the initial state is known as p_{x_0}. The model can also depend on a known (control) input u_k, so $f(x_k, u_k, v_k)$ and $h(x_k, u_k)$, but this dependence is omitted to simplify notation. The notation $s_{1:k}$ denotes the sequence $s_1, s_2, \ldots, s_k$ (s is one of the signals x, v, y, e), and n_s denotes the dimension of that signal.

In the statistical literature, a general Markov model and observation model in terms of conditional PDF's are often used

$$x_{k+1} \sim p(x_{k+1}|x_k), \qquad (9.4\text{a})$$

$$y_k \sim p(y_k|x_k). \qquad (9.4\text{b})$$

This is in a sense a more general model. For instance, (9.4) allows implicit measurement relations $h(y_k, x_k, e_k) = 0$ in (9.3b), and differential algebraic equations that add implicit state constraints to (9.3a).

9.2 Recapitulation of Nonlinear Filtering

The Bayesian approach to *nonlinear filtering* is to compute or approximate the posterior distribution for the state given the observations. The posterior is denoted $p(x_k|y_{1:k})$ for filtering, $p(x_{k+m}|y_{1:k})$ for prediction and $p(x_{k-m}|y_{1:k})$ for smoothing, respectively, where $m > 0$ denotes the prediction or smoothing lag. The theoretical derivations are based on the general model (9.4), while algorithms and discussions will be based on (9.3). Note that the Markov property of the model (9.4) implies the formulas $p(x_{k+1}|x_{1:k}, y_{1:k}) = p(x_{k+1}|x_k)$ and $p(y_k|x_{1:k}, y_{1:k-1}) = p(y_k|x_k)$, which will be used frequently.

A linearized model will turn up on several occasions, which is obtained by a first order Taylor expansion of (9.3) around $x_k = \bar{x}_k$ and $v_k = 0$:

$$x_{k+1} = f(\bar{x}_k, 0) + F(\bar{x}_k)(x_k - \bar{x}_k) + G(\bar{x}_k)v_k, \qquad (9.5a)$$
$$y_k = h(\bar{x}_k) + H(\bar{x}_k)(x_k - \bar{x}_k) + e_k, \qquad (9.5b)$$

where

$$F(\bar{x}_k) = \left.\frac{\partial f(x_k, v_k)}{\partial x_k}\right|_{x_k=\bar{x}_k, v_k=0}, \quad G(\bar{x}_k) = \left.\frac{\partial f(x_k, v_k)}{\partial v_k}\right|_{x_k=\bar{x}_k, v_k=0}, \qquad (9.5c)$$

$$H(\bar{x}_k) = \left.\frac{\partial h(x_k)}{\partial x_k}\right|_{x_k=\bar{x}_k}, \qquad (9.5d)$$

and the noise is represented by their second order moments

$$\mathrm{Cov}(e_k) = R_k, \quad \mathrm{Cov}(v_k) = Q_k, \quad \mathrm{Cov}(x_0) = P_0. \qquad (9.5e)$$

For instance, the extended Kalman filter (EKF) recursions are obtained by linearizing around the previous estimate and apply the Kalman filter equations, which gives

$$K_k = P_{k|k-1} H^T(\hat{x}_{k|k-1}) \left(H(\hat{x}_{k|k-1}) P_{k|k-1} H^T(\hat{x}_{k|k-1}) + R_k \right)^{-1}, \qquad (9.6a)$$
$$\hat{x}_{k|k} = \hat{x}_{k|k-1} + K_k(y_k - h_k(\hat{x}_{k|k-1})), \qquad (9.6b)$$
$$P_{k|k} = P_{k|k-1} - K_k H(\hat{x}_{k|k-1}) P_{k|k-1}, \qquad (9.6c)$$
$$\hat{x}_{k+1|k} = f(\hat{x}_{k|k}, 0), \qquad (9.6d)$$
$$P_{k+1|k} = F(\hat{x}_{k|k}) P_{k|k} F^T(\hat{x}_{k|k}) + G(\hat{x}_{k|k}) Q G^T(\hat{x}_{k|k}). \qquad (9.6e)$$

The recursion is initialized with $\hat{x}_{1|0} = x_0$ and $P_{1|0} = P_0$, assuming the prior $p(x_1) \sim \mathcal{N}(x_0, P_0)$. The EKF approximation of the posterior filtering distribution is then

$$\hat{p}(x_k|y_{1:k}) = \mathcal{N}(\hat{x}_{k|k}, P_{k|k}), \qquad (9.7)$$

where $\mathcal{N}(m, P)$ denotes the Gaussian density function with mean m and covariance P. The special case of a linear model is covered by (9.5) in which case $F(\bar{x}_k) = F_k$, $G(\bar{x}_k) = G_k$, $H(\bar{x}_k) = H_k$, and using these and the equalities $f(\bar{x}_k, 0) = F_k \bar{x}_k$ and $h(\bar{x}_k) = H_k \bar{x}_k$ in (9.6) gives the standard KF recursion.

The neglected higher order terms in the Taylor expansion implies that the EKF can be biased and it tends to underestimate the covariance of the state estimate. There is a variant of the EKF that also takes the second order term in the Taylor expansion into account Bar-Shalom and Fortmann (1988). This is done by adding the expected value of the second order term to the state updates and its covariance to the state covariance updates. The unscented Kalman filter (UKF) Julier and Uhlmann (2002); Julier et al. (1995) does a similar correction by using propagation of systematically chosen state points (called sigma points) through the model. Related approaches include the divided difference filter (DFF) Norgaard et al. (2000) that uses Sterling's formula to find the sigma points and the quadrature Kalman filter (QKF) Arasaratnam et al. (2007) that uses the quadrature rule in numerical integration to select the sigma points. The common theme in EKF, UKF, DDF and QKF is that the nonlinear model is evaluated in the current state estimate and for the latter ones some extra points that depend on the current state covariance.

UKF is closely related to the second order EKF Gustafsson and Hendeby (2008). Both variants improve over the EKF in certain problems and can work well as long as the posterior distribution is unimodal. Further, the algorithms are prone to diverge, and this problem is hard to mitigate or foresee by analytical methods. The choice of state coordinates is for instance crucial in EKF and UKF, see Chapter 8.9.3 in Gustafsson (2001) for one example, while this choice does not affect the performance of the PF (more than potential numerical problems).

9.2.2 Bayesian Filtering

The Bayesian solution to compute the posterior distribution, $p(x_k|y_{1:k})$, of the state vector, given past observations, is given by the general Bayesian update recursion

$$p(x_k|y_{1:k}) = \frac{p(y_k|x_k)p(x_k|y_{1:k-1})}{p(y_k|y_{1:k-1})}, \tag{9.8a}$$

$$p(y_k|y_{1:k-1}) = \int_{\mathbb{R}^{n_x}} p(y_k|x_k)p(x_k|y_{1:k-1})\, dx_k, \tag{9.8b}$$

$$p(x_{k+1}|y_{1:k}) = \int_{\mathbb{R}^{n_x}} p(x_{k+1}|x_k)p(x_k|y_{1:k})\, dx_k. \tag{9.8c}$$

This classical result Jazwinsky (1970); Van Trees (1971) is the cornerstone in nonlinear Bayesian filtering. The first equation follows directly from Bayes' law, and the other two ones follow from the law of total probability, using the model (9.4). The first equation corresponds to a measurement update, the second one is a normalization constant, and the third one corresponds to a time update.

9.2 Recapitulation of Nonlinear Filtering

The posterior distribution is the primary output from a nonlinear filter, from which standard measures as the minimum mean square (MMS) estimate $\hat{x}_k^{\text{MMS}}$ and its covariance $P_{k|k}^{\text{MMS}}$ can be extracted and compared to EKF and UKF outputs:

$$\hat{x}_{k|k}^{\text{MMS}} = \int x_k p(x_k|y_{1:k}) dx_k, \qquad (9.9a)$$

$$P_{k|k}^{\text{MMS}} = \int (x_k - \hat{x}_k^{\text{MMS}})(x_k - \hat{x}_k^{\text{MMS}})^T p(x_k|y_{1:k}) \, dx_k. \qquad (9.9b)$$

For a linear Gaussian model, the KF recursions in (9.6) also provide the solution (9.9) to this Bayesian problem. However, for nonlinear or non-Gaussian models there is in general no finite dimensional representation of the posterior distributions similar to $(\hat{x}_{k|k}^{\text{MMS}}, P_{k|k}^{\text{MMS}})$. That is why numerical approximations are needed.

9.2.3 The Point-Mass Filter

Suppose now we have a deterministic grid $\{x^i\}_{i=1}^N$ of the state space $\mathbb{R}^{n_x}$ over N points, and that we at time k based on observations $y_{1:k-1}$ have computed the relative probabilites (assuming distinct grid points)

$$w_{k|k-1}^i \propto \text{P}(x_k = x^i|y_{1:k-1}), \qquad (9.10)$$

satisfying $\sum_{i=1}^N w_{k|k-1}^i = 1$ (note that this is a relative normalization with respect to the grid points). The notation x_k^i is introduced here to unify notation with the PF, and it means that the state x_k at time k visits the grid point x^i. The prediction density and the first two moments can then be approximated by

$$\hat{p}(x_k|y_{1:k-1}) = \sum_{i=1}^N w_{k|k-1}^i \delta(x_k - x_k^i), \qquad (9.11a)$$

$$\hat{x}_{k|k-1} = \text{E}(x_k) = \sum_{i=1}^N w_{k|k-1}^i x_k^i, \qquad (9.11b)$$

$$P_{k|k-1} = \text{Cov}(x_k) = \sum_{i=1}^N w_{k|k-1}^i (x_k^i - \hat{x}_{k|k-1})(x_k^i - \hat{x}_{k|k-1})^T. \qquad (9.11c)$$

Here, $\delta(x)$ denotes the Dirac impulse function. The Bayesian recursion (9.8) now gives

$$\hat{p}(x_k|y_{1:k}) = \sum_{i=1}^{N} \underbrace{\frac{1}{c_k}p(y_k|x_k^i)w_{k|k-1}^i}_{w_{k|k}^i} \delta(x_k - x_k^i), \tag{9.12a}$$

$$c_k = \sum_{i=1}^{N} p(y_k|x_k^i)w_{k|k-1}^i, \tag{9.12b}$$

$$\hat{p}(x_{k+1}|y_{1:k}) = \sum_{i=1}^{N} w_{k|k}^i p(x_{k+1}|x_k^i). \tag{9.12c}$$

Note that the recursion starts with a discrete approximation (9.11a) and ends in a continuous distribution (9.12c). Now, to close the recursion, the standard approach is to sample (9.12c) at the grid points x^i, which computationally can be seen as a multidimensional convolution,

$$w_{k+1|k}^i = \hat{p}(x_{k+1}^i|y_{1:k}) = \sum_{j=1}^{N} w_{k|k}^j p(x_{k+1}^i|x_k^j), \quad i = 1, 2, \ldots, N. \tag{9.13}$$

This is the principle in the *point mass filter* Alspach and Sorenson (1972); Kramer and Sorenson (1988), whose advantage is its simple implementation and tuning (the engineer basically only has to consider the size and resolution of the grid). The curse of dimensionality limits the application of PMF to small models (n_x less than two or three) for two reasons: the first one is that a grid is an inefficiently sparse representation in higher dimensions, and the second one is that the multidimensional convolution becomes a real bottleneck with quadratic complexity in N. Another practically important but difficult problem is to translate and change the resolution of the grid adaptively.

9.3 The Particle Filter

9.3.1 Relation to the Point Mass Filter

The particle filter (PF) has much in common with the point mass filter (PMF). Both algorithms approximate the posterior distribution with a discrete density of the form (9.11a), and they are both based on a direct application of (9.8) leading to the numerical recursion in (9.12). However, there are some major differences:

- The deterministic grid x^i in the PMF is replaced with a dynamic stochastic grid x_k^i in the PF that changes over time. The stochastic grid is a much more efficient representation of the state space than a fixed or adaptive deterministic grid in most cases.

9.3 The Particle Filter

- The PF aims at estimating the whole trajectory $x_{1:k}$ rather than the current state x_k. That is, the PF generates and evaluates a set $\{x_{1:k}^i\}_{i=1}^N$ of N different trajectories. This affects (9.8c) as follows:

$$p(x_{1:k+1}^i|y_{1:k}) = \underbrace{p(x_{k+1}^i|x_{1:k}^i, y_{1:k})}_{p(x_{k+1}^i|x_k^i)} \underbrace{p(x_{1:k}^i|y_{1:k})}_{w_{k|k}^i} \quad (9.14)$$

$$= w_{k|k}^i p(x_{k+1}^i|x_k^i). \quad (9.15)$$

Comparing this to (9.12c) and (9.13), we note that the double sum leading to a quadratic complexity is avoided by this trick. However, this quadratic complexity appears if one wants to recover the marginal distribution $p(x_k|y_{1:k})$ from $p(x_{1:k}|y_{1:k})$, more on this in Section 9.3.3.

- The new grid is in the PF obtained by sampling from (9.12c) rather than reusing the old grid as done in the PMF. The original version of the PF Gordon et al. (1993) samples from (9.12c) as it stands by drawing one sample each from $p(x_{k+1}|x_k^i)$ for $i = 1, 2, \ldots, N$. More generally, the concept of *importance sampling* Robert and Casella (1999) can be used. The idea is to introduce a *proposal density* $q(x_{k+1}|x_k, y_{k+1})$ which is easy to sample from, and rewrite (9.8c) as

$$p(x_{k+1}|y_{1:k}) = \int_{\mathbb{R}^{n_x}} p(x_{k+1}|x_k) p(x_k|y_{1:k}) \, dx_k \quad (9.16)$$

$$= \int_{\mathbb{R}^{n_x}} q(x_{k+1}|x_k, y_{k+1}) \frac{p(x_{k+1}|x_k)}{q(x_{k+1}|x_k, y_{k+1})} p(x_k|y_{1:k}) \, dx_k. \quad (9.17)$$

The trick now is to generate a sample at random from the proposal density $x_{k+1}^i \sim q(x_{k+1}|x_k^i, y_{k+1})$ for each particle, and then adjust the posterior probability for each particle with the *importance weight*

$$p(x_{1:k+1}|y_{1:k}) = \sum_{i=1}^N \underbrace{\frac{p(x_{k+1}^i|x_k^i)}{q(x_{k+1}^i|x_k^i, y_{k+1})} w_{k|k}^i}_{w_{k+1|k}^i} \delta(x_{1:k+1} - x_{1:k+1}^i). \quad (9.18)$$

As indicated, the proposal distribution $q(x_{k+1}^i|x_k^i, y_{k+1})$ depends on the last state in the particle trajectory $x_{1:k}^i$, but also the next measurement y_{k+1}. The simplest choice of proposal is to use the dynamic model itself, $q(x_{k+1}^i|x_k^i, y_{k+1}) = p(x_{k+1}^i|x_k^i)$, leading to $w_{k+1|k}^i = w_{k|k}^i$. The choice of proposal and its actual form are discussed more thoroughly in Section 9.5.

- Resampling is a crucial step in the PF. Without resampling, the PF would break down to a set of independent simulations yielding trajectories $x_{1:k}^i$ with relative probabilities w_k^i. Since there would then be no feedback mechanism from the observations to control the simulations, they would quite soon diverge. As a result, all relative weights would tend to zero except for one that tends to one. This is called *sample depletion*, or *sample degeneracy*, or *sample impoverishment*. Note that a relative weight of one, $w_{k|k}^i \approx 1$ is not at all an indicator of how close a trajectory is to the true one since this is only a relative weight. It merely says that one sequence in the set $\{x_{1:k}^i\}_{i=1}^N$ is much more likely than all of the other ones. Resampling introduces the required information feedback from the observations, so trajectories that perform well will survive the resampling. There are some degrees of freedom in the choice of *resampling strategy* discussed in Section 9.4.1.

9.3.2 Algorithm

The PF algorithm is summarized in Algorithm 9.1. It can be seen as an algorithmic framework from which particular versions of the PF can be defined later on. It should be noted that the most common form of the algorithm combines the weight updates (9.20a,d) into one equation. Here, we want to stress the relations to the fundamental Bayesian recursion by keeping the structure of a measurement update (9.8a)–(9.12a)–(9.20a), normalization (9.8b)–(9.12b)–(9.20b), and time update (9.8c)–(9.12c)–(9.20c,d).

9.3.3 Prediction, Smoothing and Marginals

Algorithm 9.1 outputs an approximation of the trajectory posterior density $p(x_{1:k}|y_{1:k})$ and $p(x_{1:k+1}|y_{1:k})$, respectively. For a prediction problem, the simplest engineering solution is to just extract the last state x_{k+1}^i from the trajectory $x_{1:k+1}^i$ and use the particle approximation

$$\hat{p}(x_{k+1}|y_{1:k}) = \sum_{i=1}^{N} w_{k+1|k}^i \delta(x_{k+1} - x_{k+1}^i). \quad (9.19)$$

This result follows from the marginalization $p(x_{k+1}) = \int p(x_{k+1})p(x_k)\,dx_k$ using the particle approximation for $\hat{p}(x_{k:k+1}) = \sum_{i=1}^{N} w_{k+1|k}^i \delta(x_k - x_k^i)$ $\delta(x_{k+1} - x_{k+1}^i)$. This approach suffers from the depletion problem and can give a poor approximation.

A better solution is obtained by first applying the marginalization formula to the time update $p(x_{k+1}) = \int p(x_{k:k+1})dx_k = \int p(x_{k+1}|x_k)p(x_k)dx_k$. By plugging in the particle approximation $\hat{p}(x_k|y_{1:k}) = \sum_{i=1}^{N} w_{k|k}^i \delta(x_k - x_k^i)$, and scaling the result with the mixture $\sum_{i=1}^{N} q(x_{k+1}|x_k^i, y_{k+1})$ as proposal distribution according to the importance sampling principle, gives

Algorithm 9.1 Particle Filter

Choose a proposal distribution $q(x_{k+1}|x_{1:k}, y_{k+1})$, resampling strategy and the number of particles N.

Initialization: Generate $x_1^i \sim p_{x_0}, i = 1, \ldots, N$ and let $w_{1|0}^i = 1/N$.

Iteration: For $k = 1, 2, \ldots$.

1. *Measurement update:* For $i = 1, 2, \ldots, N$,

$$w_{k|k}^i = \frac{1}{c_k} w_{k|k-1}^i p(y_k|x_k^i), \qquad (9.20a)$$

where the normalization weight is given by

$$c_k = \sum_{i=1}^N w_{k|k-1}^i p(y_k|x_k^i) \qquad (9.20b)$$

2. *Estimation:* The filtering density is approximated by $\hat{p}(x_{1:k}|y_{1:k}) = \sum_{i=1}^N w_{k|k}^i \delta(x_{1:k} - x_{1:k}^i)$ and the mean (9.9a) is approximated by $\hat{x}_{1:k} \approx \sum_{i=1}^N w_{k|k}^i x_{1:k}^i$

3. *Resampling:* Optionally at each time, take N samples with replacement from the set $\{x_{1:k}^i\}_{i=1}^N$ where the probability to take sample i is $w_{k|k}^i$ and let $w_{k|k}^i = 1/N$.

4. *Time update:* Generate predictions according to the proposal distribution

$$x_{k+1}^i \sim q(x_{k+1}|x_k^i, y_{k+1}) \qquad (9.20c)$$

and compensate for the importance weight

$$w_{k+1|k}^i = w_{k|k}^i \frac{p(x_{k+1}^i|x_k^i)}{q(x_{k+1}^i|x_k^i, y_{k+1})}, \qquad (9.20d)$$

$$w_{k+1|k}^j \propto \frac{\sum_{i=1}^{N} w_{k|k}^i p(x_{k+1}^j | x_k^i)}{\sum_{i=1}^{N} q(x_{k+1}^j | x_k^i, y_{k+1})}. \quad (9.21a)$$

Note that the marginal distribution is functionally of the same form as (9.8c) in the PMF. This solution takes all paths leading to x_{k+1}^j into account, similarly to (9.13) in the PMF. It should also be noted that in the SIR-PF, $q(x_{k+1}^j | x_k^i, y_{k+1}) = w_{k|k}^i p(x_{k+1}^j | x_k^i)$, so the all weights become the same. Thus, there is no point to marginalize SIR-PF.

The *marginal particle filter* can be implemented just as Algorithm 9.1 by replacing the time update of the weights with (9.21a). Note that the complexity increases from $\mathcal{O}(N)$ in the PF to $\mathcal{O}(N^2)$ in the marginal PF, due to the new importance weight. A method with $\mathcal{O}(N \log(N))$ complexity is suggested in Klaas (2005).

The marginal particle filter has found very interesting applications in system identification, where a gradient search for unknown parameters in the model is applied Poyiadjis et al. (2005, 2006). The same parametric approach has been suggested for SLAM in Martinez-Cantin et al. (2007) and optimal trajectory planning in Sing et al. (2007).

We now turn our attention from filtering to smoothing. Though the PF appears to solve the smoothing problem for free, the inherent depletion problem of the history complicates the task, since the number of surviving trajectories with a time lag will quickly be depleted. For fixed-lag smoothing $p(x_{k-m:k}|y_{1:k})$, one can compute the same kind of marginal distributions as for the marginal particle filter leading to another compensation factor of the importance weight. However, the complexity will then be $\mathcal{O}(N^{m+1})$. Similar to the Kalman filter smoothing problem, the suggested solution Doucet et al. (2000b) is based on first running the particle filter in the usual way, and then apply a backward sweep of a modified particle filter. The set of particles generated in the forward filter is kept intact, and only the weights are modified. The backward recursion for the smoothing weights $w_{k|T}^i$, initialized at time $k = T$ with $w_{T|T}^i$, is given by

$$w_{k|T}^i = w_{k|k}^i \sum_{j=1}^{N} \frac{w_{k+1|T}^j p(x_{k+1}^j | x_k^i)}{\sum_{l=1}^{N} w_{k|k}^l p(x_{k+1}^j | x_k^l)}. \quad (9.21b)$$

This is similar to (9.21a), where all paths from $x_{k+1}^j, j = 1, \ldots, N$, to a specific x_k^i are considered. This is called the *forward filter backward smoother*. The denominator in the sum is chosen such that the weights are automatically normalized, $\sum_{i=1}^{N} w_{k|T}^i = 1$. Due to the inner summation, the complexity is $\mathcal{O}(N^2)$.

Another smoothing principle is called the *forward filter backward simulation* Godsill et al. (2004). Here, the path from a specific x_{k+1}^i is taken at random from x_k^j, $j = 1, \ldots, N$ using the accept-reject sampling principle, see Section B.2. Consider first the probability of passing from x_{k+1}^i to x_k^j given by

$$w_{k|T}^{i,j} = \frac{w_{k|k}^j p(x_{k+1}^i | x_k^j)}{\sum_{l=1}^N w_{k|k}^l p(x_{k+1}^i | x_k^l)}. \tag{9.21c}$$

This is the target distribution from which we would like to sample a j from. However, sampling from this would require quadratic complexity, since there are N^2 weights.

A clever way to avoid to compute all these N^2 weights is suggested in Douc et al. (2010). The idea is to sample from a proposal density we already have, namely the filter weights $w_{k|k}^j$. We accept the sample j with a probability that is proportional to the ratio of target and proposal densities,

$$\frac{1}{M} \frac{w_{k|T}^{i,j}}{w_{k|k}^j} = \frac{1}{M} \frac{p(x_{k+1}^i | x_k^j)}{\sum_{l=1}^N w_{k|k}^l p(x_{k+1}^i | x_k^l)}. \tag{9.21d}$$

The constant has to satisfy $M \geq p(x_{k+1}^i | x_k^j)$ for all i, j. Note that the denominator is independent of j, and can be considered as a common constant in the target and proposal distributions, and thus it does not need to be part of M. The number of backward trajectories does not need to be the same as the number of particles N, and might be smaller than N. The complexity is linear in N in average, though it should be noted that the number of rejections is not bounded in theory.

Prediction to get $p(x_{1:k+m} | y_{1:k})$ can be implemented by repeating the time update in Algorithm 9.1 m times.

9.3.4 Reading Advice

The reader may at this stage continue to Section 9.9 to see MATLAB™ code for some illustrative toy examples, or Section 16.3 to read about the result and experience in some applications, or proceed to the subsequent sections that discuss the following issues:

- The tuning possibilities and different versions of the basic PF are discussed in Section 9.4.

- The choice of proposal distribution is crucial for performance, just as in any classical sampling algorithm Robert and Casella (1999), and this is discussed in Section 9.5.

- Performance in terms of convergence of the approximation $\hat{p}(x_{1:k} | y_{1:k}) \to p(x_{1:k} | y_{1:k})$ as $N \to \infty$ and relation to fundamental performance bounds are discussed in Section 9.6.

- The particle filter is computationally quite complex, and some potential bottlenecks and possible remedies are discussed in Section 9.7.

9.4 Tuning

The number of particles N is the most immediate design parameter in the PF. There are a few other degrees of freedom discussed below. The overall goal is to avoid sample depletion, which means that only a few particles, or even only one, contribute to the state estimate. The choice of proposal distribution is the most intricate one, and it is discussed separately in Section 9.5. How the resampling strategy affects sample depletion is discussed in Section 9.4.1. The effective number of samples in Section 9.4.2 is an indicator of sample depletion in that it measures how efficiently the PF is utilizing its particles. It can be used to design proposal distributions, depletion mitigation tricks, resampling algorithms and also to choose the number of particles. It can also be used as an on-line control variable for when to resample. Some dedicated tricks are discussed in Section 9.4.3.

9.4.1 Resampling

Without the resampling step, the basic particle filter would suffer from sample depletion. This means that after a while all particles but a few ones will have negligible weights. Resampling solves this problem, but creates another one in that resampling inevitably destroys information and thus increases uncertainty by the random sampling. It is therefore of interest to start the resampling process only when it is really needed. The following options for when to resample are possible:

- The standard version of Algorithm 9.1 is termed *sampling importance resampling* (*SIR*), or *bootstrap PF*, and is obtained by resampling each time.

- The alternative is to use *importance sampling*, in which case resampling is performed only when needed. This is called *sampling importance sampling* (*SIS*). Usually, resampling is done when the effective number of samples, as will be defined in the next section, becomes too small.

As an alternative, the resampling step can be replaced with a sampling step from a distribution that is fitted to the particles after both the time and measurement update. The Gaussian particle filter (GPF) in Kotecha and Djuric (2003a) fits a Gaussian distribution to the particle cloud, after which a new set of particles is generated from this distribution. The Gaussian sum particle filter (GSPF) in Kotecha and Djuric (2003b) uses a Gaussian sum instead.

9.4.2 Effective Number of Samples

An indicator of the degree of depletion is the *effective number of samples*[1], defined in terms of the *coefficient of variation* c_v Kong et al. (1994); Liu (1996); Liu and Chen (1998) as

$$N_{\text{eff}} = \frac{N}{1 + c_v^2(w_{k|k}^i)} = \frac{N}{1 + \frac{\text{Var}(w_{k|k}^i)}{\left(\text{E}(w_{k|k}^i)\right)^2}} = \frac{N}{1 + N^2 \text{Var}(w_{k|k}^i)}. \quad (9.22\text{a})$$

The effective number of samples is thus at its maximum $N_{\text{eff}} = N$ when all weights are equal $w_{k|k}^i = 1/N$, and the lowest value it can attain is $N_{\text{eff}} = 1$, which occurs when $w_{k|k}^i = 1$ with probability $1/N$ and $w_{k|k}^i = 0$ with probability $(N-1)/N$.

A logical computable approximation of N_{eff} is provided by

$$\hat{N}_{\text{eff}} = \frac{1}{\sum_i (w_{k|k}^i)^2}. \quad (9.22\text{b})$$

This approximation shares the property $1 \leq \hat{N}_{\text{eff}} \leq N$ with the definition (9.22a). The upper bound $\hat{N}_{\text{eff}} = N$ is attained when all particles have the same weight, and the lower bound $\hat{N}_{\text{eff}} = 1$ when all the probability mass is devoted to a single particle.

The resampling condition in the PF can now be defined as $\hat{N}_{\text{eff}} < N_{\text{th}}$. The threshold can for instance be chosen as $\hat{N}_{\text{th}} = 2N/3$.

9.4.3 Tricks to Mitigate Sample Depletion

The choice of proposal distribution and resampling strategy are the two available instruments in theory to avoid sample depletion problems. There are also some simple and more practical *ad-hoc* tricks that can be tried as will be discussed below.

One important trick is to modify the noise models so the state noise and/or the measurement noise appear larger in the filter than they really are in the data generating process. This technique is called *jittering* in Fearnhead (1998), but a similar approach was introduced in Gordon et al. (1993) under the name *roughening*. Increasing the noise level in the state model (9.3a) increases the support of the sampled particles which partly mitigates the depletion problem. Further, increasing the noise level in the observation model (9.3b) implies that the likelihood decays slower for particles that do not fit the observation, and the chance to resample these increases.

---

[1] Note that the literature often defines the effective number of samples as $\frac{N}{1+\text{Var}(w_{k|k}^i)}$, which is incorrect.

In Doucet et al. (2001b), the depletion problem is handled by introducing an additional Markov Chain Monte Carlo (MCMC) step to separate the samples.

In Gordon et al. (1993), the so-called *prior editing* method is discussed. The estimation problem is delayed one time-step, so that the likelihood can be evaluated at the next time step. The idea is to reject particles with sufficiently small likelihood values, since they are not likely to be resampled. The update step is repeated until a feasible likelihood value is received. The roughening method could also be applied before the update step is invoked. The *auxiliary particle filter* Pitt and Shephard (1999) is a more formal way to sample in such a way that only particles associated with large predictive likelihoods are considered, see Section 9.5.6.

Another technique is *regularization*. The basic idea is to convolve each particle with a diffusion kernel with a certain bandwidth before resampling. This will prevent multiple copies of a few particles. One may for instance use a Gaussian kernel where the variance acts as the bandwidth. One problem in theory with this approach is that this kernel will increase the variance of the posterior distribution.

9.5 Choice of Proposal Distribution

In this section, we focus on the choice of proposal distribution, which influences the depletion problem a lot, and we will here outline available options with some comments on when they are suitable.

First, note that the general proposal distribution has the form $q(x_{1:k}|y_{1:k})$. This means that the whole *trajectory* should be sampled at each iteration, which is clearly not attractive in real-time applications. Now, the general proposal can be factorized as

$$q(x_{1:k}|y_{1:k}) = q(x_k|x_{1:k-1}, y_{1:k})q(x_{1:k-1}|y_{1:k}). \qquad (9.23)$$

The most common approximation in applications is to reuse the path $x_{1:k-1}$ and only sample the new state x_k, so the proposal $q(x_{1:k}|y_{1:k})$ is replaced by $q(x_k|x_{1:k-1}, y_{1:k})$. The approximate proposal suggests good values of x_k only, not of the trajectory $x_{1:k}$. For filtering problems this is not an issue, but for smoothing problems the second factor becomes important. Here, the idea of block sampling Doucet et al. (2006) is quite interesting.

Now, the proposal $q(x_k|x_{1:k-1}, y_{1:k})$ can due to the Markov property of the model be written as

$$q(x_k|x_{1:k-1}, y_{1:k}) = q(x_k|x_{k-1}, y_k). \qquad (9.24)$$

The following sections discuss various approximations of this proposal, and in particular how the choice of proposal depends on the *signal to noise ratio* (*SNR*). For linear Gaussian models, the SNR is in loose term defined as

9.5 Choice of Proposal Distribution

$\|Q\|/\|R\|$. That is, the SNR is high if the measurement noise is small compared to the signal noise. Here, we define SNR as the ratio of the maximal value of the likelihood and prior, respectively,

$$\text{SNR} \propto \frac{\max_{x_k} p(y_k|x_k)}{\max_{x_k} p(x_k|x_{k-1})}. \tag{9.25}$$

For a linear Gaussian model, this gives $\text{SNR} \propto \sqrt{\det(Q)/\det(R)}$.

We will in this section use the weight update

$$w_{k|k}^i \propto w_{k-1|k-1}^i \frac{p(y_k|x_k^i)p(x_k^i|x_{k-1}^i)}{q(x_k^i|x_{k-1}^i, y_k)}, \tag{9.26}$$

combining (9.20ad). The SNR thus indicates which factor in the numerator that most likely changes the weights the most.

Besides the options below that all relate to (9.24), there are many other more ad-hoc based options described in literature. For instance one idea is to run an EKF or UKF in parallel, and use the posterior Gaussian distribution from this filter as a proposal.

9.5.1 Optimal Sampling

The conditional distribution includes all information of the previous state and the current observation, and should thus be the best proposal to sample from. This conditional PDF can be written as

$$q(x_k|x_{k-1}^i, y_k) = p(x_k|x_{k-1}^i, y_k) = \frac{p(y_k|x_k)p(x_k|x_{k-1}^i)}{p(y_k|x_{k-1}^i)}. \tag{9.27a}$$

This choice gives the proposal weight update

$$w_{k|k}^i \propto w_{k-1|k-1}^i p(y_k|x_{k-1}^i). \tag{9.27b}$$

The point is that the weight will be the same whatever sample of x_k^i is generated. Put in another way, the variance of the weights is unaffected by the sampling. All other alternatives will add variance to the weights and thus decrease the effective number of samples according to (9.22a). In the interpretation of keeping the effective number of samples as large as possible, (9.27a) is the *optimal sampling*.

The drawbacks are as follows:

- It is generally hard to sample from this proposal distribution.
- It is generally hard to compute the weight update needed for this proposal distribution, since it would require to integrate over the whole state space, $p(y_k|x_{k-1}^i) = \int p(y_k|x_k)p(x_k|x_{k-1}^i)\,dx_k$.

One important special case when these steps actually become explicit is for a linear and Gaussian measurement relation, which is the subject of Section 9.5.5.

9.5.2 Prior Sampling

The standard choice in Algorithm 9.1 is to use the conditional prior of the state vector as proposal distribution,

$$q(x_k|x_{k-1}^i, y_k) = p(x_k|x_{k-1}^i), \quad (9.28a)$$

where $p(x_k|x_{k-1}^i)$ will be referred to as the prior of x_k for each trajectory. This yields

$$w_{k|k}^i \propto w_{k|k-1}^i p(y_k|x_k^i) = w_{k-1|k-1}^i p(y_k|x_k^i). \quad (9.28b)$$

This leads to the by far most common version of the PF that was originally proposed in Gordon et al. (1993). It performs well when the SNR is small, which means that the state dynamics provide accurate predictions of the state. In such cases, it is more natural to sample from the likelihood.

9.5.3 Likelihood Sampling

Consider first the factorization

$$p(x_k|x_{k-1}^i, y_k) = p(y_k|x_{k-1}^i, x_k)\frac{p(x_k|x_{k-1}^i)}{p(y_k|x_{k-1}^i)} = p(y_k|x_k)\frac{p(x_k|x_{k-1}^i)}{p(y_k|x_{k-1}^i)}. \quad (9.29a)$$

If the likelihood $p(y_k|x_k)$ is much more peaky than the prior and if it is integrable in x_k Thrun et al. (2001), then

$$p(x_k|x_{k-1}^i, y_k) \propto p(y_k|x_k). \quad (9.29b)$$

That is, a suitable proposal for the high SNR case is based on a scaled likelihood function

$$q(x_k|x_{k-1}^i, y_k) \propto p(y_k|x_k), \quad (9.29c)$$

which yields

$$w_{k|k}^i = w_{k-1|k-1}^i p(x_k^i|x_{k-1}^i). \quad (9.29d)$$

Sampling from the likelihood requires that the likelihood function $p(y_k|x_k)$ is integrable with respect to x_k Thrun et al. (2001). This is not the case when $n_x > n_y$. The interpretation in this case is that for each value of y_k, there is a manifold of possible x_k to sample from, each one equally likely.

9.5.4 Illustrations

A simple linear Gaussian model is used to illustrate the choice of proposal as a function of SNR. Figure 9.1 illustrates a high SNR case for a scalar model, where the information in the prior is negligible compared to the peaky likelihood. This means that the optimal proposal essentially becomes (a scaled version of) the likelihood.

Figure 9.2 illustrates a high SNR case for a two-dimensional state, where the observation dimension is smaller than the state space. The optimal proposal can here be interpreted as the intersection of the prior and likelihood.

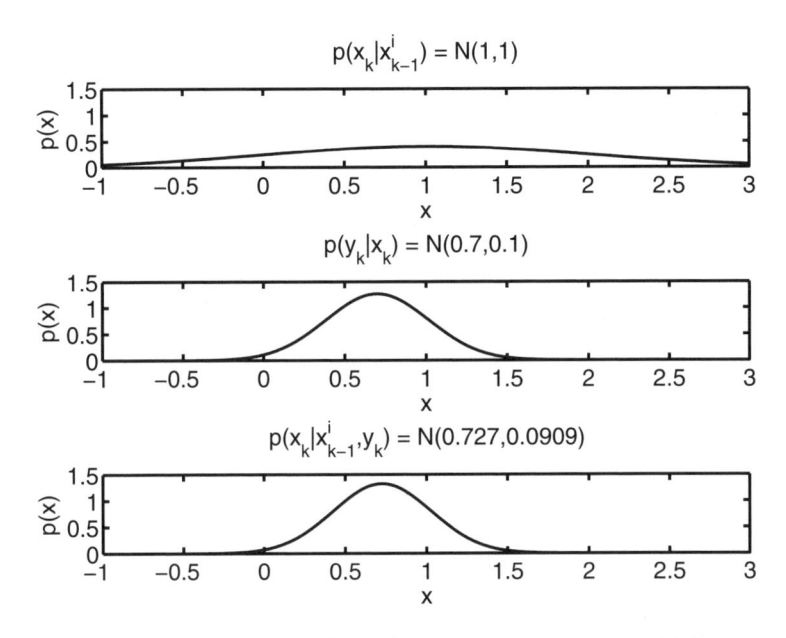

Figure 9.1: *Illustration of (9.27a) for a scalar state and observation model. The state dynamics moves the particle to $x_k = 1$ and adds uncertainty with variance 1, after which an observation $y_k = 0.7 = x_k + e_k$ is taken. The posterior in this high SNR example is here essentially equal to the likelihood.*

9.5.5 Optimal Sampling with Linearized Likelihood

The principles illustrated in Figures 9.1 and 9.2 can be used for a linearized model Doucet et al. (2000b), similar to the measurement update in the EKF (9.6ef). To simplify the notation somewhat, the process noise in (9.3a) is assumed additive $x_{k+1} = f(x_k) + v_k$. Assuming that the measurement relation (9.3b) is linearized as (9.5b) when evaluating (9.27a), the optimal proposal

230 The Particle Filter

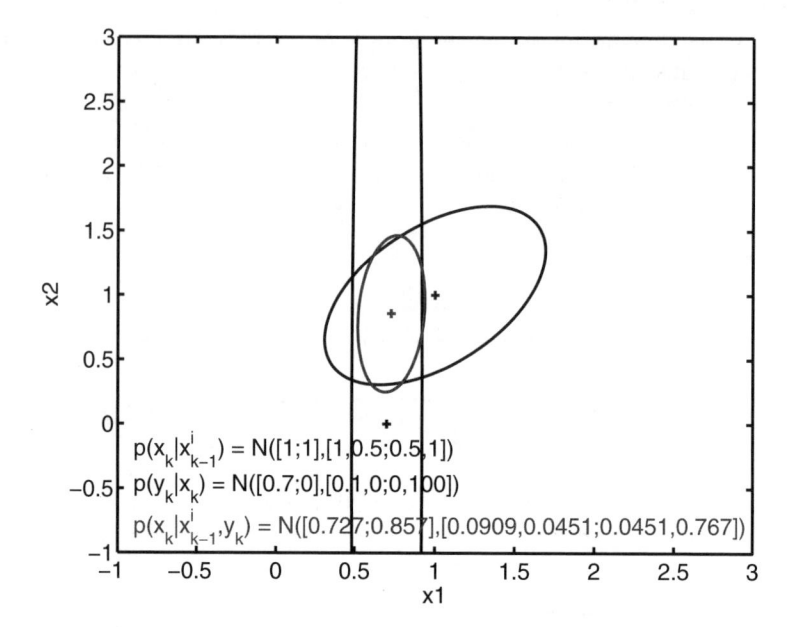

$p(x_k|x_{k-1}^i) = N([1;1],[1,0.5;0.5,1])$
$p(y_k|x_k) = N([0.7;0],[0.1,0;0,100])$
$p(x_k|x_{k-1}^i,y_k) = N([0.727;0.857],[0.0909,0.0451;0.0451,0.767])$

Figure 9.2: *Illustration of (9.27a) for a two-dimensional state and scalar observation model. The state dynamics moves the particle to $x_k = (1,1)^T$ and adds correlated noise, after which an observation $y_k = 0.7 = (1,0)x_k + e_k$ is taken. The posterior in this high SNR example corresponds roughly to the likelihood in one dimension (x_1) and the prior in the other dimension (x_2).*

can be approximated with

$$q(x_k|x_{k-1}^i, y_k) = \mathcal{N}\left(f(x_{k-1}^i) + K_k^i(y_k - \hat{y}_k^i), \left(H_k^{i,T} R_k^\dagger H^i + Q_{k-1}^\dagger\right)^\dagger\right), \quad (9.30a)$$

where † denotes pseudo-inverse. The Kalman gain, linearized measurement model and measurement prediction, respectively, are given by

$$K_k^i = Q_{k-1} H_k^{i,T} \left(H_k^i Q_{k-1} H_k^{i,T} + R_k\right)^{-1}, \quad (9.30b)$$

$$H_k^i = \left.\frac{\partial h(x_k)}{\partial x_k}\right|_{x_k = f(x_{k-1}^i)}, \quad (9.30c)$$

$$\hat{y}_k^i = h\left(f(x_{k-1}^i)\right). \quad (9.30d)$$

9.5 Choice of Proposal Distribution

The weights should thus be multiplied by the following likelihood in the measurement update,

$$p(y_k|x_{k-1}^i) = \mathcal{N}(y_k - \hat{y}_k^i, H_k^i Q_{k-1} H_k^{i,T} + R_k). \tag{9.30e}$$

The modifications of (9.30) can be motivated intuitively as follows. At time $k-1$, each particle corresponds to a state estimate with no uncertainty. The EKF recursions (9.6) using this initial value gives

$$\hat{x}_{k-1|k-1} \sim \mathcal{N}(x_{k-1}^i, 0) \Rightarrow \tag{9.31a}$$
$$\hat{x}_{k|k-1} = f(x_{k-1}^i), \tag{9.31b}$$
$$P_{k|k-1} = Q_{k-1}, \tag{9.31c}$$
$$K_k = Q_{k-1} H_k^T \left(H_k Q_{k-1} H_k^T + R_k \right)^{-1}, \tag{9.31d}$$
$$\hat{x}_{k|k} = \hat{x}_{k|k-1} + K_k(y_k - h(\hat{x}_{k|k-1})), \tag{9.31e}$$
$$P_{k|k} = Q_{k-1} - K_k H_k Q_{k-1}. \tag{9.31f}$$

The modification in Algorithm 9.1, assuming a Gaussian distribution for both process and measurement noise, is to make the following substitution in the time update

$$x_{k+1}^i = f(x_k^i) + v_k^i, \tag{9.32a}$$
$$\text{SIR}: \quad v_k^i \sim \mathcal{N}(0, Q_k), \tag{9.32b}$$
$$\text{OPT} - \text{EKF}: \quad v_k^i \in \mathcal{N}\left(K_{k+1}^i \left(y_{k+1} - h\left(f(x_k^i) \right) \right), \left(H_{k+1}^{i,T} R_{k+1}^\dagger H_{k+1}^i + Q_k^\dagger \right)^\dagger \right). \tag{9.32c}$$

and measurement update

$$\text{SIR}: \quad w_{k|k}^i = w_{k-1|k-1}^i \mathcal{N}(y_k - h(x_k^i), R_k), \tag{9.32d}$$
$$\text{OPT} - \text{EKF}: \quad w_{k|k}^i = w_{k-1|k-1}^i \mathcal{N}\left(y_k - h\left(f(x_{k-1}^i) \right), H_k^i Q_{k-1} H_k^{i,T} + R_k \right), \tag{9.32e}$$

respectively. For OPT-EKF, the SNR definition can be more precisely stated as

$$\text{SNR} \propto \frac{\|H_k^i Q_{k-1} H_k^{i,T}\|}{\|R_k\|}. \tag{9.33}$$

We make the following observations and interpretations on some limiting cases of these algebraic expressions:

- For small SNR, $K_k^i \approx 0$ in (9.30b) and $\left(H_k^{i,T} R_k^\dagger H_k^i + Q_{k-1}^\dagger \right)^\dagger \approx Q_{k-1}$ in (9.32c), which shows that the resampling (9.32c) in OPT-EKF proposal approaches (9.32b) in SIR as the SNR goes to zero. That is, for low SNR the approximation approaches prior sampling in Section 9.5.2.

- Conversely, for large SNR and assuming H_k^i invertible (implicitly implying $n_y \geq n_x$), then $\left(H_k^{i,T} R_k^\dagger H_k^i + Q_{k-1}^\dagger\right)^\dagger \approx H_k^{i,-1} R_k H_k^{i,-T}$ in (9.32c). Here, all information about the state is taken from the measurement, and the model is not used. That is, for high SNR the approximation approaches likelihood sampling in Section 9.5.3.

- The pseudo-inverse † is used consequently in the notation for the proposal covariance $\left(H_k^{i,T} R_k^\dagger H_k^i + Q_{k-1}^\dagger\right)^\dagger$ instead of inverse to accomodate the following cases:

    - Singular process noise Q_{k-1}, which is the case in most dynamic models including integrated noise.
    - Singular measurement noise R_k, to allow ficticious measurements that model state constraints. For instance, a known state constraint corresponds to infinite information in a subspace of the state space, and the corresponding eigenvector of the measurement information $H_k^i R_k^\dagger H_k^{i,T}$ will overwrite the prior information $Q_{k-1}^\dagger$.

- The EKF boils down to a KF for a linear measurement equation with Gaussian noise and with additive Gaussian process noise. That is, this is a sufficient (but not necessary) condition for the optimal proposal to exist analytically.

9.5.6 Auxiliary Sampling

The *auxiliary sampling proposal resampling filter* Pitt and Shephard (1999) uses an auxiliary index in the proposal distribution $q(x_k, i | y_{1:k})$. This leads to an algorithm that first generates a large number M (typically $M = 10N$) of pairs $\{x_k^j, i^j\}_{j=1}^M$. From Bayes' rule, we have

$$p(x_k, i | y_{1:k}) \sim p(y_k | x_k) p(x_k, i | y_{1:k-1}) \qquad (9.34a)$$
$$= p(y_k | x_k) p(x_k | i, y_{1:k-1}) p(i | y_{1:k-1}) \qquad (9.34b)$$
$$= p(y_k | x_k) p(x_k | x_{k-1}^i) w_{k-1|k-1}^i. \qquad (9.34c)$$

This density is implicit in x_k and thus not useful as an proposal density, since it requires x_k to be known. The general idea is to find an approximation of $p(y_k | x_{k-1}^i) = \int p(y_k | x_k) p(x_k | x_{k-1}^i) dx_k$. A simple though useful approximation is to replace x_k with its estimate and thus let $p(y_k | x_{k-1}^i) = p(y_k | \hat{x}_k^i)$ above. This leads to the proposal

$$q(x_k, i | y_{1:k}) = p(y_k | \hat{x}_k^i) p(x_k | x_{k-1}^i) w_{k-1|k-1}^i. \qquad (9.34d)$$

9.6 Theoretical Performance

Here, $\hat{x}_k^i = \mathrm{E}(x_k|x_{k-1}^i)$ can be the conditional mean, or $\hat{x}_k^i \sim p(x_k|x_{k-1}^i)$ a sample from the prior. The new samples are drawn from the marginalized density

$$x_k^j \sim p(x_k|y_{1:k}) = \sum_i p(x_k, i|y_{1:k}). \qquad (9.34\mathrm{e})$$

To evalute the proposal weight, first Bayes rule gives

$$q(x_k, i|y_{1:k}) = q(i|y_{1:k})q(x_k|i, y_{1:k}). \qquad (9.34\mathrm{f})$$

Here, another choice is needed. The latter proposal factor should be defined as

$$q(x_k|i, y_{1:k}) = p(x_k|x_{k-1}^i). \qquad (9.34\mathrm{g})$$

Then, this factor cancels out when forming

$$q(i|y_{1:k}) \propto p(y_k|\hat{x}_k^i)w_{k-1|k-1}^i. \qquad (9.34\mathrm{h})$$

The new weights are thus given by

$$w_{k|k}^i = w_{k-1|k-1}^{i^j} \frac{p(y_k|x_k^j)p(x_k^j|x_{k-1}^{i^j})}{q(x_k^j, i^j|y_{1:k})}. \qquad (9.34\mathrm{i})$$

Note that this proposal distribution is a product of the prior and the likelihood. The likelihood has the ability to punish samples x_k^i that give a poor match to the most current observation, unlike SIR and SIS where such samples are drawn and then immediately rejected. There is a link between the auxiliary PF and the standard SIR as pointed out in Johansen and Doucet (2008), which is useful for understanding its theoretical properties.

9.6 Theoretical Performance

The key questions here are how well the PF filtering density $\hat{p}(x_{1:k}|y_{1:k})$ approximates the true posterior $p(x_{1:k}|y_{1:k})$, and what the fundamental mean square error bounds for the true posterior are.

9.6.1 Convergence Issues

The convergence properties of the PF are well understood on a theoretical level, see the survey Crisan and Doucet (2000) and the book Moral (2004). The key question is how well a function $g(x_k)$ of the state can be approximated

$\hat{g}(x_k)$ by the PF compared to the conditional expectation $\mathrm{E}(g(x_k))$, where

$$\mathrm{E}(g(x_k)) = \int g(x_k) p(x_{1:k}|y_{1:k}) \, dx_{1:k}, \tag{9.35a}$$

$$\hat{g}(x_k) = \int g(x_k) \hat{p}(x_{1:k}|y_{1:k}) \, dx_{1:k} = \sum_{i=1}^{N} w_{k|k}^i g(x_k^i). \tag{9.35b}$$

In short, the following key results exist:

- Almost sure weak convergence

$$\lim_{N \to \infty} \hat{p}(x_{1:k}|y_{1:k}) = p(x_{1:k}|y_{1:k}), \tag{9.36}$$

in the sense that $\lim_{N \to \infty} \hat{g}(x_k) = \mathrm{E}(g(x_k))$.

- Mean square error asymptotic convergence

$$\mathrm{E}\left(\hat{g}(x_k) - \mathrm{E}(g(x_k))\right)^2 \leq \frac{p_k \|g(x_k)\|_{sup}}{N}, \tag{9.37}$$

where the supremum norm of $g(x_k)$ is used. As shown in Moral (2004) using the Feynman-Kac formula, under certain regularity and mixing conditions, the constant $p_k = p < \infty$ does not increase in time. The main condition Crisan and Doucet (2000); Moral (2004) for this result is that the unnormalized weight function is bounded. Further, most convergence results as surveyed in Crisan and Doucet (2002) are restricted to bounded functions of the state $g(x)$ such that $|g(x)| < C$ for some C. The convergence result presented in Hu et al. (2008) extends this to unbounded functions, for instance estimation of the state itself $g(x) = x$, where the proof requires the additional assumption that the likelihood function is bounded from below by a constant.

In general, the constant p_k grows polynomially in time, but does not necessarily depend on the dimension of the state space, at least not explicitly. That is, in theory we can expect the same good performance for high order state vectors. In practice, the performance degrades quickly with the state dimension due to the curse of dimensionality. However, it scales much better with state dimension than the PMF, which is one of the key reasons for the success of the particle filter.

9.6.2 Nonlinear Filtering Performance Bound

Besides the performance bound of a specific algorithm as discussed in the previous section, there are more fundamental estimation bounds for nonlinear filtering that depend only on the model and not on the applied algorithm. The

Cramér-Rao Lower Bound (CRLB) $P_{k|k}$ provides such a performance bound for any unbiased estimator $\hat{x}_{k|k}$,

$$\mathrm{Cov}(\hat{x}_{k|k}) \geq P_{k|k}^{\mathrm{CRLB}}. \tag{9.38}$$

The most useful version of CRLB is computed recursively by a Riccati equation which has the same functional form as the Kalman filter in (9.6) evaluated at the true trajectory $x_{1:k}^o$,

$$P_{k|k}^{\mathrm{CRLB}} = P_{k|k-1}^{\mathrm{CRLB}} - P_{k|k-1}^{\mathrm{CRLB}} H(x_k^o)^T (H(x_k^o) P_{k|k-1}^{\mathrm{CRLB}} H^T(x_k^o) + R_k)^{-1} H(x_k^o) P_{k|k-1}^{\mathrm{CRLB}}, \tag{9.39a}$$

$$P_{k+1|k}^{\mathrm{CRLB}} = F(x_k^o) P_{k|k}^{\mathrm{CRLB}} F^T(x_k^o) + G(x_k^o) Q_k G(x_k^o)^T. \tag{9.39b}$$

The following remarks summarize the CRLB theory with respect to the PF:

- For a linear Gaussian model

$$x_{k+1} = F_k x_k + G_k v_k, \qquad v_k \sim \mathcal{N}(0, Q_k), \tag{9.40a}$$
$$y_k = H_k x_k + e_k, \qquad e_k \sim \mathcal{N}(0, R_k), \tag{9.40b}$$

  the Kalman filter covariance $P_{k|k}$ coincides with $P_{k|k}^{\mathrm{CRLB}}$. That is, the CRLB bound is attainable in the linear Gaussian case.

- In the linear non-Gaussian case, the covariances Q_k, R_k and P_0 are replaced with the inverse intrinsic accuracies $\mathcal{I}_{v_k}^{-1}$, $\mathcal{I}_{e_k}^{-1}$ and $\mathcal{I}_{x_0}^{-1}$, respectively. Intrinsic accuracy is defined as the Fisher information with respect to the location parameter, and the inverse intrinsic accuracy is always smaller than the covariance. As a consequence of this, the CRLB is always smaller for non-Gaussian noise than for Gaussian noise with the same covariance. See Hendeby (2008) for the details.

- The parametric CRLB is a function of the true state trajectory $x_{1:k}^o$ and can thus be computed only in simulations or when ground truth is available from a reference system.

- The posterior CRLB is the parametric CRLB averaged over all possible trajectories $P_{k|k}^{\mathrm{PostCRLB}} = \mathrm{E}\bigl(P_{k|k}^{\mathrm{ParCRLB}}\bigr)$. The expectation makes its computation quite complex in general.

- In the linear Gaussian case, the parametric and posterior bounds coincide.

- The covariance of the state estimate from the particle filter is bounded by the CRLB. The CRLB theory also says that the particle filter estimate asymptotically in both the number of particles and the information in the model (basically the signal to noise ratio) attains the CRLB bound.

Consult N. Bergman and Gordon (2001) for details on these issues.

9.7 Complexity Bottlenecks

It is instructive and recommended to generate a profile report from an implementation of the particle filter. Quite often, unexpected bottlenecks are discovered that can be improved with a little extra work.

9.7.1 Resampling

One real bottleneck is the *resampling* step. This crucial step has to be performed at least regularly when N_{eff} becomes too small. Resampling strategies are overviewed in Section B.4, here we focus on the implementation of *multinomial sampling*.

The resampling can be efficiently implemented using a classical algorithm for sampling N ordered independent identically distributed variables according to Ripley (1988), commonly referred to as Ripley's method:

```
function [x,w]=resample(x,w)
% Multinomial sampling with Ripley's method
u = cumprod(rand(1,N).^(1./[N:-1:1]));
u = fliplr(u);
wc = cumsum(w);
k=1;
for i=1:N
    while(wc(k)<u(i))
        k=k + 1;
    end
    ind(i)=k;
end
x=x(ind,:);
w=ones(1,N)./N;
```

The complexity of this algorithm is linear in the number of particles N, which cannot be beaten if the implementation is done at a sufficiently low level. This is for this reason the most frequently suggested algorithm also in the particle filter literature. However, in engineering programming languages as MATLAB^{TM}, vectorized computations are often an order of magnitude faster than code based on "for" and "while" loops.

The following code also implements the resampling needed in the particle filter by completely avoiding loops.

```
function [x,w]=resample(x,w)
% Multinomial sampling with sort
u = rand(N,1);
wc = cumsum(w);
wc=wc/wc(N);
[dum,ind1]=sort([u;wc]);
ind2=find(ind1<=N);
ind=ind2-(0:N-1)';
x=x(ind,:);
w=ones(1,N)./N;
```

This implementation relies on the efficient implementation of sort. Note that sorting is of complexity $N\log_2(N)$ for low level implementations, so in theory it should not be an alternative to Ripley's method for sufficiently large N.

9.7 Complexity Bottlenecks

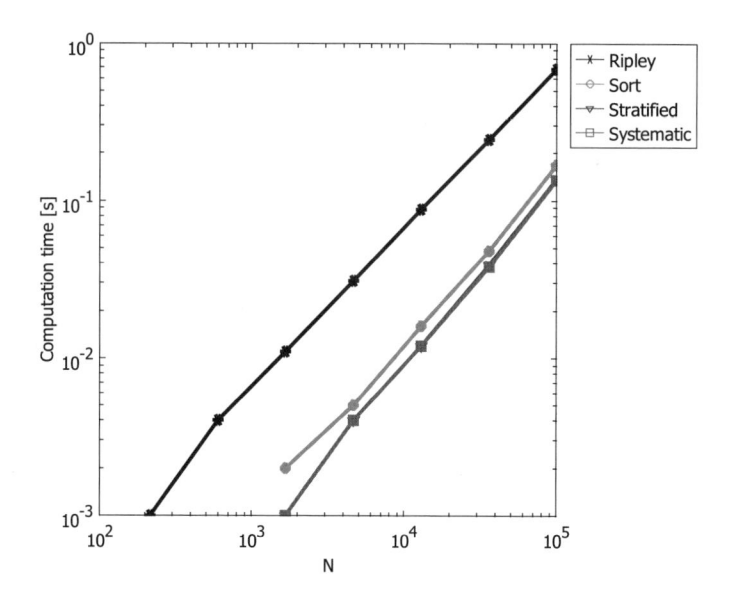

Figure 9.3: *Computational complexity in a vectorized language of different resampling algorithms.*

However, as Figure 9.3 illustrates, the sort algorithm is a factor of five faster for one instance of a vector oriented programming language. Using interpreters with loop optimization reduces this difference, but the sort algorithm is still an alternative.

Note that this code does not use the fact that wc is already ordered. The sorting gets further simplified if also the sequence of uniform numbers is ordered. This is one advantage of systematic or stratified sampling Kitagawa (1996), where the random number generation is replaced with one of the following lines:

```
% Stratified sampling
u=([0:N-1]'+(rand(N,1)))/N;
% Systematic sampling
u=([0:N-1]'+rand(1))/N;
```

Both the code based on sort and for, while are possible. Another advantage with these options is that the state space is more systematically covered, so there will not be any large uncovered volumes just by random.

9.7.2 Likelihood Evaluation and Iterated Measurement Updates

The likelihood evaluation can be a real bottleneck if not properly implemented. In case there are several independent sensors, an iterated measurement update can be performed. Denote the M sensor observations y_k^j, for $j = 1, 2, \ldots, M$. Then, independence directly gives

$$p(y_k|x_k) = \prod_{j=1}^{M} p(y_k^j|x_k). \qquad (9.41)$$

This trick is even simpler than the corresponding iterated measurement update in the Kalman filter.

However, this iterated update is not necessarily the most efficient implementation. One example is the multivariate Gaussian distribution for independent measurements

$$y_{k,j} = h_j(x_k^i) + e_{k,j}, \quad e_{k,j} \sim \mathcal{N}(0, R_{k,j}). \qquad (9.42)$$

The likelihood is given by

$$p(y_k|x_k^i) \propto e^{-0.5 \sum_{j=1}^{M} (y_{k,j} - h_j(x_k^i))^T R_{k,j}^{-1} (y_{k,j} - h_j(x_k^i))} \qquad (9.43a)$$

$$= \prod_{j=1}^{M} e^{-0.5 (y_{k,j} - h_j(x_k^i))^T R_{k,j}^{-1} (y_{k,j} - h_j(x_k^i))}. \qquad (9.43b)$$

The former equation with a sum should be used to avoid extensive calls to the exponential function. Even here, it is not trivial how to vectorize the calculations in the sum for all particles in parallel.

9.7.3 Time Update Sampling

Generating random numbers from non-standard proposals may be time consuming. Then, remembering that dithering is often a necessary practical trick to tune the PF, one should investigate proposals including dithering noise that are as simple as possible to sample from.

9.7.4 Function Evaluations

When all issues above have been dealt with, the only thing that remains is to evaluate the functions $f(x, v)$ and $h(x)$. These functions are evaluated a huge number of times, so it is worthwhile to spend time to optimize their implementation. An interesting idea is to implement these in dedicated hardware taylored to the application. This was done using analog hardware in Velmurugan et al. (2006) for an arctangens function, which is common in sensor models for bearing measurements.

9.7.5 PF versus EKF

The computational steps are compared to the Kalman filter in Table 9.1. The EKF requires only one function evaluation of $f(x, v)$ and $h(x)$ per time step, while the particle filter requires N evaluations. However, if the gradients are not available analytically in the EKF, then at least another n_x evaluations of both $f(x,v)$ and $h(x)$ are needed. These numbers increase when the step size of the numeric gradients are adaptive. Further, if the process noise is not additive, even more numerical derivatives are needed. However, the PF is still roughly a factor N/n_x more complex.

The most time consuming step in the Kalman filter is the Riccati recursion of the matrix P. Here, either the matrix multiplication FP in the time update or the matrix inversion in the measurement update are dominating for large enough models. Neither of these are needed in the particle filter. The time update of the state is the same.

The complexity of a matrix inversion using state of the art algorithms Coppersmith and Winograd (1990) is $\mathcal{O}(n_y^{2.376})$. The matrix inversion in the measurement update can be avoided by using the iterated measurement update. The condition is that the covariance matrix R_k is (block-) diagonal.

As a first order approximation for large n_x, the Kalman filter is $\mathcal{O}(n_x^3)$ from the matrix multiplication FP, while the particle filter is $\mathcal{O}(Nn_x^2)$ for a typical dynamic model where all elements of $f(x,v)$ depend on all states, for instance the linear model $f(x,v) = Fx + v$. Also from this perspective, the PF is a factor N/n_x computationally more demanding than the EKF.

Table 9.1: *Comparison of EKF in (9.6) and SIR-PF in (9.20): Main computational steps.*

Algorithm	Extended Kalman filter	Particle filter
Time update	$F = \frac{\partial f(x,v)}{\partial x}$, $G = \frac{\partial f(x,v)}{\partial v}$ $x := f(x,0)$ $P := FPF^T + GQG^T$	$v^i \sim p_v$ $x^i := f(x^i, v^i)$
Measurement update	$H = \frac{\partial h(x)}{\partial x}$ $K = PH^T(HPH^T + R)^{-1}$ $x := x + K(y - h(x))$ $P := P - KHP$	$w^i := w^i p_e(y - h(x^i))$
Estimation	$\hat{x} = x$	$\hat{x} = \sum_{i=1}^N w^i x^i$
Resampling	—	$x^i \sim \sum_{j=1}^N w^j \delta(x - x^j)$

9.8 Marginalized Particle Filter Theory

The main purpose of the *marginalized particle filter* (*MPF*) is to keep the state dimension small enough for the PF to be feasible. The resulting filter is called the MPF or the Rao-Blackwellized particle filter, and it has been known for quite some time under different names, see e.g., Andrieu and Doucet (2002);

Casella and Robert (1996); Chen and Liu (2000); Doucet et al. (2000a, 2001b); Nordlund and Gustafsson (2009); Schön et al. (2005).

The MPF utilizes possible linear Gaussian sub-structures in the model (9.3). The state vector is assumed partitioned as $x_k = ((x_k^n)^T, (x_k^l)^T)^T$, where x_k^l enters both the dynamic model and the observation model linearly. We will a bit informally refer to x_k^l as the linear state and x_k^n as the nonlinear state, respectively. MPF essentially represents x_k^n with particles, and applies one Kalman filter per particle that provides the conditional distribution for x_k^l conditioned on the trajectory $x_{1:k}^n$ of nonlinear states and the past observations.

The example below extends Example 9.1 with a speed state, releasing the assumption of having access to a speedometer, and the resulting model is linear in the speed.

─── **Example 9.2: Marginalization in 1D terrain navigation** ───

We here consider a one-dimensional terrain navigation problem. Assume here that the velocity cannot be measured, and thus has to be estimated in the filter using the model

$$x_{k+1} = x_k + u_k + \frac{T_s^2}{2} v_k, \tag{9.44a}$$

$$u_{k+1} = u_k + T_s v_k, \tag{9.44b}$$

$$y_k = h(x_k) + e_k, \tag{9.44c}$$

$$v_k \sim \mathcal{N}(0, Q_k), \tag{9.44d}$$

$$e_k \sim \mathcal{N}(0, R_k). \tag{9.44e}$$

The augmented state vector $\bar{x} = (x, u)^T$ now contains the velocity as a conditionally linear Gaussian system. That is, given a known position trajectory $x_{1:k}$, the model for the remaining state u_k is

$$u_{k+1} = u_k + T_s v_k, \tag{9.44f}$$

$$x_{k+1} - x_k = u_k + \frac{T_s^2}{2} v_k. \tag{9.44g}$$

The second equation should be interpreted as a measurement relation. That is, a linear Gaussian model (with correlated process and measurement noise), where the Kalman filter applies.

On the other hand, given a Gaussian conditional posterior of the linear state, the remaining part of the model reads

$$x_{k+1} = x_k + \hat{u}_{k|k} + \frac{T_s^2}{2} v_k, \quad \text{Cov}(\hat{u}_k) = P_{k|k} \tag{9.44h}$$

$$y_k = h(x_k) + e_k. \tag{9.44i}$$

The linear state estimate can be seen as an input with an extra process noise.

The position can be estimated using a fixed position grid, which leads to the marginalized PMF, or a stochastic position grid as in the marginalized particle filter.

The example is straightforwardly generalized to two-dimensional terrain navigation for airborne or underwater navigation. It is also almost for free to add an extra state for acceleration, which also becomes conditionally linear.

9.8.1 Model Structure

A rather general model, containing a conditionally linear Gaussian sub-structure is given by

$$x_{k+1}^n = f_k^n(x_k^n) + F_k^n(x_k^n)x_k^l + G_k^n(x_k^n)v_k^n, \tag{9.45a}$$

$$x_{k+1}^l = f_k^l(x_k^n) + F_k^l(x_k^n)x_k^l + G_k^l(x_k^n)v_k^l, \tag{9.45b}$$

$$y_k = h_k(x_k^n) + H_k(x_k^n)x_k^l + e_k. \tag{9.45c}$$

The state vector and Gaussian state noise are partitioned as

$$x_k = \begin{pmatrix} x_k^n \\ x_k^l \end{pmatrix}, \quad v_k = \begin{pmatrix} v_k^n \\ v_k^l \end{pmatrix} \sim \mathcal{N}(0, Q_k), \quad Q_k = \begin{pmatrix} Q_k^n & Q_k^{ln} \\ (Q_k^{ln})^k & Q_k^l \end{pmatrix}. \tag{9.45d}$$

Furthermore, x_0^l is assumed Gaussian, $x_0^l \sim \mathcal{N}(\bar{x}_0, \bar{P}_0)$. The density of x_0^n can be arbitrary, but it is assumed known. The underlying purpose with this model structure is that conditioned on the sequence $x_{1:k}^n$, (9.45) is linear in x_k^l with Gaussian prior, process noise and measurement noise, respectively, so the Kalman filter theory applies.

9.8.2 Algorithm Overview

The MPF relies on the following key factorization

$$p(x_k^l, x_{1:k}^n | y_{1:k}) = p(x_k^l | x_{1:k}^n, y_{1:k}) p(x_{1:k}^n | y_{1:k}). \tag{9.46}$$

These two factors decompose the nonlinear filtering task into two sub-problems:

- A Kalman filter operating on the conditionally linear, Gaussian model in (9.45) provides the *exact* conditional posterior

$$p(x_k^l | x_{1:k}^n, y_{1:k}) = \mathcal{N}\left(x_k^l; \hat{x}_{k|k}^l(x_{1:k}^{n,i}), P_{k|k}^l(x_{1:k}^{n,i})\right). \tag{9.47}$$

Here, (9.45a) becomes an extra measurement for the Kalman filter with $x_{k+1}^n - f_k^n(x_k^n)$ acting as the observation.

- A particle filter for estimating the filtering density of the nonlinear states. This involves a nontrivial marginalization step by integrating over the state space of all x_k^l using the law of total probability

$$p(x_{1:k+1}^n|y_{1:k}) = p(x_{1:k}^n|y_{1:k})p(x_{k+1}^n|x_{1:k}^n, y_{1:k}) \tag{9.48}$$

$$= p(x_{1:k}^n|y_{1:k})\int p(x_{k+1}^n|x_k^l, x_{1:k}^n, y_{1:k})p(x_k^l|x_{1:k}^n, y_{1:k})dx_k^l$$

$$= p(x_{1:k}^n|y_{1:k})\int p(x_{k+1}^n|x_k^l, x_{1:k}^n, y_{1:k})\mathcal{N}\big(x_k^l; \hat{x}_{k|k}^l(x_{1:k}^{n,i}), P_{k|k}^l(x_{1:k}^{n,i})\big)dx_k^l.$$

The intuitive interpretation of this result is that the linear state estimate acts as an extra state noise in (9.45a) when performing the particle filter time update.

The time and measurement updates of KF and PF are interleaved, so the timing is important. The information structure in the recursion is described in Algorithm 9.2. Table 9.2 summarizes the information steps in Algorithm 9.2.

Algorithm 9.2 Marginalized Particle Filter

With reference to the standard particle filter in Algorithm 9.1 and the Kalman filter, iterate the following steps for each time step:

1. PF measurement update and resampling using (9.45c) where x_k^l is interpreted as measurement noise.

2. KF measurement update using (9.45c) for each particle $x_{1:k}^{n,i}$.

3. PF time update using (9.45a) where x_k^l is intepreted as process noise.

4. KF time update using (9.45b) for each particle $x_{1:k}^{n,i}$.

5. KF extra measurement update using (9.45a) for each particle $x_{1:k}^{n,i}$.

Note that the time index appears five times in the right hand side expansion of the prior. The five steps increase each k one at the time to finally form the posterior at time $k+1$.

The posterior distribution for the nonlinear states is given by a discrete particle distribution as usual, while the posterior for the linear states is given by a Gaussian mixture:

$$p(x_{1:k}^n|y_{1:k}) \approx \sum_{i=1}^N w_{k|k}^i \delta(x_{1:k}^n - x_{1:k}^{n,i}), \tag{9.49a}$$

$$p(x_k^l|y_{1:k}) \approx \sum_{i=1}^N w_{k|k}^i \mathcal{N}\big(x_k^l; \hat{x}_{k|k}^l(x_{1:k}^{n,i}), P_{k|k}^l(x_{1:k}^{n,i})\big). \tag{9.49b}$$

Table 9.2: Summary of the information steps in Algorithm 9.2 for the marginalized PF utilizing a linear Gaussian sub-structure.

Prior	$p(x_k^l, x_{1:k}^n \mid y_{1:k})$	$= p(x_k^l \mid x_{1:k}^n, y_{1:k}) p(x_{1:k}^n \mid y_{1:k})$
1. PF TU	$p(x_{1:k}^n \mid y_{1:k})$	$\Rightarrow p(x_{1:k+1}^n \mid y_{1:k})$
2. KF TU	$p(x_k^l \mid x_{1:k}^n, y_{1:k})$	$\Rightarrow p(x_{k+1}^l \mid x_{1:k}^n, y_{1:k})$
3. KF dyn MU	$p(x_{k+1}^l \mid x_{1:k}^n, y_{1:k})$	$\Rightarrow p(x_{k+1}^l \mid x_{1:k+1}^n, y_{1:k})$
4. PF MU	$p(x_{1:k+1}^n \mid y_{1:k})$	$\Rightarrow p(x_{1:k+1}^n \mid y_{1:k+1})$
5. KF obs MU	$p(x_{k+1}^l \mid x_{1:k+1}^n, y_{1:k})$	$\Rightarrow p(x_{k+1}^l \mid x_{1:k+1}^n, y_{1:k+1})$
Posterior	$p(x_{k+1}^l, x_{1:k+1}^n \mid y_{1:k+1})$	$= p(x_{k+1}^l \mid x_{1:k+1}^n, y_{1:k+1}) p(x_{1:k+1}^n \mid y_{1:k+1})$

For a complete derivation, see Schön et al. (2005). As shown in Hendeby et al. (2007b), standard Kalman and particle filtering code can be reused when implementing the MPF. The model (9.45) can be further generalized by introducing an additional discrete mode parameter, giving a larger family of marginalized filters, see Nordlund and Gustafsson (2009).

9.8.3 Complexity Issues

In general, each Kalman filter comes with its own Riccati equation, which becomes a function $P_{k|k}^i = P_{k|k}(x_{1:k}^{n,i})$ of the whole trajectory of the nonlinear part. However, the Riccati equation is the same for all particles $P_{k|k}^i = P_{k|k}$ if the following three conditions are satisfied:

$$G_k^n(x_k^n) = G_k^n \quad \text{or} \quad F_k^n(x_k^n) = 0, \tag{9.50a}$$

$$G_k^l(x_k^n) = G_k^l, \tag{9.50b}$$

$$H_k(x_k^n) = H_k, \tag{9.50c}$$

$$\Rightarrow P_{k|k}(x_{1:k}^{n,i}) = P_{k|k}. \tag{9.50d}$$

It is easy to verify that the Riccati equations in this case only involves matrices that are the same for all trajectories $x_{1:k}^{n,i}$. This implies a significant complexity reduction.

One important special case of (9.45) in practice is a model with linear state equations with a nonlinear observation which is a function of a (small) part of the state vector,

$$x_{k+1}^n = F_k^{nn} x_k^n + F_k^{nl} x_k^l + G_k^n v_k^n, \tag{9.51a}$$

$$x_{k+1}^l = F_k^{ln} x_k^n + F_k^{ll} x_k^l + G_k^l v_k^l, \tag{9.51b}$$

$$y_k = h_k(x_k^n) \quad\quad + e_k. \tag{9.51c}$$

For instance, all applications in Section 16.3 fall into this category. In this case, step 2 in Algorithm 9.2 disappears.

The MPF appears to add quite a lot of overhead computations. It turns out, however, that the MPF is often more efficient. It may seem impossible to give any general conclusions, so application dependent simulation studies have to be performed. Nevertheless, quite realistic predictions of the computational complexity can be done with rather simple calculations, as pointed out in Karlsson et al. (2005).

Let $n_x = p + k$, where $p = \dim(x_k^n)$ and $k = \dim(x_k^l)$, respectively. That is, we both can and choose to marginalize k states and apply the PF to p states. For a linear Gaussian model, k can be choosen to any integer between 0 and n_x, where the former gives the PF and latter corresponds to the KF.

The number of flops are computed as follows. For the instruction $P_{k|k}(A_k^k)^T$, which corresponds to multiplying $P_{k|k} \sim \mathbb{R}^{k \times k}$ with $(A_k^K)^T \sim \mathbb{R}^{k \times p}$, pk^2 multiplications and $(k-1)kp$ additions are needed. The total *equivalent flop* (EF)[2] complexity is derived by Karlsson et al. (2005),

$$P_{k|k}^i = P_{k|k}(x_{1:k}^{n,i}) \Rightarrow \tag{9.52a}$$

$$\mathcal{C}(p, k, N) \approx 4pk^2 + 8kp^2 + \frac{4}{3}p^3 + 5k^3 - 5kp + 2p^2$$
$$+ (6kp + 4p^2 + 2k^2 + p - k + pc_3 + c_1 + c_2)N, \tag{9.52b}$$

$$P_{k|k}^i = P_{k|k} \Rightarrow \tag{9.52c}$$

$$\mathcal{C}(p, k, N) \approx (6kp + 4p^2 + 2k^2 + p - k + pc_3 + c_1 + c_2$$
$$+ 4pk^2 + 8kp^2 + \frac{4}{3}p^3 + 5k^3 - 5kp + 2p^2 + k^3)N. \tag{9.52d}$$

Here, the coefficient c_1 has been used for the calculation of the Gaussian likelihood, c_2 for the resampling and c_3 for the random number complexity. Note that, when $C_k = 0$ the same covariance matrix is used for all Kalman filters, which significantly reduce the computational complexity.

By requiring $\mathcal{C}(p + k, 0, N_{\text{PF}}) = \mathcal{C}(p, k, N(k))$, where N_{PF} corresponds to the number of particles used in the standard PF we can solve for $N(k)$. This gives the number of particles $N(k)$ that can be used by the MPF in order to obtain the same computational complexity as if the standard PF had been used for all states. In Figure 9.4 the ratio $N(k)/N_{\text{PF}}$ is plotted for systems with $n_x = 3, \ldots, 9$ states for the two cases. Hence, using Figure 9.4 it is possible to directly find out how much there is to gain in using the MPF from a computational complexity point of view. The figure also shows that the computational complexity is always reduced when the MPF can be used instead of the standard PF. Furthermore, as previously mentioned, the quality of the estimates will improve or remain the same when the MPF is used Doucet et al. (2001b).

---

[2]The EF complexity for an operation is defined as the number of flops that result in the same computational time as the operation.

9.8 Marginalized Particle Filter Theory

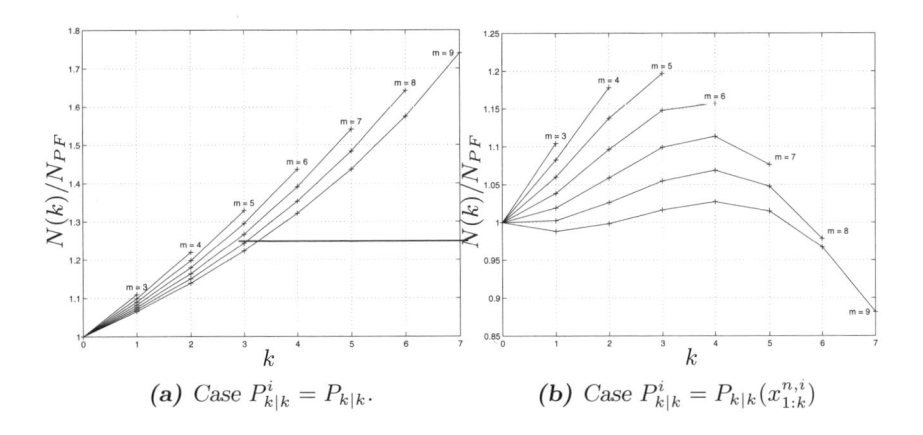

(a) Case $P_{k|k}^i = P_{k|k}$.

(b) Case $P_{k|k}^i = P_{k|k}(x_{1:k}^{n,i})$

Figure 9.4: Ratio $N(k)/N_{PF}$ for systems with $m = 3, \ldots, 9$ states for the cases $P_{k|k}^i = P_{k|k}$ (a) and $P_{k|k}^i = P_{k|k}(x_{1:k}^{n,i})$ (b), respectively. Here, $N(k)$ is defined by $\mathcal{C}(p+k, 0, N_{PF}) = \mathcal{C}(p, k, N(k))$ such that PF and MPF get the same complexity.

9.8.4 Variance Reduction

The MPF reduces the variance of the linear states which will be demonstrated below. The *law of total variance* says that

$$\text{Cov}(U) = \text{Cov}(\text{E}(U|V)) + \text{E}(\text{Cov}(U|V)). \tag{9.53}$$

Letting $U = x_k^l$ and $V = x_{1:k}^n$ gives the following decomposition of the variance of the PF:

$$\underbrace{\text{Cov}(x_k^l)}_{PF} = \text{Cov}\left(\text{E}(x_k^l|x_{1:k}^n)\right) + \text{E}\left(\text{Cov}(x_k^l|x_{1:k}^n)\right) \tag{9.54a}$$

$$= \underbrace{\text{Cov}\left(\hat{x}_{k|k}^l(x_{1:k}^{n,i})\right)}_{MPF} + \sum_{i=1}^{N} w_k^i \underbrace{P_{k|k}(x_{1:k}^{n,i})}_{KF}. \tag{9.54b}$$

Here, we recognize $(x_k^l|x_{1:k}^{n,i})$ as the Gaussian distribution, delivered by the KF, conditioned on the trajectory $x_{1:k}^{n,i}$. Now, the MPF computes the mean of each trajectory as $\hat{x}_{k|k}^l(x_{1:k}^{n,i})$ and the unconditional mean estimator is simply the mean of these,

$$\hat{x}_{k|k}^l = \sum_{i=1}^{N} w_k^i \hat{x}_{k|k}^l(x_{1:k}^{n,i}), \tag{9.55}$$

and its covariance follows from the first term in (9.54b). The first term in (9.54b) corresponds to the *spread of the mean* contribution from the Gaussian mixture, and this is the only uncertainty in the MPF.

The variance decomposition shows that the covariance for the MPF is strictly smaller than the corresponding covariance for the PF. This can also be seen as a result of Rao-Blackwell's lemma, see, e.g., Robert and Casella (1999), and the marginalization is commonly referred to as Rao-Blackwellization. This result says that the improvement in the quality of the estimate is given by the term $\mathrm{E}\left(\mathrm{Cov}(x_k^l | x_{1:k}^n)\right)$. Note that when (9.50) is satisfied, then $P_{k|k}^i = P_{k|k}$ and thus $\sum_{i=1}^N w_k^i P_{k|k}^i = P_{k|k}$. That is, the Kalman filter covariance $P_{k|k}$ is a good indicator of how much that has been gained in using the MPF instead of the PF. As a practical rule of thumb, the gain in MPF increases as the uncertainty in the linear state increases in the model. Further discussions regarding the variance reduction property of the MPF are provided for instance in Doucet et al. (2001b).

The variance reduction in the MPF can be used in two different ways:

- With the same number of particles, the variance in the estimates of the linear states can be decreased.

- With the same performance in terms of variance for the linear states, the number of particles can be decreased.

This is schematically illustrated in Figure 9.5, for the case when (9.50) is satisfied, implying that the same covariance matrix can be used for all particles. The two alternatives above are illustrated for the case a PF with 10000 particles is first applied, and then replaced by the MPF.

9.8.5 MPF Synonyms

The following names have been suggested for the filter in this section:

- MPF as is motivated by the nontrivial marginalization step (9.48).

- The *Rao-Blackwellized particle filter*, as motivated by the variance reduction in (9.54).

- The *mixture Kalman filter*, as motivated by the various mixture distributions that appear, for instance in (9.49b).

- Another logical name would be *separable particle filter* in parallel to the well established separable nonlinear least squares problem. In fact, the special case of a static problem where only (9.45c) exists falls into this class of problems. Here, the weighted least squares estimate of x_k^l is first computed as a function of $x_{1:k}^n$, which is then backsubstituted into the model with its estimation covariance to form a nonlinear least squares problem in $x_{1:k}^n$ only.

9.8 Marginalized Particle Filter Theory

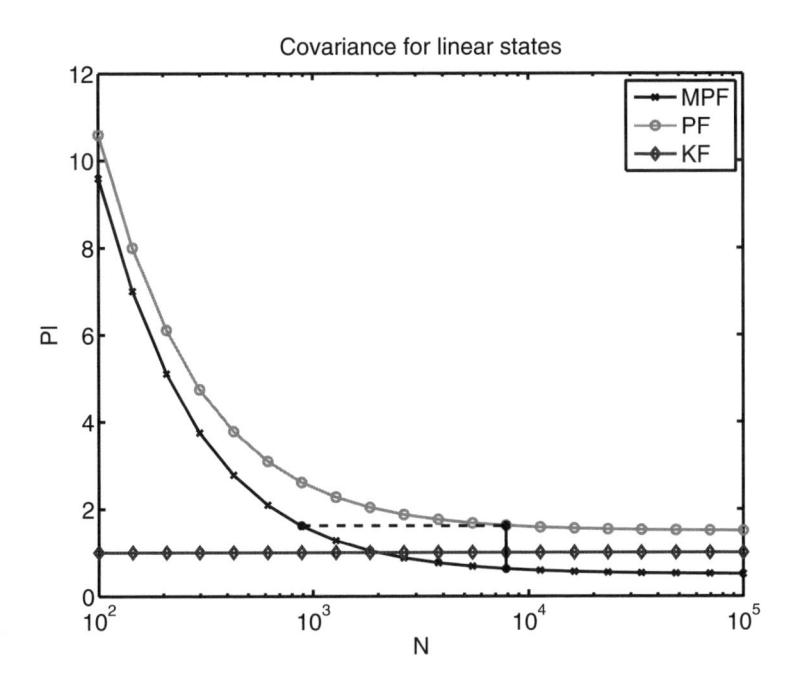

Figure 9.5: Schematic view of how the covariance of the linear part of the state vector depends on the number of particles for the PF and MPF, respectively. The gain in MPF is given by the Kalman filter covariance.

9.8.6 Illustrating Example

The aim here is to illustrate how the MPF works using the following nonlinear stochastic system.

$$x_{k+1}^n = x_k^l x_k^n + v_k^n, \tag{9.56a}$$

$$x_{k+1}^l = x_k^l + v_k^l, \tag{9.56b}$$

$$y_k = 0.2(x_k^n)^2 + e_k, \tag{9.56c}$$

where the noise is assumed white and Gaussian according to

$$v_k = \begin{pmatrix} v_k^n \\ v_k^l \end{pmatrix} \sim \mathcal{N}\left(\begin{pmatrix} 0 \\ 0 \end{pmatrix}, \begin{pmatrix} 0.25 & 0 \\ 0 & 10^{-4} \end{pmatrix} \right), \tag{9.56d}$$

$$e_k \sim \mathcal{N}(0, 1). \tag{9.56e}$$

The initial state x_0 is given by

$$x_0 \sim \mathcal{N}\left(\begin{pmatrix} 0.1 \\ 0.99 \end{pmatrix}, \begin{pmatrix} 16 & 0 \\ 0 & 10^{-3} \end{pmatrix} \right). \tag{9.56f}$$

This particular model was used in Šimandl et al. (2006), where it illustrated grid-based (point-mass) filters. Obviously, the states can be estimated by applying the standard particle filter to the entire state vector. However, a better solution is to exploit the conditionally linear, Gaussian sub-structure that is present in (9.56). The nonlinear process x_k^n is a first-order AR process, where the linear process x_k^l is the time-varying parameter. The linear, Gaussian sub-structure is used by the MPF and the resulting filtering density function at time 10, $p(x_{10}|y_{1:10})$ before the resampling step is shown in Figure 9.6 (for a particular realization). In this example 2000 particles were used, but only 100 of them are plotted in Figure 9.6 in order to obtain a clearer illustration of the result. The figure illustrates the fact that the MPF is a combination of the

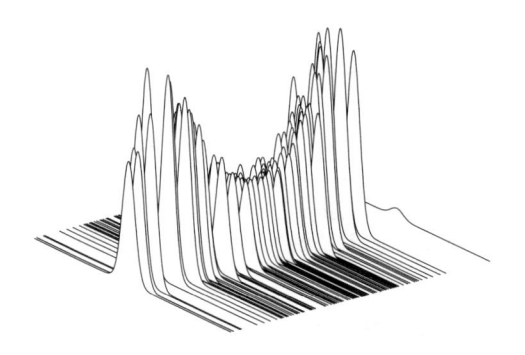

Figure 9.6: *The estimated filter PDF for system (9.56) at time 10, $p(x_{10}|y_{1:10})$ using the MPF. It is instructive to see that the linear state x_{10}^l is estimated by Gaussian densities (from the Kalman filter) and the position along the nonlinear state x_{10}^n is given by a particle (from the particle filter).*

KF and the PF. The density functions for the linear states are provided by the Kalman filters, which is evident from the fact that the marginals $p(x_k^{l,i}|y_{1:k})$ are given by Gaussian densities. Furthermore, the nonlinear state estimates are provided by the PF. Hence, the linear states are given by a parametric estimator (KF), whereas the nonlinear states are given by a nonparametric estimator (PF). In this context the MPF can be viewed as a combination of a parametric and a nonparametric estimator.

9.8.7 Marginalization of Discrete Parameters

A discrete parameter δ_k can be treated as a state in an augmented state space model, and the particle filter can be applied as usual. However, the finite state space implies that there is a finite number of possible state trajectories $\delta_{1:k}$. Though this number is exponentially growing, there are dedicated approximate algorithms for searching this space, that potentially can improve the PF performance for a finite number of particles.

Connection to Filter Banks

There is a clear connection between MPF and Kalman filters banks as was briefly introduced in (6.23). The total state space in (6.23) can be partitioned as

$$\bar{x}_k = \begin{pmatrix} x_k \\ \delta_k \end{pmatrix}. \tag{9.57}$$

Marginalization here comes quite naturally, and similar to the MPF, it is the conditional linear Gaussian model which is the key. We immediately get the Kalman filter solution

$$p(x_k|\delta_{1:k}, y_{1:k}) = \mathcal{N}(\hat{x}_{k|k}(\delta_{1:k}), P_{k|k}(\delta_{1:k})). \tag{9.58}$$

The other factor is obtained similarly to the MPF (9.48).

Mixed Problems with Linear, Nonlinear and Discrete States

Consider a state vector x_k, which can be partitioned according to

$$x_k = \begin{pmatrix} x_k^n \\ x_k^l \\ \delta_k \end{pmatrix}, \tag{9.59}$$

where x_k^l denotes the linear states, x_k^n denotes the nonlinear states, and $\delta_k \in \{1, 2, \ldots, S_k\}$ denotes the discrete state in the dynamics and measurement relation. Bayesian estimation methods, such as the particle filter, provide estimates of the filtering density function $p(x_k|y_{1:k})$. By employing the fact

$$p(x_k^l, x_{1:k}^n, \delta_{1:k}|y_{1:k}) = p(x_k^l|x_{1:k}^n, \delta_{1:k}, y_{1:k})p(x_{1:k}^n, \delta_{1:k}|y_{1:k})$$
$$= p(x_k^l|x_{1:k}^n, \delta_{1:k}, y_{1:k})p(x_{1:k}^n|\delta_{1:k}, y_{1:k})p(\delta_{1:k}|y_{1:k}), \tag{9.60}$$

the overall problem is decomposed into three sub-problems. Hence, a marginalized nonlinear filter for the general problem is characterised by

- A Kalman filter operating on the conditionally linear and Gaussian model (9.45) provides an estimate of $p(x_k^l|x_{1:k}^n, \delta_{1:k}, y_{1:k})$. As usual, the model (9.45) is linear and Gaussian conditioned on the nonlinear state sequence $x_{1:k}^n$ and the discrete mode sequence $\delta_{1:k}$.

- A marginalized nonlinear filter (e.g., PF, PMF, Gaussian sum filters, UKF) is designed for a fixed mode sequence.

- A pruning or merging scheme (IMM) for the exponentially increasing number of mode sequences, see Chapter 10.

It is very important to note that the three sub-problems mentioned above are all coupled, for example, the result from the nonlinear filter at time k is used by the Kalman filters at time k. This is further explained in the subsequent section.

---**Example 9.3: Marginalization in 1D terrain navigation**---

Consider again the one-dimensional terrain navigation problem in Example 9.2. Assume here that the measurement noise is subject to "tree-top noise", in that there is a certain probability that the terrain altitude is measured up to the tree tops. The model (9.44) is modified to

$$x_{k+1} = x_k + u_k + \frac{T_s^2}{2} v_k, \tag{9.61a}$$

$$u_{k+1} = u_k + T_s v_k, \tag{9.61b}$$

$$y_k = h(x_k) + e_k, \tag{9.61c}$$

$$v_k \sim \mathcal{N}(0, Q_k), \tag{9.61d}$$

$$e_k \sim \delta_k \mathcal{N}(0, R_k^1) + (1 - \delta_k) \mathcal{N}(\mu_k^2, R_k^2). \tag{9.61e}$$

That is, the discrete state δ_k controls which mode of a Gaussian mixture is active for each sample. Again, the example is straightforwardly generalized to two dimensions and with extra states for acceleration.

9.9 Particle Filter Code Examples

This section gives concrete MATLAB™-like code for a general SIR particle filter, and applies it to a fully annotated simulation example. Further, object oriented implementations of nonlinear filters are illustrated on a target tracking applications.

9.9.1 Terrain-Based Positioning

The following scalar state example suits three purposes. First, it enables intuitive graphical illustrations. Second, it introduces the positioning applications in the next section. Third, it should be easy to implement for interested readers for reproducing the example, and extending the code to other applications.

9.9 Particle Filter Code Examples

Consider the model

$$x_{k+1} = x_k + u_k + v_k, \tag{9.62a}$$
$$y_k = h(x_k) + e_k, \tag{9.62b}$$

where both the state and the measurement are scalar-valued. This model mimics a navigation problem in one-dimension, where u_k is a measurable velocity, v_k unmeasurable velocity disturbance, and the observation y_k measures the terrain altitude, which is known in from a *geographical information system* $h(x)$. An illustration from a real application is found in Figure 16.13. Note that the terrain altitude as a measurement relation is not one to one, since a given terrain altitude is found at many different positions. However, the observed terrain profile will after a short time be unique for the flown trajectory.

Figure 9.7 shows a trajectory, and one realization of the nonlinear function terrain profile $h(x)$, generated by the code below.

```
x=1:100;                                  % Map grid
h=20+filter(1,[1 -1.8 0.81],randn(1,100)); % Terrain altitude
N=15;
z=100+filter(1,[1 -1.8 0.81],randn(N,1));  % Measurement input
u=2*ones(N,1);                             % State input
x0=20+cumsum(u);                           % True position
y=z-interp1(x,h,x0);                       % Noisefree measurement
yn=y+1*randn(N,1);                         % Noisy measurement
plot(x0,y,'o-b',x,h,'g',x0,z-y,'go','linewidth',3)
```

The horizontal line indicates where the first measurement is taken. There are ten different intersections between the terrain profile and this observation, where the grid point just before each intersection is marked in the figure. This is clearly a problem where the posterior is multi-modal after the first measurement update.

The following code lines define the model (9.62) as an object structure:

```
m.f=inline('x+u','x','u');
m.h=inline('z-interp1(x,h,xp)','xp','h','x','z');
m.pv=ndist(0,5); m.pe=ndist(0,1);
m.p0=udist(10,90);
```

The PDF classes `ndist` and `udist` with the methods `rand` and `pdf` are assumed to be available. A script that both implements a version of the PF and also animates all the partial results is given below:

```
Np=100; w=ones(Np,1)/Np;
xp=rand(m.p0,Np);                          % Initialization
for k=1:N;
   yp=m.h(xp,h,x,z(k));                    % Measurement pred.
   w=w.*pdf(m.pe,repmat(yn(k,:),Np,1)-yp); % Likelihood
   w=w/sum(w);                             % Normalization
   subplot(3,1,1), stem(xp,Np*w/10)
   xhat(k,:)=w(:)'*xp;                     % Estimation
   [xp,w]=resample(xp,w);                  % Resampling
   subplot(3,1,2), stem(xp,Np*w)
   v=rand(m.pv,Np);                        % Random process noise
   xp=m.f(xp,u(k,:)')+v;                   % State prediction
```

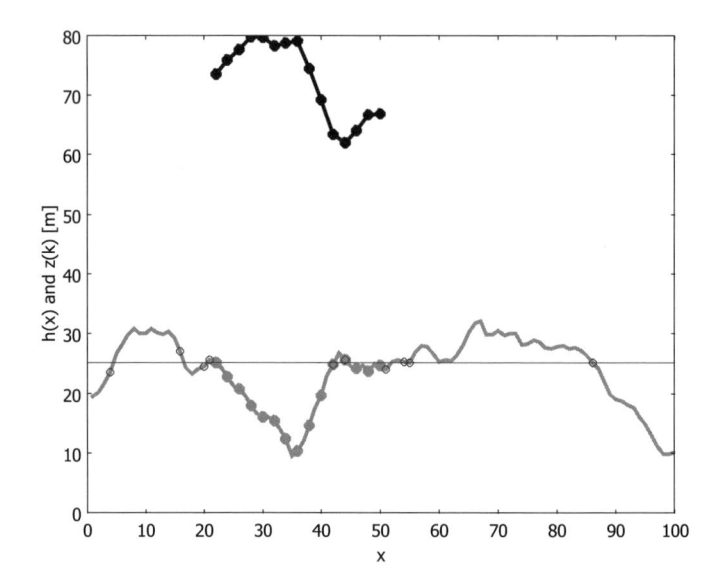

Figure 9.7: Aircraft altitude $z(x_k)$ (upper dark line) as a function of position x_k (dots on upper dark line) and nonlinear measurement relation $h(x)$ (lower gray line) for the model in (9.62). The computed terrain altitude $h(x_1)$ is also marked, and a circle is put in all grid points that give the best match to this altitude.

```
   subplot(3,1,3), stem(xp,Np*w)
end
```

Code examples of the function `resample` are given in Section 9.7.1. Figure 9.8 shows the posterior density approximation at two time instants. Figure 9.8(a) shows first the unnormalized weights after the measurement update, which with this uniform prior is just the likelihood function $p(y_1|x_0) = p(y_1)$. Then follows the particle distribution after resampling (where $w^i = 1/N$), and finally the particles after time update (which is just a translation with u_1).

Figure 9.8(b) illustrates the same thing after the 15'th measurement. The posterior is now more clustered to a unimodal distribution. Figure 9.9 shows the position error as a function of time. The break point in performance indicates when the multimodal posterior distribution becomes unimodal.

9.9.2 Target Tracking

In an object oriented implementation, simulation studies can be performed quite efficiently. The following example compares different filters for a simple

9.9 Particle Filter Code Examples

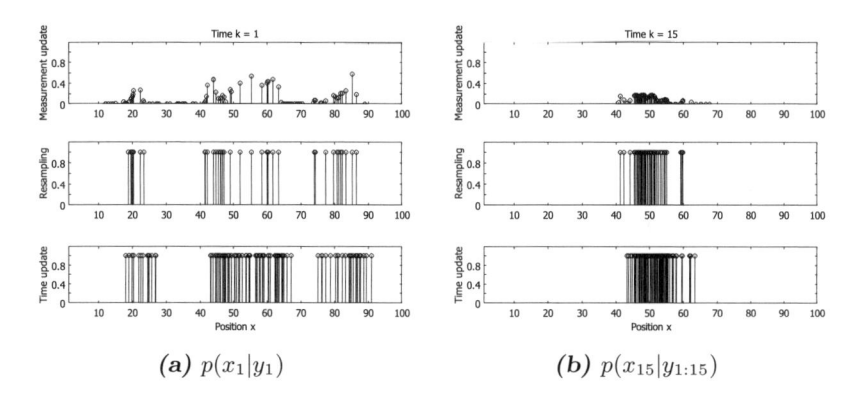

(a) $p(x_1|y_1)$

(b) $p(x_{15}|y_{1:15})$

Figure 9.8: Particle approximations of $p(x_k|y_{1:k})$ before (first subplot) and after (second subplot) resampling, respectively, together with the time update (last subplot).

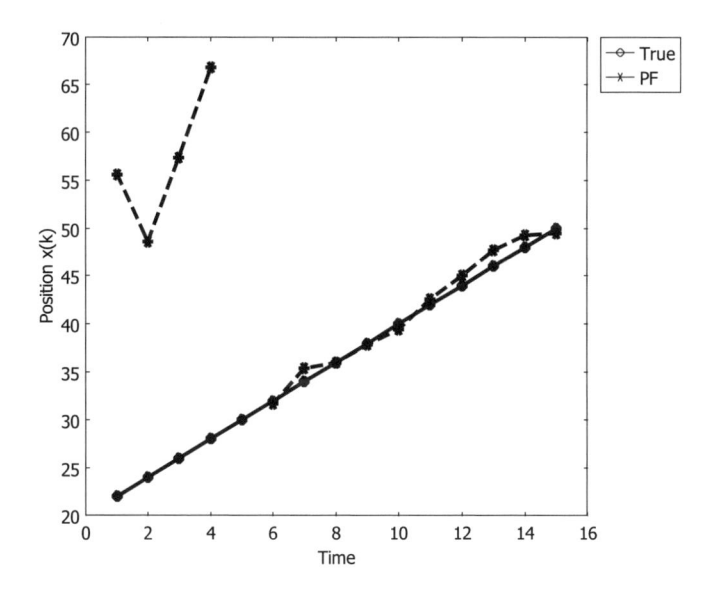

Figure 9.9: True and estimated state as a function of time. The posterior is multi-modal up to $k = 4$, which gives a large position error.

target tracking model,

$$x_{k+1} = \begin{pmatrix} I_2 & T_s I_2 \\ 0 & I_2 \end{pmatrix} x_k + \begin{pmatrix} \frac{T_s^2}{2} I_2 \\ T_s I_2 \end{pmatrix} v_k, \quad v_k \sim \mathcal{N}(0, 1 I_2), \quad x_0 = \begin{pmatrix} 0 \\ 0 \end{pmatrix},$$
(9.63a)

$$y_k = \begin{pmatrix} I_2 & 0 \end{pmatrix} x_k + e_k, \quad\quad\quad e_k \sim \mathcal{N}(0, 0.01 I_2), \quad\quad (9.63b)$$

The observation model is first linear to be able to compare to the Kalman filter that provides the optimal estimate. The example makes use of two different objects:

- Signal object where the state $x_{1:k}$ and observation $y_{1:k}$ sequences are stored, with their associated uncertainty (covariances P_k^x, P_k^y or particle representation). Plot methods in this class can then automatically provide confidence bounds.

- Model objects for linear and nonlinear models, with methods implementing simulation and filtering algorithms.

The purpose of the following example is to illustrate how little coding that is required with this object oriented approach. First, the model is loaded from an extensive example database as a linear state space model. It is then converted to the general nonlinear model structure, which does not make use of the fact that the underlying model is linear.

```
mss=exlti('cv2d');
mnl=nl(mss);
```

Now, the following state trajectories are compared:

- The true state from the simulation.

- The Cramér-Rao lower bound (CRLB) computed from the nonlinear model.

- The Kalman filter (KF) estimate using the linear model.

- The extended Kalman filter (EKF) using the nonlinear model.

- The unscented Kalman filter (UKF) using the nonlinear model.

- The particle filter (PF) using the nonlinear model.

For all except the first one, a confidence ellipsoid indicates the position estimation uncertainty.

```
y=simulate(mss,10);
xhat1=kalman(mss,y);
xhat2=ekf(mnl,y);
xhat3=ukf(mnl,y);
xhat4=pf(mnl,y,'Np',1000);
xcrlb=crlb(mnl,y);
xplot2(xcrlb,xhat4,xhat3,xhat2,xhat1,'conf',90)
```

9.9 Particle Filter Code Examples

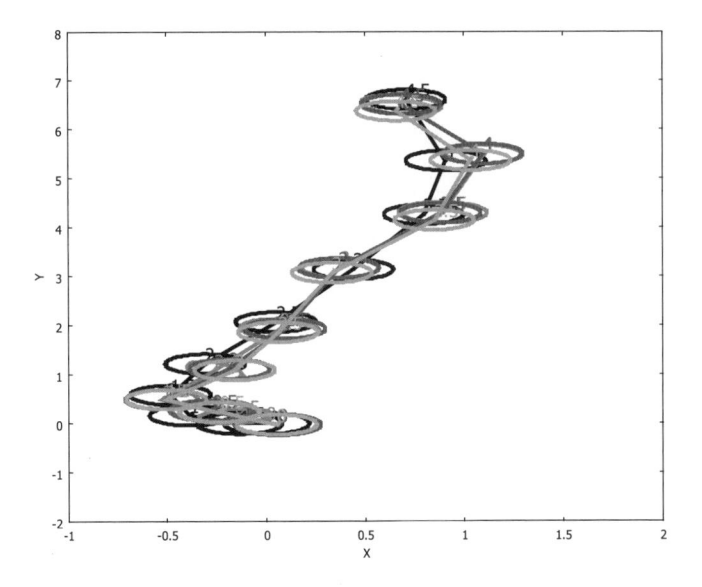

Figure 9.10: *Simulated trajectory using a constant velocity two-dimensional motion model with a position sensor, where the plots show the CRLB (darkest) and estimates from KF, EKF, UKF and PF, respectively.*

Figure 9.10 validates that all algorithms provide comparable estimates in accordance with the CRLB.

Now, consider the case of a radar sensor that provides good angle resolution but poor range. The measurement relation in model (9.63b) is changed to

$$y_k = \begin{pmatrix} \arctan\left(\frac{x_k^{(2)} - \theta^{(2)}}{x_k^{(1)} - \theta^{(1)}}\right) \\ \sqrt{\left(x_k^{(1)} - \theta^{(1)}\right)^2 + \left(x_k^{(2)} - \theta^{(2)}\right)^2} \end{pmatrix} + e_k, \quad e_k \sim \mathcal{N}\bigl(0, \operatorname{diag}(0.0001, 0.3)\bigr)$$
(9.64)

Figure 9.11 compares EKF and PF with respect to the CRLB. The PF performs well, where the covariances fitted to the particles are very similar to the CRLB. The EKF is slightly biased and too optimistic about the uncertainty, which is a typical behavior when neglecting higher order terms in the nonlinearities. However, the performance of all filters is comparable, and the nonlinear measurement relation does not in itself motivate computer intensive algorithms in this case.

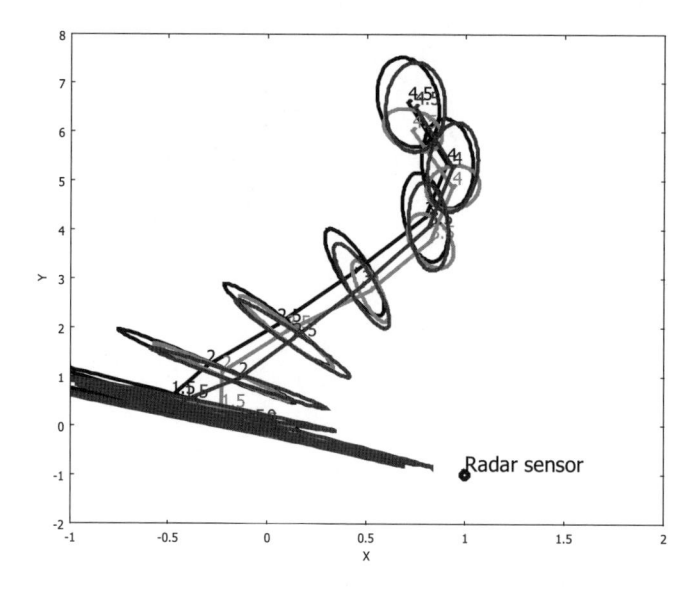

Figure 9.11: *Simulated trajectory using a constant velocity two-dimensional motion model with a radar sensor, where the plots show the CRLB (darkest) and estimates from EKF (small ellipsoids) and PF, respectively.*

9.9.3 Growth Model

The following toy example was used in the original paper Gordon et al. (1993):

$$x_{k+1} = \frac{x_k}{2} + 25\frac{x_k}{1+x_k^2} + 8\cos(k) + v_k, \quad v_k \sim \mathcal{N}(0,10), \quad x_0 \sim \mathcal{N}(5,5), \tag{9.65a}$$

$$y_k = \frac{x_k^2}{20} + e_k, \qquad\qquad e_k \sim \mathcal{N}(0,1). \tag{9.65b}$$

It has since then been used many times in the particle filter literature, and it is often claimed to be a growth model. It is included here just because it has turned into a benchmark problem. The simulation code is

```
m=exnl('pfex');
z=simulate(m,30);
zcrlb=crlb(m,z);
zekf=ekf(m,z);
zukf=ukf(m,z);
zpf=pf(m,z);
xplot(zcrlb,zpf,zekf,zukf,'conf',90,'view','cont','conftype',2)
[sqrt(mean(zcrlb.Px)) norm(z.x-zpf.x)   norm(z.x-zekf.x) norm(z.x-zukf.x)];
```

9.9 Particle Filter Code Examples

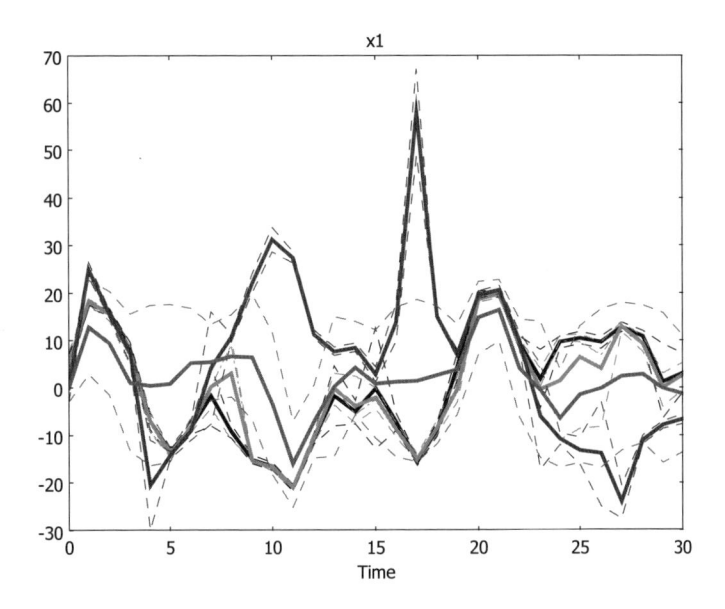

Figure 9.12: Simulated trajectory using the model (9.65), where the plots show the CRLB (darkest) and estimates from EKF, PF and UKF, respectively. Table 9.3 summarizes the performance.

Table 9.3: Mean square error performance of the estimates in Figure 9.12 for the benchmark problem in (9.65).

CRLB	PF	UKF	EKF
0.5	10	16	52

The last two lines produce the result in Figure 9.12 and Table 9.3, respectively. The conclusion from this example is that PF performs much better than the UKF which in turn is much better than the EKF. Thus, this example illustrates quite nicely the ranking of the different filters.

9.10 Summary

9.10.1 Theory

The particle filter (PF) represents the posterior filtering density $p(x_k|y_{1:k}) \approx \sum_{i=1}^{N} w_{k|k}^i \delta(x_k - x_k^i)$ with Monte Carlo samples x_k^i referred to as particles, where $w_{k|k}^i$ is the weight (probability) of each particle. The PF recursion consists of the following components:

1. *Measurement update:* For each particle, multiple the weights with the likelihood of the current observation conditioned on the particle, $w_{k|k}^i = w_{k|k-1}^i p(y_k|x_k^i)$. For a state space model with additive Gaussian noise, this becomes $w_{k|k}^i = w_{k|k-1}^i \mathcal{N}(y_k; h(x_k^i), R_k)$.

2. *Resampling:* Always (SIR) or only when needed (SIS), replace the particle set $\{x_k^i, w_{k|k}^i\}_{i=1}^N$ with a new one, taken from the original set.

3. *Time update:* For each particle, simulate a state trajectory to the next observation time using a proposal density, $x_{k+1}^i \sim q(x_{k+1}|x_{1:k}^i, y_{1:k+1})$. For a discrete state space model with Gaussian noise and SIR, this becomes $x_{k+1}^i = f(x_k^i, v_k^i)$ where $v_k^i \sim \mathcal{N}(0, Q_k)$.

The first stage of the PF design consists of modeling. The potential performance of the PF depends only on $f(x_k, v_k), h(x_k, e_k), p_v, p_e$.

The second stage consists of choosing the actual PF implementation, in order of importance:

- The number N of particles. There are two threshold effects to keep in mind for when (1) the transient and (2) stationary performance are optimized.

- Dithering: *ad-hoc* step where the covariance of the state noise v_k and measurement noise e_k, respectively, are increased.

- Choice of proposal distribution:

  - The prior $q(x_k|x_{1:k-1}^i, y_{1:k}) = p(x_k|x_{k-1}^i)$, which is the best choice for low SNR.

  - The likelihood $q(x_{k+1}|x_{1:k}^i, y_{1:k+1}) = p(y_k|x_k^i)$, which is the best choice for high SNR, but it is only possible when the likelihood function is invertible (which requires $n_x \leq n_y$). The inversion can be approximated with an EKF step.

  - The optimal choice is a combination of above, $q(x_{k+1}|x_{1:k}^i, y1:k+1) \propto p(y_k|x_k)p(x_k|x_{k-1}^i)$. It exists analytically for linear Gaussian systems but also for the case of nonlinear dynamics.

- Weight corrections as in the marginal PF to decrease weight variance, or as in the smoothing PF in off-line analysis, see Section 9.3.3.

- Resampling: how often and which resampling scheme to use, see Section 9.4.1.

9.10.2 Software

The particle filter is implemented as the method `nl.pf` in the nonlinear (NL) model class, and it has the syntax

```
xhat=pf(m,y,Property1,Value1,...)
```

Here, `m` is the NL object with model, `y` is a SIG object with measurements and `xhat` is a SIG object with estimates. The state estimate is `xhat=xhat.x` and signal estimate is `yhat=xhat.y`. The particles are saved at each time instant in the Monte Carlo field `xhat.xMC` and `xhat.yMC`, respectively. These can be illustrated setting the property `'scatter'` to `'on'` in subsequent plot commands. Alternatively, using the standard `'conf'` property, the particles are approximated with a Gaussian distribution and the corresponding confidence bound (as an ellipsoid) is shown. Both the scatter points and the Gaussian approximation can be shown simultaneously of course. The available properties are:

- `Np`: Number of particles (100).

- `k`: Prediction horizon: 0 for filter (default) and 1 for one-step ahead predictor.

- `proposal`; SIR `'sir'` (default), standard bootstrap PF (for low SNR), approximation of optimal proposal `'opt'` (for high SNR).

- `marginal`; `'off'` (default), `'mpf'` gives marginal PF.

- `smoothing`; `'off'` (default), `'ffbs'` gives forward filtering backward smoothing PF.

- `sampling`: `'simple'` (default) standard algorithm (best, but slowest in theory), `'systematic'`, `'residual'` or `'stratified'`.

- `animate` `'off'` (default) or `'on'`. There is a possibility to animate the results after each time step. Further properties control which states to animate, in which type of plot and the pause interval. This option is good for debugging, tuning and learning the PF.

10

Kalman Filter Banks

The *jump Markov linear model* (*JML*) class is essentially a linear state space model where all involved matrices are allowed to depend on a discrete *mode parameter*. The JML is defined by

$$x_{k+1} = A_k(\delta_k)x_k + B_{u,k}(\delta_k)u_k + B_{v,k}(\delta_k)v_k \tag{10.1a}$$
$$y_k = C_k(\delta_k)x_k + D_{u,k}(\delta_k)u_k + e_k \tag{10.1b}$$
$$v_k \sim \mathcal{N}(m_{v,k}(\delta_k), Q_k(\delta_k)) \tag{10.1c}$$
$$e_k \sim \mathcal{N}(m_{e,k}(\delta_k), R_k(\delta_k)), \tag{10.1d}$$
$$p(\delta_k|\delta_{k-1}) = \Pi_k^{(\delta_k,\delta_{k-1})}. \tag{10.1e}$$

Here δ_k is a discrete parameter representing the *mode* of the system (linearized mode, faulty mode etc.), and it takes on one of S different values, $\delta_k \in \{0, 1, \ldots, S-1\}$. In general, it is a hidden Markov state with a specified transition probability matrix Π_k. Here, the binary mode sequence with $S = 2$ is of special importance for its many applications.

The JML class is very flexible and incorporates a variety of problem formulations from different disciplines. The most important here is that different motion models can be used to model different operating conditions of the object, and that mixture noise distributions can be used to model outliers and different measurement conditions.

JML can be seen as a special case of the conditionally linear model studied in Section 9.8. Given the "nonlinear state" trajectory $x_{1:k}^n = \delta_{1:k}$, the model is linear and Gaussian in the "linear state" $x_k^l = x_k$. The Bayesian optimal filter is structurally the same as MPF, the main difference being that the marginalization integral is replaced with a sum due to the discrete state space of $x_{1:k}^n = \delta_{1:k}$ in JML. This discrete state space also allows for clever search algorithms, so deterministic sampling of $x_{1:k}^n = \delta_{1:k}$ is standard for JML, and stochastic sampling as in the MPF is seldom seen.

The chapter outline is as follows:

- Section 10.1 derives the general solution.

- Section 10.2 overviews on-line strategies to reduce the complexity to get recursive algorithms.

- Section 10.3 explains methods to search the space of discrete sequences off-line.

10.1 General Solution

One natural strategy for estimating (more formally, detecting is often used for discrete parameters) $\delta_{1:k}$ is the following:

- For each possible $\delta_{1:k}$, filter the data through a Kalman filter for the (conditional) known state space model (10.1).

- Choose the particular value of $\delta_{1:k}$, whose Kalman filter gives the smallest prediction errors.

- Since the number of combinations of possible sequences $\delta_{1:k}$ is S^k at time k, some logics is needed limit the search.

The resulting *filter bank* structure is illustrated in Figure 10.1.

The key tool in this chapter is a repeated application of Bayes' law to compute *a posteriori* probabilities:

$$w_{k|k} = p(\delta_{1:k}|y_{1:k}) = \frac{p(\delta_{1:k})}{p(y_{1:k})} p(y_{1:k}|\delta_{1:k}) \tag{10.2a}$$

$$= \frac{p(\delta_k|\delta_{k-1})p(\delta_{1:k-1})}{p(y_k|y_{1:k-1})p(y_{1:k-1})} p(y_k|\delta_{1:k}, y_{1:k-1}) p(y_{1:k-1}|\delta_{1:k}) \tag{10.2b}$$

$$= w_{k-1|k-1} \Pi_k^{(\delta_k,\delta_{k-1})} \frac{p(y_k|\delta_{1:k}, y_{1:k-1})}{p(y_k|y_{1:k-1})}. \tag{10.2c}$$

The Kalman filter provides the measurement prediction $\hat{y}_{k|k-1}(\delta_{1:k})$ and its associated covariance $S_{k|k-1}(\delta_{1:k})$, and the unnormalized weights can be updated according to a time update and a measurement update, respectively, as

$$w_{k|k-1} = w_{k-1|k-1} \Pi_k^{(\delta_k,\delta_{k-1})}, \tag{10.3a}$$

$$w_{k|k} \propto w_{k|k-1} \mathcal{N}\big(y_k; \hat{y}_{k|k-1}(\delta_{1:k}), S_{k|k-1}(\delta_{1:k})\big). \tag{10.3b}$$

10.1 General Solution

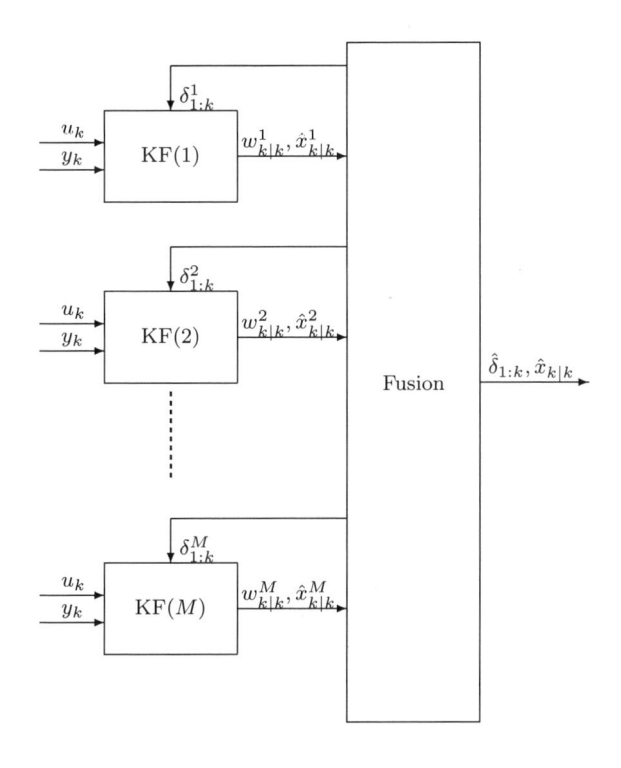

Figure 10.1: The filter bank approach.

Now, let each sequence $\delta^i_{1:k}$ be indexed by an integer i, and similarly for all Kalman filter quantities. The posterior distribution of the state can then be expressed, using the law of total probability, as a Gaussian mixture

$$p(x_k|y_{1:k}) = \sum_i p(\delta^i_{1:k}|y_{1:k}) p(x_k|\delta^i_{1:k}, y_{1:k}) \tag{10.4a}$$

$$= \sum_i w^i_{k|k} \mathcal{N}(x_k; \hat{x}^i_{k|k}, P^i_{k|k}). \tag{10.4b}$$

The minimum variance, or conditional expectation, estimator is given by

$$\hat{x}^{MV}_{k|k} = \sum_{i=1}^{S^k} w^i_{k|k} \hat{x}^i_{k|k}. \tag{10.5}$$

The conditional covariance matrix $P_{k|k}$ for x_k, given the measurements $y_{1:k}$, can be used for giving approximate confidence regions for the conditional mean. Using the *law of total variance* formula

$$\text{Cov}[X] = \text{E}_k \text{Cov}[X|k] + \text{Cov}_k \text{E}[X|k] \quad (10.6)$$
$$= \text{E}_k \text{Cov}[X|k] + \text{E}_k \left(\text{E}[X|k] \text{E}^T[X|k] \right) - \left(\text{E}_k \text{E}[X|k] \right) \left(\text{E}_k \text{E}[X|k] \right)^T$$

where index k means with respect to the probability distribution for k, the conditional covariance matrix follows, as

$$P_{k|k} = \sum_{i=1}^{S^k} w^i_{k|k} \left\{ P^i_{k|k} + \hat{x}^i_{k|k} \hat{x}^i_{k|k} \right\} - \hat{x}^{MV}_k \left(\hat{x}^{MV}_k \right)^T.$$

Thus, the *a posteriori* covariance matrix for x_k, given measurements $y_{1:k}$ is given by three terms. The first is a weighted mean of the covariance matrices for each $\delta_{1:k}$. The last two take the variation in the estimates themselves into consideration. If the estimate of x_k is approximately the same for all $\delta_{1:k}$ the first term is dominating, otherwise the variations in estimate might make the covariance matrices negligible.

As an alternative to the MV estimate, the MAP estimator can be used

$$\hat{\delta}^{MAP}_{1:k} = \arg\max_{\delta} p(\delta_{1:k}|y_{1:k}). \quad (10.7)$$

The joint MAP (JMAP) estimate of mode sequence and current state can be approximated using the assumption that the Gaussian modes are well separated by

$$\widehat{(\delta_{1:k}, x_k)}^{JMAP} = \arg\max_{(\delta_{1:k}, x_k)} p(\delta_{1:k}, x_k|y_{1:k}) \quad (10.8a)$$

$$\hat{\delta}^{JMAP}_{1:k} = \arg\max_{\delta_{1:k}} \frac{p(\delta_{1:k}, x_k|y_{1:k})}{\sqrt{|P_{k|k}(\delta_{1:k})|}} \quad (10.8b)$$

$$\leftrightarrow \arg\max_i \frac{w^i_{k|k}}{\sqrt{|P^i_{k|k}|}}. \quad (10.8c)$$

Note that the peak of a Gaussian distribution is inversely proportional to $\sqrt{|P^i_{k|k}|}$. The MAP estimate $\hat{x}^{MAP}_k$ of the state has no explicit solution and requires a numerical search algorithm.

The following sections will describe different strategies for exploring the huge space of mode sequences.

10.2 On-Line Algorithms

10.2.1 General Ideas

Interpret the exponentially increasing number of discrete sequences $\delta_{1:k}$ as a growing tree, as illustrated in Figure 10.2. The options are to *prune* or

10.2 On-Line Algorithms

merge this tree. In this section, we examine how one can discard elements in δ by cutting off branches in the tree, and lump sequences into subsets of δ by merging branches.

Thus, the basic possibilities for pruning the tree are to *cut off* branches and to *merge* two or more branches into one. That is, two state sequences are merged and in the following treated as just one. There is also a timing question: at what instant in the time recursion should the pruning be performed?

To understand this, the main steps in updating the *a posteriori* probabilities can be divided into a *time update* and a *measurement update* as follows:

- Time update:

$$p(\delta_{1:k-1}|y_{1:k-1}) \longrightarrow p(\delta_{1:k}|y_{1:k-1}) \quad (10.9a)$$
$$p(x_{k-1}|\delta_{1:k-1}, y_{1:k-1}) \longrightarrow p(x_k|\delta_{1:k}, y_{1:k-1}). \quad (10.9b)$$

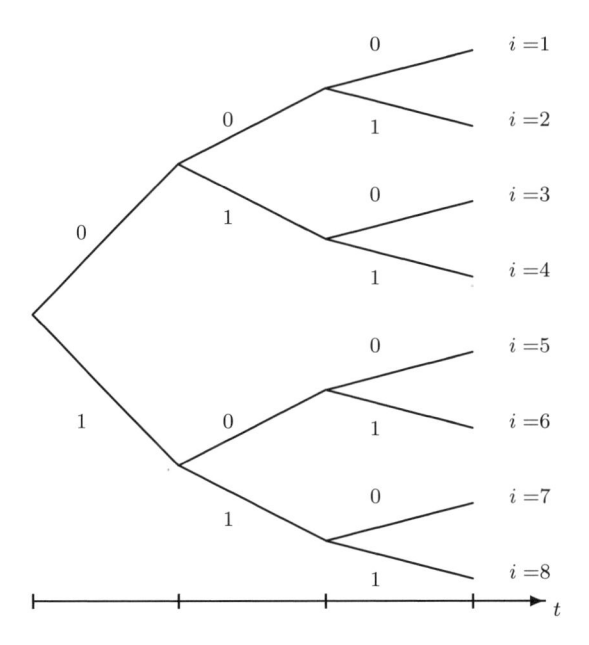

Figure 10.2: *A growing tree of discrete state sequences. For a memory of length $L = 2$, the sequences (1,5), (2,6), (3,7) and (4,8), respectively, are merged. For a memory of length $L = 1$, the sequences (1,3,5,7) and (2,4,6,8), respectively, are merged.*

- Measurement update:

$$p(\delta_{1:k}|y_{1:k-1}) \longrightarrow p(\delta_{1:k}|y_{1:k}) \quad (10.9c)$$
$$p(x_k|\delta_{1:k}, y_{1:k-1}) \longrightarrow p(x_k|\delta_{1:k}, y_{1:k}). \quad (10.9d)$$

Here, the splitting of each branch into S branches is performed in (10.9b). We define the *most probable branch* as the sequence $\delta_{1:k}$, with the largest a posteriori probability $p(\delta_{1:k}|y_{1:k})$ in (10.9c).

10.2.2 Pruning Algorithms

First, a quite general pruning algorithm is given.

In Algorithm 10.1, alternative 2a is straightforward to implement, while 2b requires some logics to keep track of the history over a sliding window, but on the other hand potentially gives better performance.

A more robust weight can be obtained by a second layer of marginalization procedure over the scale of the noise. The statistical assumption is that $P_0, Q_k(\delta_k), R_k(\delta_k)$ all contain an unknown scaling λ, where for simplicity the prior is here assumed noninformative (flat and improper). First, compute recursively the statistcs

$$D_k^i = D_{k-1}^i + \log \det S_k^i, \quad (10.12a)$$
$$V_k^i = V_{k-1}^i + (\varepsilon_k^i)^T (S_k^i)^{-1} \varepsilon_k^i. \quad (10.12b)$$

Then, a robust ML estimator, that minimizes $D_N + N \log V_N$ rather than $D_N + V_N$, is given by

$$\hat{\delta}_{1:k}^{ML} = \arg\min \left(D_k + V_k \right), \quad \text{if } \lambda \text{ is known}$$
$$\hat{\delta}_{1:k}^{ML} = \arg\min \left(D_k + (kn_y - 2) \log V_k \right), \quad \text{if } \lambda \text{ is stochastic.}$$

Exactly the same statistics are involved; the difference being only how the normalized sum of residuals appears. The corresponding weight equations are

$$w_{k|k} \propto e^{-2(D_k + V_k)}, \quad \text{if } \lambda \text{ is known} \quad (10.13a)$$
$$w_{k|k} \propto e^{-2D_k} V_k^{kn_y - 2}, \quad \text{if } \lambda \text{ is stochastic.} \quad (10.13b)$$

For change detection and segmentation purposes, $S = 2$ and $\delta_k = 0$ is the normal outcome while $\delta_k \neq 0$ corresponds to different fault modes. In such case, a much more efficient local search scheme is given below.

10.2.3 Merging Strategies

A General Merging Formula

The exact posterior density of the state vector is a mixture of S^k Gaussian distributions. The key point in merging is to replace, or approximate, a

10.2 On-Line Algorithms

Algorithm 10.1 Filter Bank Pruning: measurement updates

Given a bank of M filters for the JML model (10.1), each filter matched to a sequence $\delta^i_{1:k}$, $i = 1, 2, \ldots, M$, giving the Gaussian prediction density

$$x_k \sim \sum_{i=1}^{M} w^i_{k|k-1} \mathcal{N}(\hat{x}^i_{k|k-1}, P^i_{k|k-1}). \quad (10.10\text{a})$$

Choose the size M of the filter bank either as an integer multiple of S or as the exponential S^L matched to a finite memory L. Compute recursively

1. *Weight measurement update* for $i = 1, 2, \ldots, M$:

$$\bar{w}^i_{k|k} = w^i_{k|k-1} \mathcal{N}(y_k; \hat{y}^i_{k|k-1}, S^i_{k|k-1}), \quad (10.10\text{b})$$

$$w^i_{k|k} = \frac{\bar{w}^i_{k|k}}{\sum_{j=1}^{M} \bar{w}^j_{k|k}} \quad (10.10\text{c})$$

2. *Pruning*: Choose one of the following alternatives:

   a Sort the weights in ascending order, $w^1_{k|k} \geq w^2_{k|k} \geq \cdots \geq w^M_{k|k}$. Keep the most likely M/S first sequences. That is, remove $(S-1)M/S$ least likely sequences..

   b Keep the most likely sequence up to time $k - L + 1$ (that is, $\arg\max p(\delta_{1:k-L+1}|y_{1:k})$) and all S^{L-1} permutations of $\delta_{k-L+2:k}$.

3. *State measurement update* using the Kalman filter for $i = 1, 2, \ldots, M/S$:

$$S^i_k = C_k(\delta^i_k) P^i_{k|k-1} C^T_k(\delta^i_k) + R_k(\delta^i_k), \quad (10.10\text{d})$$

$$K^i_k = P^i_{k|k-1} C^T_k(\delta^i_k)(S^i_k)^{-1}, \quad (10.10\text{e})$$

$$\hat{y}^i_{k|k} = C_k(\delta^i_k) \hat{x}^i_{k|k-1}, \quad (10.10\text{f})$$

$$\hat{x}^i_{k|k} = \hat{x}^i_{k|k-1} + K^i_k(y_k - \hat{y}^i_{k|k-1}) \quad (10.10\text{g})$$

$$P^i_{k|k} = P^i_{k|k-1} - K^i_k S^i_k K^{i}_k. \quad (10.10\text{h})$$

Steps 4–6 are given in Algorithm 10.5.

Algorithm 10.2 Filter Bank Pruning: time updates
Continuation of Algorithm 10.1.

4 *State time update* using the Kalman filter for $i = 1, 2, \ldots, M/S$:

$$\hat{x}^i_{k+1|k} = A_k(\delta^i_k)\hat{x}^i_{k|k} + B_{u,k}(\delta^i_k)u_k \tag{10.11a}$$

$$P^i_{k+1|k} = A_k(\delta^i_k)P^i_{k|k}A^T_k(\delta^i_k) + B_{v,k}(\delta^i_k)Q_k(\delta^i_k)B^T_{v,k}(\delta^i_k). \tag{10.11b}$$

5 *Splitting:* Let each sequence split for $i = 1, 2, \ldots, M/S$ according to:

$$\delta^i_{1:k+1} = (\delta^i_{1:k}, 0), \tag{10.11c}$$

$$\delta^{i+M/S}_{1:k+1} = (\delta^i_{1:k}, 1), \tag{10.11d}$$

$$\vdots \tag{10.11e}$$

$$\delta^{i+(S-1)M/S}_{1:k+1} = (\delta^i_{1:k}, S-1). \tag{10.11f}$$

6 *Weight time update:*

$$w^i_{k+1|k} = w^i_{k|k}\Pi^{(\delta^i_{k+1},\delta^i_k)}, \quad i = 1, 2, \ldots, M. \tag{10.11g}$$

Algorithm 10.3 Filter Bank Local Pruning
As Algorithm 10.1, with

1. *Pruning*: remove all but the $M - S + 1$ most likely sequences.

2. *Splitting:* Let all sequences split according to:

$$\delta^i_{1:k} = (\delta^i_{1:k-1}, 0), \quad i = 1, 2, \ldots, M - S + 1, \tag{10.14a}$$

$$\delta^{M-S+1+s}_{1:k} = (\delta^1_{1:k-1}, s), \quad s = 1, 2, \ldots, S - 1. \tag{10.14b}$$

Some restrictions on the rules above can sometimes be useful:

- Assume a minimum segment length: let the most probable sequence split *only if it is not too young*.

- Assure that sequences are not cut off immediately after they are born: cut off the least probable sequences *among those that are older than a certain minimum life-length* (memory) L, until only M ones are left.

10.2 On-Line Algorithms

number of Gaussian distributions by one single Gaussian distribution in such a way that the first and second moments are matched. That is, a sum of L Gaussian distributions

$$p(x) = \sum_{i=1}^{L} w^i \mathcal{N}(\hat{x}^i, P^i)$$

is approximated by

$$p(x) = w \mathcal{N}(\hat{x}, P),$$

where

$$w = \sum_{i=1}^{L} w^i$$

$$\hat{x} = \frac{1}{w} \sum_{i=1}^{L} w^i \hat{x}^i$$

$$P = \frac{1}{w} \sum_{i=1}^{L} w^i \left(P^i + (\hat{x}^i - \hat{x})(\hat{x}^i - \hat{x})^T \right).$$

The second term in P is the *spread of the mean* (see (10.6)). It is easy to verify that the expectation and covariance are unchanged under the distribution approximation. When merging, all discrete information of the history is lost.

The Generalized Pseudo-Bayesian Algorithm and The Interacting Multiple Model Algorithm

The idea of the *generalized pseudo-Bayesian* (*GPB*) approach is to merge the mixture after the measurement update. It is more efficient to reverse this order, in which case the *interacting multiple model* (*IMM*) algorithm is obtained. The only difference is that merging is applied after the time update of the weights rather than after the measurement update. In this way, a lot of time updates are omitted, which usually do not contribute to performance. Computationally, IMM should be seen as an improvement over GPB. Algorithm 10.4 summarizes IMM and GPB.

The order above corresponds to IMM. For GPB, the merging step is performed after the weight time update. If the KF measurement update is the computational bottleneck, IMM is to prefer since it only has M/S measurement updates. Conversely, if the time update has higher complexity, GPB might be to prefer.

The ordering of the sequences in the splitting step is in this algorithm designed such that the sequences to be merged appear together in the merging step.

Algorithm 10.4 Filter Bank Merging (IMM): measurement updates

Given a bank of M (divisible with S) filters for the JML model (10.1), each filter matched to a sequence $\delta^i_{1:k}$, $i = 1, 2, \ldots, M$, giving the Gaussian prediction density

$$x_k \sim \sum_{i=1}^{M} w^i_{k|k-1} \mathcal{N}\big(\hat{x}^i_{k|k-1}, P^i_{k|k-1}\big). \tag{10.15a}$$

Compute recursively

1 *Weight measurement update* for $i = 1, 2, \ldots, M$:

$$\bar{w}^i_{k|k} = w^i_{k|k-1} \mathcal{N}\big(y_k; \hat{y}^i_{k|k-1}, S^i_{k|k-1}\big), \tag{10.15b}$$

$$w^i_{k|k} = \frac{\bar{w}^i_{k|k}}{\sum_{j=1}^{M} \bar{w}^j_{k|k}}. \tag{10.15c}$$

2 *Merging*: Merge the S^{L-1} sequences corresponding to the same history up to time $k - L$, so $\delta_{k-L+1:k}$ is the same in all terms in each sum below (this is automatically satisfied for the splitting enumeration below). For $j = 1, 2, \ldots, M/S$,

$$w^j_{k|k} = \sum_{i=(j-1)M+1}^{jM} w^i_{k|k} \tag{10.15d}$$

$$\hat{x}^j_{k|k-1} = \frac{1}{w^j_{k|k}} \sum_{i=(j-1)M+1}^{jM} w^i_{k|k} \hat{x}^i_{k|k-1} \tag{10.15e}$$

$$P^j_{k|k-1} = \frac{1}{w^j_{k|k}} \sum_{i=(j-1)M+1}^{jM} w^i_{k|k} \tag{10.15f}$$

$$\cdot \Big(P^i_{k|k-1} + \big(\hat{x}^i_{k|k-1} - \hat{x}^j_{k|k-1}\big)\big(\hat{x}^i_{k|k-1} - \hat{x}^j_{k|k-1}\big)^T\Big). \tag{10.15g}$$

3 *State measurement update* using the Kalman filter for $i = 1, 2, \ldots, M/S$:

$$S^i_k = C_k(\delta^i_k) P^i_{k|k-1} C^T_k(\delta^i_k) + R_k(\delta^i_k), \tag{10.15h}$$

$$K^i_k = P^i_{k|k-1} C^T_k(\delta^i_k)(S^i_k)^{-1}, \tag{10.15i}$$

$$\hat{y}^i_{k|k} = C_k(\delta^i_k) \hat{x}^i_{k|k-1}, \tag{10.15j}$$

$$\hat{x}^i_{k|k} = \hat{x}^i_{k|k-1} + K^i_k \big(y_k - \hat{y}^i_{k|k-1}\big) \tag{10.15k}$$

$$P^i_{k|k} = P^i_{k|k-1} - K^i_k S^i_k K^i_k. \tag{10.15l}$$

Steps 4–6 are given in Algorithm 10.5.

10.2 On-Line Algorithms

Algorithm 10.5 Filter Bank Merging (IMM): time updates
Continuation of Algorithm 10.4.

4 *State time update* using the Kalman filter for $i = 1, 2, \ldots, M$:

$$\hat{x}^i_{k+1|k} = A_k(\delta^i_k)\hat{x}^i_{k|k} + B_{u,k}(\delta^i_k)u_k \tag{10.16a}$$

$$P^i_{k+1|k} = A_k(\delta^i_k)P^i_{k|k}A^T_k(\delta^i_k) + B_{v,k}(\delta^i_k)Q_k(\delta^i_k)B^T_{v,k}(\delta^i_k). \tag{10.16b}$$

5 *Splitting:* Let all sequences split for $i = 1, 2, \ldots, M/S$ according to:

$$\delta^i_{1:k+1} = (\delta^i_{1:k}, 0), \tag{10.16c}$$

$$\delta^{i+M/S}_{1:k+1} = (\delta^i_{1:k}, 1), \tag{10.16d}$$

$$\vdots$$

$$\delta^{i+(S-1)M/S}_{1:k+1} = (\delta^i_{1:k}, S-1). \tag{10.16e}$$

6 *Weight time update:*

$$w^i_{k+1|k} = w^i_{k|k}\Pi^{(\delta^i_{k+1}, \delta^i_k)}, \quad i = 1, 2, \ldots, M. \tag{10.16f}$$

The hypotheses that are merged are identical up to time $N - L$. That is, we do a complete search in a sliding window of size L. In the extreme case of $L = 0$, all hypotheses are merged at the end of the measurement update. This leaves us with S time and measurement updates. Figure 10.2 illustrates how the memory L influences the search strategy.

Note that we prefer to call w^i weight factors rather than posterior probabilities. First, we do not bother to compute the appropriate scaling factors (which are never needed), and secondly, these are probabilities of merged sequences that are not easy to interprete afterwards, in contrast to the pruning alternative.

For target tracking, IMM has become a standard method, see the text books (Bar-Shalom and Fortmann, 1988) and Bar-Shalom and Li (1993). Here, there is an ambiguity in how the mode parameter should be utilized in the model, and a list of possibilities is given in Efe and Atherton (1998).

10.3 Off-Line Algorithms

10.3.1 The Expectation Maximization Algorithm

The *expectation maximization* (*EM*) algorithm (see Baum et al. (1970)), is off-line and alternates between estimating the state vector by a conditional mean $E(x_{1:k}|\delta_{1:N}, y_{1:N})$ (note that this involves a state smoother), given a mode sequence $\delta_{1:N}$, and maximizing the posterior mode sequence probability $p(\delta_{1:N}|x_{1:N}, y_{1:N})$, given the state trajectory. Application to state space models and some recursive implementations are described in the survey Krishnamurthy and Moore (1993).

10.3.2 Markov Chain Monte Carlo Algorithms

The class of *Markov chain Monte Carlo* (*MCMC*) algorithms are off-line, see B.6. The algorithm below is formulated for the case of binary δ_k, which can be represented by J change times $j_1, j_2, \ldots, j_J$. The main drawbacks

---
Algorithm 10.6 Gibbs–Metropolis Segmentation

Choose the number of changes J and marginal change time distribution.

1. Iterate Monte Carlo run i

2. Iterate *Gibbs sampler* for component j_n in $j_{1:J}$, where a random number from
$$\bar{j}_n \sim p(j_n|j_1, j_2, \ldots, j_{n-1}, j_{n+1}, j_J)$$
is taken. Denote the new candidate sequence $\overline{j_{1:J}}$. The distribution may be taken as flat, or Gaussian centered around j_n. If independent jump instants are assumed, this task simplifies to taking random numbers $\bar{j}_n \sim p(j_n)$.

3. Run the conditional Kalman filter using the sequence $\overline{j_{1:J}}$, and save the innovations ε_k and their covariances S_k.

4. The candidate $\overline{j_{1:J}}$ is accepted with probability
$$\min\left(1, \frac{p(\varepsilon_{1:N}(\overline{j_{1:J}}))}{p(\varepsilon_{1:N}(j_{1:J}))}\right).$$

That is, if the likelihood increases we always keep the new candidate. Otherwise we keep it with a certain probability which depends on its likeliness. This random rejection step is the *Metropolis step*.

After the *burn-in* (convergence) time, the distribution of change times can be computed by Monte Carlo techniques.

10.3 Off-Line Algorithms

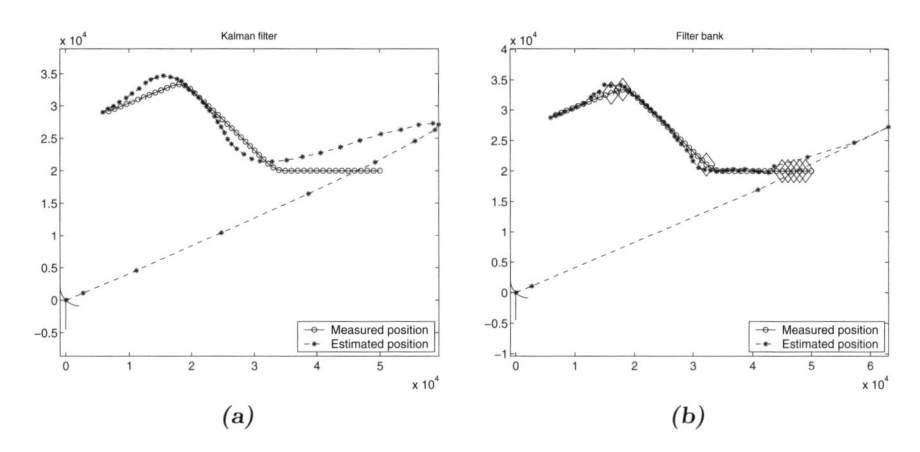

Figure 10.3: Trajectory and filtered position estimate from Kalman filter (a) and Gibbs-Metropolis algorithm with $J = 5$ (b).

with this algorithm is that J has to be specified, and the tricky choice of marginal distribution for the change times. The following example illustrates one option.

Example 10.1: Gibbs–Metropolis change detection

Consider the tracking example illustrated in Figure 10.3(a), which shows a trajectory and the result from the Kalman filter. The jump hypothesis is that the state covariance is ten times larger $Q(1) = 10\, Q(0)$. The jump sequences as a function of iterations and sub-iterations of the Gibbs sampler are shown in Figure 10.4. The iteration scheme converges to two change points at 9 and 36. For the case of overestimating the change points, these are placed at one border of the data sequence (here at the end). The improvement in tracking performance is shown in Figure 10.3(b). The distribution we would like to have in practice is the one from Monte Carlo simulations. Figure 10.5 shows the result from Kalman filter whiteness test and a Kalman filter bank. See Bergman and Gustafsson (1999) for more information.

Algorithm 10.7 removes the drawbacks in Algorithm 10.6.

The marginal distribution for the change times requires an exhaustive search of all S possibilities. However, the Gibbs sampler gives a total complexity NS of sampling instead of the exponential S^N.

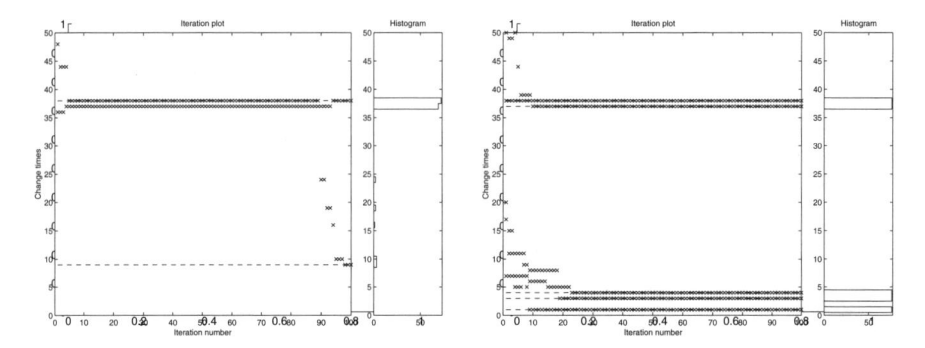

Figure 10.4: Convergence of Gibbs sequence of change times. To the left for $J=2$ and to the right for $J=5$. For each iteration, there are n sub-iterations, in which each change time in the current sequence $j_{1:J}$ is replaced by a random one. The accepted sequences $j_{1:J}$ are marked with 'x'.

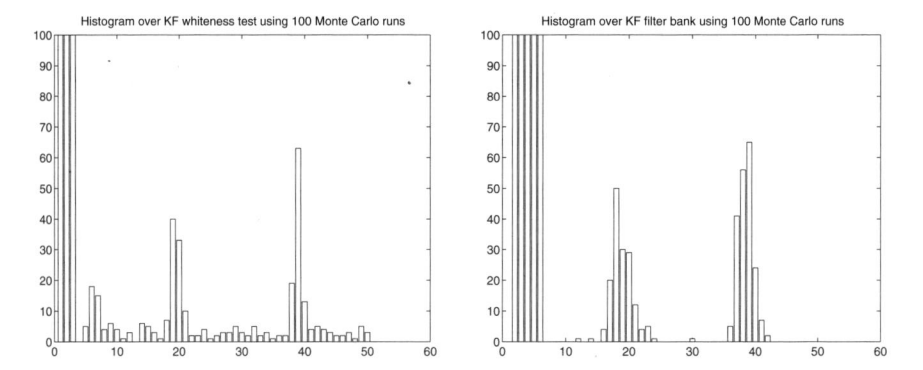

Figure 10.5: Histograms of estimated change times from 100 Monte Carlo simulations using Kalman filter whiteness test and a Kalman filter bank.

10.3 Off-Line Algorithms

Algorithm 10.7 Gibbs MCMC Segmentation

The Gibbs sequence of change times is generated by alternating taking random samples from

$$x_k^{i+1} \sim p(x_k|y_{1:N}, \delta_{1:N}^i)$$
$$\delta_{1:N}^{i+1} \sim p(\delta_{1:N}|y_{1:N}, x_{1:N}^{i+1}).$$

The first distribution is given by the conditional Kalman smoother, since

$$(x_k|\delta_{1:N}, y_{1:N}) \sim \mathcal{N}(\hat{x}_{k|N}(\delta_{1:N}), P_{k|N}(\delta_{1:N})).$$

The second distribution is

$$(\delta_k|x_{1:N}) \sim \frac{1}{\sqrt{(2\pi)^{n_v} \det(Q_k)}}$$
$$\times e^{-\frac{1}{2}(B_{v,k}^\dagger(x_{k+1}-A_k(\delta_k)x_k))^T Q_k^{-1}(\delta_k)(B_{v,k}^\dagger(x_{k+1}-A_k(\delta_k)x_k))}.$$

Here $A^\dagger$ denotes the Moore-Penrose pseudo-inverse.

10.4 Summary
10.4.1 Theory

The JML (jump Markov Linear) model is defined by

$$x_{k+1} = A_k(\delta_k)x_k + B_{u,k}(\delta_k)u_k + B_{v,k}(\delta_k)v_k$$
$$y_k = C_k(\delta_k)x_k + D_{u,k}(\delta_k)u_k + e_k$$
$$v_k \sim \mathcal{N}(m_{v,k}(\delta_k), Q_k(\delta_k))$$
$$e_k \sim \mathcal{N}(m_{e,k}(\delta_k), R_k(\delta_k)),$$
$$p(\delta_k|\delta_{k-1}) = \Pi_k^{(\delta_k, \delta_{k-1})}.$$

The general solution is structurally very similar to the marginalized particle filter with linear states x_k and nonlinear states δ_k. The main difference is that the nonlinear state has a discrete state space, which simplifies the marginalization step from an integral to a sum. The Bayesian filter consists of the following steps:

1. **Kalman filtering:** conditioned on a particular sequence $\delta_{1:k}$, the state estimation problem in (10.1) is solved by a conditional Kalman filter, which delivers the following pairs of estimates and covariances

$$\hat{x}_{k|k-1}(\delta_{1:k}), \quad P_{k|k-1}(\delta_{1:k}),$$
$$\hat{x}_{k|k}(\delta_{1:k}), \quad P_{k|k}(\delta_{1:k}),$$
$$\hat{y}_{k|k-1}(\delta_{1:k}), \quad S_{k|k-1}(\delta_{1:k}).$$

2. **Mode evaluation:** for each sequence, we can compute, up to an unknown scaling factor, the posterior probability given the measurements as weight $w_{k|k} = p(\delta_{1:k}|y_{1:k})$,

$$w_{k|k} \propto w_{k|k-1} p(y_k|\delta_{1:k}, y_{1:k-1}) = w_{k|k-1} \mathcal{N}\big(y_k - \hat{y}_{k|k-1}(\delta_{1:k}), S_{k|k-1}(\delta_{1:k})\big).$$

The weights are then normalized.

3. **Distribution:** at time k, there are S^N different sequences $\delta_{1:k}$ labeled $\delta_{1:k}^i$, $i = 1, 2, \ldots, S^N$, each with an associated weigth $w_{k|k}^i$. It follows from the theorem of total probability that the exact posterior density of the state vector is

$$p(x_k|y_{1:k}) = \sum_{i=1}^{S^N} w_{k|k}^i \mathcal{N}\big(\hat{x}_{k|k}(\delta_{1:k}^i), P_{k|k}(\delta_{1:k}^i)\big).$$

This distribution is a *Gaussian mixture* with S^N modes.

10.4 Summary

4. **Pruning and merging (on-line):** for *on-line* applications, there are two approaches to approximate the Gaussian mixture, both aiming at removing modes so only a fixed number M of modes in the Gaussian mixture are kept. The approximation strategies are *merging* and *pruning*. Pruning is simply to cut off modes in the mixture with low probability, possibly by guaranteeing a minimum lifelength L, with $M = S^L$. In merging, two or more modes are replaced by one new Gaussian distribution so that all combinations in a sliding window of length L are kept.

5. **Numerical search (off-line):** for off-line analysis, there are numerical approaches based on the EM algorithm or MCMC methods.

10.4.2 Software

The class `jml` extends the linear state space models in the `lss` class. The constructor has the syntax `mm=jml(m1,m2,...,Pi)`, and it checks that all dimensions are consistent, so `m1.nn==m2.nn` and so on. The main methods are listed below:

- `xhat=imm(mm,z)` implements the interacting multiple model (IMM) filter in Algorithm 10.4.

- `xhat=pmm(mm,z)` implements the pruning multiple model (PMM) filter in Algorithm 10.1.

- `xhat=gibbs(mm,z)` implements the Gibbs MCMC method in Algorithm 10.7.

11
Simultaneous Localization and Mapping

The area of *simultaneous localization and mapping* (*SLAM*) refers to the problem of determining ones position coordinates p_k using observations to landmarks m^j of unknown position. Landmarks (or features, or beacons) can represent any well-defined point in the world as corners, lines, markers etc. Mapping the landmark positions is thus an integrated part of the problem. For pure navigation purposes, the landmark positions can be considered as nuisance parameters. Conversely, in autonomous missions the navigation state is nuisance and the objective is to map the surroundings.

The SLAM problem is relatively new, and there is currently a fast increase in the number of research papers. The problem is almost exclusively dealt with in the robotics community in journals as IEEE Transactions on Robotics, Intenational Journal of Robotics Research, IEEE Robotics and Automotion Magazine, Autonomous Robots. The most well-cited surveys on SLAM are Bailey and Durrant-Whyte (2006), Durrant-Whyte and Bailey (2006) and Thrun et al. (2005).

The navigation state is in the SLAM literature referred to as "robot pose", where *pose* refers to position and orientation. In the first, and still most common, SLAM applications, the robot is an in-door ground vehicle and the pose is 3DoF (position and heading). For more general, for instance airborne, applications, a full 6DoF pose has to be used.

SLAM concerns the joint estimation of pose and landmark positions, and it is distinguished from standard problems as follows:

- *Localization*, or *positioning*, concerns the estimation of pose (using known landmarks) leading to a filter for the position and heading.

- *Navigation* concerns estimation of pose, velocity and perhaps other states useful for predicting the motion (using known landmarks).

- *Mapping* concerns the estimation of landmark positions from known values of pose.

- One might extend the scope of SLAM to what can be phrased *simultaneous navigation and mapping SNAM* for applications where prediction of the own platform is needed (as in control, collision avoidance, path planning), by augmenting the state vector with velocity and other states than just the pose.

Firstly, one can consider the SLAM problem as a dynamic extension of the sensor network problem in Chapter 4, where one target, rather than several, gets multiple measurements of the landmarks over time. Secondly, one can see it as a classical tracking problem with unknown parameters that include the position of the landmarks.

Section 11.1 introduces the problem and notation. The Kalman filter and particle filter approaches to SLAM are overviewed in Sections 11.2 and 11.3, respectively. Section 11.4 applies the MPF to the SLAM problem, primarily to get a working solution to the SNAM task.

11.1 Introduction

11.1.1 Problem Formulation

The following generic SLAM model will be used in this introductory discussion:

$$x_{k+1} = f(x_k, u_k, v_k), \tag{11.1a}$$

$$\mathbf{m}_{k+1} = \mathbf{m}_k, \tag{11.1b}$$

$$y_k = h(x_k, \mathbf{m}_k, u_k) + e_k. \tag{11.1c}$$

The model consists of three parts:

- A dynamic motion model with an inertial input u_k, where odometric information can be incorporated for wheeled vehicles. Further options for the input u_k are speedometer, accelerometers and gyroscopes. Here, any standard motion model from Chapter 13 can be used.

- A map memory state $\mathbf{m}$ where the position coordinates of all landmarks are stored and updated adaptively.

- A measurement model relating the current state to the map.

The concrete examples that will be discussed are ground vehicles and aircraft. One can distinguish three cases of particular interest:

2DoF The simplest one is when only horizontal position of the vehicle is unknown,

$$x_k = \begin{pmatrix} p_k^X \\ p_k^Y \end{pmatrix}. \tag{11.2}$$

For this problem, a linear model can be used which clearly simplifies algorithm derivation and analysis. Since the target state space is two-dimensional, we will refer to this case as 2DoF SLAM. 2DoF SLAM occurs for instance for a vehicle that measures range to landmarks (non-linear measurement relation), see the sensor network applications in Chapter 4.

3DoF For motion in the horizontal plane, the yaw angle ψ is often needed in the measurement equation and the target state space becomes three-dimensional (3DoF SLAM)

$$x_k = \begin{pmatrix} p_k^X \\ p_k^Y \\ \psi_k \end{pmatrix}. \tag{11.3}$$

This case applies to SLAM problems with body-fixed ('strap-down') bearing sensors that measure the angle between the vehicle and the landmarks. The yaw angle is also needed to deadreckon speed inputs. With an accurate compass, the heading state can be eliminated.

6DoF For full SLAM with a six-dimensional pose, the state vector is

$$x_k = \begin{pmatrix} p_k \\ q_k \end{pmatrix}, \tag{11.4}$$

where p_k is position (p_k^X, p_k^Y, p_k^Z) and q_k the orientation expressed in Euler angles $(\psi_k, \varphi_k, \theta_k)$ or quaternions. As in all navigation applications, derivatives (velocities, accelerations,...) of the pose states and sensor parameters may be included in the state vector.

11.1.2 Relative Measurements

In the first applications of SLAM with 3DoF pose, the observations y_k^i were taken from a scanning laser, which returns horizontal distance and bearing to a landmark. Similar observations are obtained by radar and acoustic sensors as sonar. A landmark is here defined as any obstacle that returns the emitted energy. A camera is a more challenging sensor, where sophisticated computer vision algorithms are needed to find the landmarks, which must be defined as distinct corners or patches in the images that are easy to track.

The observation model for range–bearing sensors discussed thoroughly in Chapter 3 is characteristic for SLAM applications, where only relative positions between target and landmarks are observed. To illustrate, the relation in (3.2) is

$$r_k = \sqrt{(m^X - p_k^X)^2 + (m^Y - p_k^Y)^2} + e_k^r, \tag{11.5a}$$
$$\varphi_k = \arctan 2\left(m^X - p_k^X, m^Y - p_k^Y\right) + e_k^\varphi, \tag{11.5b}$$
$$y_k = (r_k, \varphi_k)^T = h(m^X - p_k^X, m^Y - p_k^Y) + e_k = h(m - p_k) + e_k. \tag{11.5c}$$

The last equation highlights the special structure of the measurement relation. Now, (11.5) can be inverted to (see Section 3.4)

$$\bar{y}_k^1 = r_k \cos(\varphi_k) = m^X - p_k^X + \bar{e}_k^1, \tag{11.6a}$$
$$\bar{y}_k^2 = r_k \sin(\varphi_k) = m^Y - p_k^Y + \bar{e}_k^2, \tag{11.6b}$$
$$\bar{y}_k = m - p_k + \bar{e}_k. \tag{11.6c}$$

This form, which is suitable for direct application of the Kalman filter, retains the structure of relative observations.

When the sensor observations depend on the body orientation, the relative position needs to be rotated. In summary, the sensor models in Chapter 4 becomes in SLAM:

$$\text{2DoF}: \quad y_k = h(m - p_k) + e_k, \tag{11.7a}$$
$$\text{3DoF}: \quad y_k = h\big(R(\psi_k)(m - p_k)\big) + e_k, \tag{11.7b}$$
$$\text{6DoF}: \quad y_k = h\big(R(q_k)(m - p_k)\big) + e_k. \tag{11.7c}$$

11.1.3 Landmark Association

We here discuss the problem when there are many landmarks y_k^i detected which need to be associated with each landmark m^j in the memory at each time. Only the 2DoF case is discussed below for notational convenience.

The observation vector contains in general an unordered list of relative observations

$$y_k^i = h(m_X^j - X, m_Y^j - Y) + e_k^i. \tag{11.8}$$

The association of observation i to landmark j is denoted $j = c_k^i$. This problem falls into the general association problem described in Chapter 5.3. The auction algorithm can be used to compute c_k^i for each observed landmark. The *RANSAC* (RANdom SAmpling Concensus) algorithm offers a more robust alternative, where small subsets taken at random from a larger set of observations are evaluated, and the set that gives the best overall fit to the previously observed landmarks is chosen. Note that RANSAC throws away a majority of the landmark observations y_k^i for the sake of robustness.

Once the association is done, the observation vector can be sorted and the batch model

$$y_k = h(\mathbf{m} - p_k) + e_k \tag{11.9}$$

can be used. The auction algorithm uses only information from the state estimate and observations to compute the statistical distance between observation and each landmark.

Vision-based landmark observations may generally lead to an *implicit nonlinear measurement* of the kind:

$$h(y_k^i, x_k, m^j) = 0. \tag{11.10}$$

This observation model is treated in detail in Section 3.7.

11.1.4 Problem Intuition

The following example is suitable for understanding the problem and to get intuition for the various issues. It will also be used throughout this chapter.

Example 11.1: SLAM introduction

First, a sensor network is set up exactly as in Chapter 4. There are eigth ($M = 8$) landmarks, each one providing range (for instance using time of arrival, TOA) to the target.

```
m0=[0 1 1 1 0 1 -1 0 -1 -1 -1 -1 0 -1 1]';  % True sensor positions
p0=[0.9 -0.4]';                              % Initial target position
M=length(m0)/2;                              % Number of sensors
smod=exsensor('toa',M);                      % Sensor model
smod.x0=p0;                                  % Initial state
smod.th=m0(:);                               % Sensor grid
sigmath=0.1;                                 % Define perturbed network for later use
m=m0+sigmath*randn(2*M,1);                   % Perturbed sensor positions
Pmm=sigmath^2*eye(2*M);                      % Covariance of sensor positions
```

To generate the simulation model, a trick is used that combines the sensor model from the sensor network and a standard motion model's dynamics in the NL constructor. The motion model is a polar velocity model with acceleration and angular rate states (the `'cvpva2d'` option). There are $n_x = 6$ states, $n_y = M = 8$ measurements and $n_\theta = 2M = 16$ parameters in the model.

```
N=8;
fs=1;
R=1e-3*eye(M);                    % Measurement noise
mmod=exnl('ctpva2d',fs);          % Motion model
tmod=nl(mmod.f,smod.h,[6 0 M 2*M]);% Total model
tmod.fs=fs;
tmod.x0=[p0' 2*pi/N pi/2 -0.02 2*pi/N]'; % Initial state
tmod.xlabel=mmod.xlabel;
tmod.th=m;
tmod.pe=R;
y=simulate(tmod,4*N);             % Four laps
xplot2(y)
hold on
plot(smod,'linewidth',2,'fontsize',18)
axis([-1.2 1.2 -1.2 1.2])
```

The turn rate is constant. The acceleration is set to a small negative number to give a spiral rather than a circle. Figure 11.1 shows the sensor network and the trajectory on top of each other.

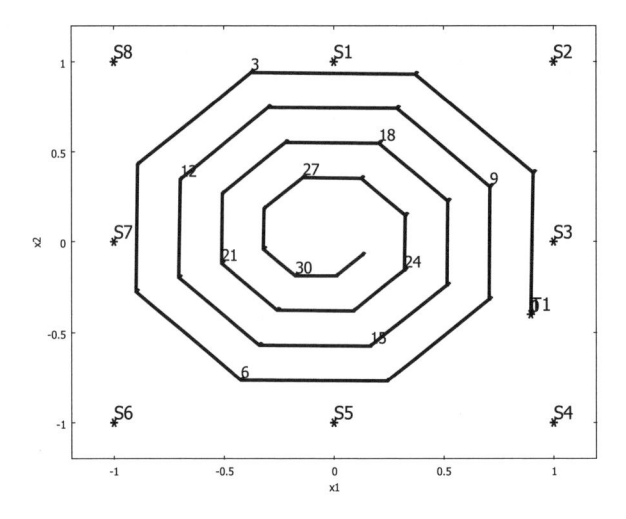

Figure 11.1: Eight landmarks and one target moving in a spiral.

The underlying problems in SLAM can be phrased in terms of observability. For illustration, the 2DoF ground robot application in the example above is used.

- There is a two-dimensional unobservable subspace, since the absolute position cannot be recovered from relative measurements $h(p-m)$ without an earth fixed reference point. For the 3DoF SLAM problem, also the initial yaw angle is unknown and the unobservable subspace becomes three-dimensional. A global reference sensor may for instance be a compass, that removes ambiguity in orientation.

- By initializing the pose in an assumingly known point, observability is obtained, at least initially. The initial reference point is arbitrary and can for instance be taken as the origin: $x_0 = (p_0^X, p_0^Y) = (0,0)$ in 2DoF SLAM and $x_0 = (p_0^X, p_0^Y, \psi_0) = (0,0,0)$ in 3DoF SLAM. Even so, the global reference frame will become diffuse as time increases.

- The cross-correlations between the landmarks are important for splitting up the information in the parts that correspond to relative position and heading, relative landmark positions and the unobservable subspace.

- When the landmarks have not been visible for a while, the relative uncertainty between pose and landmark location increases. Put in other words, using the initialization $(X_0, Y_0, \psi_0) = (0,0,0)$, the landmark covariance matrix becomes rank deficient and loses three degrees of freedom in the limit, which can be seen as a long term drift.

- To eliminate long term drift, a so called *loop closure* is needed. This means that the same set of landmarks are observed after having been out of sight for some while. Loop closure should instantaneously decrease state covariance.

It is important to realize that the pose and landmark positions are too dependent on each other to allow for any approximate *ad-hoc* algorithm. Any approach based on separating the mapping and localization problems will fail, since

$$p(x_k, \mathbf{m}|y_{1:k}) \neq p(x_k|y_{1:k})p(\mathbf{m}|y_{1:k}). \qquad (11.11)$$

In particular, the *certainty equivalence* principle, which iterates between estimating the state given the previous map estimate and estimating the map given the estimated state,

$$\hat{x}_k = \arg\max_{x} p(x_k|y_{1:k}, \hat{\mathbf{m}}_{k-1}), \qquad (11.12a)$$

$$\hat{\mathbf{m}}_k = \arg\max_{\mathbf{m}} p(\mathbf{m}|y_{1:k}, \hat{x}_k), \qquad (11.12b)$$

does not promise anything.

--- **Example 11.2: Certainty equivalence tracking** ---

The following code uses the simulation model for EKF filtering.

```
ekfmod=tmod;                        % Create estimation model
ekfmod.x0=[p0' 0 0 0 0]';           % Initial state
ekfmod.x0=y.x(1,:)';
ekfmod.px0=1000*diag([0 0 1 1 1]);  % Known initial position for SLAM
   observ.
ekfmod.pv=1e-3*diag([0 0 0 0 1 1]);
ekfmod.th=m0;                       % True positions
xhat1=ekf(ekfmod,y);                % EKF for true sensor positions
ekfmod.th=m;                        % Perturbed positions
ekfmod.P=Pmm;
xhat2=ekf(ekfmod,y);                % EKF for perturbed sensor positions
xplot2(xhat1,xhat2,'conf',90,'linewidth',2,'fontsize',18)
hold on
plot(smod,'linewidth',2,'fontsize',18)
axis([-1.2 1.2 -1.2 1.2])
```

Figure 11.2 shows the estimated state trajectory using the true and perturbed vector of landmark positions, respectively. With the true values, the trajectory is more or less perfect, indicating that the tracking problem is quite easy given the measurement noise and multitude of geometrically well placed sensors. With perturbed positions, the performance degrades quickly. Note that the covariance is basically the same in both cases, since it only depends on the landmark geometry and the model.

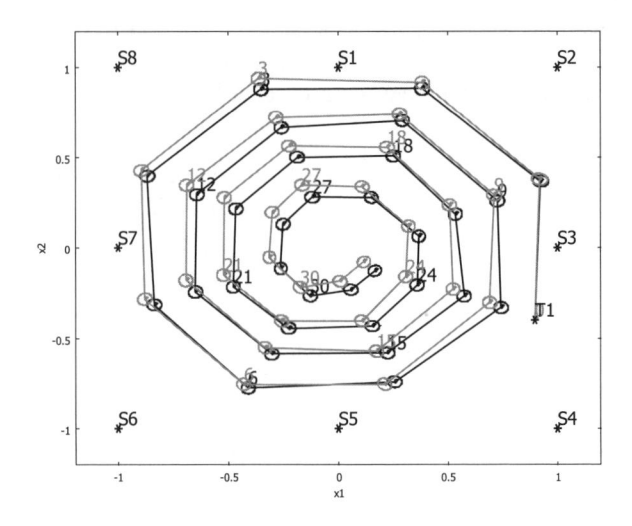

Figure 11.2: *EKF for tracking without mapping using the true (black) and perturbed (gray) landmark positions* **m**, *respectively.*

11.1.5 Parameter Estimation Approach

In a parameter estimation approach, the landmark locations **m** are considered as unknown parameters in a parametric state space model

$$x_{k+1} = f(x_k, u_k, v_k), \tag{11.13a}$$

$$y_k = h(x_k, \mathbf{m}, u_k) + e_k. \tag{11.13b}$$

The NLS algorithm can then be applied to iteratively refine the parameter estimate $\hat{\mathbf{m}}$ based on the innovation sequence and its gradient for each iterate $\hat{\mathbf{m}}^i$. For Gaussian parameter independent measurement noise, NLS coincides with maximum likelihood estimation. This approach is perhaps more conceptual. One of its drawbacks is that it requires a good starting point for **m** to converge.

―― **Example 11.3: NLS SLAM** ――――――――――――――――――――

The code below first estimates the parameters **m** using the NLS algorithm called by the `nl.estimate` method. Then, the calibrated model is used in the EKF filter similar to Example 11.2.

```
nlsmod=tmod;              % Create estimation model
nlsmod.x0=[p0' 0 0 0 0]'; % Initial state
nlsmod.x0=y.x(1,:)';
nlsmod.px0=1000*diag([0 0 1 1 1]);   % Known initial position gives ⇒
    observ.
nlsmod.pv=1e-3*diag([0 0 0 0 1 1]);
nlsmod.th=m;
nlsmod.P=Pmm;
```

11.1 Introduction

```
modhat=estimate(nlsmod,y,'x0mask',[0 0 1 1 1 →
    1],'ctol',1e-2,'gtol',1e-2,'disp','on');
-----------------------------------------------
Iter#           Cost    Grad. norm     BT#   Alg
-----------------------------------------------
    0       1.262e+002       -          -    rgn
    1       5.353e+001   1.771e+003     1    rgn
    2       3.330e+001   7.535e+002     1    rgn
    3       2.859e+001   7.321e+001     1    rgn
    4       2.421e+001   3.014e+002     1    rgn
    5       2.317e+001   8.447e+001     1    rgn
    6       2.297e+001   1.252e+001     1    rgn
    7       2.291e+001   7.313e+000     1    rgn
    8       2.289e+001   4.214e+000     1    rgn
    9       2.288e+001   2.860e+000     1    rgn
Relative difference in the cost function < opt.ctol.
for k=1:8
    Pth{k}=ndist(modhat.th(2*k-1:2*k),mhat.P(2*k-1:2*k,2*k-1:2*k));
end
plot2(Pth{:},'legend','off')
sfcodeoff
hold on
plot(smod,'linewidth',2,'fontsize',18)
hold off
```

Figure 11.3 shows the sensor network, the calibrated sensor locations and the estimated state trajectory using EKF and the estimated landmark positions, respectively. The sensor locations get most of the information from the first measurements, since the initial state is assumed known. That explains why the radial uncertainty to the position at time one has decreased the most for all landmarks.

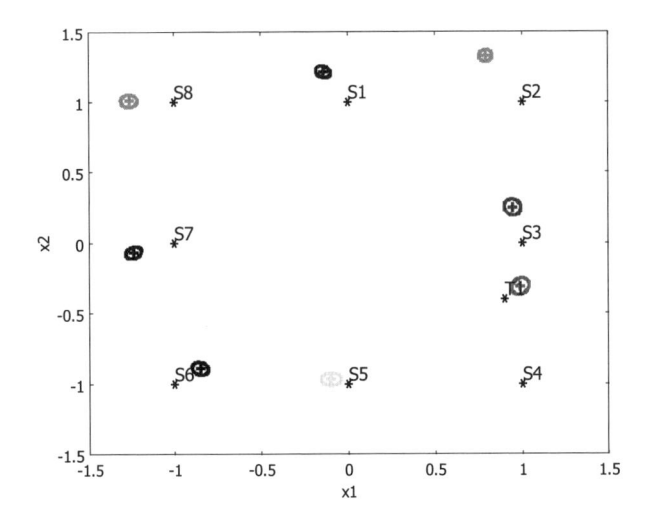

Figure 11.3: NLS estimated landmark positions $\hat{m}_{32|32}$.

11.1.6 Nonlinear Filtering Approach

Using the augmented state vector

$$z_k = \begin{pmatrix} x_k \\ m^1 \\ \vdots \\ m^{n_m} \end{pmatrix} = \begin{pmatrix} x_k \\ \mathbf{m} \end{pmatrix}, \qquad (11.14)$$

the SLAM model falls into the general nonlinear filtering framework, with the following main principles:

- The EKF/UKF approach as described in Section 11.2. The non-standard problem is to overcome the dimensionality of the covariance matrix, since it turns out that the cross-correlations between the landmark positions are very important to keep track of. The information filter resolves this problem, but creates other ones.

- The PF approach leading to the so called FastSLAM algorithm, as described in Section 11.3. The marginalization concept is used to eliminate the landmarks from the PF. The big advantage is that conditioned on the state trajectory $x_{1:k}$, each landmark is represented with a Gaussian distribution, where the conditional cross correlations are zero.

11.2 Kalman Filter Approach

First, the basic algorithm is derived in Section 11.2.1 by direct application of the KF/EKF recursions to the augmented state space model. Then a couple of examples are provided in Section 11.2.2. Sections 11.2.3 and 11.2.4 give the EKF SLAM algorithm in standard and information forms, respectively.

11.2.1 Basic Algorithm

The EKF SLAM algorithm for the linearized model can be summarized as applying the EKF to the model

$$x_{k+1} = A_k x_k + B_k v_k, \qquad \text{Cov}(v_k) = Q_k \qquad (11.15a)$$
$$\mathbf{m}_{k+1} = \mathbf{m}_k, \qquad (11.15b)$$
$$\mathbf{y}_k = C_k^x x_k + C_k^{\mathbf{m}}(c_k^{1:I_k}) \mathbf{m}_k + e_k, \qquad \text{Cov}(e_k) = R_k. \qquad (11.15c)$$

Here, the association indices $(c_k^{1:I_k})$ are implicit in how the matrix $C_k^{\mathbf{m}}(c_k^{1:I_k})$ is constructed based on the association decisions. Basically, $C_k^{\mathbf{m}}$ is a sparse matrix. An association $j = c_k^i$ implies that column j of $C_k^{\mathbf{m}}$ is non-zero. All unassociated landmarks lead to zero columns in $C_k^{\mathbf{m}}$. Exploiting this sparseness, the complexity for the measurement update is $\mathcal{O}(n_y^3 + (n_x + n_m)n_y^2 + (n_x + n_m)^2 n_y)$ which becomes essentially quadratic in the number of observed landmarks n_m for large n_m, see Törnqvist (2008). If all landmarks are observed all the time, then $n_y = n_m$ and the complexity becomes cubic. In the sequel, the association is often surpressed in notation, but this structure is important to keep in mind.

The EKF propagates the following statistics recursively in time

$$\hat{z}_{k|k-1} = \begin{pmatrix} \hat{x}_{k|k-1} \\ \hat{\mathbf{m}}_{k|k-1} \end{pmatrix}, \quad P_{k|k-1} = \begin{pmatrix} P_{k|k-1}^{xx} & P_{k|k-1}^{x\mathbf{m}} \\ P_{k|k-1}^{\mathbf{m}x} & P_{k|k-1}^{\mathbf{mm}} \end{pmatrix}. \tag{11.16}$$

Exploiting the block structure, certain insights for motivating sub-optimal algorithms can be made. The basic form of the EKF recursion gets a measurement update

$$S_k = C_k^x P_{k|k-1}^{xx} C_k^{xT} + C_k^{\mathbf{m}} P_{k|k-1}^{\mathbf{mm}} C_k^{\mathbf{m}T}$$
$$+ C_k^{\mathbf{m}} P_{k|k-1}^{\mathbf{m}x} C_k^{xT} + C_k^x P_{k|k-1}^{x\mathbf{m}} C_k^{\mathbf{m}T} + R_k, \tag{11.17a}$$

$$K_k^x = \left(C_k^x P_{k|k-1}^{xx} + C_k^{\mathbf{m}} P_{k|k-1}^{\mathbf{m}x} \right)^T S_k^{-1}, \tag{11.17b}$$

$$K_k^{\mathbf{m}} = \left(C_k^x P_{k|k-1}^{x\mathbf{m}} + C_k^{\mathbf{m}} P_{k|k-1}^{\mathbf{mm}} \right)^T S_k^{-1}, \tag{11.17c}$$

$$\varepsilon_k = y_k - C_k^x \hat{x}_{k|k-1} - C_k^{\mathbf{m}} \hat{\mathbf{m}}_{k|k-1}, \tag{11.17d}$$

$$\hat{z}_{k|k} = \hat{z}_{k|k-1} + \begin{pmatrix} K_k^x \\ K_k^{\mathbf{m}} \end{pmatrix} \varepsilon_k, \tag{11.17e}$$

$$P_{k|k} = P_{k|k-1} - \begin{pmatrix} K_k^x \\ K_k^{\mathbf{m}} \end{pmatrix} S_k \begin{pmatrix} K_k^x \\ K_k^{\mathbf{m}} \end{pmatrix}^T, \tag{11.17f}$$

and a time update

$$\hat{z}_{k+1|k} = \begin{pmatrix} A_k & 0 \\ 0 & I \end{pmatrix} \hat{z}_{k|k}, \tag{11.18a}$$

$$P_{k+1|k} = \begin{pmatrix} A_k P_{k|k}^{xx} A_k^T + B_k Q_k B_k^T & A_k P_{k|k}^{x\mathbf{m}} \\ P_{k|k}^{\mathbf{m}x} A_k^T & P_{k|k}^{\mathbf{mm}} \end{pmatrix}. \tag{11.18b}$$

11.2.2 Examples

The first example computes the CRLB for the SLAM problem. This is a good tool to investigate the information in a given problem. It can be used to decide if a given sensor deployment, sensor performance and motion model are insufficient for a good SLAM solution.

Example 11.4: EKF SLAM CRLB

So far, a parametric state space model has been used, where the landmark positions are the parameters θ. To put the parameters into an augmented state vector, a new model has to be defined. The code below concatenates the existing function $f(x_k, \theta)$ with the identity mapping. A new simulation is performed, that gives an identical trajectory. The point is that when computing the CRLB, the true state trajectory is needed, and the state is 22 dimensional for this SLAM problem.

```
Ts=1/fs;
fpar=['x(7:',num2str(7+2*M-1),',:)'];
B=[0 0;0 0; Ts^2/2 0; 0 Ts^2/2; Ts 0; 0 Ts];
h=['['];
for k=1:M
    h=[h,['sqrt((x(1,:)-x(',num2str(6+2*k-1),',:)).^2+...
        (x(2,:)-x(',num2str(6+2*k),',:)).^2)']];
    if k<M
        h=[h,';'];
    end
end
h=[h,']'];
f=['[',char(tmod.f),';',fpar,']'];
slammod=nl(f,h,[6+2*M 0 M 0]);   % Total model
slammod.fs=fs;
slammod.xlabel={tmod.xlabel{:},tmod.thlabel{:}};
slammod.x0=[tmod.x0;m0];          % True positions
slammod.pe=1*R;
y2=simulate(slammod,4*N);
slammod.x0=[tmod.x0;m];           % Perturbed positions
P0=1*diag([0 0 1 1 1 1]);         % Known initial state for SLAM observ.
slammod.px0=1*diag([diag(P0);diag(Pmm)]);
Q=zeros(6+2*M);
Q(1:6,1:6)=1e-2*B*B';
slammod.pv=Q;
xcrlb=crlb(slammod,y2);
hold on
xplot2(xcrlb,'conf',90,'linewidth',2,'fontsize',18)
title('CRLB for EKF-SLAM')
for k=1:8
    Pmmk=squeeze(xcrlb.Px(1,6+2*k-1:6+2*k,6+2*k-1:6+2*k));
    mhatk=xcrlb.x(1,6+2*k-1:6+2*k)';
    Pth{k}=ndist(mhatk,Pmmk);
end
plot2(Pth{:},'legend','off')
```

Figure 11.4 shows the sensor network, and the CRLB for filtered trajectory $\hat{x}_{k|k}^{1:2}$ and landmark positions at the end $\hat{x}_{32|32}^{6+2m-1:6+2m}$, for $m = 1, 2, \ldots, 8$.

It is interesting to study the evolution of the state uncertainty. Initially, the uncertainty is set to zero by convention to get observability. The uncertainty then grows for a while because of the process noise. After a few steps ($k = 4$), it starts to decrease again. This effect is due to adaptive learning of the landmark positions.

The actual performance of a SLAM algorithm can often be far from the CRLB bound. However, the next example shows that for this problem, the

11.2 Kalman Filter Approach

CRLB for EKF-SLAM

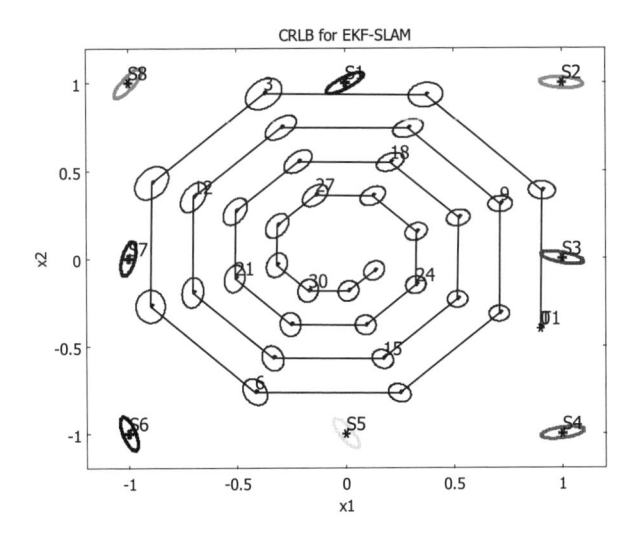

Figure 11.4: *CRLB for SLAM. The plot shows the covariance matrix lower bound for the filtered trajectory $\hat{x}_{k|k}$ and landmark positions at the end $\hat{m}^{(k)}_{32|32}$, for $k = 1, 2, \ldots, 8$. Each ellipsoid is centered around the true value.*

performance is close to CRLB.

---**Example 11.5: EKF SLAM using original model**---

Given the code in Example 11.4, the result from EKF SLAM is now easily computed.

```
xekfslam=ekf(slammod,y2);
hold on
xplot2(xekfslam,'conf',90,'linewidth',2,'fontsize',18)
title('EKF-SLAM')
for k=1:8
    Pmmk=squeeze(xekfslam.Px(1,6+2*k-1:6+2*k,6+2*k-1:6+2*k));
    mhatk=xekfslam.x(1,6+2*k-1:6+2*k)';
    Pth{k}=ndist(mhatk,Pmmk);
end
plot2(Pth{:},'legend','off')
```

Figure 11.5 shows the sensor network, the learned sensor locations and the estimated state trajectory using EKF.

Here, the performance comes close to the CRLB. It should be noted that some careful tuning of Q is needed. It reflects not only the state noise (which happens to be zero for this simulation), but also linearization errors. Choosing it too large gives divergence and too small gives a larger spiral.

EKF-SLAM

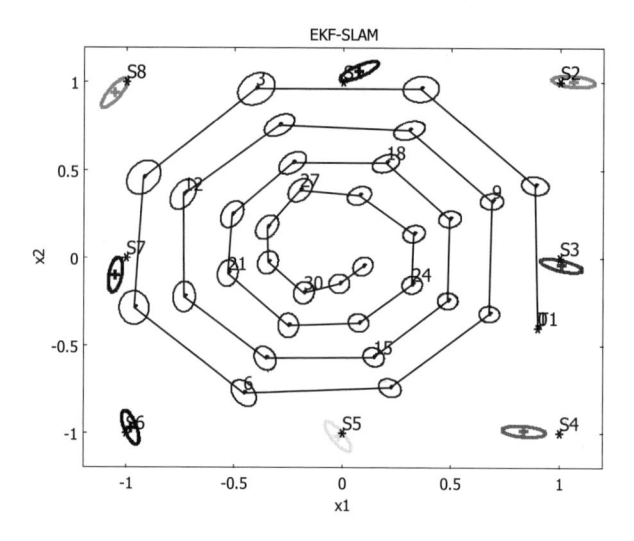

Figure 11.5: *EKF SLAM estimate of trajectory $\hat{x}_{k|k}$ and landmark positions $\hat{m}_{32|32}^{(k)}$ (final estimate) using the correct nonlinear model.*

The good performance of EKF SLAM above is due to the high signal to noise ratio and that the model is able to perfectly predict the whole trajectory if initialized with the true values. The next example investigates a more realistic case where there are model errors. This model cannot simulate an error-free trajectory, and the process noise has to be used as a tool for tuning the model and linearization errors.

── **Example 11.6: EKF SLAM for linear state model** ──

The model (11.15) is linear, so let us here try a linear state space model with constant velocity. The model is here defined from scratch.

```
Ts=1/fs;
f1=['x(1,:)+x(3,:)*',num2str(Ts)];
f2=['x(2,:)+x(4,:)*',num2str(Ts)];
f3='x(3,:)';
f4='x(4,:)';
fpar=['x(5:',num2str(5+2*M-1),',:)'];
B=[Ts^2/2 0; 0 Ts^2/2; Ts 0; 0 Ts];
h=['['];
for k=1:M
   h=[h,['sqrt((x(1,:)-x(',num2str(4+2*k-1),',:)).^2+...
       (x(2,:)-x(',num2str(4+2*k),',:)).^2)']];
   if k<M
      h=[h,';'];
   end
end
h=[h,']'];
f=['[',f1,';',f2,';',f3,';',f4,';',fpar,']'];
slammod=nl(f,h,[4+2*M 0 M 0]);   % Total model
slammod.fs=fs;
slammod.xlabel={'pX','pY','vX','vY',tmod.thlabel{:}};
```

11.2 Kalman Filter Approach

```
slammod.x0=[p0(:);0;2*pi/N;m];
P0=1*diag([0 0 1 1]);              % Known initial state for SLAM observ.
slammod.px0=1*diag([diag(P0);diag(Pmm)]);
Q=zeros(4+2*M);
Q(1:4,1:4)=1e1*B*B';
slammod.pv=Q;
slammod.pe=1*R;
xekfslam=ekf(slammod,y);
hold on
xplot2(xekfslam,'conf',90,'linewidth',2,'fontsize',18)
title('EKF-SLAM')
for k=1:8
    Pmmk=squeeze(xekfslam.Px(1,4+2*k-1:4+2*k,4+2*k-1:4+2*k));
    mhatk=xekfslam.x(1,4+2*k-1:4+2*k)';
    Pth{k}=ndist(mhatk,Pmmk);
end
plot2(Pth{:},'legend','off')
```

Figure 11.6 shows the result, which should be compared to the excellent results in Figure 11.5. The model cannot catch the constant turn rate, so obviously the result is worse now as expected. However, the learning is interesting to study. The final phase of the spiral looks quite good.

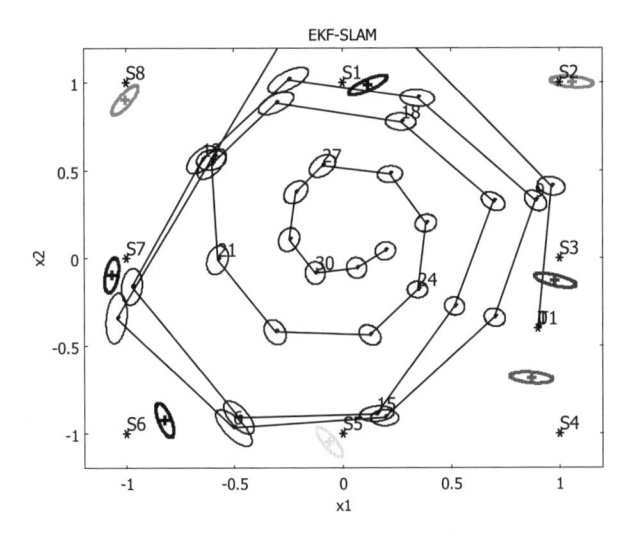

Figure 11.6: EKF SLAM estimate of trajectory $\hat{x}_{k|k}$ and landmark positions $\hat{m}_{32|32}^{(k)}$ (final estimate) using an approximate constant velocity linear model.

It should be remarked that tuning is here even harder than in Example 11.5, and strange results can be obtained with a too small or large Q.

11.2.3 EKF SLAM on Standard Form

The standard form of EKF SLAM follows directly from the recursions above. Algorithm 11.1 summarizes the computations in block component form.

Note that all elements in $P_{k|k}^{\mathbf{mm}}$ are affected by the measurement update, so the sparse time update leading to $P_{k|k-1}^{\mathbf{mm}} = P_{k-1|k-1}^{\mathbf{mm}}$ does not help in mitigating the complexity that makes real-time implementation impossible when the number of landmarks increases to hundreds or thousands. The full correlation structure needs to be taken care of, and there is no obvious sparse structure. The information filter finds such a sparse representation, though.

11.2.4 EKF-SLAM on Information Form

As a rule of thumb, the information filter is computationally to prefer to the standard form when the number of measurements is larger than the process noise dimension, $n_y > n_v$, see Section 7.3.1. This certainly holds for SLAM, where $n_v = 2$ typically in both the 2DoF and 3DoF SLAM formulations. The particular structure of the dynamics where most states are constant in time can be utilized for massive savings in complexity.

First, the sufficient statistic ι and information matrix $\mathcal{I}$ are propagated by the information filter described in Section 7.1.4:

$$\iota_{k|l} = \mathcal{I}_{k|l} \hat{z}_{k|l}, \tag{11.22a}$$

$$\mathcal{I}_{k|l} = P_{k|l}^{-1} = \begin{pmatrix} P_{k|l}^{xx} & P_{k|l}^{x\mathbf{m}} \\ P_{k|l}^{\mathbf{m}x} & P_{k|l}^{\mathbf{mm}} \end{pmatrix}^{-1} = \begin{pmatrix} \mathcal{I}_{k|l}^{xx} & \mathcal{I}_{k|l}^{x\mathbf{m}} \\ \mathcal{I}_{k|l}^{\mathbf{m}x} & \mathcal{I}_{k|l}^{\mathbf{mm}} \end{pmatrix} \tag{11.22b}$$

for $l = k-1$ and $l = k$, respectively. The measurement update (11.17) on information form is given by

$$\iota_{k|k} = \iota_{k|k-1} + C_k^T R_k^{-1} y_k, \tag{11.23a}$$

$$\mathcal{I}_{k|k} = \mathcal{I}_{k|k-1} + C_k^T R_k^{-1} C_k. \tag{11.23b}$$

or by writing out the block components

$$\iota_{k|k}^x = \iota_{k|k-1}^x + C_k^{xT} R_k^{-1} y_k, \tag{11.24a}$$

$$\iota_{k|k}^{\mathbf{m}} = \iota_{k|k-1}^{\mathbf{m}} + C_k^{\mathbf{m}T} R_k^{-1} y_k, \tag{11.24b}$$

$$\mathcal{I}_{k|k}^{xx} = \mathcal{I}_{k|k-1}^{xx} + C_k^{xT} R_k^{-1} C_k^x, \tag{11.24c}$$

$$\mathcal{I}_{k|k}^{x\mathbf{m}} = \mathcal{I}_{k|k-1}^{x\mathbf{m}} + C_k^{xT} R_k^{-1} C_k^{\mathbf{m}}, \tag{11.24d}$$

$$\mathcal{I}_{k|k}^{\mathbf{mm}} = \mathcal{I}_{k|k-1}^{\mathbf{mm}} + C_k^{\mathbf{m}T} R_k^{-1} C_k^{\mathbf{m}}. \tag{11.24e}$$

It is important to note that $C_k^{\mathbf{m}}$ has a very sparse structure, where all columns corresponding to non-observed landmarks are zero. That implies that all correction terms in the measurement update have a very sparse structure.

Algorithm 11.1 EKF SLAM on standard form

Given the model in (11.15).
Initialization:

$$\hat{z}_{1|0} = \begin{pmatrix} \hat{x}_{1|0} \\ \hat{\mathbf{m}}_{1|0} \end{pmatrix}, \quad P_{1|0} = \begin{pmatrix} P^{xx}_{1|0} & P^{x\mathbf{m}}_{1|0} \\ P^{\mathbf{m}x}_{1|0} & P^{\mathbf{mm}}_{1|0} \end{pmatrix}. \tag{11.19a}$$

Time recursion:

1. Associate a map landmark $j = c_k^i$ to each observed landmark i, and construct the matrix $C_k^\mathbf{m}$ in (11.15). This step includes data gating for outlier rejection and track handling to start and end landmark tracks.

2. Measurement update:

$$S_k = C_k^x P^{xx}_{k|k-1} C_k^{xT} + C_k^\mathbf{m} P^{\mathbf{mm}}_{k|k-1} C_k^{\mathbf{m}T}$$
$$+ C_k^\mathbf{m} P^{\mathbf{m}x}_{k|k-1} C_k^{xT} + C_k^x P^{x\mathbf{m}}_{k|k-1} C_k^{\mathbf{m}T} + R_k, \tag{11.20a}$$

$$K_k^x = \left(C_k^x P^{xx}_{k|k-1} + C_k^\mathbf{m} P^{\mathbf{m}x}_{k|k-1} \right)^T S_k^{-1}, \tag{11.20b}$$

$$K_k^\mathbf{m} = \left(C_k^x P^{x\mathbf{m}}_{k|k-1} + C_k^\mathbf{m} P^{\mathbf{mm}}_{k|k-1} \right)^T S_k^{-1}, \tag{11.20c}$$

$$\varepsilon_k = y_k - C_k^x \hat{x}_{k|k-1} - C_k^\mathbf{m} \hat{\mathbf{m}}_{k|k-1}, \tag{11.20d}$$

$$\hat{x}_{k|k} = \hat{x}_{k|k-1} + K_k^x \varepsilon_k, \tag{11.20e}$$

$$\hat{\mathbf{m}}_{k|k} = \hat{\mathbf{m}}_{k|k-1} + K_k^\mathbf{m} \varepsilon_k, \tag{11.20f}$$

$$P^{xx}_{k|k} = P^{x}_{k|k-1} - K_k^x S_k K_k^{x,T}, \tag{11.20g}$$

$$P^{\mathbf{mm}}_{k|k} = P^{\mathbf{mm}}_{k|k-1} - K_k^\mathbf{m} S_k K_k^{\mathbf{m},T}, \tag{11.20h}$$

$$P^{x\mathbf{m}}_{k|k} = P^{x\mathbf{m}}_{k|k-1} - K_k^x S_k K_k^{\mathbf{m},T}. \tag{11.20i}$$

3. Time update:

$$\hat{x}_{k+1|k} = A_k \hat{x}_{k|k}, \tag{11.21a}$$

$$\hat{\mathbf{m}}_{k+1|k} = \hat{\mathbf{m}}_{k|k}, \tag{11.21b}$$

$$P^{xx}_{k+1|k} = A_k P^{xx}_{k|k} A_k^T + B_k Q_k B_k^T, \tag{11.21c}$$

$$P^{\mathbf{mm}}_{k+1|k} = P^{\mathbf{mm}}_{k|k}, \tag{11.21d}$$

$$P^{x\mathbf{m}}_{k+1|k} = A_k P^{x\mathbf{m}}_{k|k}. \tag{11.21e}$$

For instance, only the elements $\mathcal{I}^{(i,j)}$ are changed, where both i and j are observed landmark indices.

The derivation utilizes the splitted time update in one step for the dynamics and one part for the inputs in Algorithm 7.2. To derive the time update, assume first a noise-free dynamic model with $Q_k = 0$ and define the corresponding intermediate information matrix $\bar{\mathcal{I}}$. From (11.18b),

$$\bar{\mathcal{I}}_{k+1|k} = \begin{pmatrix} A_k & 0 \\ 0 & I \end{pmatrix}^{-1} \begin{pmatrix} P^{xx}_{k|k} & P^{xm}_{k|k} \\ P^{mx}_{k|k} & P^{mm}_{k|k} \end{pmatrix}^{-1} \begin{pmatrix} A_k & 0 \\ 0 & I \end{pmatrix}^{-T} \quad (11.25a)$$

$$= \begin{pmatrix} A_k^{-1} \mathcal{I}^{xx}_{k|k} A_k^{-T} & A_k^{-1} \mathcal{I}^{xm}_{k|k} \\ \mathcal{I}^{mx}_{k|k} A_k^{-T} & \mathcal{I}^{mm}_{k|k} \end{pmatrix}. \quad (11.25b)$$

The time update can now be written on inverse form as

$$\mathcal{I}^{-1}_{k+1|k} = \bar{\mathcal{I}}^{-1}_{k+1|k} + \begin{pmatrix} B_k \\ 0 \end{pmatrix} Q_k \begin{pmatrix} B_k^T & 0 \end{pmatrix}, \quad (11.25c)$$

where the *matrix inversion lemma*

$$(A + BCD)^{-1} = A^{-1} - A^{-1} B (DA^{-1} B + C^{-1})^{-1} DA^{-1},$$

gives

$$\mathcal{I}_{k+1|k} = \bar{\mathcal{I}}_{k+1|k} - \bar{\mathcal{I}}_{k+1|k} \begin{pmatrix} B_k \\ 0 \end{pmatrix} \left(\begin{pmatrix} B_k^T & 0 \end{pmatrix} \bar{\mathcal{I}}_{k+1|k} \begin{pmatrix} B_k \\ 0 \end{pmatrix} + Q_k^{-1} \right)^{-1} \begin{pmatrix} B_k^T & 0 \end{pmatrix} \bar{\mathcal{I}}_{k+1|k}. \quad (11.25d)$$

Utilizing the block form, we get

$$M_k = \left(B_k^T \bar{\mathcal{I}}^{xx}_{k+1|k} B_k + Q_k^{-1} \right)^{-1}, \quad (11.26a)$$

$$\mathcal{I}^{xx}_{k+1|k} = \bar{\mathcal{I}}^{xx}_{k+1|k} - \bar{\mathcal{I}}^{xx}_{k+1|k} B_k M_k B_k^T \bar{\mathcal{I}}^{xx}_{k+1|k}, \quad (11.26b)$$

$$\mathcal{I}^{xm}_{k+1|k} = \bar{\mathcal{I}}^{xm}_{k+1|k} - \bar{\mathcal{I}}^{xx}_{k+1|k} B_k M_k B_k^T \bar{\mathcal{I}}^{xm}_{k+1|k}, \quad (11.26c)$$

$$\mathcal{I}^{mm}_{k+1|k} = \bar{\mathcal{I}}^{mm}_{k+1|k} - \bar{\mathcal{I}}^{mx}_{k+1|k} B_k M_k B_k^T \bar{\mathcal{I}}^{xm}_{k+1|k}. \quad (11.26d)$$

Note that the matrix M_k that has to be inverted is only $n_v \times n_v$. A rank n_v matrix is thus subtracted from each block information matrix. Note, however, that this subtraction still affects all elements in the information matrix. The important thing to note is that since information can never be negative, many small elements in the information matrix can be approximated with zeros, and this sound approximation can be used to decrease the complexity.

11.2 Kalman Filter Approach

It remains to find the update for $\iota_{k|k}$. First, using definition (11.22) twice and the time update in (11.18a) once,

$$\iota_{k+1|k} = \mathcal{I}_{k+1|k}\hat{z}_{k+1|k}$$
$$= \mathcal{I}_{k+1|k}\begin{pmatrix} A_k & 0 \\ 0 & I \end{pmatrix}\hat{z}_{k|k}$$
$$= \mathcal{I}_{k+1|k}\begin{pmatrix} A_k & 0 \\ 0 & I \end{pmatrix}\mathcal{I}_{k|k}^{-1}\iota_{k|k}. \quad (11.27a)$$

To avoid explicit inversion of the information matrix, use the original time update in (11.18b)

$$\mathcal{I}_{k+1|k}^{-1} = \begin{pmatrix} A_k & 0 \\ 0 & I \end{pmatrix}\mathcal{I}_{k|k}^{-1}\begin{pmatrix} A_k^T & 0 \\ 0 & I \end{pmatrix} + \begin{pmatrix} B_k Q_k B_k^T & 0 \\ 0 & 0 \end{pmatrix}. \quad (11.28a)$$

By pre-multiplying this equation with $\mathcal{I}_{k+1|k}$ we get

$$I = \mathcal{I}_{k+1|k}\begin{pmatrix} A_k & 0 \\ 0 & I \end{pmatrix}\mathcal{I}_{k|k}^{-1}\begin{pmatrix} A_k^T & 0 \\ 0 & I \end{pmatrix} + \mathcal{I}_{k+1|k}\begin{pmatrix} B_k Q_k B_k^T & 0 \\ 0 & 0 \end{pmatrix}. \quad (11.28b)$$

By combining the results,

$$\iota_{k+1|k} = \left(I - \mathcal{I}_{k+1|k}\begin{pmatrix} B_k Q_k B_k^T & 0 \\ 0 & 0 \end{pmatrix}\right)\begin{pmatrix} A_k^{-T} & 0 \\ 0 & I \end{pmatrix}\iota_{k|k}$$
$$= \begin{pmatrix} I - \mathcal{I}_{k+1|k}^{xx} B_k Q_k B_k^T A_k^{-T} & 0 \\ -\mathcal{I}_{k+1|k}^{mx} B_k Q_k B_k^T A_k^{-T} & I \end{pmatrix}\iota_{k|k}. \quad (11.29)$$

The recursions are summarized in Algorithm 11.2.

The following remarks are important:

- Independent landmark observations imply that R_k is block-diagonal, so the measurement update can be implemented sequentially in the number of observed landmarks, where each update is a low rank correction.

- Each measurement update using observation y_k^i only affects n_y^i elements in the information state $\iota_{k|k}^m$ and an $n_y^i \times n_y^i$ sub-matrix in the information matrix $I_{k|k}^{mm}$. That is, the measurement update scales linearly in the number of observations.

- The time update is a rank(n_v) correction to the information matrix. However, it affects all elements in the information state and matrix, respectively. That means that the time update scales quadratically in the number of landmarks, which will eventually be a problem. An approximation can be based on the following reasoning: If a landmark has not been visible for a long while, its information $\mathcal{I}^{xm^j}$ should be close

Algorithm 11.2 EKF SLAM on information form

Given the model in (11.15).
Initialization:

$$\mathcal{I}_{1|0} = P_{1|0}^{-1}, \tag{11.30a}$$

$$\iota_{1|0} = \mathcal{I}_{1|0}\hat{z}_{1|0}. \tag{11.30b}$$

Time recursion:

1. Associate a map landmark $j = c_k^i$ to each observed landmark j, and construct the matrix $C_k^{\mathbf{m}}$ in (11.15). This step includes data gating for outlier rejection and track handling to start and end landmark tracks.

2. Measurement update:

$$\iota_{k|k}^x = \iota_{k|k-1}^x + C_k^{xT} R_k^{-1} y_k, \tag{11.31a}$$

$$\iota_{k|k}^{\mathbf{m}} = \iota_{k|k-1}^{\mathbf{m}} + C_k^{\mathbf{m}T} R_k^{-1} y_k, \tag{11.31b}$$

$$\mathcal{I}_{k|k}^{xx} = \mathcal{I}_{k|k-1}^{xx} + C_k^{xT} R_k^{-1} C_k^x, \tag{11.31c}$$

$$\mathcal{I}_{k|k}^{x\mathbf{m}} = \mathcal{I}_{k|k-1}^{x\mathbf{m}} + C_k^{xT} R_k^{-1} C_k^{\mathbf{m}}, \tag{11.31d}$$

$$\mathcal{I}_{k|k}^{\mathbf{mm}} = \mathcal{I}_{k|k-1}^{\mathbf{mm}} + C_k^{\mathbf{m}T} R_k^{-1} C_k^{\mathbf{m}}. \tag{11.31e}$$

3. Time update:

$$\bar{\mathcal{I}}_{k+1|k}^{xx} = A_k^{-1} \mathcal{I}_{k|k}^{xx} A_k^{-T}, \tag{11.32a}$$

$$\bar{\mathcal{I}}_{k+1|k}^{x\mathbf{m}} = A_k^{-1} \mathcal{I}_{k|k}^{x\mathbf{m}}, \tag{11.32b}$$

$$M_k = \left(B_k^T \bar{\mathcal{I}}_{k+1|k}^{xx} B_k + Q_k^{-1}\right)^{-1}, \tag{11.32c}$$

$$\mathcal{I}_{k+1|k}^{xx} = \bar{\mathcal{I}}_{k+1|k}^{xx} - \bar{\mathcal{I}}_{k+1|k}^{xx} B_k M_k B_k^T \bar{\mathcal{I}}_{k+1|k}^{xx}, \tag{11.32d}$$

$$\mathcal{I}_{k+1|k}^{x\mathbf{m}} = \bar{\mathcal{I}}_{k+1|k}^{x\mathbf{m}} - \bar{\mathcal{I}}_{k+1|k}^{xx} B_k M_k B_k^T \bar{\mathcal{I}}_{k+1|k}^{x\mathbf{m}}, \tag{11.32e}$$

$$\mathcal{I}_{k+1|k}^{\mathbf{mm}} = \bar{\mathcal{I}}_{k+1|k}^{\mathbf{mm}} - \bar{\mathcal{I}}_{k+1|k}^{\mathbf{m}x} B_k M_k B_k^T \bar{\mathcal{I}}_{k+1|k}^{x\mathbf{m}}, \tag{11.32f}$$

$$\iota_{k+1|k}^x = \left(I - \mathcal{I}_{k+1|k}^{xx} B_k Q_k B_k^T A_k^{-T}\right) \iota_{k+1|k}^x, \tag{11.32g}$$

$$\iota_{k+1|k}^{\mathbf{m}} = \iota_{k|k}^{\mathbf{m}} - \mathcal{I}_{k|k}^{\mathbf{m}x} B_k Q_k B_k^T A_k^{-T} \iota_{k|k}^x \tag{11.32h}$$

4. Estimation:

$$P_{k|k} = \mathcal{I}_{k|k}^{-1}, \tag{11.33a}$$

$$\hat{x}_{k|k} = P_{k|k}^{xx} \iota_{k|k}^x + P_{k|k}^{x\mathbf{m}} \iota_{k|k}^{\mathbf{m}}, \tag{11.33b}$$

$$\hat{\mathbf{m}}_{k|k} = P_{k|k}^{\mathbf{m}x} \iota_{k|k}^x + P_{k|k}^{\mathbf{mm}} \iota_{k|k}^{\mathbf{m}}. \tag{11.33c}$$

to zero (covariance close to infinity) and there is no need to decrease it further, which would affect only the last decimal points. Therefore, the information for these landmarks could be set to a small number of even zero until loop closure occurs.

- These two remarks imply a masking effect, which implies that not all the map needs to be in the cache memory. A submap corresponding to the most nearby landmarks can be masked out in the information filter, yielding a low-dimensional information matrix which can at any time be merged with the full information matrix.

The bottleneck of the EKF SLAM algorithm on information form is the estimation step. The pose estimate is needed for state linearization and data gating, so it is an important step that cannot be circumvented. The problem is the inversion of the information matrix. However, there are decent approximate solutions sufficient for the purpose of linearization and data gating. A first alternative is to use a local map of landmarks when computing (11.33a–b). In this way, the information matrix is replaced with a much smaller matrix. Guidelines for choosing the local map include:

- The most nearby elements provided in a local map.

- The most informative one, where counters of the number of observations and other tricks can be used.

- Exploit approximate sparseness of the information matrix, by replacing small values with zeros.

Another approach is based on parallel algorithms, where a local map is initiated regularly and fused with the current state estimate. At some regular intervals, the local map is fused with the global one.

The preferred alternative, however, is based on a simple gradient search, see Thrun et al. (2005). First note the definition $\iota_{k|k} = \mathcal{I}_{k|k} z_{k|k}$. This can be seen as a linear model $\mathbf{y} = \mathbf{H}x$ and a least mean square (LMS) algorithm can be used to iteratively find a more accurate solution starting with the previous estimate:

$$\bar{z}^{(0)} = \hat{z}_{k-1|k-1}, \tag{11.34a}$$

$$\bar{z}^{(k+1)} = \bar{z}^{(k)} + \mu \mathcal{I}_{k|k}\left(\iota_{k|k} - \mathcal{I}_{k|k} z^{(k)}_{k|k}\right), \quad k = 0, 1, 2, \ldots, K-1 \tag{11.34b}$$

$$\hat{z}_{k|k} = \bar{z}^{(K)}. \tag{11.34c}$$

Here, μ is a suitable step length, and the iterations are terminated at step K which is either pre-defined or determined by a convergence criteria.

11.2.5 Practical Aspects

The following items relate to the problem of handling the landmark vector:

- *Adding new landmarks.* Initializing with a new landmark m_{new} with no prior information is done by state augmentation in the information filter

$$\bar{z} = \begin{pmatrix} z \\ m^{\text{new}} \end{pmatrix} \Rightarrow \bar{\mathbf{m}} = \begin{pmatrix} \mathbf{m} \\ 0 \end{pmatrix}, \quad \bar{\mathcal{I}} = \begin{pmatrix} \mathcal{I} & 0 \\ 0 & 0 \end{pmatrix}. \quad (11.35)$$

The EKF time update takes care of proper initialization of all cross correlations with this new landmark.

- *Data association.* Standard gating methods from target tracking, where each detected landmark is compared to each landmark in the map, may be used but are generally not recommended. An incorrect association may have drastic consequences on performance. In *batch gating*, the idea is to gate all or a large number of observed landmarks jointly to exploit the relative structure in the map. The idea in *appearance signatures* is to use additional landmarks associated to detected landmarks as color and local structure and texture for vision sensors and reflectance for radar. Also *multiple hypothesis test data association (MHT)* is conceivable if computational resources allow.

- *Partitioned updates.* The idea is to initialize regularly new filters with local landmarks, and update the global landmark vector at a slower update rate.

- *Convergence.* It is clear that the measurement update in the information filter can only increase the information matrix. The question is if the time update can destroy information. Algebraically, the time update subtracts information. However, this information loss has low rank. More precisely, it has rank n_v, since the matrix M_k has rank n_v. For 2DoF SLAM, $n_v = 2$. It is intuitively clear that this matrix compensates for the relative uncertainty between pose and all the landmarks, leaving the relative landmark information unchanged. That is, the information of relative landmark position can only increase over time. In the limit when all landmarks are seen many times, the information tends to infinity and the covariance matrix P^{mm} goes to zero.

11.2.6 Summary of EKF SLAM

The EKF approach to SLAM has the following properties:

- EKF SLAM is a rather straightforward application of the EKF to the SLAM problem and should be rather easy to get running on smaller examples.

11.2 Kalman Filter Approach

- EKF SLAM scales cubically in state dimension just as the EKF, which is manageable in most cases (in contrast to the exponential complexity of the FastSLAM algorithm in the next section). This is one of the strengths of EKF SLAM, allowing for a full navigation state.

- EKF SLAM scales quadratically in the number of observed landmarks if the sparseness of the C matrix is exploited. However, the complexity is cubic if all landmarks are observed all the time.

- EKF SLAM on information form scales only quadratically in landmark space, since the measurement update only affects a sub-matrix and the time update is a low rank correction. However, to extract the state estimate requires a matrix inversion that has a cubic cost in state dimension. Decent approximations to state estimation exist, that do not affect global performance of the information filter (except for possible effects of increased linearization error and incorrect associations).

- EKF SLAM is quite sensitive to incorrect associations. Batch data gating is one mitigating action. The RANSAC algorithm picks out a subset of landmarks leading to more robust algorithms at a moderate decrease in information.

- Track handling is a complex issue for the standard filter form, but relatively easy to implement for the information form. Each row in the information state corresponds to one landmark, which can be removed or new ones added without affecting all other statistics. The same holds for the information matrix.

There is a strong duality between the standard and information forms of the Kalman filter. The good properties in the measurement update in the information filter is compensated by getting back the problems in the time update. What favors the information form for SLAM can eventually be summarized in the following rules of thumb:

- *It is easier to set elements in $\mathcal{I}$ to zero than elements in P to infinity.* The numeric algebra simplifies using sparse matrices compared to matrices with very large or infinite elements.

- *It is more sensible to approximate the time update than the measurement update.* A small change in the time update can be interpreted as a change in the motion model, and this is anyhow often *ad-hoc*, while the measurement equation is more exact and should not be approximated.

The next section describes the FastSLAM approach, which basically has opposite properties to EKF SLAM.

11.3 The FastSLAM Algorithm

The main task in localization, navigation and tracking is to estimate the position and heading of the object under consideration. The particle filter has proven to be an enabling technology for many applications of this kind, in particular when the observations are complicated nonlinear functions of the position and heading, see the survey Gustafsson et al. (2002) for a list of such cases.

The FastSLAM algorithm, as described in Montemerlo et al. (2002), has proven to be an enabling technology for such applications. FastSLAM can be seen as a special case of MPF, where the landmark vector $\mathbf{m}_k$ containing the coordinates for all landmarks used in the mapping can be interpreted as the linear part x^l of the state vector. The main difference is that the landmark vector is a constant parameter of increasing dimension over time, rather than a time-varying state with a dynamic evolution over time. The derivation is analogous to marginalized particle filter in Section 9.8, and uses the factorizations

$$p(x_{1:k}, \mathbf{m}|y_{1:k}) = p(\mathbf{m}|x_{1:k}, y_{1:k})p(x_{1:k}|y_{1:k}), \qquad (11.36a)$$

$$p(\mathbf{m}|x_{1:k}, y_{1:k}) = \prod_{j=1}^{n_m} p(\mathbf{m}^j|x_{1:k}, y_{1:k}). \qquad (11.36b)$$

The second factorization is a key one. It says that each landmark position can be treated separately, conditioned on the trajectory, and this will save a lot of computations. For instance, this relaxes the requirement to keep track of the full correlation structure of the landmarks, which is one of the bottlenecks in the EKF SLAM.

The fastSLAM algorithm was originally devised to solve the SLAM problem for mobile robots, where the dimension of the state vector is small, typically consisting of three states (horizontal position and a heading angle), see Thrun et al. (2005). This implies that all platform states can be estimated by the PF.

11.3.1 Conditional Mapping

The goal here is to solve the mapping problem given by the first factor

$$p(\mathbf{m}|x_{1:k}, y_{1:k}), \qquad (11.37)$$

in (11.36). We will constrain the problem to cases where the observation is linear in the landmark positions.

To make the notation more specific, each observed landmark l is associated with a map landmark $j = c_l$, with the inverse association mapping $l = \bar{c}_j$, according to the model

$$0 = h^0(y_k^l, x_k) + h^1(y_k^l, x_k)m^{c_l} + e_k^l, \quad \text{Cov}(e_k^l) = R_k^l. \qquad (11.38)$$

11.3 The FastSLAM Algorithm

This formulation covers:

- Approximate Taylor expansions as detailed in Section 3.7, where an implicit observation model $h^0(y_k, x_k, \mathbf{m}) = 0$ is linearized with respect to the landmark positions $\mathbf{m}$. Note that one does not have to linearize with respect to target pose in contrast to EKF SLAM.
- Bearing and range measurements, where $h(y_k^l, x_k)$ has two rows per landmark in 2DoF SLAM.
- Bearing-only measurements coming from a radar ray or computed by projection theory for image frames. Here, $h(y_k^l, x_k)$ has only one row per landmark in 2DoF SLAM.

The WLS estimate is given by summing up the information relating to map landmark c_l

$$\hat{m}^j = \Big(\underbrace{\sum_{k=1}^{N} h^{1T}(y_k^{\bar{c}_j}, x_k)(R_k^{\bar{c}_j})^{-1} h^1(y_k^{\bar{c}_j}, x_k)}_{\mathcal{I}_N^j} \Big)^{-1}$$

$$\underbrace{\sum_{k=1}^{N} -h^{1T}(y_k^{\bar{c}_j}, x_k)(R_k^{\bar{c}_j})^{-1} h^0(y_k^{\bar{c}_j}, x_k)}_{\iota_N^j}. \quad (11.39)$$

In this sum, h^0 and h^1 are empty if the map landmark j does not get an associated observation landmark at time k.

Under a Gaussian noise assumption, the posterior distribution is Gaussian

$$p\big(m^j | y_{1:N}, x_{1:N}\big) = \mathcal{N}\big((\mathcal{I}_N^j)^{-1} \iota_N^j, (\mathcal{I}_N^j)^{-1}\big). \quad (11.40)$$

This distribution is exact if the measurement model is exact and the noise Gaussian, otherwise it is just an approximation. However, for mapping there is probably no need for more sophisticated posterior representations than the Gaussian. The main shortcoming with a Gaussian approximation is its inability to model multimodal posteriors, but the landmark location posterior distribution is unimodal by nature for most sensors.

A recursive version of the WLS estimate is provided by the Kalman filter. Here, we again favorize the information form for its simpler book-keeping for large maps,

$$\iota_k^j = \iota_{k-1}^j + h^{1T}(y_k^{\bar{c}_j}, x_k) R_k^{-1} h^0(y_k^{\bar{c}_j}, x_k), \quad (11.41a)$$
$$\mathcal{I}_k^j = \mathcal{I}_{k-1}^j + h^{1T}(y_k^{\bar{c}_j}, x_k) R_k^{-1} h^1(y_k^{\bar{c}_j}, x_k), \quad (11.41b)$$
$$\hat{m}^j = (\mathcal{I}_k^j)^{-1} \iota_k^j. \quad (11.41c)$$

An important consequence of this result is that the likelihood for a new observation of map landmark l can be computed by

$$p(y_k^{\bar{c}_j}|y_{1:k-1}, x_{1:k}) =$$
$$\mathcal{N}\left(-h^0(y_k^{\bar{c}_j}, x_k); h^1(y_k^{\bar{c}_j}, x_k)\hat{m}_{k-1}^l, R_k^{\bar{c}_j} + h^1(y_k^{\bar{c}_j}, x_k)(\mathcal{I}_k^j)^{-1}h^{1T}(y_k^{\bar{c}_j}, x_k)\right).$$
(11.42)

This follows from standard results for linear estimation.

11.3.2 Marginalization

The goal here is to solve the mapping problem given by the second factor $p(x_{1:k}|y_{1:k})$ in (11.36). Here, the marginalization trick

$$p(x_{1:k+1}|y_{1:k}) = p(x_{1:k}|y_{1:k})p(x_{k+1}|x_{1:k}, y_{1:k})$$
$$= p(x_{1:k}|y_{1:k})\int p(x_{k+1}, \mathbf{m}|x_{1:k}, y_{1:k})d\mathbf{m}$$
$$= p(x_{1:k}|y_{1:k})\int p(x_{k+1}|\mathbf{m}, x_{1:k}, y_{1:k})p(\mathbf{m}|x_{1:k}, y_{1:k})d\mathbf{m}$$
$$= p(x_{1:k}|y_{1:k})\int p(x_{k+1}|\mathbf{m}, x_{1:k}, y_{1:k})\mathcal{N}(\mathbf{m}; \hat{\mathbf{m}}, \text{Cov}(\mathbf{m}))d\mathbf{m}.$$

This is completely analogous to (9.48) in Section 9.8. Intuitively, the equation above says that the map uncertainty can be interpreted as an extra process noise in the time update of the particle filter.

11.3.3 The Algorithm

Algorithm 11.3 summarizes the steps in FastSLAM as a special case of the MPF in Algorithm 9.2. From Chapter 9 the proposal densities in what in SLAM literature is called FastSLAM 1.0 and 2.0 correspond to SIR and the optimal proposal, respectively, see Section 9.5.

11.3.4 Summary of FastSLAM

Algorithm 11.3 has the following properties:

- FastSLAM is surprisingly simple to implement. Each particle corresponds to one state trajectory with an associated map consisting of mean and covariance of each map landmark. There are no cross-correlations thanks to the conditioning on pose history.

- FastSLAM scales linearly in landmark dimension, since each landmark has its own WLS filter.

11.3 The FastSLAM Algorithm

Algorithm 11.3 FastSLAM

Given the motion model $p(x_{k+1}|x_k)$ and an observation model of the form

$$h^0(y_k^l, x_k) + h^1(y_k^l, x_k)m^{c_l} + e_k^l = 0.$$

Initialize N particles:

$$x_1^{(i)} \sim p_0(x), \quad i = 1, 2, \ldots, N,$$

Time recursion:

1. Perform data association that assigns a map landmark $c_l^{(i)}$ to each observed landmark l and each particle i. Initialize new map landmarks if necessary.

2. Compute the importance weights according to

$$\omega_k^{(i)} = \prod_l \mathcal{N}\big(h^0(y_k^l, x_k^{(i)}) + h^1(y_k^l, x_k^{(i)})\hat{m}_{k-1}^{c_l^{(i)}},$$

$$R_k^l + h^1(y_k^l, x_k^{(i)})(\mathcal{I}_k^{c_l^{(i)}})^{-1} h^{1T}(y_k^l, x_k^{(i)})\big).$$

where the product is taken over all observed landmarks l, and normalize $\bar{\omega}_k^{(i)} = \omega_k^{(i)} / \sum_{j=1}^N \omega_k^{(j)}$.

3. Draw N new particles with replacement (resampling) according to

$$\Pr(x_k^{(i)} = x_k^{(j)}) = \bar{\omega}_k^{(j)}, j = 1, \ldots, N.$$

4. Pose time update as in FastSLAM 1.0

$$x_{k+1}^{(i)} \sim p(x_{k+1}|x_{1:k}^{(i)}) = p(x_{k+1}|x_k^{(i)}).$$

or FastSLAM 2.0

$$x_{k+1}^{(i)} \sim p(x_{k+1}|x_{1:k}^{(i)}, y_{1:k+1}) = \frac{1}{C} p(x_{k+1}|x_k^{(i)}) p(y_{k+1}|x_{k+1}).$$

5. Map measurement update:

$$\iota_k = \iota_{k-1} + h^{1T}(y_k, x_k) R_k^{-1} h^0(y_k, x_k),$$
$$\mathcal{I}_k = \mathcal{I}_{k-1} + h^{1T}(y_k, x_k) R_k^{-1} h^1(y_k, x_k),$$
$$p(\mathbf{m}^{(i)}|x_{1:k}^{(i)}, y_{1:k}) = \mathcal{N}\left((\mathcal{I}_k^{(i)})^{-1} \iota_k^{(i)}, (\mathcal{I}_k^{(i)})^{-1}\right).$$

- As the standard PF, FastSLAM scales badly in the state dimension. This limits its application to mainly 3DoF localization problems with three states.

- FastSLAM has a built-in *association diversity* that makes it relatively robust to local incorrect associations. The reason is that each particle has its unique association history, and the particle cloud can handle multiple association hypotheses automatically.

- Also *landmark handling* is easy to implement for FastSLAM. Since each map landmark is treated independently, new ones can be introduced and old ones removed without affecting the other ones.

- The main problem with FastSLAM is how to utilize *loop closure*. There are two reasons for this:

  - Revisiting an old feature/landmark gives important information of the whole map, but the map's correlation structure is not kept track of.

  - As an alternative to update the complete map, one could adjust the complete pose trajectory, and then rely on that each landmark is correctly represented with respect to the trajectory. However, the particle filter is quite bad on estimating the trajectory $x_{1:k}$ and the reason is depletion caused by the resampling.

The next section describes the marginalized FastSLAM approach, that combines all the good landmarks of the EKF SLAM and FastSLAM.

11.4 Marginalized FastSLAM

The aim is to solve the SLAM problem when the state dimension of the platform is too large to be estimated by the PF. This section provides a problem formulation and introduces the necessary notation.

The marginalized particle filter (MPF) enables estimation of velocity, acceleration and sensor error models by utilizing linear substructures in the model, which is fundamental for performance in applications as surveyed in Schön et al. (2006). The MPF as described in Schön et al. (2005) splits up the state vector x into one 'nonlinear' part x^n where the particle filter is applied, and another 'linear' part x^l where the Kalman filter is applied.

Concepts and Notation

The total state vector to be estimated at time k is

$$x_k = \left((x_k^n)^T \quad (x_k^l)^T \quad \mathbf{m}^T \right)^T, \qquad (11.45)$$

11.4 Marginalized FastSLAM

where x_k^n denotes the states of the platform that are estimated by the particle filter, and x_k^l denotes the states of the platform that are linear-Gaussian given information about x_k^n. These states together with the map **m** are estimated using Kalman filters.

The *key* factorization, which allows us to solve this problem successfully is

$$p(x_{1:k}^n, x_k^l, \mathbf{m}|y_{1:k}) = \prod_{j=1}^{M_k} \underbrace{p(m^j|x_{1:k}^n, x_k^l, y_{1:k})}_{\text{WLS}} \underbrace{p(x_k^l|x_{1:k}^n, y_{1:k})}_{\text{KF}} \underbrace{p(x_{1:k}^n|y_{1:k})}_{\text{PF}}. \quad (11.46)$$

A general model structure that extends MPF in Section 9.8 to SLAM is given by

$$x_{k+1}^n = f_k^n(x_k^n) + A_k^n(x_k^n)x_k^l + B_k^n(x_k^n)v_k^n, \quad (11.47\text{a})$$

$$x_{k+1}^l = f_k^l(x_k^n) + A_k^l(x_k^n)x_k^l + B_k^l(x_k^n)v_k^l, \quad (11.47\text{b})$$

$$m_{k+1}^j = m_k^j, \quad (11.47\text{c})$$

$$y_{1,k} = h_{1,k}(x_k^n) + C_k(x_k^n)x_k^l + e_{1,k}, \quad (11.47\text{d})$$

$$y_{2,k}^{(j)} = h_{2,k}(x_k^n) + C_{j,k}(x_k^n)m_k^j + e_{2,k}^{(j)}, \quad (11.47\text{e})$$

where $j = 1, \ldots, M_k$ and the noise for the platform states is assumed white and Gaussian distributed with

$$v_k = \begin{pmatrix} v_k^n \\ v_k^l \end{pmatrix} \sim \mathcal{N}(0, Q_k), \quad Q_k = \begin{pmatrix} Q_k^n & Q_k^{nl} \\ (Q_k^{nl})^T & Q_k^l \end{pmatrix}. \quad (11.47\text{f})$$

The measurement noise is assumed white and Gaussian distributed according to

$$e_{1,k} \sim \mathcal{N}(0, R_{1,k}), \quad (11.47\text{g})$$

$$e_{2,k}^j \sim \mathcal{N}(0, R_{2,k}^j), \quad j = 1, \ldots, M_k. \quad (11.47\text{h})$$

Finally, x_0^k is Gaussian,

$$x_0 \sim \mathcal{N}(\bar{x}_0, \bar{P}_0), \quad (11.47\text{i})$$

and the density for x_0^p can be arbitrary, but it is assumed known.

There are two different measurement models, (11.47d) and (11.47e), where the former only measures quantities related to the platform, whereas the latter will also involve the map states.

Algorithm

The FastSLAM algorithm first presented in Montemerlo et al. (2002) is here extended to include both the map states *and* the states corresponding to a linear Gaussian sub-structure in Algorithm 11.4, see Törnqvist et al. (2009) for more details and examples.

Algorithm 11.4 Marginalized FastSLAM

Initialize N particles:

$$x_{1|0}^{n,(i)} \sim p(x_{1|0}^{n}), \quad x_{1|0}^{l,(i)} = \bar{x}_{1|0}^{l}, \quad P_{1|0}^{l,(i)} = \bar{P}_{1|0}, \quad i = 1, \ldots, N,$$

Time recursion:

1. If there is a new map related measurement available perform data association for each particle, otherwise proceed to step 3.

2. Compute the importance weights according to

$$\omega_k^{(i)} = p(y_k | x_{1:k}^{n,(i)}, y_{1:k-1}), \qquad i = 1, \ldots, N,$$

   and normalize $\omega_k^{(i)} = \omega_k^{(i)} / \sum_{j=1}^{N} \omega_k^{(j)}$.

3. Draw N new particles with replacement (resampling) according to, for each $i = 1, \ldots, N$

$$\Pr(x_{k|k}^{n,(i)} = x_{k|k}^{n,(j)}) = \tilde{\omega}_k^j, \qquad j = 1, \ldots, N.$$

   The resampling also involves copying the associated Kalman filter state and covariance associated with each particle.

4. If there is a new map related measurement, perform map estimation and management (detailed below), otherwise proceed to step 6.

5. Particle filter prediction and Kalman filter (for each particle $i = 1, \ldots, N$)

   (a) Kalman filter measurement update,

$$p(x_k^l | x_{1:k}^n, y_{1:k}) = \mathcal{N}(x_k^l; \hat{x}_{k|k}^{l,(i)}, P_{k|k}^{(i)}),$$

   where $\hat{x}_{k|k}^{l,(i)}$ and $P_{k|k}^{(i)}$ are given by the Kalman filter.

   (b) Time update for the nonlinear particles,

$$x_{k+1|k}^{n,(i)} \sim p(x_{k+1|k} | x_{1:k}^{n,(i)}, y_{1:k}).$$

   (c) Kalman filter time update,

$$p(x_{k+1}^l | x_{1:k+1}^n, y_{1:k}) = \mathcal{N}(x_{k+1}^l | \hat{x}_{k+1|k}^{l,(i)}, P_{k+1|k}^{(i)}),$$

   where $\hat{x}_{k+1|k}^{k,(i)}$ and $P_{k+1|k}^{(i)}$ are given by the Kalman filter.

11.4.1 Summary of MFastSLAM

Algorithm 11.4 has the following properties:

- MFastSLAM is more complex than FastSLAM to implement, while the difference is perhaps smaller than expected. Each particle corresponds to one *pose* trajectory with one Gaussian distribution representing the other navigation states (velocities and so on) and an associated map consisting of mean and covariance of each map landmark. There are no cross-correlations.

- MFastSLAM scales linearly in landmark dimension.

- As the standard MPF, MFastSLAM scales nicely with state dimension in most navigation and positioning applications.

- MFastSLAM is as robust to local incorrect associations as FastSLAM.

- MFastSLAM shares the difficulty to utilize loop closure with FastSLAM.

11.5 Summary

11.5.1 Theory

The SLAM model consists of a motion model for the state x_k, a map memory for features/landmarks $\mathbf{m}_k$ and a measurement relation:

$$x_{k+1} = f(x_k, u_k, v_k),$$
$$\mathbf{m}_{k+1} = \mathbf{m}_k,$$
$$y_k^l = h(x_k, \mathbf{m}_k^{c_l}, u_k) + e_k, \ l \in I_k.$$

Here, I_k is the set of landmarks at time k and c_l is the association mapping from observation index l to map index c_l. A good SLAM algorithm is characterized by the following properties:

- Robust to association errors.
- Flexible to nonlinear motion models and measurement relations.
- The fundamental information when having a *loop closure* must be fully utilized.
- It should scale nicely with both the state and landmark dimensions.

There are three conceptually different approaches:

1. *Parameter estimation:* The map vector $\mathbf{m}$ is seen as a parameter that is iteratively estimated by running a conditional nonlinear filter (EKF, PF) at each iteration. This is suitable for batch-processing and when a good prior value of the vector $\mathbf{m}$ is available. Using gradient search techniques, this approach scales nicely in both state and landmark dimensions. Loop closure does not pose any difficulties.

2. *EKF-SLAM:* The map vector $\mathbf{m}$ is augmented to the state vector and the EKF is applied using a linearized model. EKF-SLAM handles loop closure, but is sensitive to linearization errors and more importantly it is inherently sensitive to association errors.

   The standard EKF has cubic complexity $\mathcal{O}\big((n_x + n_m)^3\big)$ in the time update. The particular structure with constant parameters imply here that the time update is only $\mathcal{O}\big((n_x)^3\big)$, which is independent of landmark dimension. However, the measurement update affects all elements in the covariance matrix. This is mitigated with the information filter, where only the elements in the information matrix corresponding to the observed landmarks are affected. The drawback is that all elements of the information matrix are affected in the time update (there are ideas for how to approximate this) and that computation of the state estimate requires inversion of the information matrix (again approximations are available).

11.5 Summary

3. *FastSLAM:* The map vector $\mathbf{m}$ is augmented to the state vector, and the MPF is applied using $x_k^l = \mathbf{m}$. FastSLAM represents the pose with a set of particles, where each particle has its own global map $\mathbf{m}^i$ of independent map landmark locations. The PF approach has a built-in association diversity, since each particle handles one hypothesis of the complete association history $c_{1:k}^i$. It is difficult to utilize loop closure in FastSLAM.

   FastSLAM scales well with landmark dimension, since each landmark is estimated with one WLS algorithms. As the standard PF, FastSLAM is restricted to small state dimensions. Using a second layer of marginalization for the dynamic state, the MFastSLAM algorithm offers a solution that scales well in both state and landmark dimensions.

Thus, there is so far no approach that satisfies all criteria for a good SLAM algorithm.

11.5.2 Software

There are no dedicated multi-purpose SLAM algorithms implemented so far. For simple simulation experiments, the NLS algorithm in `nl.estimate` and the EKF in `nl.ekf` can be used as illustrated in the examples in this chapter.

Part III
Practice

12

Modeling

Physical modeling often leads to a mixture of continuous-time state dynamics and discrete-time observations,

$$\dot{x}(t) = a(x(t), u(t); \theta) + v(t), \qquad (12.1a)$$
$$y(t_k) = c(x(t_k), u(t_k); \theta) + e(t_k). \qquad (12.1b)$$

The time update in filters needs a prediction model for how the state evolves from one observation to the next, so the continuous-time model needs to be either simulated or discretized. These cases correspond to numerical and analytical solutions, respectively, to the ordinary differential equation in (12.1a). The analytical approach is in focus here. In general, the analytical discretization can also affect the measurement relation, leading to the standard model for nonlinear filtering

$$x_{k+1} = f(x_k, u_k; \theta) + v_k \qquad (12.2a)$$
$$y_k = h(x_k, u_k; \theta) + e_k. \qquad (12.2b)$$

Further, in a Kalman filtering framework, a linear model of the form

$$x_{k+1} = F(\theta)x_k + G(\theta)u_k + v_k \qquad (12.3a)$$
$$y_k = H(\theta)x_k + J(\theta)u_k + e_k. \qquad (12.3b)$$

is needed. Another problem is that the physical model may have unknown or partially known parameters θ that need to be determined.

This chapter discusses the following issues:

- Discretization of linear models in Section 12.1.

- Discretization of nonlinear models in Section 12.2.
- Discretization of the continuous-time process noise v_k in Section 12.3.
- Choice of coordinate system to faciliate discretization of nonlinear models and to minimize the effects of linearization errors in Section 12.4.
- Modeling the time variations of the sensor measurement error in Section 12.5.
- The option to upsample or downsample the dynamics in Section 12.6.
- Calibration, or gray-box identification, of unknown parameters from a sequence of test data in Section 12.7.

12.1 Discretizing Linear Models

This section provides a summary of sampling *linear time-invariant* (*LTI*) continuous-time systems on state space form

$$\dot{x}(t) = Ax(t) + Bu(t),$$
$$y(t) = Cx(t) + Du(t),$$

to a discrete-time system

$$x_{k+1} = Fx_k + Gu_k,$$
$$y_k = Hx_k + Ju_k,$$

The solution for a *zero-order hold* (*ZOH*) assumption, where the input $u(\tau) = u_k$ is assumed piecewise constant over the interval $t = kT \leq \tau < t + T = (k+1)T$, is given by multiplying with the integrating factor

$$\frac{d}{dt}\left(e^{-At}x(t)\right) = e^{-At}\dot{x}(t) - Ae^{-At}x(t) = e^{-At}Bu(t).$$

The matrix exponential is defined by a Taylor expansion

$$e^{AT} = I + AT + A^2\frac{T^2}{2!} + A^3\frac{T^3}{3!} + \ldots$$

Note that the matrix exponential commutes, so $Ae^{-At} = e^{-At}A$. Integrating both sides from t to $t + T$ gives

$$\left[e^{-At}x(t)\right]_t^{t+T} = \int_t^{t+T} e^{-A\tau}Bu(\tau)d\tau$$

12.1 Discretizing Linear Models

The solution is

$$x(t+T) = e^{AT}x(t) + \int_0^T e^{A\tau}Bu(t+T-\tau)d\tau = e^{AT}x(t) + \int_0^T e^{A\tau}d\tau Bu(t).$$

From this, we can identify

$$x_{k+1} = Fx_k + Gu_k,$$
$$F = e^{AT},$$
$$G = \int_0^T e^{A\tau}d\tau B = \left(IT + A\frac{T^2}{2!} + A^2\frac{T^3}{3!} + \ldots\right)B.$$

Using the definition of the matrix exponential, the sampled system can be computed in one step as

$$\exp\left(\begin{bmatrix} AT & BT \\ 0 & 0 \end{bmatrix}\right) = \begin{bmatrix} F & G \\ 0 & I \end{bmatrix}.$$

This follows directly from the definition. There is a similar formula for *first-order hold* (*FOH*), where the rate of change is assumed constant over each sample period. For both ZOH and FOH, the output relation is unchanged, so $H = C$ and $J = D$.

The matrix exponential can be computed efficiently using the following equality

$$e^A = \left(e^{A/2^j}\right)^{2^j}.$$

The idea is that the outer power is computed by squaring the exponential matrix j times, where j is determined to be large enough to get a rapid decay in the Taylor expansion of $e^{A/2^j}$.

The *bilinear transformation* is another alternative to ZOH and FOH common in signal processing. It maps the left half plane onto the unit circle, so stability is always preserved. The conversion between continuous and discrete-time is done using the following operator formalism:

$$s \approx \frac{2}{T}\frac{z-1}{z+1},$$
$$z \approx \frac{2/T + s}{2/T - s}.$$

Applying this to a LTI state space model gives

$$\frac{2}{T}\frac{z-1}{z+1}x(t) \approx sx(t) = Ax + Bu$$

which can be written in standard form as

$$M = (I_{n_x} - \tfrac{T}{2}A)^{-1},$$
$$F = M(I_{n_x} + \tfrac{T}{2}A),$$
$$G = \tfrac{T}{2}MB,$$
$$H = CM,$$
$$J = D + HG.$$

The conversion back again is given by

$$A = \tfrac{2}{T}\left(I_{n_x} + F\right)^{-1},$$
$$\bar{M} = \left(I_{n_x} - \tfrac{T}{2}F\right),$$
$$B = \tfrac{2}{T}\bar{M}G,$$
$$C = H\bar{M},$$
$$D = J - HG.$$

12.2 Discretizing Nonlinear Models

This section considers a continuous-time nonlinear model with discrete time observations:

$$\dot{x}(t) = a(x(t); \theta) + v(t) \tag{12.4a}$$
$$y(t_k) = c(x(t_k); \theta) + e(t_k). \tag{12.4b}$$

The input $u(t)$ in (12.1a) and (12.1b) is here omitted for clarity.

There are basically two alternatives in passing from a continuous nonlinear model to a discrete linear model:

- First linearize (12.4a) by a first order Taylor expansion, and then apply the sampling formulas in Section 12.1.

- First try to discretize (12.4a) by analytically solving an integral formula, and then linearize.

These approaches will be referred to as *discretized linearization* and *linearized discretization*, respectively.

12.2.1 Discretized Linearization

A Taylor expansion of each component of the state in (12.4a) around the current estimate $\hat{x}$ yields

$$\dot{x}_i = a_i(\hat{x}) + a'_i(\hat{x})(x - \hat{x}) + \frac{1}{2}(x - \hat{x})^T a''_i(\xi)(x - \hat{x}) + v_{i,t}, \tag{12.5}$$

12.2 Discretizing Nonlinear Models

where ξ is a point in the neighborhood of x and $\hat{x}$. Here a'_i denotes the derivative of the ith row of a with respect to the state vector x, and a''_i is the exptected value of the second derivative. We will not investigate transformations of the state noise yet. It will be assumed that the discrete-time counterpart has a covariance matrix $\bar{Q}$. The following example applies discretized linearization to a standard motion model, see also Table 13.4.

---**Example 12.1: Target tracking: coordinated turns**---

There are two main alternatives to model the dynamics of a *coordinated turn* (circular movement) using either Cartesian or polar velocity, which are here compared. The dynamics for the state vector using Cartesian velocity, $(x^{(1)}, x^{(2)}, v^{(1)}, v^{(2)}, \omega)^T$, is given by:

$$\begin{aligned}\dot{x}^{(1)} &= v^{(1)} \\ \dot{x}^{(2)} &= v^{(2)} \\ \dot{v}^{(1)} &= -\omega v^{(2)} \\ \dot{v}^{(2)} &= \omega v^{(1)} \\ \dot{\omega} &= 0.\end{aligned} \qquad (12.6)$$

If the turn rate ω is piecewise constant, the dynamics describe a trajectory that consists of circular movements. The derivative in the Taylor expansion (12.5) is

$$A = a'_{cv}(x) = \begin{pmatrix} 0 & 0 & 1 & 0 & 0 \\ 0 & 0 & 0 & 1 & 0 \\ 0 & 0 & 0 & -\omega & -v^{(2)} \\ 0 & 0 & \omega & 0 & v^{(1)} \\ 0 & 0 & 0 & 0 & 0 \end{pmatrix}. \qquad (12.7)$$

Second, with the state vector with polar velocity, $(x^{(1)}, x^{(2)}, v, h, \omega)^T$, the state dynamics become:

$$\begin{aligned}\dot{x}^{(1)} &= v\cos(h) \\ \dot{x}^{(2)} &= v\sin(h) \\ \dot{v} &= 0 \\ \dot{h} &= \omega \\ \dot{\omega} &= 0.\end{aligned} \qquad (12.8)$$

The derivative in the Taylor expansion (12.5) is now

$$A = a'_{pv}(x) = \begin{pmatrix} 0 & 0 & \cos(h) & -v\sin(h) & 0 \\ 0 & 0 & \sin(h) & v\cos(h) & 0 \\ 0 & 0 & 0 & 0 & 0 \\ 0 & 0 & 0 & 0 & 1 \\ 0 & 0 & 0 & 0 & 0 \end{pmatrix}. \tag{12.9}$$

The discrete-time system is now specified by the matrix exponential, $F = e^{AT}$.

12.2.2 Linearized Discretization

A different and more accurate approach is to first discretize (12.4a). In some rare cases, of which tracking with constant turn rate is one example, the state space model can be discretized exactly by solving the sampling formula

$$x(t+T) = x(t) + \int_t^{t+T} a(x(\tau))d\tau \tag{12.10}$$

analytically. The solution can be written

$$x(t+T) = f(x(t)). \tag{12.11}$$

The following example illustrates the principle of exact sampling for one very common motion model. This model is also summarized in Table 13.4.

──── **Example 12.2: Exact sampling of coordinated turn model** ────

Consider the tracking example with $x = (x^{(1)}, x^{(2)}, v, h, \omega)^T$. The analytical solution of (12.10) using (12.8) is

$$x^{(1)}(t+T) = x^{(1)}(t) + \frac{2v(t)}{\omega(t)}\sin(\frac{\omega(t)T}{2})\cos(h(t) + \frac{\omega(t)T}{2})$$
$$x^{(2)}(t+T) = x^{(2)}(t) + \frac{2v(t)}{\omega(t)}\sin(\frac{\omega(t)T}{2})\sin(h(t) + \frac{\omega(t)T}{2})$$
$$v(t+T) = v(t) \tag{12.12}$$
$$h(t+T) = h(t) + \omega(t)T$$
$$\omega(t+T) = \omega(t).$$

This defines the function $x_{k+1} = f(x_k, T)$. Note that this discretization allows a time varying sample interval, which is convenient in filtering problems with irregular sampling. The alternate state coordinates $x = (x^{(1)}, x^{(2)}, v^{(1)}, v^{(2)}, \omega)^T$ give, with (12.6) in (12.10),

$$\begin{aligned}
x^{(1)}(t+T) &= x^{(1)}(t) + \frac{v^{(1)}(t)}{\omega(t)}\sin(\omega(t)T) - \frac{v^{(2)}(t)}{\omega(t)}(1-\cos(\omega(t)T)) \\
x^{(2)}(t+T) &= x^{(2)}(t) + \frac{v^{(1)}(t)}{\omega(t)}(1-\cos(\omega(t)T)) + \frac{v^{(2)}(t)}{\omega(t)}\sin(\omega(t)T) \\
v^{(1)}(t+T) &= v^{(1)}(t)\cos(\omega(t)T) - v^{(2)}(t)\sin(\omega(t)T) \quad (12.13) \\
v^{(2)}(t+T) &= v^{(1)}(t)\sin(\omega(t)T) + v^{(2)}(t)\cos(\omega(t)T) \\
\omega(t+T) &= \omega(t).
\end{aligned}$$

For both models, care should be taken in the implementation for the case of straight motion, when $\omega(t) \approx 0$. These calculations are quite straightforward to compute using symbolic computation programs.

12.2.3 Numerical Simulation by Fast Sampling

If exact sampling is not possible to do analytically, we can get a good approximation of an exact sampling by numerical integration. This can be implemented using the discretized linearization approach in the following way. Suppose we have a function [x,P]=tu(x,P,T) for the time update, implementing $x_{k+1} = A(x_k, T)x_k$. Then the fast sampling method, in MATLAB™ formalism,

```
for i=1:M;
    [x,P]=tu(x,T/M);
end;
```

provides a more accurate time update for the state. In the limit, we can expect that the resulting state update converges to $f(x)$. An advantage with numerical integration is that it provides a natural approximation of the state noise, see the next section. Section 12.6.1 describes fast sampling in more detail.

A further unusual yet sound alternative is based on using numerical integration routines. There are many good simulation tools for nonlinear ordinary differential equations, like ode45 in MATLAB™. Note that the continuous time Ricatti equation $\dot{P} = AP + PA + Q$ can be used as the ODE for the covariance matrix.

12.3 Discretizing State Noise

The state noise in (12.4a) has hitherto been neglected. In this section, we discuss different ideas on how to approximately sample the state noise $\bar{Q}$ in the discrete-time model.

There are basically five different alternatives for computing a time discrete state noise covariance from a time continuous one,

$$\dot{x}(t) = a(x(t)) + v(t), \quad \text{Cov}(v(t)) = Q,$$
$$x(t+T) = f(x(t)) + \bar{v}(t), \quad \text{Cov}(\bar{v}(t)) = \bar{Q},$$

The alternatives are listed below. All expressions are normalized with T, so that one and the same Q can be used independently of the sampling interval.

a. $v(t)$ is assumed to be continuous white noise with variance Q.

$$\bar{Q}_a = \int_0^T e^{a'(x(kT+\tau))} Q e^{\left(a'(x(kT+\tau))\right)^T} d\tau \tag{12.14a}$$

This integral can seldom be solved analytically. However, using the fast sampling idea in Section 12.2.3, a numerical approximation is obtained by the Riemann sum $\sum_{m=1}^{M} A(x_{kT+mT/M}, T/M) Q A^T(x_{kT+mT/M}, T/M)$.

b. $v(t) = v_k$ is assumed to be a stochastic variable which is constant in each sample interval and has variance Q/T.

$$\bar{Q}_b = \frac{1}{T} \int_0^T e^{a'(x(kT+\tau))} d\tau Q \int_0^T e^{\left(a'(x(kT+\tau))\right)^T \tau} d\tau \tag{12.14b}$$

c. $v(t)$ is assumed to be a sequence of Dirac impulses active immediately after a sample is taken. Loosely speaking, we assume $\dot{x} = a(x) + \sum_k v_k \delta_{kT-t}$ where v_k is discrete white noise with variance TQ,

$$\bar{Q}_c = T e^{a'(x(kT))T} Q e^{(a'(x(kT)))^T T} \tag{12.14c}$$

d. $v(t)$ is assumed to be white noise such that its total influence during one sample interval is TQ,

$$\bar{Q}_d = TQ \tag{12.14d}$$

e. $v(t)$ is assumed to be a discrete white noise sequence with variance TQ. That is, we assume that the noise enters immediately after a sample time, so $x(t+T) = f(x(t) + v(t))$,

$$\bar{Q}_e = T f'(x) Q \big(f'(x)\big)^T. \tag{12.14e}$$

All but the first methods correspond to more or less *ad hoc* assumptions on the state noise. Figure 12.1 shows examples of noise realizations corresponding to assumptions a–c.

It is impossible to say *a priori* which assumption is the most logical one. Instead, one should investigate the alternatives (12.14a)–(12.14e) by Monte Carlo simulations and determine the importance of this choice for a certain trajectory.

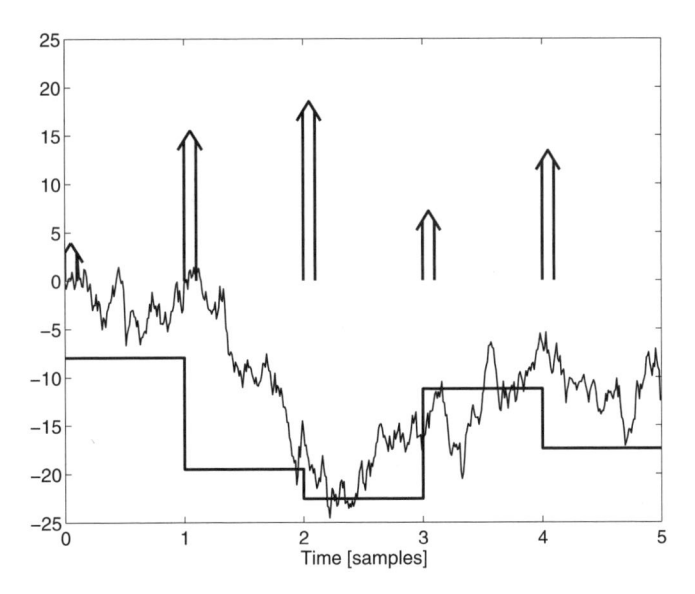

Figure 12.1: *Examples of assumptions on the state noise. The arrow denotes impulses in continuous-time and pulses in discrete time.*

12.4 Linearization Error and Choice of State Coordinates

The error we make when linearizing $A = a'_x(x)$ or $F = f'_x(x)$ depends only on the size of the second order rest term $(x - \hat{x})^T (a^{(i)})''(\xi)(x - \hat{x})$ and $(x - \hat{x})^T (f^{(i)})''(\xi)(x - \hat{x})$, respectively. This error propagates to the covariance matrix update and for discretized linearization also to the state update. This observation implies that we can use the same analysis for linearized discretization and discretized linearization. The Hessian a''_i of component i of $a(x)$ will be used for notational convenience, but the same expressions holds for f''_i. Let $\tilde{x}(t+T) = x(t+T) - \hat{x}(t+T|t)$ denote the state prediction error. Then we have

$$4\|\dot{\tilde{x}}^{(i)}\|_2^2 = \|(x - \hat{x})^T (a^{(i)})''(\xi)(x - \hat{x})\|_2^2. \tag{12.15}$$

Care must be taken when comparing different state vectors because $\|x - \hat{x}\|_2$ is not an invariant measure of state error in different coordinate systems, as the following example illustrates.

─── **Example 12.3: Target tracking: best state coordinates** ───

In target tracking with coordinated turns, we want to determine which of the following two coordinate systems is likely to give the smallest linearization error:

$$x = (x^{(1)}, x^{(2)}, v^{(1)}, v^{(2)}, \omega)^T$$
$$x = (x^{(1)}, x^{(2)}, v, h, \omega)^T.$$

For notational simplicity, we will drop the time index in this example. For these two cases, the state dynamics are given by

$$a_{cv}(x) = (v^{(1)}, v^{(2)}, -\omega v^{(2)}, \omega v^{(1)}, 0)^T$$
$$a_{pv}(x) = (v\cos(h), v\sin(h), 0, \omega, 0)^T,$$

respectively.

Suppose the initial state error is expressed in $\|\tilde{x}_{cv}\|_2^2 = (\tilde{x}^{(1)})^2 + (\tilde{x}^{(2)})^2 + (\tilde{v}^{(1)})^2 + (\tilde{v}^{(2)})^2 + \tilde{\omega}^2$. The problem is that the error in heading angle is numerically much smaller than in velocity. We need to find the scaling matrix such that $\|\tilde{x}_{pv}\|_2^2$ is a representative metric. Using

$$v^{(1)} = v\cos(h)$$
$$v^{(2)} = v\sin(h),$$

we have

$$(v^{(1)} - \hat{v}^{(1)})^2 + (v^{(2)} - \hat{v}^{(2)})^2 = v^2 + \hat{v}^2 - 2v\hat{v}\cos(h - \hat{h})$$
$$= v^2 + \hat{v}^2 - 2v\hat{v} + 2v\hat{v}(1 - \cos(h - \hat{h}))$$
$$= (v - \hat{v})^2 + 4v\hat{v}\sin^2\left(\frac{h - \hat{h}}{2}\right)$$
$$\approx (v - \hat{v})^2 + v^2(h - \hat{h})^2.$$

The approximation is accurate for angular errors less than, say, 40° and relative velocity errors less than, say, 10%. Thus, a weighting matrix

$$J_{pv} = \text{diag}(1, 1, 1, v, 1)$$

should be used in (12.15) for polar velocity and the identity matrix J_{cv} for Cartesian velocity. Only then can the norms be compared to each other.

The error in linearizing state variable i is

$$4\|\dot{\tilde{x}}^{(i)}\|_2^2 = \|(x - \hat{x})^T J J^{-1}(a^{(i)})''(\xi) J^{-1} J(x - \hat{x})\|_2^2$$
$$\leq \|J(x - \hat{x})\|_2^4 \|J^{-1}(a^{(i)})''(\xi) J^{-1}\|_2^2.$$

The Frobenius norm defined by $\|X\|_F^2 = \sum (x^{(ij)})^2$, where the sum is taken over all elements $x^{(ij)}$ of X, will be used to bound the 2-norm. Using the

12.4 Linearization Error and Choice of State Coordinates

inequality $\frac{1}{n}\|X\|_F^2 \leq \|X\|_2^2 \leq \|X\|_F^2$, the linearization error can be bounded by

$$4\|\dot{\tilde{x}}^{(i)}\|_2^2 \leq \|J(x-\hat{x})\|_2^4 \|J^{-1}(a^{(i)})''(\xi)J^{-1}\|_F^2. \quad (12.16)$$

The total error is then

$$4\|\dot{\tilde{x}}\|_2^2 = \sum_{i=1}^{5} 4\|\dot{\tilde{x}}^{(i)}\|_2^2. \quad (12.17)$$

The above calculations show that if we start with the same initial error in two different state coordinate systems, (12.16) quantizes the linearization error, which can be upper bounded by taking the Frobenius norm instead of two-norm, on how large the error in the time update of the extended Kalman filter can be.

The remaining task is to give explicit expressions of the rest terms in the Taylor expansion, first given in Gustafsson and Isaksson (1996). For discretized linearization, the Frobenius norm of the Hessian for the state transition function $f(x)$ is

$$\sum_{i=1}^{5} \|(a_{cv}^{(i)})''(x)\|_F^2 = 4T^2$$

for coordinates $(x^{(1)}, x^{(2)}, \dot{x}^{(1)}, \dot{x}^{(2)}, \omega)^T$, and

$$\sum_{i=1}^{5} \|J_{pv}^{-1}(a_{pv}^{(i)})''(x)J_{pv}^{-1}\|_F^2 = (1+\frac{2}{v})T^2,$$

for coordinates $(x^{(1)}, x^{(2)}, v, h, \omega)^T$. Here we have scaled the result with T^2 to get the integrated error during one sampling period, so as to be able to compare it with the results below. For linearized discretization, the corresponding norms are

$$\sum_{i=1}^{5} \|(f_{cv}^{(i)})''\|_F^2 \approx 4T^2 + T^4\left(1+v^2\right) \quad (12.18)$$

$$+v^2T^6\left(\frac{1}{9}(\omega T)^0 - \frac{1}{240}(\omega T)^2 + \frac{1}{12600}(\omega T)^4 - \frac{1}{1088640}(\omega T)^6 + O\left((\omega T)^8\right)\right)$$

and

$$\sum_{i=1}^{5} \|J_{pv}^{-1}(f_{pv}^{(i)})''J_{pv}^{-1}\|_F^2 \approx T^2\left(1+\frac{2}{v}\right) + T^4\left(\frac{1}{2}+\frac{v}{2}\right) \quad (12.19)$$

$$+v^2T^6\left(\frac{1}{9}(\omega T)^0 - \frac{1}{240}(\omega T)^2 + \frac{1}{12600}(\omega T)^4 - \frac{1}{1088640}(\omega T)^6 + O\left((\omega T)^8\right)\right).$$

The approximation here is that ω^2 is neglected compared to $4v$. The proof is straightforward, but the use of symbolic computation programs as MAPLE[TM] or MATHEMATICA[TM] is recommended.

Note first that linearized discretization gives an additive extra term that grows with the sampling interval, but vanishes for small sampling intervals, compared to discretized linearization. Note also that

$$\frac{\sum_{i=1}^{5} \|J_{pv}^{-1}(a_{pv}^{(i)})''(x)J_{pv}^{-1}\|_F^2}{\sum_{i=1}^{5} \|(a_{cv}^{(i)})''(x)\|_F^2} = \frac{1}{4} + \frac{1}{2v},$$

$$\lim_{T \to 0} \frac{\sum_{i=1}^{5} \|J_{pv}^{-1}(f_{pv}^{(i)})''(x)J_{pv}^{-1}\|_F^2}{\sum_{i=1}^{5} \|(f_{cv}^{(i)})''(x)\|_F^2} = \frac{1}{4} + \frac{1}{2v}.$$

That is, the continuous-time result is consistent with the discrete time result. We can thus conclude the following:

- The formulas imply an upper bound on the rest term that is neglected in the EKF. Other weighting matrices can be used in the derivation to obtain explicit expressions for how e.g. an initial position error influences the time update. This case, where $J = \text{diag}(1,1,0,0,0)$, is particularly interesting, because now both weighting matrices are exact, and no approximation will be involved.

- The Frobenius error for Cartesian velocity has an extra term v^2T^4 compared to the one for polar velocity. For $T = 3$ and $\omega = 0$ this implies twice as large a bound. For $\omega = 0$, the bounds converge as $T \to \infty$.

- We can summarize the results in the plot in Figure 12.2 illustrating the ratio of (12.18) and (12.19) for $\omega = 0$ and three different velocities. As noted above, the asymptotes are 4 and 1, respectively. Note the huge peak around $T \approx 0.1$. For $T = 5$, Cartesian velocity is only 50% worse.

12.5 Sensor Noise Modeling

There is a kind of fundamental bias–variance trade-off in many kinds of common low-quality sensors. This stems from the fact that the measurement error rarely consists of just independent noise, but also some strongly correlated noise that can be interpreted as a time-varying offset. This can be investigated in the following way.

12.5.1 A Bias-Variance Plot

Suppose we can collect multiple measurements of a constant physical value s. A perfect sensor would give $y_k = s$ for all k, but in reality there are several

12.5 Sensor Noise Modeling

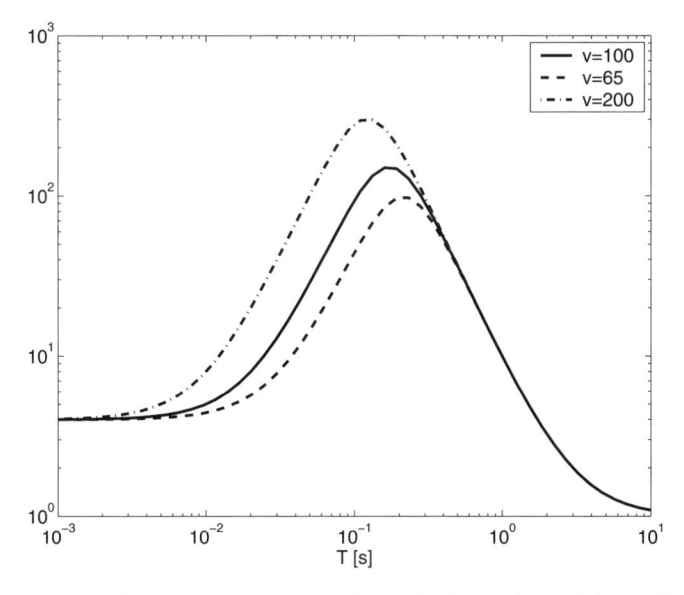

Figure 12.2: Ratio of the upper bounds (12.18) and (12.19) versus sampling interval. Cartesian velocity has a larger upper bound for all sampling intervals.

kinds of noise added to y_k. By averaging over an interval of length N, we get an estimate of the signal value

$$\hat{s} = \frac{1}{N} \sum_{k=1}^{N} y_k. \tag{12.20}$$

Such an experiment is simple to perform, and the estimation error variance can be estimated from multiple time intervals in a longer measurement series,

$$\hat{s}_m = \frac{1}{N} \sum_{k=1}^{N} y_{(m-1)N+k}, \quad m = 1, 2, \ldots, M \tag{12.21a}$$

$$\hat{\lambda}_m(N) = \frac{1}{N-1} \sum_{k=1}^{N} \left(y_{(m-1)N+k} - \hat{s}_m \right)^2, \tag{12.21b}$$

$$\hat{\lambda}(N) = \frac{1}{M} \sum_{m=1}^{M} \hat{\lambda}_m(N). \tag{12.21c}$$

As we will see, a plot of $\hat{\lambda}(N)$ as function of N reveals fundamental properties of the noise process in the sensor.

12.5.2 Allan Variance

An established measure of bias and variance for sensors is the *Allan variance*. It was originally defined for characterizing clock errors, but are today used more widely, for instance for inertial sensors, see the IEEE standard Board (R2008). It is defined as

$$\hat{s}_m = \frac{1}{N} \sum_{k=1}^{N} y_{(m-1)N+k}, \quad m = 1, 2, \ldots, M \quad (12.22a)$$

$$\hat{\lambda}(N) = \frac{1}{N-1} \sum_{m=1}^{N-1} (\hat{s}_{m+1} - \hat{s}_m)^2. \quad (12.22b)$$

The measure $\hat{\lambda}(N)$ is called the *Allan variance*. Clearly, the two measures in (12.21) and (12.22) are closely related. We use (12.21) in the sequel since it is somewhat easier to analyze.

12.5.3 Offset Models

To understand the theoretical properties of the bias-variance measure in (12.21), assume first that each sample has independent measurement noise,

$$y_k = s + e_k, \quad \mathrm{Cov}(e_k) = R. \quad (12.23)$$

Then, the bias-variance measure becomes $\hat{\lambda}(N) = \frac{R}{N}$, which decays monotonously in N.

Next, assume that the sensor has a time-varying offset subject to a random walk,

$$b_{k+1} = b_k + v_k, \qquad \mathrm{Cov}(v_k) = Q, \qquad (12.24a)$$
$$y_k = s + b_k + e_k, \qquad \mathrm{Cov}(e_k) = R. \qquad (12.24b)$$

Then, the bias-variance measure is easily derived from the sample average,

$$\bar{y}(N) = \frac{1}{N} \sum_{k=1}^{N} y_k = s + \frac{1}{N} \sum_{k=1}^{N} \left(e_k + \sum_{l=1}^{k} (b_1 + v_l) \right), \quad (12.24c)$$

$$\hat{\lambda}(N) = \frac{R}{N} + \frac{Q(N+1)(2N+1)}{6N} \approx \frac{R}{N} + \frac{QN}{3}. \quad (12.24d)$$

Note that the initial state b_1 of the offset does not affect the variance. In the derivation, the equality $\sum_{k=1}^{N} k^2 = N(N+1)(2N+1)/6$ is needed. Figure 12.3 illustrates this bias-variance function. Thus, it shows the following behavior for random walk offsets:

- Initially, averaging improves the signal estimate.

12.5 Sensor Noise Modeling

- For long enough averaging intervals, the variance starts to grow.
- There is an optimal integration length, which gives the minimum variance of the signal estimate.

Low-quality sensors often show this behavior.

Further, assume that there is a regularization forgetting in the bias dynamics,

$$b_{k+1} = ab_k + v_k, \qquad \text{Cov}(v_k) = Q, \qquad (12.25a)$$
$$y_k = s + b_k + e_k, \qquad \text{Cov}(e_k) = R. \qquad (12.25b)$$

The offset dynamics becomes a stochastic process with stationary variance $Q/(1-a^2)$. For a finite averaging window, the sample average and bias-variance measure can be calculated as

$$\bar{y}(N) = \frac{1}{N}\sum_{k=1}^{N} y_k = s + \frac{1}{N}\sum_{k=1}^{N}\left(e_k + \sum_{l=1}^{k} a^{k-l}(b_1 + v_l)\right), \qquad (12.25c)$$

$$\hat{\lambda}(N) = \frac{R}{N} + \frac{1}{N^2}\sum_{k=1}^{N}\underbrace{\sum_{n=0}^{N-k-1} a^{2n}}_{\frac{1-a^{2(N-k-1)}}{1-a^2}} Q, \qquad (12.25d)$$

$$= \frac{R}{N} + \frac{Q}{N^2(1-a^2)}\left(N - \frac{a^{2N}-1}{a^2-1}\right). \qquad (12.25e)$$

In this case, the bias-variance function is bounded.

12.5.4 Offset Modeling

The offset models (12.23), (12.24) and (12.25) are all appropriate for filtering applications. The basic idea is to augment the state vector with the offset b_k and use the dynamics in (12.24) and (12.25), respectively. Choosing the appropriate offset model is a standard *system identification* and *model selection* problem. For each model, the free parameters R, Q and a can be estimated from the bias-variance curve using the *nonlinear least squares* (NLS) method, as described in Section 3.2. The *parsimonious principle* should be used when selecting between the various standard models, which means that the best model gives a good compromise between curve fitting and complexity.

In short, the NLS system identification procedure for estimating $\theta = (Q, R)$ for a random walk offset model is

$$\hat{\theta} = \arg\min_{\theta} \sum_{N}\left(\hat{\lambda}(N) - \frac{R}{N} - \frac{Q(N+1)(2N+1)}{6N}\right)^2. \qquad (12.26)$$

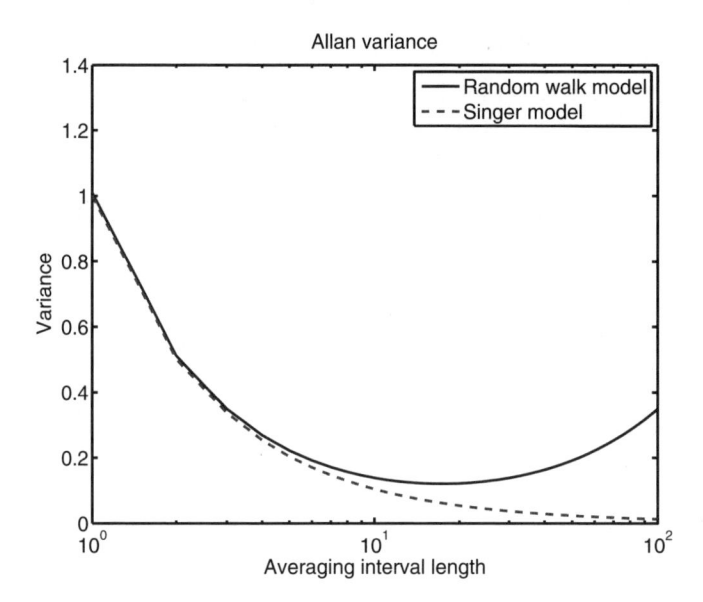

Figure 12.3: *Bias-variance plot for different offset models.*

It should here be pointed out that the Kalman filter presents a much more consequent approach. Let $x_k = (s, b_k)$, initialize with $x_0 = (0,0)^T$, apply the linear Kalman filter and then use dynamic system version of NLS, as will be presented in Section 12.7, for optimizing the sensor error parameters in an outer loop. This approach should make use of the data in a more efficient way.

12.6 Choice of Sampling Interval

The sampling interval in a filter is often routinely taken as the same as for the sensor. In multi-rate applications, where different sensors have different sampling rates, the fastest sensor dictates the filter's sampling interval. This is definitely a natural choice. However, there might be reasons for actually sampling faster, or even slower, than this default choice. We focus first on faster sampling and its benefits.

12.6.1 Up-Sampling the Dynamics

Let T_f be the sampling interval for the dynamic model, and T_h the sampling interval for the measurement equations. Assume that $T_h = MT_f$, where M

12.6 Choice of Sampling Interval

is an integer. The model is then

$$x(lT_f + T_f) = f(x(lT_f), v_l), \qquad l = 1, 2, \ldots, MN. \qquad (12.27)$$
$$y(kT_h) = h(x(kMT_f), e_k), \qquad k = 1, 2, \ldots, N. \qquad (12.28)$$

A practical advantage is the possibility to get an accurate discretization of continuous-time models. Given a continuous-time model $\dot{x}(t) = a(x(t), v(t))$, where $v(t)$ is a continuous-time Gaussian process with intensity Q, the discrete-time model can be approximated using *Euler sampling*,

$$x(lT_f + T_f) = x(lT_f) + T_f a(x(lT_f), v_l), \quad l = 1, 2, \ldots, MN. \qquad (12.29)$$

Here, v_l is a usual Gaussian noise process with covariance $T_f Q$. The approximation is arbitrarily good for small values of T_f. This might be a preferable implementation even when there is an exact discretization of the dynamics, as is the case for the coordinated turn models in Table 13.4. Note that the fast Euler sampling also gives a correct description of the noise model, which is not the case for the sampled models in Table 13.4, where the noise model is approximated.

Besides accurate sampling of continuous-time models, possible advantages of fast sampling include:

- The estimated state trajectory is interpolated to a smoother function. This gives nicer illustrations for slowly sampled systems.

- The influence of sensor time delays and time jitter as described in Section 15.3.1, that are not integer multiples of the sensor sample interval T_h, can be simulated, using the state values on a denser grid. In the case the time delay is known, the measurement update can be invoked at a more accurate time using the fast sampling.

- Collision avoidance and other decision support systems may require denser sampling to catch the inter-sample behavior of the trajectory.

It is hard to give any general rule of thumbs to decide if the faster sampling is needed in either state simulation or filtering. Numerical simulation studies are as usual recommended on a trial and error basis when the sensor sampling is considerably slower than the motion dynamics.

12.6.2 Down-Sampling the Dynamics

The basic idea of down-sampling is best illustrated for a linear model,

$$x_{k+1} = F_k x_k + G_k v_k,$$
$$y_k = H_k x_k + e_k.$$

This can be decimated to half the sample rate by stacking two adjacent measurements as

$$x_{2k+2} = F_{2k+1}F_{2k}x_{2k} + F_{2k+1}G_{2k}v_{2k} + G_{2k+1}v_{2k+1} \quad (12.30a)$$

$$\begin{pmatrix} y_{2k} \\ y_{2k+1} \end{pmatrix} = \begin{pmatrix} H_{2k} \\ H_{2k+1}F_{2k} \end{pmatrix} x_{2k} + \begin{pmatrix} e_{2k} \\ H_{2k+1}G_{2k}v_{2k} + e_{2k+1} \end{pmatrix}. \quad (12.30b)$$

By substituting $\bar{x}_l = x_{2k}$ and $\bar{y}_l = \left(y_{2k}^T, y_{2k+1}^T\right)^T$, we get the new linear model

$$\bar{x}_{l+1} = \bar{F}_l \bar{x}_l + \bar{v}_l, \quad (12.30c)$$
$$\bar{y}_l = \bar{H}_l \bar{x}_l + \bar{e}_l. \quad (12.30d)$$

Here,

$$\bar{F}_l = F_{2k+1}F_{2k}, \quad (12.30e)$$

$$\bar{H}_l = \begin{pmatrix} H_{2k} \\ H_{2k+1}F_{2k} \end{pmatrix} \quad (12.30f)$$

$$\bar{Q}_l = F_{2k+1}G_{2k}Q_{2k}G_{2k}^T F_{2k+1}^T + G_{2k+1}Q_{2k+1}G_{2k+1}^T, \quad (12.30g)$$

$$\bar{R}_l = \begin{pmatrix} R_{2k} & 0 \\ 0 & R_{2k+1} \end{pmatrix} \quad (12.30h)$$

$$\bar{S}_l = \mathrm{E}\left(\bar{e}_l \bar{v}_l^T\right) = \begin{pmatrix} 0 & 0 \\ H_{2k+1}G_{2k}Q_{2k}G_{2k}^T F_{2k+1}^T & 0 \end{pmatrix}. \quad (12.30i)$$

This model runs at half the rate of the original one. The decimation can be repeated an integer number of times. In the limit, we get a batch formulation of the filter problem, where only x_0 is unknown.

There is a similar re-formulation for nonlinear models. For instance, the model

$$x_{k+1} = f(x_k, v_k), \quad (12.31a)$$
$$y_k = h(x_k, e_k), \quad (12.31b)$$

can be written with half the rate as

$$x_{2k+2} = f(f(x_{2k}, v_{2k}), v_{2k+1}), \quad (12.31c)$$

$$\begin{pmatrix} y_{2k} \\ y_{2k+1} \end{pmatrix} = \begin{pmatrix} h(x_{2k}, e_{2k}) \\ h(f(x_{2k}, v_{2k}), e_{2k+1}) \end{pmatrix}. \quad (12.31d)$$

One immediate drawback is that decimation introduces correlation between process and measurement noise ($\bar{S}$ in the linear case). This can be handled by the Kalman filter, see Section 7.2.3. Further, with a different assumption of the process noise (impulsive just before the sample time), this dependence can be avoided. After all, the process noise is in most cases only instrumental.

For a linear model, there is probably not much to gain with down-sampling. For nonlinear models, it can resolve problems when using nonlinear transformations (as the unscented transform) and when designing the proposal distribution in the particle filter. One possible use, which is actually common in practice, is for filter initialization from a very uncertain prior. An EKF or UKF initialized with very vague prior can easily diverge without a dedicated initialization procedure. The standard approach is to estimate the initial state using the first few measurements. This is indeed an *ad-hoc* version of down-sampling as described here.

One motivation for down-sampling in the measurement update in the UKF is that the output distribution becomes more unimodal and Gaussian the more informative the measurement equation is. This option is particularly interesting for the practically important sub-class of problems with linear dynamics and nonlinear observation models. Note that the down-sampled model (12.31) gets quite a simple form in this case.

The possible advantage of using decimated models in the particle filter can be motivated as follows. For high SNR, the dynamical model $p(x_k|x_{k-1})$ is a quite poor proposal density. The likelihood function $p(y_k|x_k)$ is more peaky than the prior, and should be used as the proposal, see Section 9.5. However, if $n_x > n_y$, the likelihood cannot be used as a proposal alone, since it is not invertible. A necessary condition for getting an invertible likelihood function is that the decimation factor D satisfies $Dn_y \geq n_x$. The idea is to choose D large enough so the likelihood becomes invertible.

12.7 Calibration of Dynamical Systems

Calibrating a dynamical model of a system is often a necessary prerequisite in nonlinear filtering. Consider the following nonlinear model with uncertain parameters θ:

$$x_{k+1} = f(x_k, u_k, v_k; \theta), \tag{12.32}$$

$$y_k = h(x_k, u_k, e_k; \theta). \tag{12.33}$$

In nonlinear filtering, x_0, v_k, e_k are considered as nuisance, and the state sequence $x_{1:N}$ is to be estimated. In calibration, the state sequence is nuisance, and the estimation problem can be recast into the standard form of Chapter 3,

$$\mathbf{y} = \mathbf{h}(x_0, \theta, \mathbf{e}, \mathbf{v}). \tag{12.34}$$

This is a nonlinear estimation problem, where nonlinear least squares (NLS) applies. Compared to NLS for estimating parameters in a nonlinear static model as described in Section 3.2, the time-varying state introduces an extra difficulty.

The terminology *model calibration* is motivated by the fact that quite an accurate initial value of x_0 and θ is needed to get a fairly good state trajectory to start with. That is, the problem is to calibrate, or fine-tune, already good parameters rather than estimating them from scratch. This problem is also called *gray-box identification*.

The NLS method described below also applies, with the same notation, to a continuous-time model with discrete-time observations

$$\dot{x}(t) = a(x(t), u(t), v(t), \theta), \tag{12.35a}$$
$$y(t_k) = c(x(t_k), u(t_k), \theta) + e(t_k), \tag{12.35b}$$
$$x(0) = x_0. \tag{12.35c}$$

All the parameters that will be estimated are denoted η. Hence, if all the parameters and the initial conditions should be estimated, we have the following definition of η,

$$\eta = \begin{pmatrix} \theta \\ x_0 \end{pmatrix}. \tag{12.36}$$

The calibration problem can be formulated in terms of a nonlinear least-squares problem,

$$\hat{\eta} = \arg\min_\eta V(\eta, \mathbf{y}), \tag{12.37a}$$

$$V(\eta, \mathbf{y}) = \frac{1}{2} \sum_{t=1}^{N} \varepsilon^2(t_k, \eta). \tag{12.37b}$$

Here, the residual

$$\varepsilon(t_k, \eta) = y(t_k) - \hat{y}(t_k, \eta), \tag{12.38}$$

denotes the prediction error and $\hat{y}(t_k, \eta)$ is the one-step ahead prediction of the output. The notation $\varepsilon(t_k, \eta)$ is used to be able to account for non-equidistant sampling. Analogously to Section 3.2.1, define

$$\varepsilon(\eta) = \begin{pmatrix} \varepsilon^T(t_1, \eta) & \varepsilon^T(t_2, \eta) & \cdots & \varepsilon^T(t_N, \eta) \end{pmatrix}^T. \tag{12.39}$$

Using the residual vector $\varepsilon(\eta)$ the cost function (12.37b) can be written

$$V(\eta, \mathbf{y}) = \frac{1}{2} \varepsilon^T(\eta) \varepsilon(\eta). \tag{12.40}$$

12.7.1 Nonlinear Least-Squares

The gradient of the cost function is given by

$$\frac{\partial V(\eta)}{\partial \eta} = \sum_{k=1}^{N} \frac{\partial \varepsilon(t_k, \eta)}{\partial \eta} \varepsilon(t_k, \eta) = \sum_{k=1}^{N} \left(-\frac{\partial \hat{y}(t_k, \eta)}{\partial \eta}\right) \varepsilon(t_k, \eta)$$

$$= \sum_{k=1}^{N} \psi(t_k, \eta) \varepsilon(t_k, \eta) = J(\eta) \varepsilon(\eta), \quad (12.41)$$

where ψ is the gradient of $-\hat{y}(t_k, \eta)$,

$$\psi(t_k, \eta) = -\frac{\partial \hat{y}(t_k, \eta)}{\partial \eta}, \qquad \psi(t_k, \eta) \in \mathbb{R}^{n_\eta} \quad (12.42)$$

and the Jacobian $J(\eta)$ is

$$J(\eta) = \frac{\partial \varepsilon^T(\eta)}{\partial \eta} = \begin{pmatrix} \psi(t_1, \eta) & \psi(t_2, \eta) & \cdots & \psi(t_N, \eta) \end{pmatrix} \quad (12.43)$$

The Hessian of the cost function,

$$\frac{\partial^2 V(\eta)}{\partial \eta \partial \eta^T} = \sum_{k=1}^{N} \psi(t_k, \eta) \psi^T(t_k, \eta) - \sum_{k=1}^{N} \varepsilon(t_k, \eta) \frac{\partial \psi(t_k, \eta)}{\partial \eta}$$

$$= J(\eta) J^T(\eta) - \sum_{k=1}^{N} \varepsilon(t_k, \eta) \frac{\partial \psi(t_k, \eta)}{\partial \eta}, \quad (12.44)$$

is commonly approximated by

$$H(\eta) \approx J(\eta) J^T(\eta). \quad (12.45)$$

A suitable finite-difference approximation for the derivative is given by the central-difference formula,

$$\psi(t_k, \eta) \approx -\frac{\partial y(t_k, \eta)}{\partial \eta_i} \approx -\frac{y(t_k, \eta + he_i) - y(t_k, \eta - he_i)}{2h}, \quad (12.46)$$

where h is a small number and e_i is the ith unit vector. Typically h is chosen as

$$h = \sqrt{u}, \quad (12.47)$$

where u is the unit-roundoff (the value for u in double-precision computations is about 10^{-15}). In order to form the central-difference approximations (12.46) we need

$$\{x(t_k, \eta \pm he_i)\}_{i=1}^{n_\eta}, \quad (12.48)$$

which are obtained by solving $2n_\eta$ differential equations.

12.7.2 AUV Steering Dynamics

The steering dynamics from rudder angle δ_r to lateral speed v, yaw rate r and yaw angle ψ can from physical relations be shown to be given by

$$\begin{pmatrix} \dot{v} \\ \dot{r} \\ \dot{\psi} \end{pmatrix} = \begin{pmatrix} a_{11} & a_{12} & 0 \\ a_{21} & a_{22} & 0 \\ 0 & 1 & 0 \end{pmatrix} \begin{pmatrix} v \\ r \\ \psi \end{pmatrix} + \begin{pmatrix} b_1 \\ b_2 \\ 0 \end{pmatrix} \delta_r. \quad (12.49)$$

The model calibration task is to estimate the parameters

$$\theta = \begin{pmatrix} a_{11} & a_{12} & a_{21} & a_{22} & b_1 & b_2 \end{pmatrix}^T \quad (12.50)$$

in the continuous-time physical model structure (12.49), given observations of $v(t), r(t), \delta_r(t)$ at the non-uniform sampling times t_k, $k = 1, 2, \ldots, N$.

For the parameter identification, data collected during a sea-trial mission with a HUGIN 4500 vehicle are used. Prior to applying the identification procedure, the measurements were outlier filtered and smoothed.

A complete estimation approach is as follows:

1. **Estimation of initial discrete-time model:** In order to get good initial values to the NLS optimization, a discrete-time transfer function is first estimated. The single input multiple output (SIMO) model structure follows from (12.49) as

$$\begin{pmatrix} v(kT) \\ r(kT) \end{pmatrix} = G(z)\delta_r(kT). \quad (12.51)$$

This is identified using the following steps:

(a) Resample the data regularly at a sampling frequency of 1 Hz.

(b) Estimate a high order finite impulse response (FIR) model, where $G(z)$ has only zeros, no poles. This is done by the least squares method.

(c) Simulate noise-free measurements using the FIR model and true input.

(d) Estimate $G(z)$ as an ARX(2,2,1) model using the least squares method.

The ARX model could be estimated directly in step (b) above. The point with (b) and (c) is that a so called output error model is obtained, which is much better suited for simulation than an ARX model. The resulting transfer function is

$$Y_1(z) = \frac{1.48 \cdot z + 1.02}{z^2 + 3.63 \cdot z + 0.338},$$

$$Y_2(z) = \frac{-35.5 \cdot z - 11.2}{z^2 + 3.63 \cdot z + 0.338}$$

The identification and tex code above are generated by

12.7 Calibration of Dynamical Systems

```
[N,ny,nu]=size(z);
mtf=estimate(tf([ny ny 1 nu ny]),zip,'MC',0);
tex(mtf,'decimals',3)
```

2. **Inverse sampling and change of state coordinates:** The discrete-time transfer function is converted to a continuous-time state space model with the structure (12.49) using the following steps:

    (a) Convert discrete-time transfer function to discrete-time state space model.

    (b) Apply inverse sampling to get a continuous-time model.

    (c) Make a change of state variables to $x := C^{-1}x$ so that the C matrix becomes the identity matrix in the new coordinates.

    The resulting model is already on the required form with $H = I$,

    $$\dot{x}(t) = \begin{pmatrix} 1.555 & 0.187 \\ -44.925 & -5.183 \end{pmatrix} x(t) + \begin{pmatrix} 1.476 \\ -35.475 \end{pmatrix} u(t)$$

    $$y(t) = \begin{pmatrix} 1.000 & 0.000 \\ 0.000 & 1.000 \end{pmatrix} x(t)$$

    The conversions and resulting tex code are generated by

    ```
    s=ss(mtf);
    s=modred(s,ny);
    sc1=d2c(s);
    sc2=transform(sc1,sc1.C);
    sc1==sc2
        ans = true
    tex(sc2,'decimals',3)
    ```

3. **NLS calibration:** The Gauss-Newton algorithm is applied to the NLS cost function, using a continuous-time ODE solver to produce the simulated $\hat{y}(t_k)$ at non-uniformly sampled data. The ODE solver handles nonlinear continuous-time ODE's, though the model here is linear in the states.

    First, the nonlinear model structure is defined (though the model is linear, the more general nonlinear model is used here):

    ```
    nn=[2 1 2 6];          % [nx nu ny nth]
    f='[th(1)*x(1)+th(2)*x(2)+th(5)*u; th(3)*x(1)+th(4)*x(2)+th(6)*u]';
    th=[sc2.A(1,1) sc2.A(1,2) sc2.A(2,1) sc2.A(2,2) sc2.B(1) sc2.B(2)]' ;
    h='x';
    m=nl(f,h,nn);
    m.th = th;
    m.pe=[];
    m.x0=z.y(1,:)';    %x0=y1
    m
    NL object
        dx/dt = [th(1)*x(1)+th(2)*x(2)+th(5)*u;
                 th(3)*x(1)+th(4)*x(2)+th(6)*u]
            y = x
    ```

```
x0' = [-0.031           6]
th' = [0.13       1.1      -0.47      -2.7      1.3
       -2.6]
```

Estimation with a simulation is done by

```
mhat=estimate(m,z,'disp',1,'miter',1,'x0mask',zeros(1,m.nn(1)))
```

The `disp` option generates the information in Table 12.1.

Table 12.1: NLS iterations

Iteration	Cost	Grad. norm
0	$5.602 \cdot 10^2$	–
1	$5.531 \cdot 10^2$	$4.890 \cdot 10^3$
2	$5.525 \cdot 10^2$	$3.965 \cdot 10^3$

Finally, the linear model and its tex code is generated by

```
mhatl=nl2ss(mhat,zhat1(1))    % 'linearize' nl object to ss
tex(mhatl(:,1),'decimals',3)  % only first real input
```

The iterations of the Gauss-Newton algorithm are shown in Table 12.1. In this case, the NLS cost (12.40) decreased only marginally, though the resulting model looks quite different from the initialization:

$$\dot{x}(t) = \begin{pmatrix} 1.529 & 0.190 \\ -44.970 & -5.355 \end{pmatrix} x(t) + \begin{pmatrix} 1.490 \\ -35.586 \end{pmatrix} u(t)$$

$$y(t) = \begin{pmatrix} 1.000 & 0.000 \\ 0.000 & 1.000 \end{pmatrix} x(t)$$

4. **Validation by simulation.**

   Figure 12.4 shows the simulation of the identified discrete-time model, final continuous-time model and measurements.

5. **Cross validation.** A standard step in system identification to prevent overfitting observed data, is to validate the model on a new data set not used for estimation. An example with excellent performance is shown in Figure 12.5.

   The simulation is done by:

```
mhat.pe=[];   % Remove measurement noise
u=sig(z.u,z.t);
zhat1=simulate(mhat,u);
plot(zhat1,z,'conf',0)
```

12.7 Calibration of Dynamical Systems

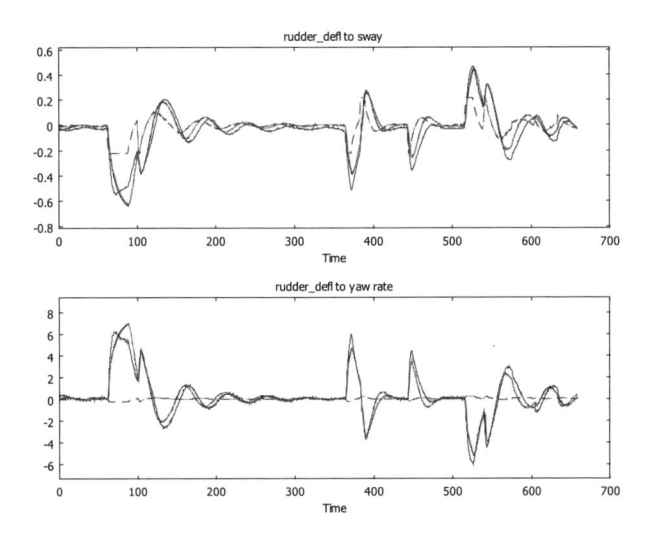

Figure 12.4: Simulation of identified discrete-time model, final continuous-time model and measurements.

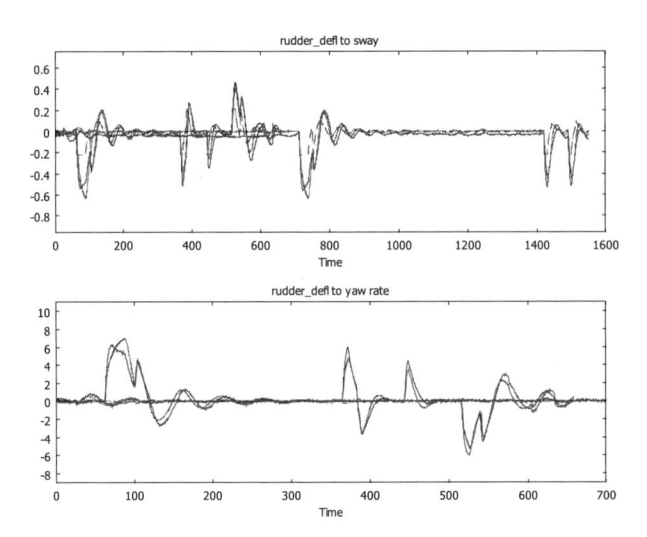

Figure 12.5: As Figure 12.4, but using cross validation on a longer data set with data not used in the estimation.

12.8 Summary

12.8.1 Theory

Sampling a continuous-time nonlinear model to a discrete-time linear model concerns passing from the upper left corner to the lower right corner below:

Classification	Nonlinear	Linear
Continuous-time	$\dot{x} = a(x,u) + v$	$\dot{x} = Ax + Bu + v$
	$y = c(x,u) + e$	$y = Cx + Du + e$
Discrete-time	$x_{k+1} = f(x,u) + v$	$x_{k+1} = Fx + Gu + v$
	$y = h(x,u) + e$	$y = Hx + Ju + e$

There are two options:

- Discetized linearization: The model is first linearized using

$$A = a'_x(x,u), \quad B = a'_u(x,u), \quad C = c'_x(x,u), \quad D = c'_u(x,u),$$

and then the linear model is sampled. Sampling a continuous-time linear model can be done with the zero-order hold formulas, $F = e^{AT}$, $G = \int_0^T e^{At} dt B$, $H = C$ and $J = D$. First-order hold and bilinear transformation are two other alternatives.

- Linearized discretization: The model is first sampled by solving the integral

$$x(t+T) = f(x(t), u(t)) = \int_t^{t+T} a(x(\tau), u(\tau)) d\tau,$$

and then this discrete-time nonlinear model is linearized using $F = f'_x(x_k, u_k)$ and $G = f'_u(x_k, u_k)$.

The first approach is always applicable, while the second alternative is recommended whenever it is possible to do analytically.

Sampling the state noise is in practice not so critical, and the simple choice $\bar{Q} = TQ$ is often sufficient. If the process noise can be written $B_v v(t)$ so the state noise covariance is $B_v Q B_v^T$, then the discrete-time state noise can be taken as $G_v Q T G_v^T$, where $G_v = \int_0^T e^{At} dt B_v$.

Fast sampling is an interesting alternative. The idea is to replace integrals with Riemann sums, and for each term in the sum use the discretized linearization approach.

For model calibration, the nonlinear least squares method can be used to fine-tune the partially known parameters in either the continuous or discrete-time model. Unknown initial states should here be considered as parameters. The algorithm is conceptually simple:

12.8 Summary

1. Simulate the state trajectory using the current parameter estimate.

2. Compute the output for the trajectory and compare to test data to get a residual vector.

3. Perturb the parameters in all directions, and observe how the residual vector changes.

4. Apply a steepest descent algorithm or Gauss-Newton algorithm to minimize the residual vector in the least squares (or maximum likelihood) sense.

12.8.2 Software

The following functions and methods are useful for the theory in this chapter:

- fp=numgrad(f,ind,varargin) computes numerical gradients using a central difference. varargin is the list of arguments to the inline function or string f, and ind is the index to the argument for which the gradient is computed. For example, let $f(x,u) = x^T x + ux^{(2)}$, then the gradient with respect to x and u, respectively, is computed by

```
x=[1 1]'; u=2;  f='[x''*x+u*x(2)]',  fx=numgrad(f,1,x,u)
x=[1 1]'; u=2;  f='[x''*x+u*x(2)]',  fu=numgrad(f,2,x,u)
```

- fp=numhess(f,ind,varargin) computes numerical Hessian, with the same syntax as above.

- md=c2d(mc,opt) converts a continuous-time LTI model to discrete-time LTI model. Here, the options are 'zoh', 'foh' or 'bil'. The inverse operation is md=d2c(mc,opt)

- mss=nl2ss(mnl,x) approximates a nonlinear model with a linear model using the linearization point specified in the SIG object x (this object must have only one time instant).

- mhat=estimate(m,y) calibrates the fields m.th and m.x0 using the data in y using the NLS method. Here, y can be a cell with several data sets.

- nls is the function that the method nl.estimate calls.

13
Motion Models

For both navigation and tracking of objects, a good model for describing the motion of the object is fundamental. This chapter will survey a number of general and application specific models in primarily continuous time form but the sampled discrete time models are also provided whenever they exist in a simple form. Mathematically, the models are in the form

$$\dot{x}(t) = a(t, x(t), u(t), w(t); \theta), \qquad (13.1a)$$
$$x(t+T) = f(t, x(t), u(t), w(t); \theta, T). \qquad (13.1b)$$

The standard discrete time model corresponds to $x_k = x(kT)$ or regular sampling. The notation in (13.1) implicitely also covers the case of irregular sampling. Here a and f specify the continuous and discrete time model, respectively, $u(t)$ is the input (an assumption of intersample behaviour is needed to compute f from a), $w(t)$ denotes process noise and θ are possible free parameters in the model.

There are two types of model for state estimation:

- Kinematic models that model unknown inputs as process noise. This can be seen as a kind of general purpose black-box structures.

- Application specific force models, where known inputs to the system are incorporated.

The chapter starts with an extensive overview of kinematic models, where each section is split up into the following subsections:

- Motion in 1D, 2D and 3D.

- Pure kinematics compared to sensed velocity as input.
- Continuous time and discrete time models.

Then, a number of motion models are listed for automotive vehicles, aircraft, surface and underwater vessels. A similar survey of motion models, in particular for target tracking, is provided in Li and Jilkov (2003).

13.1 Translational Kinematics

13.1.1 Motion in One Dimension

Consider motion in only one dimension $X(t)$. The main option for the state vector is to include velocity (*constant velocity model, CV*) and acceleration (*constant acceleration model, CA*). With only $X(t)$ in the state vector, the consequent name is *constant position model, (CP)*. Such models are rather trivial, and the results for the translational kinematics is summarized in Table 13.1 with $n = 1$.

Table 13.1: *Discretization as in Section 12.1 for translational kinematics models in nD, where $p(t)$ denotes the position X (1D), X,Y (2D) and X,Y,Z (3D), respectively. Another interpretation is $p(t) = \psi(t)$. The signal $w(t)$ is process noise for a pure kinematic model and a motion input signal in position, velocity and acceleration, respectively, for the case of using sensed motion as an input rather than as a measurement.*

Name	State	Continuous/discrete time
CP	$x(t) = p(t)$	$\dot{x}(t) = w(t)$
		$x(t+T) = x(t) + Tw(t)$
CV	$x(t) = \begin{pmatrix} p(t) \\ v(t) \end{pmatrix}$	$\dot{x}(t) = \begin{pmatrix} 0_n & I_n \\ 0_n & 0_n \end{pmatrix} x(t) + \begin{pmatrix} 0_n \\ I_n \end{pmatrix} w(t)$
		$x(t+T) = \begin{pmatrix} I_n & TI_n \\ 0_n & I_n \end{pmatrix} x(t) + \begin{pmatrix} \frac{T^2}{2}I_n \\ TI_n \end{pmatrix} w(t)$
CA	$x(t) = \begin{pmatrix} p(t) \\ v(t) \\ a(t) \end{pmatrix}$	$\dot{x}(t) = \begin{pmatrix} 0_n & I_n & 0_n \\ 0_n & 0_n & I_n \\ 0_n & 0_n & 0_n \end{pmatrix} x(t) + \begin{pmatrix} 0_n \\ 0_n \\ I_n \end{pmatrix} w(t)$
		$x(t+T) = \begin{pmatrix} I_n & TI_n & \frac{T^2}{2}I_n \\ 0_n & 0_n & I_n \\ 0_n & 0_n & 0_n \end{pmatrix} x(t) + \begin{pmatrix} \frac{T^3}{6}I_n \\ \frac{T^2}{2}I_n \\ TI_n \end{pmatrix} w(t)$

The continuous time model for a chain of integrators follows directly from the definition of the state vector. To sample such models, the series expansion of the exponential occuring in sampling, see Section 12.1, is useful,

$$e^{AT} = I + AT + \frac{T^2}{2}A^2 + \dots. \tag{13.2}$$

13.1 Translational Kinematics

The point is that A is nilpotent for integrators, so the $(m+1)$'th term becomes zero for an integrator of order m.

13.1.2 Motion in Higher Dimensions

The key point with translational kinematics is that motion is independent in each dimension. That means that the 1D model is copied to each dimension, and with a particular choice of state vector, the same notation can be used to cover all dimensions, as illustrated in Table 13.1.

13.1.3 Alternative Discretization Schemes

Section 12.1 provides three alternatives for discretizing a LTI system, and these methods can be applied to integrators. Table 13.2 compares the *zero-order hold* (piece-wise constant input) assumption with *first-order hold* (piece-wise linear input) and *bilinear interpolation* (bandlimited input). It is interesting to note that all other assumptions than piece-wise constant noise gives a process noise "leakage" into the measurement equation. The total measurement noise is $R + JQJ^T$, while the cross-correlation is JQ.

Table 13.2: Discretization as in Section 12.1 for similar to Table 13.1 for the state $x(t) = (p(t),\ v(t))^T$, but comparing different sampling assumptions for the double integrator.

Continuous time	$A = \begin{pmatrix} 0_n & I_n \\ 0_n & 0_n \end{pmatrix}$	$B = \begin{pmatrix} 0_n \\ I_n \end{pmatrix}$	$C = (I_n,\ 0_n)$	$D = 0_n$	
ZOH	$F = \begin{pmatrix} I_n & TI_n \\ 0_n & I_n \end{pmatrix}$	$G = \begin{pmatrix} \frac{T^2}{2} I_n \\ TI_n \end{pmatrix}$	$H = (I_n,\ 0_n)$	$J = 0_n$	
FOH	$F = \begin{pmatrix} I_n & TI_n \\ 0_n & I_n \end{pmatrix}$	$G = \begin{pmatrix} T^2 I_n \\ TI_n \end{pmatrix}$	$H = (I_n,\ 0_n)$	$J = \frac{T^2}{6} I_n$	
BIL	$F = \begin{pmatrix} I_n & TI_n \\ 0_n & I_n \end{pmatrix}$	$G = \begin{pmatrix} \frac{T^2}{4} I_n \\ \frac{T}{2} I_n \end{pmatrix}$	$H = (I_n,\ \frac{T}{2} I_n)$	$J = \frac{T^2}{2} I_n$	

Inputs or Outputs?

A non-trivial question is whether a measured motion state should be considered as a measurement or an input. Assume for instance that the acceleration a^x is measured. The options are

- Consider the accelerometer as a measurement $y = a^x + e^a$, and use the triple integrator model that includes a^x as one state.

- Consider the accelerometer as an input $u = a^x$, and use the double integrator model that does not include a^x in the state vector. The measurement noise now takes the role as process noise $w = e^a$.

Similar questions occur for all kind of models and motion sensors. The latter option is the standard choice for positioning, since the state vector is smaller. This is related to the *Luenberger observer*, where measured states are removed. However, if the motion state is needed for prediction, the first alternative is the only option.

13.1.4 Singer Model

The CP, CV and CA models all assume a random walk input to the highest derivative in the state vector. Such continuous time random walk models are a bit unphysical in a mathematical sense. For instance, the continuous time noise has spectrum $\Phi(\omega) = \sigma^2$ and thus it has infinite power. A more practical problem is that the state diverges in long time simulations. One sound approach to mitigate these problems is to assume that the noise is low-pass filtered. A first order low-pass filter for a CA model is

$$\dot{a}(t) = -\alpha a(t) + v(t). \tag{13.3}$$

In this way, the power is limited, and a simulation is due to the feedback term $-\alpha a(t)$ controlled back to $a(t) \approx 0$. The Singer CA model in nD is thus given by

$$\dot{x}(t) = \begin{pmatrix} 0_n & I_n & 0_n \\ 0_n & 0_n & I_n \\ 0_n & 0_n & -\alpha I_n \end{pmatrix} x(t) + \begin{pmatrix} 0_n \\ 0_n \\ I_n \end{pmatrix} w(t). \tag{13.4}$$

The discrete time dynamics using ZOH sampling can be shown to be

$$F(T) = \begin{pmatrix} I_n & T I_n & \frac{\alpha T - 1 + e^{-\alpha T}}{\alpha^2} I_n \\ 0_n & I_n & \frac{1 - e^{-\alpha T}}{\alpha} I_n \\ 0_n & 0_n & e^{-\alpha T} I_n \end{pmatrix}. \tag{13.5}$$

The process noise becomes a complicated but feasible integral expression $G = \int_0^T F(t) dt B$.

A nice interpretation is that α can be used to interpolate between a CV and a CA model. For $\alpha \to 0$, the CA model is attained, and for $\alpha \to \infty$ the CV model is approached.

There are many variants of the Singer model to get more realistic models for the acceleration, see the survey Li and Jilkov (2003). Note also that the Singer principle applies to CP, CV as well as higher order integrators.

13.2 Rotational Kinematics

Rotational kinematics is needed in most navigation applications in general and in particular the ones where an IMU is used. Besides, it can be used

13.2 Rotational Kinematics

Table 13.3: *Coordinate notation for rotations of a body in local coordinate system (x, y, z) relative to an earth fixed coordinate system. Note that for instance $\omega_x = \dot{\phi}$ holds only for one-dimensional rotation around the x-axis.*

Motion components	Rotation Euler angle	Angular speed
Longitudinal forward motion x	Roll ϕ	ω_x
Lateral motion y	Pitch θ	ω_y
Vertical motion z	Yaw ψ	ω_z

in tracking applications where an extended target model is used. The notation used throughout this section is summarized in Table 13.3. General comprehensive references in this area are Britting (1971); Salychev (1998); Titterton and Weston (1997).

13.2.1 Motion in 2D

For 2D motion, the convention is that only yaw is considered. The constraints are that the rigid body is moving in the horizontal plane only, with $z = 0$, $\phi = 0$ and $\theta = 0$. The yaw kinematics is simply given by

$$\dot{\psi}(t) = w(t) \tag{13.6a}$$

or as double integrated noise

$$\dot{\psi}(t) = \omega_z(t), \tag{13.6b}$$
$$\dot{\omega}_z(t) = w(t), \tag{13.6c}$$

or any higher order integrator. The same continuous time and discrete time models as in Tables 13.1 and 13.2 can be used, with $p(t) = \psi(t)$ and $n = 1$.

13.2.2 Motion in 3D

There is a huge step in going from rotations in 2D to 3D. A full derivation and details would be outside the scope of this book. However, the complete set of equations is given to give a hint of what is needed to implement the model for simulation and filtering.

The basic choice here is whether the simplest option of roll, pitch and yaw angles are to be used in the state vector, or if the more general unit quaternion representation should be used.

Euler Angle Representation

The Euler angle describes the rotation of one coordinate system with respect to another one. The most common case is to determine orientation of a body relative to earth coordinates, and this is the mental model used here if not otherwise states. For that reason, no additional notation is used to denote the two implicit coordinate systems.

The rotational kinematics is needed whenever vectors relating to the sensor have to be rotated. One important example encountered in the next section is for rotating accelerometer sensed forces in a body fixed sensor to world coordinates. For instance, the gravitational vector $\mathbf{g} = (0, 0, g)^T$ given in world coordinates is sensed by the accelerometer as the force $Q\mathbf{g}$. This can be obtained by three subsequent rotations, or as one rotation, using the following rotation matrices

$$Q = Q_\phi^x Q_\theta^y Q_\psi^z \tag{13.7}$$

$$= \begin{pmatrix} 1 & 0 & 0 \\ 0 & \cos\phi & \sin\phi \\ 0 & -\sin\phi & \cos\phi \end{pmatrix} \begin{pmatrix} \cos\theta & 0 & -\sin\theta \\ 0 & 1 & 0 \\ \sin\theta & 0 & \cos\theta \end{pmatrix} \begin{pmatrix} \cos\psi & \sin\psi & 0 \\ -\sin\psi & \cos\psi & 0 \\ 0 & 0 & 1 \end{pmatrix}$$

$$= \begin{pmatrix} \cos\theta\cos\psi & \cos\theta\sin\psi & -\sin\theta \\ \sin\phi\sin\theta\cos\psi - \cos\phi\sin\psi & \sin\phi\sin\theta\sin\psi + \cos\phi\cos\psi & \sin\phi\cos\theta \\ \cos\phi\sin\theta\cos\psi + \sin\phi\sin\psi & \cos\phi\sin\theta\sin\psi - \sin\phi\cos\psi & \cos\phi\cos\theta \end{pmatrix}.$$

We will here use the clock-wise rotation convention, motivated by navigation and geo-location applications. For instance, heading is defined clock-wise relative north. A more mathematical convention is the right-hand rule. The latter is also common in literature, and leads to similar rotation matrices but with different sign permutations.

Rotation matrices have the following properties:

- Q is orthogonal, so the inverse rotation matrix is given by $Q^{-1} = Q^T$.

- Subsequent rotations are computed as a product of rotation matrices, often indexed as $Q^{(3,2)}Q^{(2,1)}$. This means that a vector in coordinate system 1 is first rotated to coordinate system 2, and then further rotated to coordinate system 3. An example when this is needed is when the coordinate systems 1–3 relate to the world, body and sensor, respectively.

- Note the important role of the order of rotations. For another order than the xyz rule, a different kinematic model is obtained.

To derive the kinematic equations, the rotation matrix is applied to each unit vector, and the result is differentiated:

$$\begin{pmatrix} \omega_x \\ \omega_y \\ \omega_z \end{pmatrix} = \begin{pmatrix} \dot\phi \\ 0 \\ 0 \end{pmatrix} + Q_\phi^x \begin{pmatrix} 0 \\ \dot\theta \\ 0 \end{pmatrix} + Q_\phi^x Q_\theta^y \begin{pmatrix} 0 \\ 0 \\ \dot\psi \end{pmatrix} \tag{13.8}$$

13.2 Rotational Kinematics

Solving for the angle derivatives, we get

$$\begin{pmatrix} \dot{\phi} \\ \dot{\theta} \\ \dot{\psi} \end{pmatrix} = \begin{pmatrix} 1 & -\sin(\phi)\tan(\theta) & \cos(\phi)\tan(\theta) \\ 0 & \cos(\phi) & \sin(\phi) \\ 0 & -\sin(\phi)\sec(\theta) & \cos(\phi)\sec(\theta) \end{pmatrix} \begin{pmatrix} \omega_x \\ \omega_y \\ \omega_z \end{pmatrix} \quad (13.9)$$

The model above is linear in turn rates ω. An approximate model linear in the angles can be derived for a small angle assumption.

As usual for trigonometric functions, Euler angles suffer from the 2π ambiguity. There are also singular points at 'the poles' of the sphere, where a reparametrization is needed. The core problem occurs for $\theta = \pm 90$ degrees. A stalling aircraft has a pitch angle that approaches 90 degrees, but what is then yaw and roll?

Quaternion Representation

A unit quaternion is defined by a vector with four real numbers as $q = (q^0, q^1, q^2, q^3)^T$, as an extension to complex numbers. To represent an angle in 3D, the vector is constrained to the unit sphere, so $q^T q = 1$. The *quaternion representation* does not suffer from the weaknesses of the Euler angle, and are often used as internal states in navigation. The drawback is more notation and complex formulas, less intuition and the fact that the state contains an algebraic constraint that has to be fulfilled all the time.

The first relation concerns transforming Euler angles to quaternions,

$$q^0 = \cos(\phi/2)\cos(\theta/2)\cos(\psi/2) + \sin(\phi/2)\sin(\theta/2)\sin(\psi/2), \quad (13.10a)$$
$$q^1 = \sin(\phi/2)\cos(\theta/2)\cos(\psi/2) - \cos(\phi/2)\sin(\theta/2)\sin(\psi/2), \quad (13.10b)$$
$$q^2 = \cos(\phi/2)\sin(\theta/2)\cos(\psi/2) + \sin(\phi/2)\cos(\theta/2)\sin(\psi/2), \quad (13.10c)$$
$$q^3 = \cos(\phi/2)\cos(\theta/2)\sin(\psi/2) - \sin(\phi/2)\sin(\theta/2)\cos(\psi/2). \quad (13.10d)$$

The kinematics model corresponding to (13.9) is given without derivation as

$$\dot{q} = -\frac{1}{2}S(\omega)q = -\frac{1}{2}\bar{S}(q)\omega \quad (13.11a)$$

where S and $\bar{S}$ are skew-symmetric matrices

$$S(\omega) = \begin{pmatrix} 0 & -\omega_x & -\omega_y & -\omega_z \\ \omega_x & 0 & -\omega_z & \omega_y \\ \omega_y & \omega_z & 0 & -\omega_x \\ \omega_z & -\omega_y & \omega_x & 0 \end{pmatrix}, \quad (13.11b)$$

$$\bar{S}(q) = \begin{pmatrix} -q_1 & -q_2 & -q_3 \\ q_0 & q_3 & -q_2 \\ -q_3 & q_0 & q_1 \\ q_2 & -q_1 & q_0 \end{pmatrix}. \quad (13.11c)$$

That is, the model is bilinear in quaternions and angular speed, which can be expressed as either

- a linear input model with nonlinear state dynamics $\dot{q} = -\frac{1}{2}\bar{S}(q)\omega$, or
- a linear state model with nonlinear input $\dot{q} = -\frac{1}{2}S(\omega)q$.

Using the sampling formula for the assumption that ω is piece-wise constant between sampling instants, ZOH sampling (see Section 12.1) gives

$$q(t+T) = e^{-\frac{1}{2}S(\omega(t))T} q(t). \qquad (13.12a)$$

The terms in a series expansion can be identified as

$$q(t+T) = \left(\cos\left(\frac{\|\omega(t)T\|}{2}\right) I_4 + \frac{\sin\left(\frac{\|\omega(t)\|T}{2}\right)}{\|\omega(t)\|} S(\omega(t)) \right) q(t) \qquad (13.12b)$$

$$\approx \left(I_4 + \frac{T}{2} S(\omega(t)) \right) q(t). \qquad (13.12c)$$

The latter small angle approximation is identical to a Euler forward approximation and will be used to find explicit but approximate discrete time models.

Letting ω be part of the state vector and assuming a white noise torque result in the following model

$$\begin{pmatrix} \dot{q}(t) \\ \dot{\omega}(t) \end{pmatrix} = \begin{pmatrix} +\frac{1}{2} S(\omega(t)) q(t) \\ w(t) \end{pmatrix}. \qquad (13.13)$$

There is no known closed form discretized model here. However, the approximate form can be discretized using the chain rule to

$$\begin{pmatrix} q(t+T) \\ \omega(t+T) \end{pmatrix} \approx \underbrace{\begin{pmatrix} I_4 + \frac{T}{2} S(\omega(t)) & \frac{T^2}{2} \bar{S}(q(t)) \\ 0_{3\times 4} & I_3 \end{pmatrix}}_{F(t)} \begin{pmatrix} q(t) \\ \omega(t) \end{pmatrix} \qquad (13.14)$$

$$+ \begin{pmatrix} \frac{T^3}{4} \bar{S}(q(t)) \\ T I_3 \end{pmatrix} v(t). \qquad (13.15)$$

The rotation matrix can be expressed in the quaternions as

$$Q = \begin{pmatrix} q_0^2 + q_1^2 - q_2^2 - q_3^2 & 2q_1q_2 - 2q_0q_3 & 2q_0q_2 + 2q_1q_3 \\ 2q_0q_3 + 2q_1q_2 & q_0^2 - q_1^2 + q_2^2 - q_3^2 & -2q_0q_1 + 2q_2q_3 \\ -2q_0q_2 + 2q_1q_3 & 2q_2q_3 + 2q_0q_1 & q_0^2 - q_1^2 - q_2^2 + q_3^2 \end{pmatrix} \qquad (13.16)$$

13.3 Rigid-Body Kinematics

To compute the relation between quaternion and Euler angle, the fact that the two rotation matrices must be the same can be used, which gives

$$\psi = \tan^{-1}\left(\frac{Q^{12}}{Q^{11}}\right) = \tan^{-1}\left(\frac{2q_1q_2 - 2q_0q_3}{2q_0^2 + 2q_1^2 - 1}\right), \quad (13.17a)$$

$$\theta = -\sin^{-1}(Q^{13}) = -\sin^{-1}(2q_1q_3 + 2q_0q_2), \quad (13.17b)$$

$$\phi = \tan^{-1}\left(\frac{Q^{23}}{Q^{33}}\right) = \tan^{-1}\left(\frac{2q_2q_3 - 2q_0q_1}{2q_0^2 + 2q_3^2 - 1}\right). \quad (13.17c)$$

13.3 Rigid-Body Kinematics

A general form allowing for using both accelerometer and gyroscope as measurements is given by

$$\begin{pmatrix} \dot{p} \\ \dot{v} \\ \dot{a} \\ \dot{q} \\ \dot{\omega} \\ \dot{b}^{\text{acc}} \\ \dot{b}^{\text{gyro}} \end{pmatrix} = \begin{pmatrix} 0 & I & 0 & 0 & 0 & 0 & 0 \\ 0 & 0 & I & 0 & 0 & 0 & 0 \\ 0 & 0 & 0 & 0 & 0 & 0 & 0 \\ 0 & 0 & 0 & -\frac{1}{2}S(\omega) & 0 & 0 & 0 \\ 0 & 0 & 0 & 0 & 0 & 0 & 0 \\ 0 & 0 & 0 & 0 & 0 & 0 & 0 \\ 0 & 0 & 0 & 0 & 0 & 0 & 0 \end{pmatrix} \begin{pmatrix} p \\ v \\ a \\ q \\ \omega \\ b^{\text{acc}} \\ b^{\text{gyro}} \end{pmatrix} + \begin{pmatrix} 0 & 0 & 0 & 0 \\ 0 & 0 & 0 & 0 \\ 1 & 0 & 0 & 0 \\ 0 & 0 & 0 & 0 \\ 0 & 1 & 0 & 0 \\ 0 & 0 & 1 & 0 \\ 0 & 0 & 0 & 1 \end{pmatrix} \begin{pmatrix} v^a \\ v^\omega \\ v^{\text{acc}} \\ v^{\text{gyro}} \end{pmatrix}. $$

$$(13.18)$$

Bias states for the accelerometer b^{acc} and gyro b^{gyro} have been added as well. This is common in high-end navigation systems to improve dead-reckoning and estimation performance in the end. The model contains 22 states, but there are only 21 free states due to the quaternion constraint $q^T q = 1$.

The model can be discretized using the same idea as in (13.14). The measurement relations for accelerometer and gyroscope can then be stated as

$$y^{\text{acc}} = Q(q)(a - \mathbf{g}) + b^{\text{acc}} + e^{\text{acc}}, \qquad e^{\text{acc}} \sim \mathcal{N}(0, R^{\text{acc}}), \quad (13.19a)$$

$$y^{\text{gyro}} = \omega + b^{\text{gyro}} + e^{\text{gyro}}, \qquad e^{\text{gyro}} \sim \mathcal{N}(0, R^{\text{gyro}}), \quad (13.19b)$$

It is here implicit that the quaternion describes the rotation between an earth fixed and a body fixed coordinate system. The sensors are body fixed, the platform's motion is described relative to earth, and $\mathbf{g} = (0, 0, g)^T$ is the gravitation vector in earth fixed coordinates. The linearized measurement relation becomes, assuming $a = 0$ to simplicity and first differentiating the last column of (13.16) to get H_q^{acc},

$$H_q^{\text{acc}} = -2 \cdot g \begin{pmatrix} q_2 & q_3 & q_0 & q_1 \\ -q_1 & -q_0 & q_3 & q_2 \\ q_0 & -q_1 & -q_2 & q_3 \end{pmatrix}, \tag{13.20a}$$

$$H^{\text{acc}} = \begin{pmatrix} 0_{3\times 9}, & H_q^{\text{acc}}, & 0_{3\times 3}, & I_{3\times 3}, & 0_{3\times 3} \end{pmatrix}, \tag{13.20b}$$

$$H^{\text{gyro}} = \begin{pmatrix} 0_{3\times 13} & I_3 & 0_{3\times 3} & I_3 \end{pmatrix}. \tag{13.20c}$$

By defining $H_{q,i}^{\text{acc}} = dQ_{:,i}(q)/dq$, the generalization of (13.20b) in case of $a = 0$ can be written

$$H^{\text{acc}} = \begin{pmatrix} 0_{3\times 6}, & Q(q), & H_{q,1}^{\text{acc}}a, & H_{q,2}^{\text{acc}}a, & H_{q,3}^{\text{acc}}(a-\mathbf{g}), & 0_{3\times 3}, & I_{3\times 3}, & 0_{3\times 3} \end{pmatrix}. \tag{13.20d}$$

13.4 Constrained Kinematic Models

A coordinated turn means that the object will follow a circular path as long as no external force (process noise) is present. This circle is first parametrized in world coordinates with turn rate and then in the next section in local body fixed coordinates using yaw rate or (inverse) curve radius.

13.4.1 Coordinated Turns in 2D using World Coordinates

Two versions of the coordinated turn model using world coordinates are presented in Example 12.1 in continuous time, and the discretized versions are given in Example 12.2. Table 13.4 summarizes the models.

13.4.2 Coordinated Turns in 2D using Local Coordinates

Various models can be constructed based on the following equations of motion:

$$\dot{\psi} = \frac{v_x}{R} = v_x R^{-1}, \tag{13.21a}$$

$$a_y = \frac{v_x^2}{R} = v_x^2 R^{-1} = v_x \dot{\psi}, \tag{13.21b}$$

$$a_x = -v_y \frac{v_x}{R} = -v_y v_x R^{-1} = -v_y \dot{\psi}. \tag{13.21c}$$

The state vector may contain $x = (v^x, v^y, a^x, a^y, \psi, R^{-1})$, where R^{-1} is the inverse curve radius (assuming circular motion). Depending on the sensor configuration, certain states may be eliminated using the measurement as input trick.

Odometry as described in Section 13.5 may be incorporated in the model by augmenting the state vector with position X, Y in world coordinates similar to (13.30).

13.4 Constrained Kinematic Models

Table 13.4: Coordinated turn models in Cartesian or polar velocity in continuous time and the corresponding exact ZOH sampled model as given in Examples 12.1 and 12.2. The discrete time dynamics are given by $F = e^{A(x)T}$ and $F = f'_x(x)$, respectively.

	Cartesian velocity
$a(x)$	$\dot{X} = v^X$ $\dot{Y} = v^Y$ $\dot{v}^X = -\omega v^Y$ $\dot{v}^Y = \omega v^X$ $\dot{\omega} = 0$
$A(x)$	$\begin{pmatrix} 0 & 0 & 1 & 0 & 0 \\ 0 & 0 & 0 & 1 & 0 \\ 0 & 0 & 0 & -\omega & -v^Y \\ 0 & 0 & \omega & 0 & v^X \\ 0 & 0 & 0 & 0 & 0 \end{pmatrix}$
$f(x)$	$X(t+T) = X + \frac{v^X}{\omega}\sin(\omega T) - \frac{v^Y}{\omega}(1-\cos(\omega T))$ $Y(t+T) = Y + \frac{v^X}{\omega}(1-\cos(\omega T)) + \frac{v^Y}{\omega}\sin(\omega T)$ $v^X(t+T) = v^X \cos(\omega T) - v^Y \sin(\omega T)$ $v^Y(t+T) = v^X \sin(\omega T) + v^Y \cos(\omega T)$ $\omega(t+T) = \omega$
	Polar velocity
$a(x)$	$\dot{X} = v\cos(h)$ $\dot{Y} = v\sin(h)$ $\dot{v} = 0$ $\dot{h} = \omega$ $\dot{\omega} = 0$
$A(x)$	$\begin{pmatrix} 0 & 0 & \cos(h) & -v\sin(h) & 0 \\ 0 & 0 & \sin(h) & v\cos(h) & 0 \\ 0 & 0 & 0 & 0 & 0 \\ 0 & 0 & 0 & 0 & 1 \\ 0 & 0 & 0 & 0 & 0 \end{pmatrix}$
$f(x)$	$X(t+T) = X + \frac{2v}{\omega}\sin(\frac{\omega T}{2})\cos(h + \frac{\omega T}{2})$ $Y(t+T) = Y + \frac{2v}{\omega}\sin(\frac{\omega T}{2})\sin(h + \frac{\omega T}{2})$ $v(t+T) = v$ $h(t+T) = h + \omega T$ $\omega(t+T) = \omega$

Speed Sensor as Input

As one particular example, a state space model for the yaw kinematics using known forward speed v_x is

$$x = \begin{pmatrix} \psi \\ R^{-1} \end{pmatrix}, \quad u = v_x, \quad \dot{x} = f(x, u) + w = \begin{pmatrix} v_x R^{-1} \\ 0 \end{pmatrix} + w \quad (13.22)$$

The model can be used for filtering if some additional information is provided, for instance a vision system measuring the curve radius.

A Fusion Model for Lateral Dynamics

The measurements under consideration are

- $y_t^1 = \dot{\psi}$ from yaw rate sensor.
- $y_t^2 = a_{y,m}$ from lateral accelerometer.
- $y_t^3 = a_{x,m}$ from longitudinal accelerometer.

The gyro signal is subject to an offset and scale factor error

$$y_t^1 = (1 + \delta_{\text{sc,gyro}})\dot{\psi}_t + \delta_{o,\text{gyro}} + e_t^1$$

Here $\delta_{\text{sc,gyro}}$ is the scale factor error in the gyro, which enters the measurement nonlinearly. A linearization around the current estimate yields

$$y^1 + \hat{\delta}_{\text{sc,gyro}}\hat{\dot{\psi}} = (1 + \hat{\delta}_{\text{sc,gyro}})\dot{\psi} + \hat{\dot{\psi}}\delta_{\text{sc,gyro}} + \delta_{o,\text{gyro}}.$$

A good working approximation might be to use

$$y^1 = \dot{\psi} + \hat{\dot{\psi}}\delta_{\text{sc,gyro}} + \delta_{o,\text{gyro}}. \quad (13.23a)$$

The speed can be measured as $v_{x,m} = \omega r_{nom} = v_x - \omega \delta_r$. Thus, for the accelerometers, we have

$$y_t^2 = v_x \dot{\psi}_t + \delta_{o,\text{acc},y} = (v_{x,m} + \omega \delta_r)\dot{\psi}_t + \delta_{o,\text{acc}}, \quad (13.23b)$$
$$y_t^3 = a_x + \delta_{o,\text{acc},x}. \quad (13.23c)$$

Again, there is a nonlinear scaling factor error due to absolute wheel radius. A linearization as above is necessary. Note that the two scale factors are linearly independent when the velocity is changing. Table 13.5 summarizes the slowly time-varying parameters that must be estimated (in order of relative importance). See Section 13.5 for more information on the sensor models, and Section 16.2.2 for an application.

13.4 Constrained Kinematic Models

Table 13.5: *Offsets in the lateral motion model (13.23).*

δ_r	Relative difference in actual and nominal radius
$\delta_{o,\text{acc}}$	Accelerometer offset $[m/s^2]$
$\delta_{o,\text{gyro}}$	Gyro offset $[rad/s]$
$\delta_{sc,\text{gyro}}$	Gyro scale factor relative error

Road-Aligned Turn Model

Road geometry can be approximated with constant jerk, leading to a linear relation for inverse curve radius $R^{-1} = c_0 + c_1 x$. According to (14.35d), the yaw rate then becomes linear in time when the speed is constant corresponding to smooth steering wheel maneuvers. The coordinates x and y denote the position in the curved coordinate system, which is attached to the road according to Figure 13.1. The model below has been developed independently in Eidehall and Gustafsson (2004) and Klotz et al. (2004).

First, we can obtain a relationship between the relative course to the road ψ_{sc} and the absolute course in world coordinates ψ_{abs} by differentiating ψ_{sc} w.r.t. time,

$$\psi_{\text{sc}} = \psi_{\text{abs}} + \psi_{\text{sc}} \quad \Rightarrow \tag{13.24a}$$

$$\dot\psi_{\text{sc}} = \dot\psi_{\text{abs}} + \dot\psi_{\text{sc}} = \dot\psi_{\text{abs}} + \frac{v}{R} = \dot\psi_{\text{abs}} + c_0 v, \tag{13.24b}$$

where R is the current road radius, v the velocity and ψ_{lane} denotes the angle between the lane and some fix reference. We also have

$$\dot y_{\text{sc}} = \sin(\psi_{\text{sc}})v \approx \psi_{\text{sc}} v. \tag{13.25}$$

Assuming that the road has constant width $\dot W = 0$ and jerk $\dot c_1 = 0$, the continuous-time motion equations for the host vehicle states and the road states can be written

$$\dot x = v, \tag{13.26a}$$
$$\dot v = 0, \tag{13.26b}$$
$$\dot y = 0, \tag{13.26c}$$
$$\dot W = 0, \tag{13.26d}$$
$$\dot y_{\text{sc}} = v \sin(\psi_{\text{sc}}), \tag{13.26e}$$
$$\dot\psi_{\text{sc}} = v c_0 + \dot\psi_{\text{abs}}, \tag{13.26f}$$
$$\dot c_0 = v c_1, \tag{13.26g}$$
$$\dot c_1 = 0. \tag{13.26h}$$

A further extension of this model to three-dimensional road models is possible by including a road curvature R_y around the lateral y-axis. Roads are constructed under constraints on the visible horizon with a parabolic design rule $z = x^2/(2R_y)$. A linear model is plausible for the variation in R_y^{-1} here as well, similar to the for R_z around the z-axis. Such a 3D model can be used by ACC systems for fuel economy and collision avoidance system by warning for possible vehicles still out of sight.

The discrete-time dynamics for the last five states is

$$W(t+T) = W + w_4, \tag{13.27a}$$

$$y_{sc}(t+T) = y_{sc} + vT_s\psi_{sc} + v^2\frac{T_s^2}{2}c_0 + v^3T_s^3c_1/6 + vT^2\dot{\psi}_{abs}/2 + w_5,$$

$$\psi_{sc}(t+T) = \psi_{sc} + vT_sc_0 + v^2T_s^2c_1/2 + T_s\dot{\psi}_{abs} + w_6, \tag{13.27b}$$

$$c_0(t+T) = c_0 + vT_sc_1 + w_7, \tag{13.27c}$$

$$c_1(t+T) = c_1 + w_8. \tag{13.27d}$$

The variables $\{w_i\}_{i=1}^8$ are white Gaussian process noise, with covariance matrices Q_{host} and Q_{obj} for the host and the tracked vehicle's state, respectively.

13.5 Odometric Models

Odometry concerns computation of relative position for wheeled vehicles, and involves formulas for dead-reckoning wheel speeds on one common axle. The raw signals are here angular velocities of the wheel, measured in cars with the ABS sensors. Equations (14.34) and (14.35d) state that

$$v_x^m = \frac{\omega_3 r_3 + \omega_4 r_4}{2} \tag{13.28a}$$

$$\dot{\psi}^m = v_x^m \frac{2}{B} \frac{\frac{\omega_3 r_3}{\omega_4 r_4} - 1}{\frac{\omega_3 r_3}{\omega_4 r_4} + 1} = \frac{\omega_3 r_3 - \omega_4 r_4}{B}. \tag{13.28b}$$

The superscript m indicates a measured value, where 3 is the rear left wheel and 4 the rear right wheel. The principle of dead-reckoning under the assumption that

- there is no lateral velocity,
- both speed and yaw rate are measured and considered as inputs,

gives the integral

$$\psi(t) = \psi(0) + \int_0^t \dot{\psi}(t)dt, \tag{13.29a}$$

$$X(t) = X(0) + \int_0^t v^x(t)\cos(\psi(t))dt, \tag{13.29b}$$

$$Y(t) = Y(0) + \int_0^t v^x(t)\sin(\psi(t))dt. \tag{13.29c}$$

13.5 Odometric Models

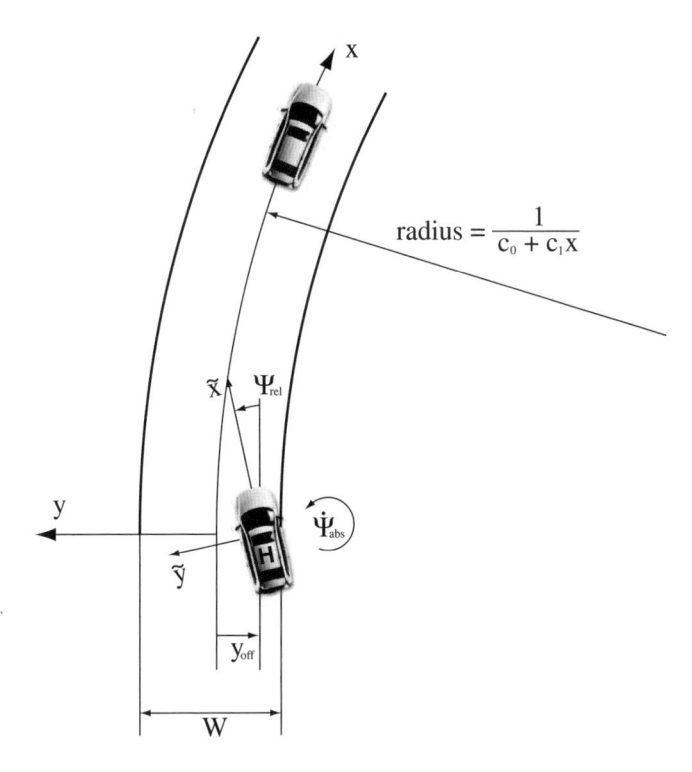

Figure 13.1: The coordinate systems used in deriving the dynamic motion model. Here, (x, y) denotes the position in a curved coordinate system, which is attached to and follows the road. Furthermore, $(\tilde{x}, \tilde{y})$ denotes the position in a coordinate system, which is attached to the moving host vehicle. Figure from Eidehall and Gustafsson (2004).

The corresponding model, having the above as the solution, is given by

$$x(t) = \begin{pmatrix} X(t) \\ Y(t) \\ \psi(t) \end{pmatrix}, \quad \dot{x}(t) = \begin{pmatrix} v^x(t) \cos(\psi(t)) \\ v^x(t) \sin(\psi(t)) \\ \dot{\psi}(t) \end{pmatrix}. \quad (13.30a)$$

Assuming piecewise constant speed and yaw rate, the discrete time formula is given by

$$X(t+T) = X(t) + \frac{2v^x(t)}{\dot{\psi}(t)} \sin(\frac{\dot{\psi}(t)T}{2}) \cos(\psi(t) + \frac{\dot{\psi}(t)T}{2}) \quad (13.31a)$$
$$\approx X(t) + v^x(t)T\cos(\psi(t)), \quad (13.31b)$$
$$Y(t+T) = Y(t) + \frac{2v^x(t)}{\dot{\psi}(t)} \sin(\frac{\dot{\psi}(t)T}{2}) \sin(\psi(t) + \frac{\dot{\psi}(t)T}{2}) \quad (13.31c)$$
$$\approx Y(t) + v^x(t)T\sin(\psi(t)). \quad (13.31d)$$

The latter expressions are based on small changes in yaw during one sample interval.

The virtual speed and yaw rate sensor in (13.28) can be used in (13.31) to give a position $(X(t; r_3, r_4), Y(t; r_3, r_4))$ that depends on the values of wheel radii r_3 and r_4 that are used. Further sources of error come from wheel slip in longitudinal and lateral direction. To support the plain odometric relation, a model based filter can be used. The standard idea of augmenting the state vector with unknown offsets works if only enough sensor information is available.

13.6 Vehicle Models

Table 13.6 summarizes the notation in this section. The state vector contains at least
$$x = (v_x, v_y, \psi, \dot{\psi}, \phi, \theta)^T$$
and the measurement vector at least
$$y = (\omega_1, \omega_2, \omega_3, \omega_4, T_{\text{eng}}, \omega_{\text{eng}}, P_{\text{brake}}, a_x, a_y, a_z, \dot{\psi})^T.$$

13.6.1 Longitudinal Dynamics

Assume that this vehicle has mass m, normal force F_z and the wheel has moment of inertia J and radius r. The basic dynamic equations of motion are:
$$m\dot{v}_x = F_z \mu_x(s) \quad (13.32)$$
$$J\dot{\omega} = rF_z\mu_x(s) + T_{\text{traction}} - T_{\text{brake}} \quad (13.33)$$

Here $\mu_x(s)$ is the normalized traction force F_x/F_z. The slip definition is
$$s = \frac{\omega r - v_x}{v_x} = \frac{\omega r}{v_x} - 1 \quad (13.34)$$

13.6 Vehicle Models

Table 13.6: Notation for vehicle models

Car dynamics	Yaw ψ, roll ϕ and pitch θ. Velocities v_x, v_y, v_z. Accelerations a_x, a_y, a_z
Road	Yaw (direction) Ψ, roll (banking) Φ and pitch (inclination) Θ.
Driveline	Engine torque T_{eng} and speed ω_{eng} computed from injection signal or manifold pressure by lookup table. Gearbox with ratios k_i. Inertia $J_1 + k^2 J_2$.
Body constants	Mass m, inertias J_x, J_y, J_z around center of gravity, distances L_r, L_f to rear and front wheel axle, respectively. Wheel base B. Air drag F_{drag} and roll resistance F_{roll}.

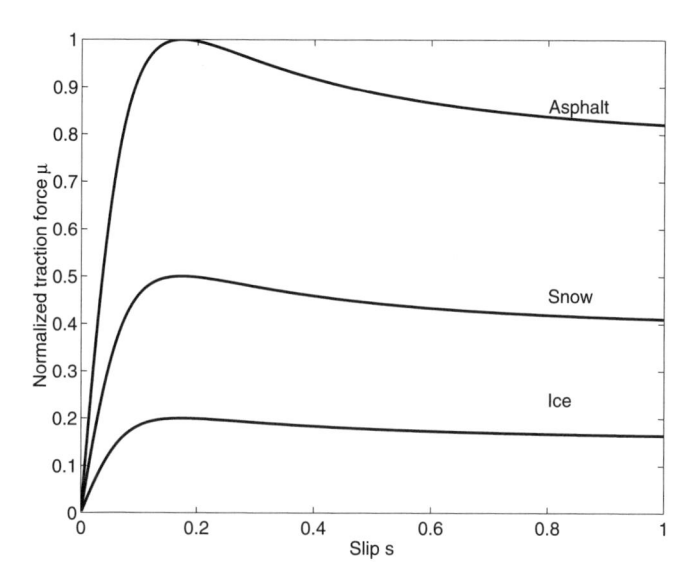

Figure 13.2: Schematic view of the slip curve.

This static relation works fine for our purposes, while an extension using a first order dynamical model is presented in Bernard and Clover (1995), which is suitable for simulation. The knowledge of the μ–s relation is critical, and will be devoted its own section. The principal form of this relation is shown in Figure 13.2, and the best possible friction is attained for the slip yielding the peak value of this curve. The basic control strategy is to reduce the engine torque and thus μ_x so as to never pass the peak value.

The chain rule applied to (13.34), using (13.33), gives the dynamics for the slip as

$$\dot{s} = \frac{r}{v_x}\dot{\omega} - \frac{\omega r}{v_x^2}\dot{v}_x$$
$$= \frac{r}{Jv_x}(rF_z\mu_x(s,\mu_0) + T_{\text{traction}} - T_{\text{brake}}) - \frac{\omega r}{mv_x^2}F_z\mu_x(s,\mu_0) \qquad (13.35)$$

That is, each tire is modeled as a first order nonlinear ordinary differential equation.

A driveline model from the engine torque reduction signal to the tire force T_{traction} can be taken as

$$T_{\text{traction}} = G(p)T_{\text{engine}} \qquad (13.36)$$

where $G(p)$ is a linear transfer function, here expressed in the differentiating operator p. We will assume that a torque reduction algorithm is given,

$$T_{\text{engine}} = \min(u,1)T_{\text{commanded}}, \quad u > 0, \qquad (13.37)$$

where u is the reduction factor (could be discrete). The minimum operator is included so that the controller can be active all the time and normally it aims at increasing the engine torque ($u > 1$) so that the slip would approach its reference value. Common principles for torque reduction include controlling the throttle or ignition angle or to use fuel blocking.

Slip Models

The so called *"magic tire formula"* Bakker et al. (1987) is the best known and most cited parametric model of the slip curve. It is defined as

$$\mu(s) = D\sin\left(C\tan^{-1}\left(B((1-E)s + \frac{E}{B}\tan^{-1}(Bs))\right)\right). \qquad (13.38)$$

This model is widely accepted for its flexibility, and used for simulation and curve fitting to test bench data. A suitable set of initial values is $B = 14$, $C = 1.3$, $D = \mu_{max}$ and $E = -0.2$.

Many other alternative slip functions have been suggested, see Liu and Sun (1995), Alvarez et al. (2001), Harned et al. (1969) and Szostak et al. (1988). For example, the rational function $\mu = ks/(as^2 + bs + 1)$ is suggested in Kiencke and Daiss (1994) for estimation during ABS braking. A theoretical model for friction, aimed for control purposes, is developed in de Wit et al. (1995) based on fundamental friction relations Armstrong-Hélouvry (1991). The model was later applied to slip curves in de Wit and Horowitz (1999); de Wit and Tsiotras (1999). Their μ–s model is dynamic, as opposed to the static one in (13.38),

$$\dot{z} = v_x - \frac{\sigma_0 |sv_x|}{g(sv_x)} z, \tag{13.39a}$$

$$F_x = (\sigma_0 z + \sigma_1 \dot{z} + \sigma_2 v_x) F_z, \tag{13.39b}$$

$$g(sv_x) = \mu_C + (\mu - \mu_C) e^{-|sv_x/v_s|^{0.5}} \tag{13.39c}$$

Here z is the state in the model and σ_i, s_C, v_s are parameters. They refer to the original model (which has a power 2 instead of 0.5 above) as the *LuGre* model.

13.6.2 Bicycle Model

Assume that spring–damper dynamics are much faster than road dynamics, so

$$\psi \approx \Psi, \quad \phi \approx \Phi, \quad \theta \approx \Theta. \tag{13.40}$$

The underlying mechanical equations belong to three categories: lateral, longitudinal and inertia.

$$F_y = \frac{v_x^2}{R_z} = v_x \dot{\psi} \tag{13.41a}$$

$$F_x = m \dot{v}_x \tag{13.41b}$$

$$J_z \ddot{\psi} = T_z. \tag{13.41c}$$

The dynamical equations for a front-wheel driven bicycle model are as follows:

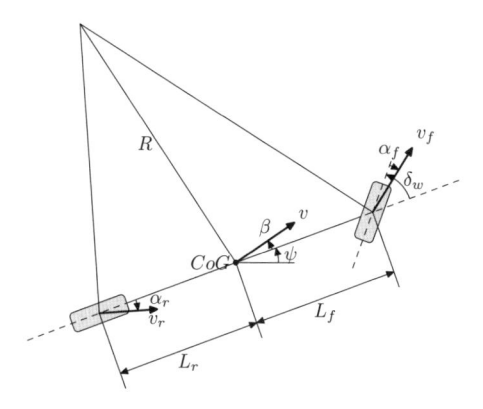

Figure 13.3: *Notation for the lateral dynamics of vehicles.*

$$\ddot{\omega}_f(J_1 + k^2 J_2) = T_{\text{eng}} - F_{f,x} r_w \tag{13.42a}$$
$$F_{f,x} = f(\omega, v_{f,x}, T_{\text{eng}}, N_{f,z}) \tag{13.42b}$$
$$F_x = m\ddot{x} = F_{f,x} + F_{r,x} - F_{\text{drag}} - F_{\text{roll}} - mg\sin\Theta \tag{13.42c}$$
$$F_y = m\ddot{y} = v_x\dot{\psi} + mg\sin\Phi + F_{f,y} + F_{r,y} \tag{13.42d}$$
$$F_z = m\ddot{z} = mg\cos(\Phi)\cos(\Theta) - v_x\dot{\Theta} - v_y\dot{\Phi} \tag{13.42e}$$
$$J_z\ddot{\psi} = F_{f,y}L_f - F_{r,y}L_r. \tag{13.42f}$$

13.7 Aircraft Dynamics

Complete dynamic models of aircraft are rare, and linearized partial models are used in academic publications. We here provide linearized discrete-time dynamics of two aircraft, where the roll dynamics are neglected.

13.7.1 F-16

The vertical dynamics of the *F-16 aircraft* is used in many simulation studies, due to the fact that the model has been public for a long time. The state, inputs and outputs for this application are summarized in Table 13.7.

Table 13.7: F-16 notation

Inputs	Outputs	States
u_1: spoiler angle $[0.1 deg]$	y_1: relative altitude $[m]$	x_1: altitude $[m]$
u_2: forward acceleration $[m/s^2]$	y_2: forward speed $[m/s]$	x_2: forward speed $[m/s]$
u_3: elevator angle $[deg]$	y_3: pitch angle $[deg]$	x_3: pitch angle $[deg]$
		x_4: pitch rate $[deg/s]$
		x_5: vertical speed $[deg/s]$

The numerical values below are taken from Maciejowski (1989) (given in

13.7 Aircraft Dynamics

continuous time) sampled with 10 Hz:

$$A = \begin{pmatrix} 1 & 0.0014 & 0.1133 & 0.0004 & -0.0997 \\ 0 & 0.9945 & -0.0171 & -0.0005 & 0.0070 \\ 0 & 0.0003 & 1.0000 & 0.0957 & -0.0049 \\ 0 & 0.0061 & -0.0000 & 0.9130 & -0.0966 \\ 0 & -0.0286 & 0.0002 & 0.1004 & 0.9879 \end{pmatrix},$$

$$B_u = \begin{pmatrix} -0.0078 & 0.0000 & 0.0003 \\ -0.0115 & 0.0997 & 0.0000 \\ 0.0212 & 0.0000 & -0.0081 \\ 0.4150 & 0.0003 & -0.1589 \\ 0.1794 & -0.0014 & -0.0158 \end{pmatrix},$$

$$B_d = \begin{pmatrix} 0 \\ 1 \\ 0 \\ 0 \\ 0 \end{pmatrix}, \quad B_f = \begin{pmatrix} -0.0078 & 0.0000 & 0.0003 & 0 & 0 & 0 \\ -0.0115 & 0.0997 & 0.0000 & 0 & 0 & 0 \\ 0.0212 & 0.0000 & -0.0081 & 0 & 0 & 0 \\ 0.4150 & 0.0003 & -0.1589 & 0 & 0 & 0 \\ 0.1794 & -0.0014 & -0.0158 & 0 & 0 & 0 \end{pmatrix},$$

$$C = \begin{pmatrix} 1 & 0 & 0 & 0 & 0 \\ 0 & 1 & 0 & 0 & 0 \\ 0 & 0 & 1 & 0 & 0 \end{pmatrix}, \quad D_f = \begin{pmatrix} 0 & 0 & 0 & 1 & 0 & 0 \\ 0 & 0 & 0 & 0 & 1 & 0 \\ 0 & 0 & 0 & 0 & 0 & 1 \end{pmatrix}.$$

The disturbance is assumed to act as an additive term to the forward speed. The model is available as m=exlti('airc') in *Signals and Systems Lab*.

13.7.2 Gripen

We here summarize a similar linearized discrete-time model for the lateral dynamics of the Swedish fighter *Gripen*, see Glad and Ljung (2000) for more details. The states and measurements are somewhat different, and the notation is summarized in Table 13.8. A linear model for Mach 0.6, altitude 500

Table 13.8: Gripen notation

Inputs	Outputs	States
u_1: aileron command [rad]	$y_1 = x_1$	$x_1 = v_y$ lateral speed [m/s]
u_2: rudder command [rad]	$y_2 = x_2$	$x_2 = \dot{\phi}$: roll rate [rad/s]
	$y_3 = x_3$	$x_3 = \dot{\psi}$ yaw rate [rad/s]
		$x_4 = \phi$: roll angle [rad]
		$x_5 = \psi$: yaw angle [rad]
		x_6: aileron angle [rad]
		x_7: rudder angle [rad]

Table 13.9: *The SNAME notation for marine vessels*

Motion components	Forces	Velocities	Pose
surge (x-direction)	X	u	x
sway (y-direction)	Y	v	y
heave (z-direction)	Z	w	z
roll (rotation about x)	K	p	ϕ
pitch (rotation about y)	M	q	θ
yaw (rotation about z)	N	r	ψ

m and angle of attack of 0.04 rad is given by

$$A = \begin{pmatrix} -0.3 & 8.1 & -201.0 & 9.8 & 0.0 & -12.5 & 17.1 \\ -0.2 & -2.5 & 0.6 & -0.0 & 0.0 & 107.0 & 7.7 \\ 0.0 & -0.1 & -0.5 & 0.0 & 0.0 & 4.7 & -8.0 \\ 0.0 & 1.0 & 0.0 & 0.0 & 0.0 & 0.0 & 0.0 \\ 0.0 & 0.0 & 1.0 & 0.0 & 0.0 & 0.0 & 0.0 \\ 0.0 & 0.0 & 0.0 & 0.0 & 0.0 & -20.0 & 0.0 \\ 0.0 & 0.0 & 0.0 & 0.0 & 0.0 & 0.0 & -20.0 \end{pmatrix} \quad (13.43a)$$

$$B = \begin{pmatrix} 0.0 & -2.1 \\ -31.7 & 0.0 \\ 0.0 & 1.5 \\ 0.0 & 0.0 \\ 0.0 & 0.0 \\ 20.0 & 0.0 \\ 0.0 & 20.0 \end{pmatrix} \quad (13.43b)$$

$$C = \begin{pmatrix} 0.0 & 0.0 & 0.0 & 1.0 & 0.0 & 0.0 & 0.0 \\ 0.0 & 0.0 & 0.0 & 0.0 & 1.0 & 0.0 & 0.0 \end{pmatrix}. \quad (13.43c)$$

The model is available as `m=exlti('gripen')` in *Signals and Systems Lab*.

13.8 Underwater Vehicle Dynamics

Six independent coordinates are required to completely describe the position and orientation of an underwater vehicle. For marine vehicles it is common to use the SNAME notation, see SNAME (1950), summarized in Table 13.9.

The state coordinates are grouped into two vectors, where

$$x^{pr} = \begin{pmatrix} x & y & z & \phi & \theta & \psi \end{pmatrix}^T, \quad (13.44)$$

denotes the position and rotation pr, and

$$x^{vq} = \begin{pmatrix} u & v & w & p & q & r \end{pmatrix}^T \quad (13.45)$$

13.8 Underwater Vehicle Dynamics

denotes velocity and angular rates vq, respectively. The position coordinates

$$x^{\text{pos}} = \begin{pmatrix} x & y & z \end{pmatrix}^T \tag{13.46}$$

are decomposed in an Earth-centered and Earth-fixed frame (ECEF). However, for local navigation it is convenient to use a local North East Down (NED) coordinate frame instead. Linear and angular velocities are decomposed in the body fixed frame.

The nonlinear dynamic equations of motion can be expressed in a compact form, see Fossen (2002), as

$$\dot{x}^{pr} = J(x^{pr})x^{vq} \tag{13.47a}$$
$$M\dot{x}^{vq} + C(x^{vq})x^{vq} + D(x^{vq})x^{vq} + g(x^{pr}) = \tau + w, \tag{13.47b}$$

where $J(x^{pr})$ is a nonlinear transformation matrix for the kinematics, M is the inertia matrix of the vehicle including added mass, $C(x^{vq}) = C_{RB}(x^{vq}) + C_A(x^{vq})$ is the centrifugal and coriolis matrix, $D(x^{vq})$ is the hydrodynamic damping matrix, $g(x^{pr})$ is the vector of gravity and buoyant forces, τ is the control input vector of forces and moments, and w is a vector of environmental disturbances.

The vehicle dynamics are affected by currents. Taking this into account, a more accurate model can be given as

$$M\dot{x}^{vq} + C_{RB}(x^{vq})x^{vq} + C_A(x^{vq} - c^{vq})(x^{vq} - c^{vq})$$
$$+ D(x^{vq} - c^{vq})(x^{vq} - c^{vq}) + g(x^{pr}) = \tau + w, \tag{13.48}$$

where c^{vq} is the velocity of the surrounding fluid. In open waters the current is often assumed to be irrotational and can be written as

$$c^{vq} = \begin{pmatrix} u_c & v_c & w_c & 0_{1\times 3} \end{pmatrix}. \tag{13.49}$$

In this model, the current velocity is assumed to be very small, and (13.47) will be used. Underwater currents will however nearly always be present and need to be handled. A common method is to model the current as a slowly varying bias.

Equation (13.47) is not practical for controller or observer design. For slender and symmetric vehicles it is possible to separate the system into three noninteracting (or lightly interacting) systems Fossen (2002); Jalving (1994); Ni (2001). The three subsystems and their state variables are:

- Speed: $u(t)$
- Steering: $v(t), r(t), \psi(t)$
- Diving: $w(t), q(t), \theta(t), z(t)$

Table 13.10: Decoupled subsystems of an underwater vehicle

Subsystem	State variables	Control inputs
Speed	$u(t)$	$n(t)$
Steering	$v(t), r(t), \psi(t)$	$\delta_r(t)$
Diving	$w(t), q(t), \theta(t), z(t)$	$\delta_s(t)$

The subsystems with states and control inputs for a cruciform tail configuration, are summarized in Table 13.10.

In the following, simplified models of the speed, steering and diving subsystems will be given, based on Fossen (2002); Jalving (1994). The SNAME notation in Table 13.9 will be used for denoting forces and moments.

More general information about marine vehicle dynamics can be found in for instance Fossen (2002).

Speed Subsystem

Neglecting interactions from other parts of the system, the surge subsystem can be modeled as

$$(m - X_{\dot{u}})\dot{u} - X_u u - X_{|u|u}|u|u = \tau_1(n) + T_{\text{loss}}, \quad (13.50)$$

where m and $X_{\dot{u}}$ denote vehicle mass and added mass respectively, X_u is linear damping, $X_{|u|u}$ is quadratic damping, and T_{loss} contains coupling terms and environmental disturbances. For low speed applications $X_{|u|u} \approx 0$. The propeller dynamics can be modeled as

$$\tau_1 = T_{|n|n}|n|n + T_{|n|u}|n|u \quad (13.51)$$

where $T_{|n|n}$ and $T_{|n|u}$ are propeller coefficients and n is the propeller revolution.

The HUGIN AUV speed controller usually tries to maintain a constant propeller revolution rate instead of trying to maintain a constant surge speed u_0. This is to optimize the battery operation. As a consequence the surge speed u will drop somewhat during maneuvers due to some coupling between the surge and the steering subsystem.

Steering Subsystem

Under the assumption of nearly constant speed $u \approx u_0$, the vehicle dynamics in sway and yaw can be simplified to:

$$m\dot{v} + mu_0 r = Y \quad (13.52\text{a})$$
$$I_z \dot{r} = N, \quad (13.52\text{b})$$

where

$$Y = Y_{\dot{v}}\dot{v} + Y_{\dot{r}}\dot{r} + Y_v v + Y_r r + Y_\delta \delta_r \quad (13.52c)$$
$$N = N_{\dot{v}}\dot{v} + N_{\dot{r}}\dot{r} + N_v v + N_r r + N_\delta \delta_r. \quad (13.52d)$$

For small roll and pitch angles we can assume that

$$\dot{\psi} \approx r. \quad (13.53)$$

The steering subsystem can then be rearranged as

$$\begin{pmatrix} m - Y_{\dot{v}} & -Y_{\dot{r}} & 0 \\ -N_{\dot{v}} & I_z - N_{\dot{r}} & 0 \\ 0 & 0 & 1 \end{pmatrix} \begin{pmatrix} \dot{v} \\ \dot{r} \\ \dot{\psi} \end{pmatrix}$$
$$+ \begin{pmatrix} -Y_v & -Y_r + mu_0 & 0 \\ -N_v & -N_r & 0 \\ 0 & -1 & 0 \end{pmatrix} \begin{pmatrix} v \\ r \\ \psi \end{pmatrix} = \begin{pmatrix} Y_\delta \\ N_\delta \\ 0 \end{pmatrix} \delta_r. \quad (13.54a)$$

Writing the above equation in state space form yields

$$\begin{pmatrix} \dot{v} \\ \dot{r} \\ \dot{\psi} \end{pmatrix} = \begin{pmatrix} a_{11} & a_{12} & 0 \\ a_{21} & a_{22} & 0 \\ 0 & 1 & 0 \end{pmatrix} \begin{pmatrix} v \\ r \\ \psi \end{pmatrix} + \begin{pmatrix} b_1 \\ b_2 \\ 0 \end{pmatrix} \delta_r. \quad (13.54b)$$

The steering dynamics calibrated from real data presented in Fauske et al. (2007) is given by

$$\theta = (0.1273, 1.1069, -0.4665, -2.6960, 1.2888, -2.5942)^T, \quad (13.55)$$

and is available as m=exnl('auv') (nonlinear model object) or m=exlti('auv') (linear model object in state space form) in *Signals and Systems Lab*.

To calculate the horizontal position, kinematic differential equations are needed

$$\dot{x} = u \sin \psi + v \cos \psi \quad (13.56a)$$
$$\dot{y} = u \cos \psi - v \sin \psi. \quad (13.56b)$$

Depth Dynamics

A neutrally buoyant vehicle dives using its stern planes and forward motion. The vehicle depth dynamics can be simplified to

$$m(\dot{w} - u_0 q) = Z \quad (13.57a)$$
$$I_y \dot{q} = M, \quad (13.57b)$$

where Z and M are forces and moments due to added mass, damping and stern plane deflection. Expanding the terms gives the model

$$\begin{pmatrix} m - Z_{\dot{w}} & -Z_{\dot{q}} & 0 & 0 \\ -M_{\dot{w}} & I_y - M_{\dot{q}} & 0 & 0 \\ 0 & 0 & 1 & 0 \\ 0 & 0 & 0 & 1 \end{pmatrix} \begin{pmatrix} \dot{w} \\ \dot{q} \\ \dot{\theta} \\ \dot{z} \end{pmatrix}$$

$$+ \begin{pmatrix} -Z_w & mu_0 - Z_q & 0 & 0 \\ -M_w & -M_q & \overline{BG}_z W & 0 \\ 0 & -1 & 0 & 0 \\ -1 & 0 & u_0 & 0 \end{pmatrix} \begin{pmatrix} w \\ q \\ \theta \\ z \end{pmatrix} = \begin{pmatrix} Z_\delta \\ M_\delta \\ 0 \\ 0 \end{pmatrix} \delta_s, \quad (13.58)$$

where $W = mg$ and $\overline{BG}_z$ is a moment caused by a difference between the center of gravity and center of buoyancy.

Writing the above equation in state space form yields

$$\begin{pmatrix} \dot{w} \\ \dot{q} \\ \dot{\theta} \\ \dot{z} \end{pmatrix} = \begin{pmatrix} c_{11} & c_{12} & c_{13} & 0 \\ c_{21} & c_{22} & c_{23} & 0 \\ 0 & 1 & 0 & 0 \\ 1 & 0 & -u_0 & 0 \end{pmatrix} \begin{pmatrix} w \\ q \\ \theta \\ z \end{pmatrix} + \begin{pmatrix} d_1 \\ d_2 \\ 0 \\ 0 \end{pmatrix} \delta_s. \quad (13.59)$$

The heave speed, w, is usually low, so the above model can be simplified to

$$\begin{pmatrix} \dot{q} \\ \dot{\theta} \\ \dot{z} \end{pmatrix} = \begin{pmatrix} c_{22} & c_{23} & 0 \\ 1 & 0 & 0 \\ 0 & -u_0 & 0 \end{pmatrix} \begin{pmatrix} q \\ \theta \\ z \end{pmatrix} + \begin{pmatrix} d_2 \\ 0 \\ 0 \end{pmatrix} \delta_s. \quad (13.60)$$

13.9 Summary

13.9.1 Theory

The following guidelines apply when choosing a suitable model for state estimation:

- The level of detail should be matched to the SNR. With very precise measurements with respect to the object's maneuverability, a complex model can be motivated. On the other hand, with large measurement uncertainty, a very simple model often works at least as well as a complex one. One can here contrast the two problems of tracking walking humans and satellites.

- As linear models as possible are to be prefered when the EKF is to be used. The choice of coordinate vector is here of importance, since some choices can give models that are more nonlinear than others.

- For numerical approximate filters as the PF, the choice of coordinates is not important for the performance.

- Discrete time models should be prefered when they exist in nonlinear filtering, to avoid invoking time-consuming numerical ODE solvers.

The models can be divided into the following categories:

- *Kinematic models.* A kinematic model describes the motion of an object as a function of forces, but does not attempt to model the forces. In simpler words, the forces are modelled using the process noise, and possible known control inputs are not used. The main alternatives are:

  - *Translation kinematics.* Usually, independent motion in all dimensions, where Newton's force law $F = ma$ gives the model

  $$x_t^{(1)} = X_t, \tag{13.61a}$$
  $$\dot{x}_t^{(1)} = x_t^{(2)} = v_t^X, \tag{13.61b}$$
  $$\dot{x}_t^{(2)} = w_t^X = a_t^X, \tag{13.61c}$$

  for movements along the X axis. The choice of coordinates is often crucial (for instance Cartesian or polar, world frame or local Cartesian).

  - *Rotational kinematics* describes the orientation of the object. The coordinates are either Euler angles or *quaternions*.

  - *Rigid body kinematicsrigid body kinematics* combines translational and rotational kinematics for describing motion in both orientation and position.

- *Constrained kinematics.* Constrained motion is often difficult to fit into the model framework, unless a smart choice of coordinates is possible. The simplest case of motion in the plane (2D motion) can be modelled using the states X, Y, ψ for position X, Y and heading ψ, respectively, and possibly derivatives of these. Coordinated turn models is another example of motion models, where the motion is constrained to circular trajectories.

- *Application specific force models.* A force model complements a kinematic model with application specific models for describing the forces and torques acting on the object. As an example, suppose a DC motor with input u_t generates the force $F_t = ku_t$ in (13.61), then the model can be written

$$\dot{x}_t^{(2)} = a_t^X = \frac{k}{m}u_t + \frac{k}{m}w_t. \qquad (13.62)$$

Here, the process noise is assumed to model uncertainty in the control input. There are application specific models for aircraft, vehicles, ships, underwater vessels and so on.

13.9.2 Software

The most common kinematic blackbox models are available in discrete time:

- Linear models in `'m=exlti(ex)'`, where `ex='cxnD'`. Here n stands for the dimension that can be 1,2 or 3, and x stands for the number of integrators and it can take on the values p (position, 0 integrators), v (velocity, 1), a (acceleration, 2) or j (jerk, 3). The model includes a position measurement.

- Coordinated turn models with Cartesian or polar velocity are obtained with `m=exnl(ex)`, where `ex='ctcv2D'`, `'ctpv2D'`. These include range and bearing measurements.

- The example library `m=exmotion(ex)` generates just the motion models as NL objects. The user can then append sensor models from the library `m=exsensor(ex)`. For instance,

```
m=exmotion('ctcv2d');
s=exsensor('radar',1);
ms1=addsensor(m,s);
ms2=exnl('ctcv2D')
```

give the same models, but the first alternative has the largest flexibility. A selection of available motion models is summarized below.

13.9 Summary

exmotion(fun)	nD	Description
ctcv2d	2D	Coordinated turn model, Cartesian velocity
ct, ctpv2d	2D	Coordinated turn model, polar velocity
ctpva2d	2D	Coordinated turn model, polar velocity and acc
cv2d	2D	Cartesian velocity linear model
imu2d	2D	Dead-reckoning of acceleration and yaw rate
imukin2d	2D	Two-dimensional inertial model with aX, aY, wX as inputs
imukin2dbias	2D	As imukin2d but with 3 bias states for the inertial measurements
imukin3d	3D	Three-dimensional inertial model with a, w as the 6D input
imukin3dbias	3D	As imukin3d but with 6 bias states for the inertial measurement

- Many dynamical models described in this chapter are available as examples in exlti(ex), as for instance 'airc', 'gripen' and 'auv'.

14

Sensors and Sensor Near Processing

This chapter describes the most common sensors for typical sensor fusion applications in navigation and tracking.

The most fundamental physical relations are surveyed, but the focus is on sensor-near signal processing for extracting as much information as possible from each sensor. Conventional sensors as thermometer returns a temperature related signal, and there is no need for pre-processing such signals. However, already such a simple device as an accelerometer contains much more information than the acceleration. First, the spectrum of the signal reveals important vibration modes of the platform, that can be related to motion. For instance, the wheel angular speed might be seen superpositioned to the accelerometer spectrum. Further, a three-dimensional accelerometer can be used as an inclinometer when at stand-still, since the gravity vector also affects the sensor. These are two examples of *virtual sensors*. The most challenging sensors from our sensor fusion perspective are ranging sensors, wheel speed sensors and wireless radio measurements, and these will be allocated the most space in this sensor survey.

14.1 Ranging Sensors

The ranging sensors radar, lidar, sonar and IR work on different frequency bands, but otherwise have a lot in common. The main principle is that a signal or pulse $s(t)$ is emitted at a certain direction θ, and the returned pulse $y(t)$ is measured. The received signal can be expressed as $y(t) = h * s(t - 2R/c)$, where $*$ denotes convolution, R is the range, c the speed of the media, $2R/c$ the time of flight and $h(t)$ is the dynamic distortion caused by scattering and multipath effects. Pre-processing involves pre-filtering and thresholding the returned pulse, which is conventionally done in hardware. The thresholding gives the approximate time of flight for the signal. The result is a set of range–bearing pair R_k, θ_k. The returned pulse is quite information rich, so a modern trend is to perform more advanced analysis in either hardware or software.

14.1.1 The Frequency Spectrum

Table 14.1: *Electromagnetic and audio spectrum. In the name abbreviations, F stands for frequency, LMH for low, medium and high respectively, and ESUV for extremely, super, ultra and very, respectively.*

Name	Frequency [Hz]	Wavelength [m]	Application
ELF 1	$3 - 30$	$10^8 - 10^7$	Communication with submarines
SLF 2	$30 - 300$	$10^7 - 10^6$	Communication with submarines
ULF 3	$300 - 3000$	$10^6 - 10^5$	Communication within mines
VLF 4	$3 \cdot 10^3 - 3 \cdot 10^4$	$10^5 - 10^4$	Submarine communication, avalanche beacons, wireless heart rate monitors, geophysics
LF 5	$3 \cdot 10^4 - 3 \cdot 10^5$	$10^4 - 10^3$	Navigation, time signals, AM longwave broadcasting
MF 6	$3 \cdot 10^5 - 3 \cdot 10^6$	$10^3 - 100$	AM (Medium-wave) broadcasts
HF 7	$3 \cdot 10^6 - 3 \cdot 10^7$	$100 - 10$	Shortwave broadcasts, amateur radio and over-the-horizon aviation communications
VHF 8	$3 \cdot 10^7 - 3 \cdot 10^8$	$10 - 1$	FM, television broadcasts and line-of-sight ground-to-aircraft and aircraft-to-aircraft communications
UHF 9	$3 \cdot 10^8 - 3 \cdot 10^9$	$1 - 0.1$	television broadcasts, microwave ovens, mobile phones, wireless LAN, Bluetooth, GPS and two-way radio
SHF 10	$3 \cdot 10^9 - 3 \cdot 10^{10}$	$0.1 - 0.01$	microwave devices, wireless LAN, most modern radars
EHF 11	$3 \cdot 10^{10} - 3 \cdot 10^{11}$	$0.01 - 0.001$	Radio astronomy, high-speed microwave radio relay
	$3 \cdot 10^{11}$	10^{-3}	Far IR
	$3 \cdot 10^{12}$	10^{-4}	Thermal IR
	$3 \cdot 10^{14}$	10^{-6}	Near IR
	$6 \cdot 10^{14}$	$5 \cdot 10^{-7}$	Visible light: red $(4 \cdot 10^{-7} \text{m}, 7.5 \cdot 10^{14} \text{Hz})$ to violet $(7 \cdot 10^{-7} \text{m}, 4.3 \cdot 10^{14} \text{Hz})$
UV	$3 \cdot 10^{15} - 3 \cdot 10^{16}$	$10^{-7} - 10^{-8}$	
X-rays	$3 \cdot 10^{16} - 3 \cdot 10^{18}$	$10^{-8} - 10^{-10}$	
gamma	$3 \cdot 10^{18} -$	$10^{-10} -$	

Table 14.1 summarizes the electromagnetic spectrum and how it is used by different sensors and applications. Above 300 GHz, the absorption of electromagnetic radiation by the earth's atmosphere effectively attenuates higher frequencies. This high attenuation lasts up to the infrared and optical window frequency ranges. The ELF, SLF, ULF, and VLF bands overlap the audio spectrum, which is approximately 20–20,000 Hz. In the audio spectrum, sound is transmitted by atmospheric compression rather than electromagnetic

14.1.2 Radar

The term RADAR appeared 1941 as an acronym for Radio Detection and Ranging, when it replaced the previously used RDF (Radio Direction Finding). The term is now a standard word, radar, without capitalization. The first patents appeared in 1904 on detection and ranging, respectively. German, French, American and British researchers made major contributions to the development at the time before the Second World War.

Principles

A radar emits electromagnetic waves which reflect (scatter) from any large change in the dielectric or diamagnetic constants. One distinguishing property between different radar principles is how the emitted wave looks like:

- *CW radar* (continuous wave) was the first system used. It transmits a sinusoid which is also mixed with the returned signal to provide the Doppler shift which corresponds to range rate. All computations can be realized in analogue electronics.

- *FM-CW radar* (frequency modulated contionuous wave) sweeps the frequency according to a sawtooth (or sine), which makes it possible to also compute the range.

- *PD radar* (pulse Doppler) transmits a short pulse of duration T_p with *pulse repetition frequency* (PRF) f_{PRF}. Each pulse is modulated with a sine wave of frequency f, so the range rate can be computed from the frequency shift Δ_f as

$$\dot{R} = \frac{cf_{PRF}(\Delta_f + n\pi)}{4\pi f}, \quad \Delta_f \in [-\pi, \pi], \quad n \in \mathcal{Z}. \tag{14.1}$$

It is the 2π ambiguity in the phase shift that gives rise to possible aliasing: The delivered range rate is the true range rate modulo $cf_{PRF}/4f$. The remedy is to increase the PRF, which may decrease accuracy.

The range is computed from the returned pulse which is thresholded at a delay of T_d as

$$R = \frac{cT_d}{2}. \tag{14.2}$$

Here, another kind of aliasing occurs due to ambiguity in which transmitted pulse to associate the detected peak with. In general,

$$R = \frac{c(T_d + n/f_{PRF})}{2}, \quad T_d \in [0, 1/f_{PRF}], \quad n \in \mathcal{Z}. \tag{14.3}$$

The consequence is that the range modulo $cT_d/2 = c/(2f_{PRF})$ is delivered. For this reason, the PRF should be chosen small. Typical values are shown in Table 14.2

Table 14.2: *Typical values on range and range rate ambiguity.*

PRF	Velocity Ambiguity	Range Ambiguity
Low (2kHz)	30m/s	75km
Medium (12kHz)	180m/s	12.5km
High (200kHz)	3000m/s	750m

The resolution in bearing and azimuth depends on the radar loob. Basically, it is inversely proportional to the size of the antenna. The returned energy decays quickly with distance. The radar equation gives a R^4 decay:

$$P_r = \frac{P_t G_t A_r \sigma F^4}{(4\pi)^2 R_t^2 R_r^2} \tag{14.4}$$

where the notation is given in Table 14.3.

Table 14.3: *Notation in the radar equation (14.4).*

P_t	transmitter power
G_t	gain of the transmitting antenna
A_r	effective aperture (area) of the receiving antenna
σ	radar cross section, or scattering coefficient, of the target
F	pattern propagation factor
R_t	distance from the transmitter to the target
R_r	distance from the target to the receiver

Surface Radar Illustrations

To illustrate the pre-processing in the radar, a surface radar mounted on a boat will be studied in some detail, using the results of the master thesis Dahlin and Mahl (2007).

A scanning radar emits energy in all bearing angles, and the returned energy is displayed in polar coordinates on a monitor. Figure 14.1 illustrates one such image in comparison to the actual map from a sea chart. The marine standard EDICS describes how such a radar image should be overlaid on the sea chart, as done in Figure 14.2. To get a good match, both position and heading have to be known.

Traditional radar monitors are black and white, which requires to digitalize the returned pulse response. The threshold is here the critical design

14.1 Ranging Sensors

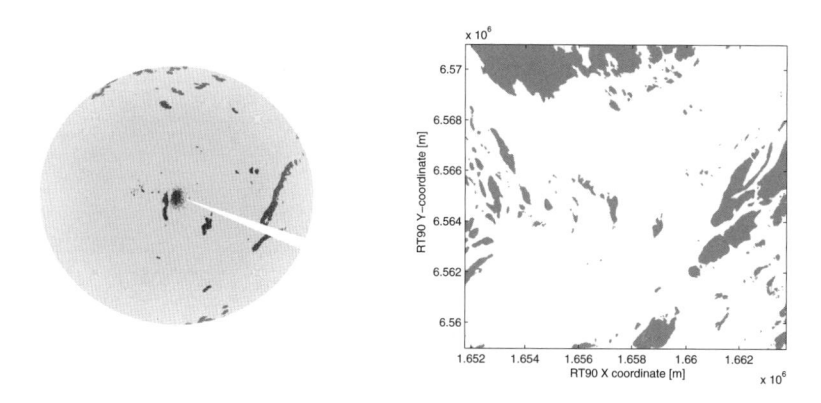

Figure 14.1: Radar video frame and corresponding part of sea chart.

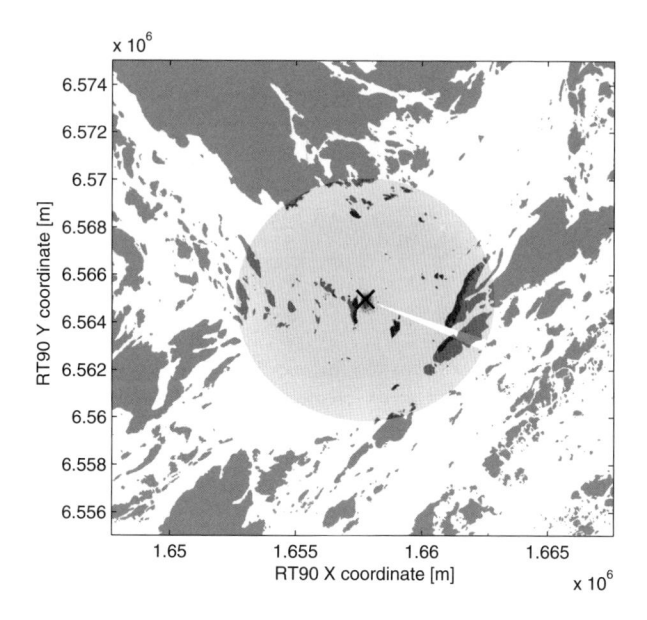

Figure 14.2: The radar overlay principle.

parameter, and this might have to be adapted in different weather conditions for example.

Figure 14.3 shows two cases where the pulse response has one unique and easily distinguished peak. The one in Figure 14.3(b) is easy to localize in time, while the one in Figure 14.3(d) has a longer duration due to the larger island that returns energy at all parts of land.

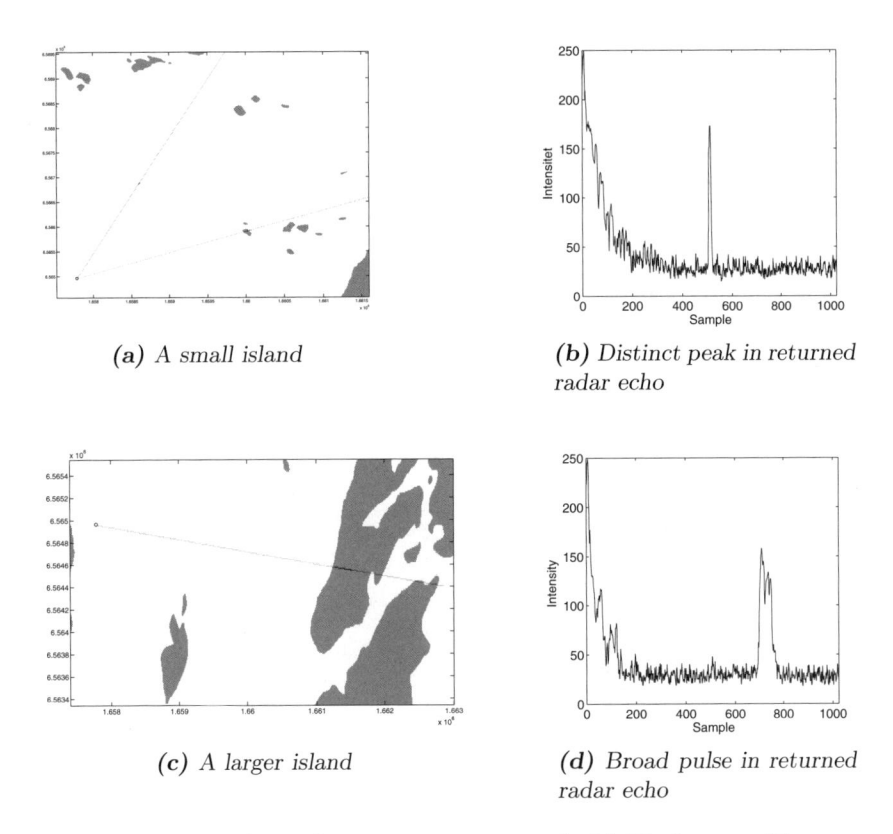

(a) A small island

(b) Distinct peak in returned radar echo

(c) A larger island

(d) Broad pulse in returned radar echo

Figure 14.3: Radar pulse energy response. (a–b) Ideal case with one small object leading to a narrow pulse response. (c–d) Case with one large object leading to a broad pulse response.

Figure 14.4 shows three other examples of impulse responses exhibiting multiple peaks for several islands on the same bearing, and in one case from another boat in front of an island.

The conclusion is that the thresholding device in the radar should:

- Allow for multiple echoes.

- Compute the returned energy in each echo, from which extended target

14.1 Ranging Sensors

models can be used and topological information in geographic information systems fully utilized. It can also faciliate association over time.

- Be well above the noise floor to avoid excessive clutter.

- Be larger or blind at small distances to avoid detection due to near-field back-scattering.

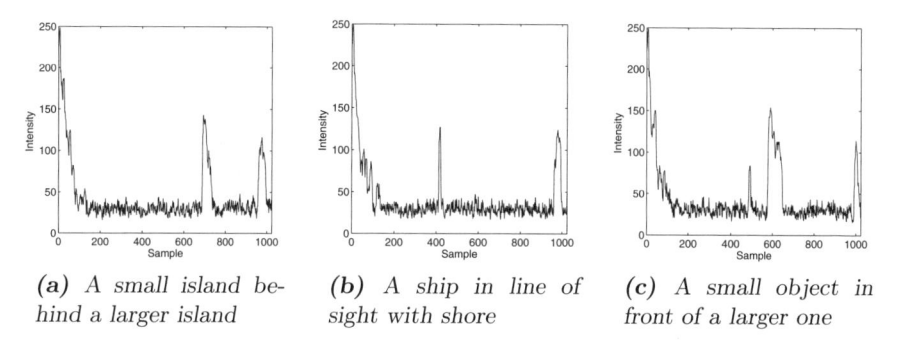

(a) A small island behind a larger island

(b) A ship in line of sight with shore

(c) A small object in front of a larger one

Figure 14.4: Radar pulse energy response in case of ambiguities.

14.1.3 Laser Radar

A laser-based system used for measuring the distance to an object or the object's speed is called *laser radar*. Laser radar is sometimes called *ladar* (laser detection and ranging), laser radar, or *lidar* (light detection and ranging).

Laser radar systems have been investigated over several decades primarily for military applications, see for instance Jelalian (1992). Laser radars are, just as conventional radars, mainly used for remote sensing and mapping. Depending on the application, the chosen wavelength is usually in the 0.5-10 μm range (from visible to thermal infrared). As in microwave radar technology, the range to object and background is often obtained by measuring the time-of-flight for a modulated laser beam from the transmitter to the object and back to the receiver. Some unique features in laser radar systems are high angular, range and velocity resolution.

In this section, the two main measurement techniques for laser radar, coherent and direct-detection, are described shortly. The main signal properties of laser radars and models of the returned signals are presented. For further information we recommend Kamerman (1993, 1997); Shapiro et al. (1981); Stone et al. (2004).

Coherent and Direct Detection

Two main detection schemes can be identified in laser radar systems; *coherent detection* and *direct detection*. The difference lies in how the returning signal is processed in the detector. In coherent detection, the phase information is preserved. To achieve coherent detection, the returned laser beam is mixed with the local oscillator beam. Such systems are common for aerosol measurements, and velocity and vibration measurements with very high accuracy. Compared to direct-detecting systems the sensitivity can be up to 1000 times higher. The received signal's phase is also saved, which means that more detailed information of the sensed object can be collected. The high Doppler resolution gives the opportunity to estimate the speed of slowly moving objects or vibrating surfaces in one laser pulse (i.e., one measurement).

In direct-detection systems, the phase information is lost since only the amplitude of the returning signal is collected in the detector. Direct-detection laser radar systems are less complex and are common in 3D imaging applications. An object's velocity can be measured with a direct-detecting system too, using two or more measurements.

Imaging Techniques

With 3D laser radar a new dimension is added to active imaging. In addition to intensity and angular coordinates, also range is included in the image. There are several principles for 3D imaging laser radar; *scanning laser radar*, *staring laser radar* and *gated viewing*. Staring and gated viewing systems are usually direct-detection systems, while scanning can be of both detection types. A scanning or staring laser radar usually gives both an intensity and a range image of the scene.

The first laser radar systems were *single point sensors* and in the 1980's the single point sensors were combined with a device for beam deflection, i.e., rotating mirrors, to achieve *scanning systems*. This was the first type of 3D imaging laser radar. A straight-forward method to acquire 3D information of a scene is to scan with a single point detector laser radar. With every laser pulse, a very small part of the scene is illuminated and the time-of-flight of the reflected pulse is stored. Some detectors give a time-resolved pulse response (full waveform), whereas other detectors only give the time for the pulse return (above a certain threshold). With some systems it is possible to store first and last echo, or even more returns, each echo representing a different object range. There are also line array scanners, where an array of point sensors are used. An advantage of a scanning system is the possibility to achieve high angular resolution. The main disadvantage is the long data acquisition time, which prevents the capture of moving objects.

The development of Focal Plane Array (FPA) detectors with timing capability in each pixel has made non-scanning 3D imaging laser radars feasible.

These *staring systems* (flash laser radars) enable the capture of a complete 3D image with just one transmitted laser pulse. With such a system, the frame rate can be increased to video rate (50 Hz or 60 Hz), data from moving objects can be obtained. The sensor has the same size as an ordinary camera, excluding the laser source, which can be fit to the application, and the data acquisition platform, normally a computer. In short range systems, the laser can be incorporated into the camera unit itself.

With the *gated viewing* (GV) technique, also called burst illumination, the sensor can be a simple camera constructed for the laser wavelength, but the shutter gain is synchronized with an illuminating short pulse laser. This enables image collection at certain range interval in the scene. With an adjustable delay setting, corresponding to the time-of-flight back and forth to a desired range, the opening of the camera shutter is controlled. This exposes the camera only for a desired range slice, with the depth corresponding to the shutter open time (usually in the region of nanoseconds). The delay can be changed through a predefined program, resulting in a number of slices representing different ranges, i.e., a 3D volume. The set-back of this system is the power inefficiency, since every range slice image requires a total scene illumination. The advantage is the low cost and robustness, since rather simple components can be used. It has been shown that a rather small set of gated images can give high resolution 3D images, if the depth information is taken into account Andersson (2005).

Signal Properties

From a signal processing point of view, the benefits and drawbacks with laser radar measurements can be summarized as:

- A laser radar system returns range and intensity data with high angular precision. The data are, however, noisy and there are sometimes artifacts at borders of objects. This is a type of blurring phenomenon, resulting in object samples at erroneous ranges behind or in front of the object. The returned intensity values in the image are a function of the object's surface properties, which can be used to distinguish different materials. On the other hand, the returned intensity is a function of all objects that the laser beam has illuminated. This means that for partly obscured objects the returned intensity for an object can vary severely over the surface.

- The active illumination with a laser results in complete independence of ambient light conditions (such as day or night), and hence the image contrast is very robust in that respect. On the other hand, the illumination can be detected.

- The short wavelength makes it possible to collect data of high resolution.

Details of the object can be acquired, which is powerful in recognition applications. On the other hand, when large objects or scenes are measured a lot of data are collected. This requires fast hardware, large storage capabilities and fast algorithms.

- Due to the short wavelength, laser radars are more sensitive than conventional radars to atmospheric conditions with high attenuation, like fog, but less sensitive to rain and snow. This drawback can be partly compensated for by the gating technique.

- Sparse structures can be penetrated, which adds the ability to collect data from objects that are partly hidden behind vegetation, camouflage nets, curtains, or Venetian blinds.

- The laser radar does not penetrate dense structures, as tree stems, metal surfaces, roofs, and walls. Those object types do not transmit the particular wavelength. This means that data are only collected from the parts of the objects that are in the line of sight from the sensor. This effect is called *self-occlusion* and a *2.5D* representation of the scene is collected.

- When an object is measured with these types of a laser radars, a 3D coordinate is retrieved in each sample. This means that data can be projected to an arbitrary view. It is only a 2.5D representation of the scene, due to the self-occlusion. To achieve a full 3D representation, 2.5D images collected from different positions are combined. That process is called *registration*, see Besl and McKay (1992); Tolt et al. (2006).

- Laser radar systems are affected by speckle noise. The speckle noise is generated by the interference of different parts of the laser beam on the object's surface. Diffuse and semi-diffuse surfaces, like vegetation and non-glossy painted surfaces, give rise to speckles which in turn cause amplitude and phase modulation of the returning signal. Coherent detection systems are more sensitive to speckle-induces noise than direct-detection systems. Therefore, a coherent-detection imaging system will have a larger fraction of drop-out pixels compared to a direct-detecting system.

Signal Models

The different detection principles give returning signals with different level of detail and therefore the signal models differ. The model of the returning signal from a coherent detection system is complex-valued. For an object consisting of one, homogeneous surface, that moves with constant, radial velocity, the received signal is modeled:

$$S_{cd}(t) = A(t)\exp\bigl(-i(\omega_d t + \gamma(t))\bigr),$$
$$\gamma(t) = \mu\sin(\omega_v t + \gamma_v),$$

where $A(t)$ describes the speckle-induced modulation, modulations due to turbulence and other atmospheric effects, $\omega_d/2\pi$ is the doppler-induced frequency, μ is the modulation index, and $\omega_v/2\pi$ and γ_v are the vibration-induced frequency and phase, respectively.

In a direct-detection laser radar, the intensity estimate $\hat{I}$ and the range estimate $\hat{R}$ are calculated from the received pulse $S_{dd}(t)$. If the sensor elements in the detector are assumed to be statistically independent, the intensity values in sensor element (x,y) can be modeled as (see Gerwe and Idell (2003))

$$\hat{I}(x,y) = I(x,y) + n_I, \tag{14.5a}$$
$$I(x,y) \sim \text{Po}(S_{dd}(x,y,t)), \tag{14.5b}$$
$$n_I \sim \mathcal{N}(m,\sigma_I^2), \tag{14.5c}$$

where $\text{Po}(s)$ is a Poisson distributed random variable with mean s, and n_I is a Gaussian random variable with mean m and standard deviation σ_I. The mean m models a deterministic bias in the amplifier-part of the receiver, see Snyder and Hammoud (1993) for details. This model is also used for passive imaging, like visual and infrared cameras, see Snyder and Hammoud (1993). The uncertainty in intensity value is $\Delta I = \hat{I} - I^0$, where I^0 is the true value. The uncertainty in intensity value in pixel (x,y) can be modeled as, see Gerwe and Idell (2003),

$$\Delta I(x,y) + \sigma_I^2 \sim \text{Po}\bigl(S_{dd}(x,y,t) + \sigma_I^2\bigr)$$

The range estimate can be modeled as, see Grönwall et al. (2007),

$$\hat{R} = R_0 + \Delta R, \tag{14.6a}$$
$$\Delta R \sim \mathcal{N}\bigl(b(\theta),\sigma_R^2(\theta)\bigr), \tag{14.6b}$$

where R_0 is the true but unknown range and the ΔR is the uncertainty in the range estimate. The bias b and variance σ_R^2 vary with the object's shape, θ, and signal-to-noise ratio. If an optimal signal detector, like the matched filter, is used in the receiver the bias is close to zero.

14.1.4 Sonar

SONAR (SOund Navigation And Ranging) can be either active and passive. SONAR uses either infrasonic or ultrasonic sound, and the counterpart to acoustics in air is in water called ultraacoustics. The performance of SONAR is either noise limited or reverbation (scattering) limited.

The main applications are detection of targets, navigation, and underwater communication. Already Leonardo da Vinci demonstrated that detection of ships is possible by putting a tube into the water. The first patents appeared shortly after the Titanic disaster in 1912, and the first system for detecting icebergs was reported in 1914.

In active SONAR, the distance is measured by $R = ct/2$, where t is the time until an echo is received and c is the speed of sound in water. This speed can be approximated by

$$c(T, S, z) = \theta_1 + \theta_2 T + \theta_3 T^2 + \theta_4 T^3 + \theta_5 (S - 35) \tag{14.7a}$$
$$+ \theta_6 z + \theta_7 z^2 + \theta_8 T(S - 35) + \theta_9 T z^3, \tag{14.7b}$$
$$\theta = \big(1448.96, 4.591, -5.304 \cdot 10^{-2}, 2.374 \cdot 10^{-4}, 1.340, \tag{14.7c}$$
$$1.630 \cdot 10^{-2}, 1.675 \cdot 10^{-7}, -1.025 \cdot 10^{-2}, -7.139 \cdot 10^{-13}\big)^T, \tag{14.7d}$$

where T, S, and z are temperature in degrees Celsius, salinity in parts per thousand and depth in meters, For instance, $c = 1550.744$ m/s for $T = 25°C$, $S = 35$, $z = 1000$ m.

The speed in water is thus about five times faster than the speed in air, which is approximately given by

$$c_{air} = 331.3 \sqrt{1 + \frac{T}{273.15}} \approx 331.3 + 0.6T, \tag{14.8}$$

where T is in Celsius.

14.2 Physical Sensors

The sensors in this section have in common that they measure a physical quantity directly. It includes inertial sensors and geophones.

14.2.1 Accelerometer

An archetypal accelerometer consists of a proof mass anchored to a frame by a suspension beam acting as a spring-damper system. The proof mass is displaced during acceleration by Newton's force law. Today, *micromachined devices*, or *micro-electrical-mechanical systems* (*MEMS*) are the by far most common type of accelerometers by their low price and small size, see Yazdi et al. (1998). Here, the automotive market has been a driver for development, see Fleming (2001).

There are some different physical principles for how to measure the beam deflection. First, the beam can be connected to a piezoresistor or capacitor.

14.2 Physical Sensors

Table 14.4: *MEMS accelerometer specifications (from Yazdi et al. (1998))*

Parameter	Automotive	Navigation
Range	± 2g (ADAS) or ± 50g (airbag)	± 1g
Resolution	< 10mg (ADAS) or < 100mg (airbag)	< 4μg
Frequency range	0–400Hz	0–100Hz

Second, a resonant vibrating beam can be used with the principle that the resonance frequency shifts during acceleration. The primary mechanical noise source is due to Brownian motion of gas molecules around the proof mass. Typical specifications are summarized in Table 14.4.

A stationary three-axis accelerometer can be used as an *inclinometer*, where the specific force due to the gravity field is sensed by the accelerometer and used to find the downward direction. This is a good complement to the *compass* function of magnetometers to find the orientation of a stationary platform.

14.2.2 Gyroscopes

Classical gyroscopes consisted of a spinning wheel and are today rarely used in practice. High-precision fiber-optic and ring laser gyros are still too expensive for mass produced devices, and are reserved for high performance navigation applications. MEMS gyros have the by far largest volumes, and can today be found in a range of products as vehicle stability systems and navigation systems, video-camera stabilization, virtual reality, computer games and robotics.

All MEMS sensors use vibrating mechanical structures to sense angular rate utilizing the Coriolis effect. The vibrating structure can be a ring, fork or disc. Some typical data are summarized in Table 14.5, for three categories of quality.

The rotation of the earth affects the gyroscope measurements. This can be utilized to turn an accurate three-axis gyroscope into a *compass*. The requirement on the gyroscope is that the drift is considerably smaller than 15 degrees per hour (the angular rate of the earth).

14.2.3 Geophones

Geophones, also known as *seismometers*, consist of a mass flexibly connected to earth, and a means to register its movements. Advanced intruments may also contain a feedback loop that keeps the mass in the same position relative to the body to avoid saturation. The classical use is for detecting earth quakes and explosions and in seismic surveys for oil and gas. However, geophones are

Table 14.5: MEMS gyroscope specifications (from Yazdi et al. (1998))

Parameter	Rate grade	Tactical grade	Inertial grade
Range °/sec	50–1000	>500	>400
Angle random walk °/h	>0.5	0.05–0.5	<0.001
Bias drift °/h	10–1000	0.1–10	<0.01
Scale factor accuracy °/sec	0.1–1	0.01–0.1	<0.001
Frequency range	0–70Hz	0–100Hz	0–100Hz

very versatile and cheap tracking sensors.

14.2.4 Magnetometer

Magnetometers have like many other sensors been miniaturized recently and are now part of many consumer products as cellular phones. The most common principle is to use a *Hall sensor* to sense the magnetic field.

The most common application of magnetometers is a *compass*. However, it can also be used for sensing magnetic objects and also metallic objects since these deflect the earth magnetic field. That means that a metallic object can be seen as a magnetic dipole. The magnetometer sensor model follows from Maxwell's equations. Basically, a magnetic dipole is specified by a vector $\mathbf{m}$, and it affects the magnetometer signal with the three-dimensional vector

$$y_k = \frac{\mu_0}{4\pi |\mathbf{r}_k|^5} \left((\mathbf{r}_k^T \mathbf{m}) \mathbf{r}_k - |\mathbf{r}_k|^2 \mathbf{m} \right), \tag{14.9}$$

where $\mathbf{r}_k$ is the vector between the sensor and dipole and μ_0 a known constant.

The model can be simplified somewhat by assuming a planar geometry. Suppose the (x,y) plane is parallel to $\mathbf{m} = m(1,0,0)^T$ and $\mathbf{r}_k = (x,y,0)^T$. Then y_k^z is unaffected. A transformation of y_k^x, y_k^y from Cartesian to polar coordinates gives the new measurement vector

$$\bar{y}_k^1 = \sqrt{(y_k^x)^2 + (y_k^y)^2} = \frac{\mu_0}{4\pi |\mathbf{r}_k|^3} \sqrt{1 + 3\sin^2(\theta_k)}, \tag{14.10a}$$

$$\bar{y}_k^2 = \frac{y_k^x}{y_k^y} = \frac{x + x^2 + y^2}{y}. \tag{14.10b}$$

Here, θ is the angle between $\mathbf{r}$ and $\mathbf{m}$. Note that the magnitude does not reveal the direction of motion by itself. The phase contains information of direction of speed due to the x term in the numerator.

Figure 14.5 shows two examples. Clearly, the magnitude of the magnetic field is different for a large and a small vehicle. Note here that the smaller car is passing closer to the sensor than the larger vehicle, giving a relatively larger measurement. The direction of travel can be seen from the phase difference of the colored curves. Note here that the sensor is not in the same plane as

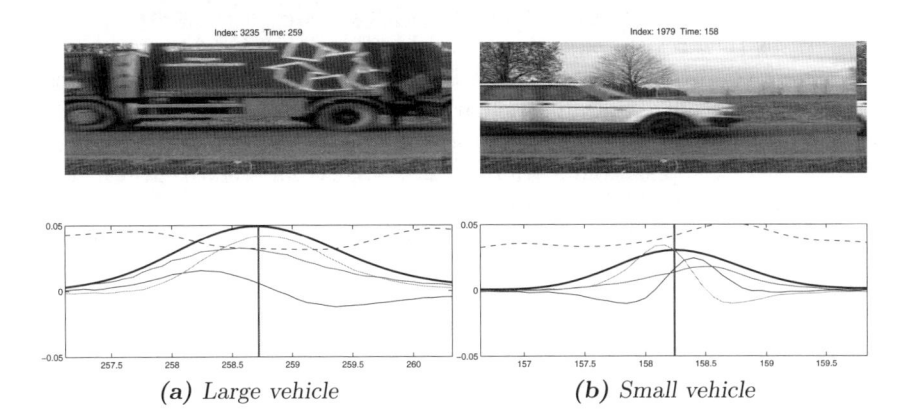

Figure 14.5: *Example of two vehicles and magnetometer measurements. The sensor is seen in the lower center of the images. Legend: thick black is the magnitude, blue y_k^x, green y_k^y, red y_k^z. Dashed line shows accelerometer signal.*

the dipole centrum, as it is located beneath the road surface. That is, the assumption above is not satisfied in this experiment.

14.3 Wheel Speed Sensors

The wheel speed is one of the most information rich signals in all kind of applications for wheeled vehicles. This section focuses on wheel speed sensors such as found in cars equipped with ABS. The section is based on the survey Gustafsson, which is the proper reference to this material.

14.3.1 Introduction

Rotational speed sensors are used in a variety of applications and are known by many different names: tachometers, revolution-counters, RPM gauges, etc. There are two different principles, (i) *rotary encoders* that provide the absolute angle and (ii) *incremental rotary encoders* that give angular speed. The basic technology in such sensors is based on conductive tracks, optical reflections or magnetic field variation. The last two types are the most common today, where the latter is the dominant technology in rough environments such as automotive applications. The principle for rotational speed sensors is that a toothed wheel is attached to the rotating shaft. A magnet attached to one side causes a variation in the magnetic field that can be sensed by a *Hall sensor*, and the variation is converted to a square wave signal where each edge corresponds to one edge of the toothed wheel. The time between two or more edges is then registered and converted into angular speed.

Some applications, such as volume knobs in home audio equipment and speedometers, do not require particularly high resolution while other applications require varying degrees of accuracy. Examples include fault detection (see the gear box application in Ahlberg and Sundström (2009)), health condition monitoring and virtual sensors computing physical quantities such as tire–road friction and tire pressure as over-viewed in Gustafsson et al. (2001b), and yaw rate and absolute velocity Gustafsson et al. (2001a). We will describe the computational process in more detail, with a focus on sampling, quantization effects, multi-domain signal processing aspects and the pre-processing needed to learn the manufacturing errors in the toothed wheel for very high accuracy applications.

For specific examples, our attention is directed to the particular rotational speed sensors found in cars equipped with *ABS (anti-locking brake systems)*. Since its introduction thirty years ago, these have found many new applications such as *electronic stability control (ESC)*, *traction control systems*, *tire pressure monitoring systems (TPMS)*, and *odometry* (navigation system support) to mention a few. See the survey Gustafsson (2009) for details and more examples.

14.3.2 Basic Functionality

The rotational speed sensor works as follows. A *toothed wheel (cog wheel)*, see Figure 14.6, with N_{cog} cogs is mounted on a rotating shaft. Initially we assume that the toothed wheel is ideal, so the angle between each tooth is $2\pi/N_{\text{cog}}$. The sensor gives a sinusoidal signal with varying amplitude and frequency, which is converted in a comparator to a square wave with constant amplitude. Each edge in this signal corresponds to an edge of a tooth. The time when tooth k is passing is denoted t_k. The corresponding angle is denoted $\varphi_k = k\frac{2\pi}{N_{\text{cog}}}$. The angle can also be given as a function of time $\varphi(t_k)$. Here, we encounter the problems of having two different domains: the angle domain and the time domain. Since the square wave is synchronized with the angle, the angle domain is to be prefered in many cases. One can say that the angle is sampled equidistantly in angle, where the *sample interval* is $\alpha = \frac{2\pi}{N_{\text{cog}}}$. The different domains are summarized in Table 14.6.

The *angular speed* can now be approximated as

$$\hat{\omega}_k = \frac{2\pi}{N_{\text{cog}}(t_k - t_{k-1})}. \qquad (14.11)$$

This is the basic formula that will be analyzed in the following. Note that the absolute angular speed (the direction of rotation is not observable) is measured in the cog domain, not the time domain. The main computational problems that perturb the result are:

14.3 Wheel Speed Sensors

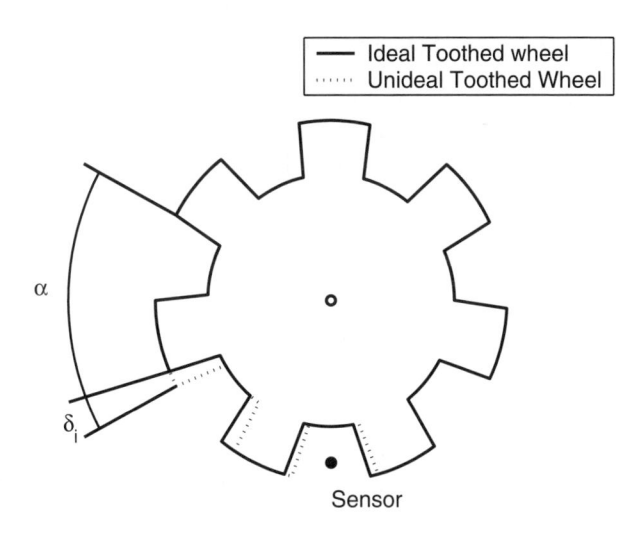

Figure 14.6: Ideal toothed wheel (solid) has the angle $\alpha = 2\pi/N_{\text{cog}}$ radians between each cog. Real toothed wheels are always non-ideal with a small cog offset δ_i.

Table 14.6: Definitions of the different domains

Domain	Index	Angle	Angular speed	Index interval
Time	t	$\varphi(t)$	$\omega(t)$	–
Sample index	n	$\varphi(nT_s)$	$\omega(nT_s)$	T_s
Timer index	m	$\varphi(mT_c)$	$\omega(mT_c)$	T_c
Angle index	k	$k\alpha$	–	$\alpha = \frac{2\pi}{N_{\text{cog}}}$

- Sampling time effects. The sensor provides a square wave signal that corresponds to equidistant sampling in the angular domain. However, most applications are based on equidistant sampling in the time domain, defined by the sampling instants $t = nT_s$, where T_s is the sampling interval. Approximations are needed to convert from one domain to the other.

- Quantization effects. The exact times t_k cannot be determined. A counter based on a fast internal clock with cycle time T_c is used to approximate t_k with $m_k T_c$, where m_k is the number of clock cycles, but there is still a time jitter.

- Non-ideal teeth. As indicated in Figure 14.6, small manufacturing errors δ_i, $i = 1, 2, \ldots, N_{\text{cog}}$, produce small angular deviations in the tooth positions that perturb the computed angular speed signal (14.11).

14.3.3 Sampling and Resampling

As noted previously, the basic expression (14.11) provides angular speed equidistantly in the angle domain. It is, however, in most cases more natural to interpolate the values to pre-specified points in other domains.

Interpolation and Resampling

First consider interpolation to the time domain. Any standard *interpolation* method can be used, such as linear interpolation

$$\hat{\omega}(t) = \frac{(t_{k+1} - t)\hat{\omega}_k + (t - t_k)\hat{\omega}_{k+1}}{t_{k+1} - t_k}, \quad (14.12)$$

where $t_k \leq t \leq t_{k+1}$. In particular, *equidistant sampling* is possible by letting $t = nT_s$, where T_s is the sampling interval. Note, however, that the sampling theorem requires the continuous time signal to be bandlimited to avoid aliasing. Even if the true $\omega(t)$ is bandlimited, the sampling process in (14.11) may induce *aliasing*. A remedy to this is given in Section 14.3.3.

Multi-Domain Signal Processing

There may be other domains of interest. For instance, shafts are often connected to other shafts through gears, and then there are at least two different angle domains. One should be careful when defining the time sequences for the cog events. We will use superindices to distinguish different cog domains, so $t_m^{(2)}$ denotes the time when the angle is $m\alpha^{(2)}$ in the second domain, corresponding to the m'th edge of the toothed wheel. If the sensor is located in one angle domain, samples equidistant in this domain can be converted to the other one using the gearing ratio r and the formula

$$\varphi^{(2)}(t_m^{(2)}) = \varphi_m^{(2)} = m\alpha^{(2)}, \quad (14.13a)$$

$$t_m^{(2)} = \frac{\left(m\alpha^{(2)} - kr\alpha^{(1)}\right)t_{k+1}^{(1)} + \left((k+1)r\alpha^{(1)} - m\alpha^{(2)}\right)t_k^{(1)}}{\alpha^{(1)}}, \quad (14.13b)$$

where k is chosen such that $kr\alpha^{(1)} \leq m\alpha^{(2)} \leq (k+1)r\alpha^{(1)}$. Now, the basic formula (14.11) can be applied using the time instants defined above.

Synchronous Averaging

Synchronous averaging is a technique to monitor vibrations in shafts. The idea is to average the angular speed over the revolutions. This can be done numerically in a simple way by putting the angular speeds in a matrix with N_{cog} columns, where each row corresponds to equidistant angular sampling of one complete revolution. By averaging over the rows, time-varying deterministic components and measurement noise will be attenuated, and only the superimposed vibration modes are left. The principle is described in Hochman and Sadok (2004). In Gustafsson, one application is summarized, where a whole gear box is monitored for faults using one rotational speed sensor and the known gearing ratios.

Integrated Sampling

To motivate the idea, consider first a standard sampling example where a continuous time signal $y(t)$ is to be sampled to $y_k \leftrightarrow y(t_k)$ where $t_k = kT_s$. A standard sampler includes a hold circuit that integrates the signal over a short time Δ, so

$$y_k = \frac{1}{\Delta} \int_{t-\Delta}^{t} y(\tau)d\tau.$$

Integrated sampling, on the other hand, integrates over the whole sampling period,

$$y_k = \frac{1}{T_s} \int_{t-T_s}^{t} y(\tau)d\tau.$$

There are certain disadvantages with the latter from a sampling theory perspective, but there are obvious practical advantages. First, the averaging over the sampling interval corresponds to a (non-ideal) anti-aliasing filter, so the aliasing effect is (partially) mitigated automatically. Second, the integral can be recovered exactly

$$\int_{0}^{kT_s} y(\tau)d\tau = \sum_{k} y_k T_s, \qquad (14.14)$$

which follows from the elementary properties of the integral. The most immediate drawback of integrated sampling is that it gives a positive bias if the signal is decreasing and a negative bias if it is increasing. If the signal is linearly increasing/decreasing in the sample interval, which consitutes a first order approximation, it can be written $y(t) = y((k-1)T_s) + (t-(k-1)T_s)\dot{\omega}$, where $\dot{\omega}$ denotes the true angular acceleration, and the integrated sampling will give the value $y_k = y(kT_s) - \dot{\omega}T_s/2$.

For a rotational speed sensor, there is no way to include a continuous-time anti-aliasing filter, so integrated sampling is the only way to mitigate frequency folding. The basic formula (14.11) is then replaced with

$$\hat{\omega}(nT_s) = \frac{2\pi(k_n - k_{n-1})}{N_{\text{cog}}(t_{k_n} - t_{k_{n-1}})}, \qquad (14.15)$$

where t_{k_n} denotes the time when the last cog k_n passed before time $t = nT_s$. The integral property (14.14) is preserved in this formula, which means that the angle can be recovered by summing $\hat{\omega}(nT_s)$ over time (except for the facts that the initial condition might not be known and the direction of speed might change unnoticed). All in all, integrated sampling should be preferred generally. We will return to the formula (14.15) in the next section.

Frequency Analysis

Frequency analysis is a classical tool in signal processing to detect resonances and dynamic effects in linear systems. The *discrete Fourier transform* (*DFT*) is the main tool here. For the computed wheel speed, it is defined as

$$\Omega_1(f) = \sum_{n=1}^{N} \hat{\omega}(nT_s)e^{-2\pi fn}, \quad f = i/(NT_s). \tag{14.16}$$

The advantage with this formulation is that the computationally efficient *FFT* (*fast Fourier transform*) algorithm can be used, and the minor drawback is that the integrated sample principle has to be used. The alternative is to approximate the Fourier integral by a *Riemann sum* in the following way

$$\Omega_2(f) = \int \omega(t)e^{-2\pi ft} dt \tag{14.17a}$$

$$\approx \sum_{k=1}^{N} \hat{\omega}_k (t_k - t_{k-1}) e^{-2\pi f t_k} \tag{14.17b}$$

$$= \frac{2\pi}{N_{\text{cog}}} \sum_{k=1}^{N} e^{-2\pi f t_k}, \quad f = i/(NT_s), \tag{14.17c}$$

using $\hat{\omega}_k$ defined in (14.11). Note here that the integrand collapses since $\hat{\omega}_k(t_k - t_{k-1}) = 2\pi/N_{\text{cog}}$ by definition. Instead, the time indexes t_k enter the exponential function and thus propagate the frequency information.

In practice, the difference between (14.16) and (14.17) is small, and (14.16) is to be preferred in off-line applications due to the speed of the FFT algorithm while (14.17) is best in on-line applications due to its simplicity.

There is, however, an interesting definition of the DFT in the cog domain. The idea is to transform the time intervals

$$\Omega_3(f) = \sum_{k=1}^{N} (t_k - t_{k-1}) e^{-2\pi f t_k}, \quad f = i/(NT_s). \tag{14.18}$$

This can be seen as (a scaled version of) the DFT of the inverse wheel speed. We will refer to the magnitude $|\Omega_3(f)|$ as the "*cog spectrum*". Note, that this definition is confusingly similar to (14.16) and (14.17) above, but its properties are different as will be demonstrated later on.

14.3.4 Quantization Effects

We now focus on how the time in the basic expression (14.11) is measured and on the induced quantization effects.

Timer Clock

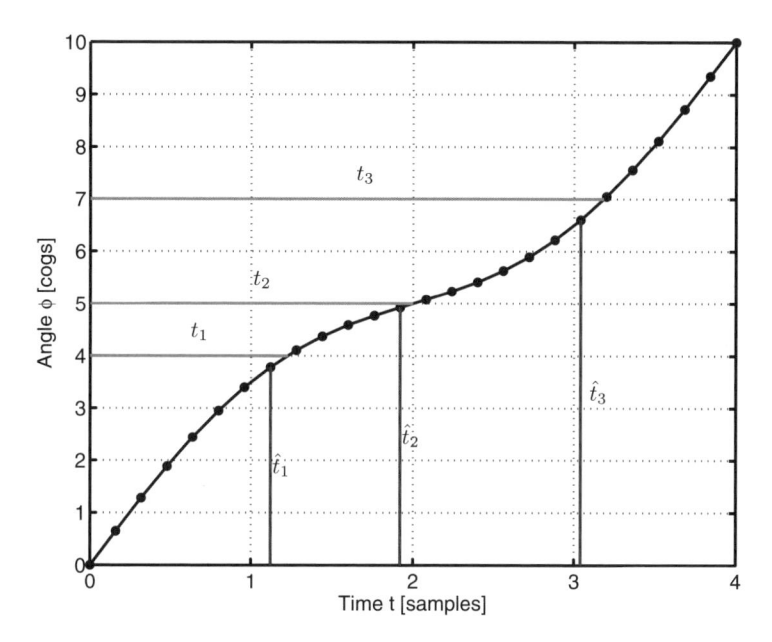

Figure 14.7: Illustration of the different event domains: angle index k, sample index n and timer index m (dots on the line). When a new sample interval starts, the sensor waits for the next pulse from the cogs that defines time t_k and angle $\varphi_k = 2\pi k/N_{cog}$. The clock value is read as an approximation $\hat{t}_k$ of t_k. The rotational speed can then be computed according to (14.11).

In practice the exact time instants t_k are not available. Instead, a *counter* triggered by a fast oscillator/clock is used to measure time. Let the duration of one clock cycle be T_c. To explain the *quantization* problems associated with this, consider Figure 14.7. As soon as a pulse is received at time $t_n \geq nT_s$, the value m_n of the clock counter is read. This value corresponds to cog number k_n. The cog time t_n is then approximated as $\hat{t}_n = m_n T_c$. Then, the computed angular velocity corresponding to integrated sampling in (14.15) is

$$\hat{\omega}(nT_s) = \frac{2\pi(k_n - k_{n-l})}{N_{\text{cog}}(\hat{t}_{k_n} - \hat{t}_{k_{n-l}})} = \frac{2\pi(k_n - k_{n-1})}{N_{\text{cog}}(m_n - m_{n-1})T_c}. \qquad (14.19)$$

As a side note, one must treat counter overflow properly when evaluating $m_n - m_{n-1}$. The difference has to be taken modulo the maximum counter value.

Statistical Quantization Analysis

Formula (14.19) compared to (14.15) suffers from *quantization noise* due to clock *jitter*. We have

$$t_i - \hat{t}_i \in [n_i, n_i + 1)T_c, \tag{14.20}$$

for $i = 1, 2$. It is reasonable to assume that the error is uniformly distributed between 0 and 1, that is

$$\frac{t_i - \hat{t}_i}{T_c} \in U(0, 1). \tag{14.21}$$

Since the difference $\hat{t}_k - \hat{t}_{k-1}$ contains the difference of two uniform distributions (an independence assumption is always justified in practice), it becomes an unbiased estimate of $t_k - t_{k-1}$ with triangular error distribution. The variance of $U(0, 1)$ is $1/12$, so the variance of the difference is twice that value, which implies

$$\text{Var}\left(\hat{t}_k - \hat{t}_{k-1}\right) = \frac{T_c^2}{6}. \tag{14.22}$$

It follows that the variance of the inverse angular frequency is

$$\text{Var}\left(\frac{1}{\hat{\omega}}\right) = \frac{T_c^2}{6}\left(\frac{N_{\text{cog}}}{2\pi(k_n - k_{n-1})}\right)^2 \approx \frac{T_c^2}{6\omega^2 T_s^2}. \tag{14.23}$$

The variance of $\hat{\omega}$ in (14.19) depends on its expected value. A first order Taylor expansion using the notation $\hat{\omega} = \omega + \tilde{\omega}$, where ω is the mean and $\tilde{\omega}$ is a zero mean stochastic variable, gives

$$\frac{1}{\hat{\omega}} = \frac{1}{\omega + \tilde{\omega}} = \frac{1}{\omega\left(1 + \frac{\tilde{\omega}}{\omega}\right)} \approx \frac{1}{\omega}\left(1 - \frac{\tilde{\omega}}{\omega}\right) \Rightarrow \tag{14.24a}$$

$$\tilde{\omega} = \omega\left(\frac{1}{\omega} - \frac{1}{\hat{\omega}}\right) \tag{14.24b}$$

This first order expansion indicates that $\tilde{\omega}$ is zero-mean with variance

$$\text{Var}(\hat{\omega}) = \text{E}(\tilde{\omega})^2 = \omega^2 \text{Var}\left(\frac{1}{\hat{\omega}}\right) \approx \frac{T_c^2}{6T_s^2}. \tag{14.25}$$

The mean square error is a trade-off between bias and variance. We have

$$\text{MSE} = \text{E}\big((\omega(nT_s) - \hat{\omega}(nT_s))^2\big) = (\omega(nT_s) - \text{E}[\hat{\omega}(nT_s)])^2 + \text{Var}\big(\hat{\omega}(nT_s)\big). \tag{14.26}$$

Using the linear approximation $\omega(t) = \omega((k-1)T_s) + (t-(k-1)T_s)\dot\omega((k-1)T_s)$ of the true signal, we get the concrete formula for the *bias–variance trade-off*

$$\text{MSE} = \frac{\dot\omega^2 T_s^2}{4} + \frac{T_c^2}{6T_s^2}. \qquad (14.27)$$

There is thus a trade-off between getting a small bias and small variance when choosing the sample interval T_s. A rule of thumb is thus that the sampling interval T_s should be small compared to the maximum rate of change of $\omega(t)$ to avoid a large bias, and the clock interval T_c should be small compared to T_s to avoid excessive variance. What "small" is quantitatively depends on the accuracy the application requires.

The bias and variance will propagate to all subsequently defined quantities, as for example the frequency domain expressions (14.16)–(14.18).

Saturation Effect

There is also an important *saturation effect* here. If the angular speed is too small compared to the sample interval, then (14.19) collapses, since no pulse will be available. The condition for this not to happen follows from (14.19) setting $k_n - k_{n-1} = 1$ and using $(m_n - m_{n-1})T_c \approx T_s$ as

$$\omega \geq \frac{2\pi}{N_{\text{cog}} T_s}. \qquad (14.28)$$

This gives a lower bound on the sampling interval for a given minimum speed.

14.3.5 Cog Error Compensation

The wheel speed computed as in (14.19) is sufficient for most applications. However, in high-precision applications, where very accurate wheel speed information is needed, more sophisticated signal processing is required. The general principles are described mainly in words below, see Persson et al. (2002) and Schwarz et al. (1997) for further information.

Principles

The main reason for the limited precision in (14.19) is the cog errors $\{\delta_k\}_{k=1}^{2N_{\text{cog}}}$ illustrated in Figure 14.6. The true angle to the kth edge on the toothed wheel can be written as $\frac{2\pi}{N_{\text{cog}}} - \delta_{k \bmod 2N_{\text{cog}}}$. With known cog errors, a constant angular speed can be computed by

$$\omega_k = \left(\frac{2\pi}{N_{\text{cog}}} - \delta_{k \bmod N_{\text{cog}}}\right) \frac{1}{t_k - t_{k-1}}. \qquad (14.29)$$

A simple turn-around is to measure the time for one complete revolution,

$$\hat{\omega}_k = \frac{2\pi}{N_{\text{cog}}(t_k - t_{k-N_{\text{cog}}})}. \tag{14.30}$$

The idea is that the sum or errors $\sum_{i=1}^{N_{\text{cog}}} \delta_i = 0$ is zero and cancels out over one revolution. However, for a changing true speed, this introduces a significant bias that might not be tolerated in high-accuracy applications.

A better idea is to estimate the cog errors and replace $\delta_{k \bmod N_{\text{cog}}}$ in (14.29) with this estimate. Once the sequence $\{\hat{\delta}_k\}_{k=1}^{2N_{\text{cog}}}$ is estimated, the angular speed in (14.11) can be estimated by

$$\hat{\omega}_k = \left(\frac{2\pi}{N_{\text{cog}}} - \hat{\delta}_{k \bmod 2N_{\text{cog}}}\right) \frac{1}{t_k - t_{k-1}}. \tag{14.31}$$

An Example

Figure 14.8 illustrates the problem using time stamps t_k from an ABS sensor with $N_{\text{cog}} = 48$ cogs, where both edges are used and $T_c = 4 \cdot 10^{-7}$. The test drive was deliberately designed with almost constant speed. The upper plot shows the cog spectrum as defined in (14.18), where the bias (0 Hz) component corresponding to the speed has been removed. The spectrum shows various vibration modes as will be described later, but more importantly there are $2N_{\text{cog}}$ distinct strong peaks corresponding to the Fourier series of the sequence $\{\delta_k\}_{k=1}^{2N_{\text{cog}}}$. These dominate the rather weak signal components, and further frequency analysis clearly requires compensation. The possibility of highlighting cog errors is a key property of the cog spectrum in (14.18).

The lower plot in Figure 14.8 shows the estimated cog errors in an errorbar plot with standard deviations computed as the inverse Fourier series of the peaks in the upper plot. The standard deviation is much smaller (almost invisible confidence interval), so the errors are significant. It should be remarked that this approach only works when the speed is constant. A better way to compute these errors is to use synchronous averaging as described in Section 14.3.3.

Figure 14.9 shows the result of cog error compensation in the frequency domain. In this case, where the true speed is almost constant, the cog and time spectra look similar. However, the more the speed varies, the more important this pre-processing is.

14.3.6 Applications

Examples of applications of the wheel speed information found in the literature are illustrated in Figure 14.10 and include the following approaches:

14.3 Wheel Speed Sensors

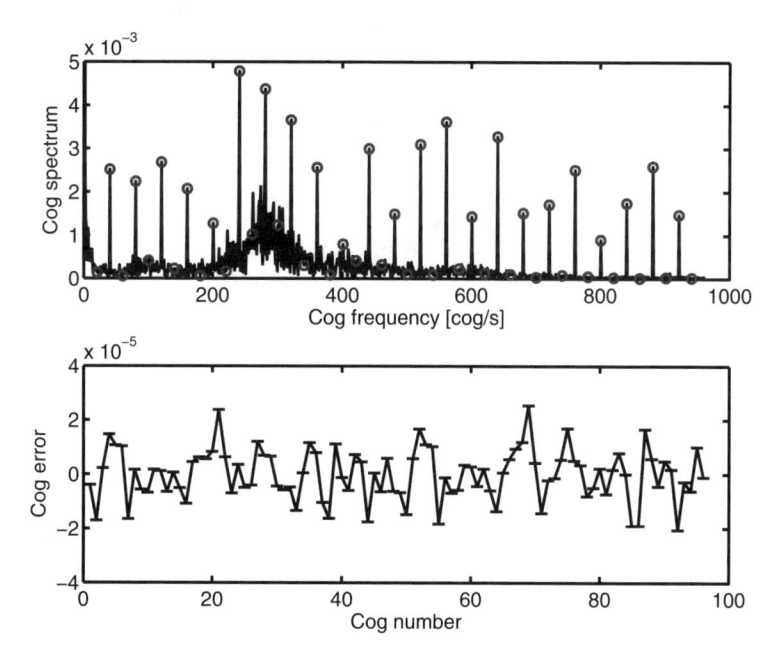

Figure 14.8: *Cog spectrum* $DFT(t_k - t_{k-1})$ *where the signal component from speed variations is concealed in the* $2N_{cog}$ *peaks caused by the cog errors. The lower plot shows the estimated cog errors* $\hat{\delta}_i$, $i = 1, 2, \ldots, 2N_{cog}$ *as an errorbar plot, where the standard deviation is shown for each cog error.*

- Temporal information (Fourier transformed angular velocity)

$$\Omega(f) = \int \omega(t) e^{-2\pi f t} dt, \tag{14.32}$$

reveals, in the different frequency bands, information about tire-road friction, tire pressure, tire condition, surface texture and wheel balance information as well as wheel suspension information (spring-damper condition). A model-based alternative to the Fourier transform is to use an AR(2) model, which can be motivated by a damper-spring model of the tire modes. In the frequency domain, the AR(2) model yields

$$\Omega(f) = \frac{1}{(i2\pi f)^2 + 2\eta f_0(i2\pi f) + f_0^2}, \tag{14.33}$$

where η is the damper constant and f_0 the eigen-frequency of the spring. This model holds only locally in the frequency band according to Table 14.7, so frequency selective estimation algorithms are needed.

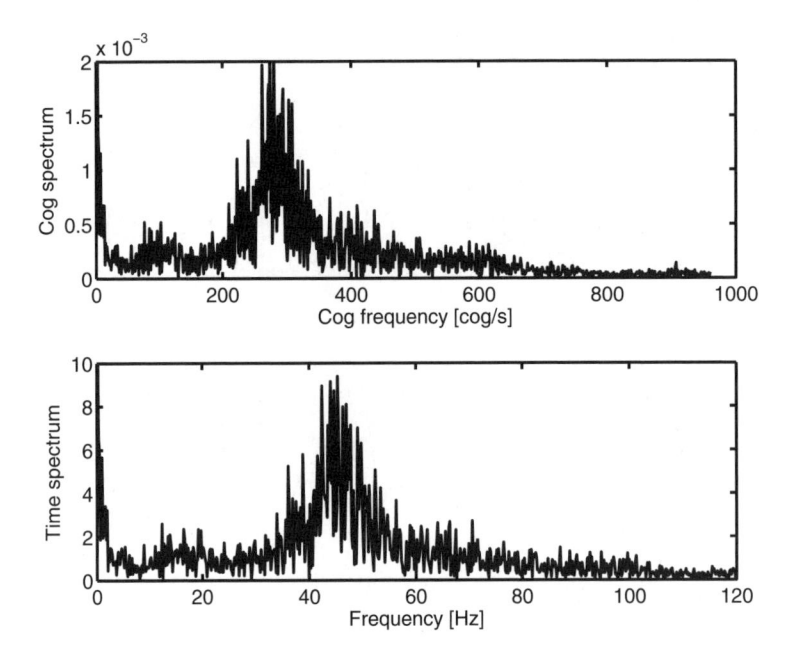

Figure 14.9: *Upper plot: Cog spectrum as in Figure 14.8, where the time stamps are compensated with the estimated cog errors. Lower plot: Spectrum of the speed estimate in (14.31).*

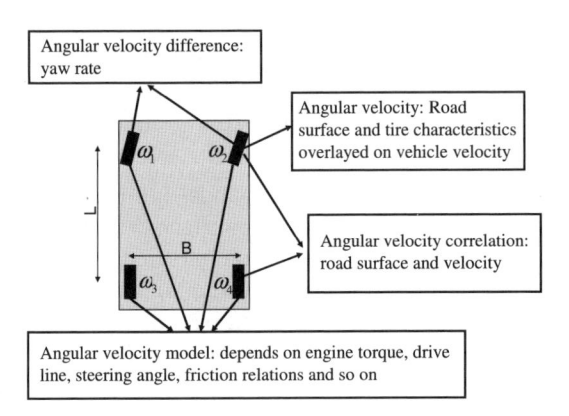

Figure 14.10: *Wheel speed applications by spectral analysis, by comparing wheel speeds in the lateral or longitudinal direction, or by model based approaches.*

14.3 Wheel Speed Sensors

- Spatial information (correlation of velocities) reveals absolute velocity when comparing two wheels on different axles and same side, and yaw rate information comparing two wheels on the same axle.

- Model-based approaches to friction estimation (longitudinal and lateral slip models), including handling conditions as aqua planning and rough road detector. Furthermore, dynamic state estimation and navigation using dead-reackoning are based on wheel speeds.

Development and algorithms in these areas are found by patent database searches, and very few scientific publications exist. This is still an open research field.

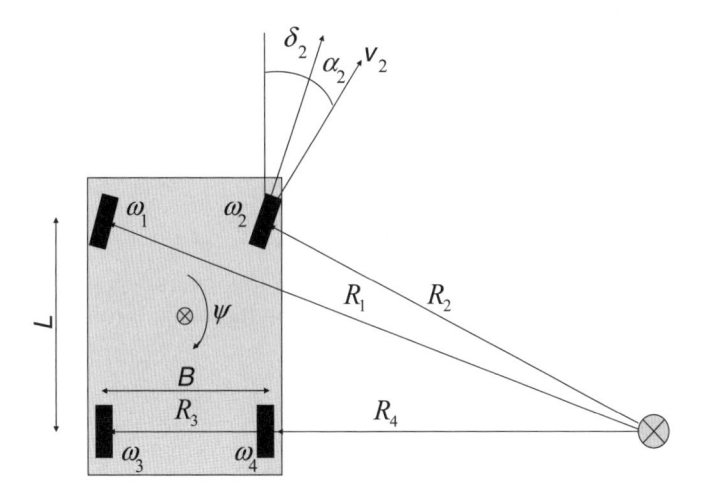

Figure 14.11: *Notation for lateral dynamics and curve radius relations.*

Table 14.7: *Frequency spectrum (14.16) (or (14.17)) for the wheel speed signal where different physical sources are given, with approximate limits in Hz. Here, mode refers to oscillation modes in the tire.*

0–10	10–15	15–30	30–60	60–80	80–100	100–
Speed	Mode 1	Noise	Mode 2	Noise	Mode 3	Noise
Damper	Narrow-band noise components					

14.3.7 Virtual Sensors

Several important dynamical vehicle states can be computed for the angular speeds if only the wheel radius on all wheels where exactly the same (and in some cases known), as the following sections illustrate. This is here called *virtual sensors*. However, the wheel radii differ as a combination of different load on each wheel, different tire inflation pressure, perhaps even differently worn tires and other factors. The consequence is that the virtual sensors will be biased with an offset. In many applications, this offset can be included in the set of unknown states and thus estimated.

Velocity

The velocity in the center of an axle can be computed as the average of the wheel speeds. For instance, on a front-wheel driven four-wheeled vehicle, the speed should be computed on the rear axle to avoid longitudinal traction slip

$$v_x = \frac{v_3 + v_4}{2} = \frac{\omega_3 r_3 + \omega_4 r_4}{2}. \tag{14.34}$$

Yaw Rate

Yaw rate $\dot{\psi}$ at the center point of the rear axle can be computed from

$$\dot{\psi} = \frac{v_x}{R} = v_x R^{-1}, \tag{14.35a}$$

where R is defined as $(R_3 + R_4)/2$. Note that the inverse curve radius should always be used to avoid singularities on straight roads. From Figure 14.11, we have

$$\frac{v_3}{v_4} = \frac{R_3}{R_4} = \frac{R + B/2}{R - B/2}, \tag{14.35b}$$

leading to

$$R^{-1} = \frac{2 \frac{v_3}{v_4} - 1}{B \frac{v_3}{v_4} + 1} = \frac{2 \frac{\omega_3 r_3}{\omega_4 r_4} - 1}{B \frac{\omega_3 r_3}{\omega_4 r_4} + 1}. \tag{14.35c}$$

That is, the yaw rate can, for a front-wheel driven car, be computed as

$$\dot{\psi} = \frac{\omega_3 r_3 + \omega_4 r_4}{2} \frac{2 \frac{\omega_3 r_3}{\omega_4 r_4} - 1}{B \frac{\omega_3 r_3}{\omega_4 r_4} + 1}, \tag{14.35d}$$

provided that the wheel radii are known and that the angular speeds of the rear wheels are measured.

For computations using wheel speeds on a steered axle, the wheel base B should be replaced by $B \cos(\delta)$, where δ denotes the steering angle.

Longitudinal Slip

One definition of *longitudinal slip* (there are several similar) at a driven wheel w is

$$s = \frac{\omega_w r_w - v_w}{v_w}, \tag{14.36}$$

14.3 Wheel Speed Sensors

where $\omega_w r_w$ is the circumferential velocity, and v_w is the absolute velocity. The circumferential velocity is during traction larger and during braking smaller than the absolute velocity due to slip. Slip is during normal driving in the order of a few parts per thousand.

A first problem in computing the slip (14.36) is that the wheel radii are not the same, so we cannot compute the wheel velocities directly. Consider the front left wheel as a slipping driven wheel, while the rear left wheel is non-driven and taken as a velocity reference (assuming straight ahead driving). In this case, the slip s_l on the left side is

$$s_l = \frac{\omega_1 r_1}{\omega_3 r_3} - 1 \qquad (14.37)$$

14.3.8 Slip Computation in Curves

While driving in a curve, all four wheels have different angular velocities. This is easily seen from Figure 14.11, where the different distances from each wheel to the momentary center of motion are defined.

Consider again the virtual speed and yaw rate sensor in (14.34) and (14.35d), respectively. Geometrical relations can be used to compute the local velocity at any point of the car, in particular the driven front wheels as

$$v_1 = \dot{\psi} R_1 = \dot{\psi} \sqrt{\left(R + \frac{B}{2}\right)^2 + L^2}$$

$$= v_x \sqrt{\left(1 + \frac{R^{-1}B}{2}\right)^2 + (R^{-1}L)^2}, \qquad (14.38a)$$

$$v_2 = \dot{\psi} R_2 = \dot{\psi} \sqrt{\left(R - \frac{B}{2}\right)^2 + L^2}$$

$$= v_x \sqrt{\left(1 - \frac{R^{-1}B}{2}\right)^2 + (R^{-1}L)^2}. \qquad (14.38b)$$

As usual, the expression is rewritten to be a function of inverse curve radius. These expressions give correction terms for transforming the speed v_x to the driven wheels. Note that these factors are both one for straight ahead driving.

We have here assumed that the car is front wheel driven. However, similar expressions for how to compensate for cornering are easily derived for rear wheel driven cars as well.

Absolute Velocity

Absolute velocity can be computed from the wheel axle distance L and time delay between disturbances passing both front and rear wheel according to the following principle:

$$\hat{\tau} = \arg\max_{\tau} \mathrm{E}\left[\omega_{\text{front}}(t)\omega_{\text{rear}}(t-\tau)\right], \qquad (14.39a)$$

$$\hat{v} = \frac{L}{\hat{\tau}}. \qquad (14.39b)$$

This works when the velocity is constant over the estimation time of the correlation. Open problems are how to make it velocity adaptive and to get sufficient resolution (better than the uncertainty in wheel radius, which is about a few percent).

14.4 Wireless Network Measurements

This section overviews measurements available in wireless networks. Details regarding practical implementations and standards are adopted from the papers Drane et al. (1998); Krizman et al. (1997); Zhao (2001, 2002).

14.4.1 General Observation Models

Denote the two-dimensional mobile position at time t by $p_t = (X_t, Y_t)^T$, and the known base station positions by $p^i = (X_i, Y_i)^T$. The generic measurements relative reference point i, y_t^i are subject to uncertainties e_t^i and typical measurements are summarized below. All measured times are automatically multiplied by the speed of light to get a measure in meters rather than in nano seconds.

- **Received signal strength:** $y_t^i = h_{RSS}(|p_t - p^i|) + e_t^i$ is averaged over fast fading, and depends on distance and slow fading. A model that solely depends on the relative distance is the so called Okumura-Hata model described in Y.Okumura et al. (1968), stating that $y_t^i = K - 10\alpha \log_{10}(|p_t - p^i|) + e_t^i$, where $\mathrm{Std}(e_t^i) \approx 4 - 12$ dB depending on the environment (desert to dense urban). It is also possible to utilize a predicted or measured spatial digital map with received signal strength values as discussed below.

- **Time of arrival:** $y_t^i = |p_t - p^i| + e_t^i = h_{TOA}(p_t, p^i) + e_t^i$ is for example estimated in the uplink in GSM at multiple base stations upon request from the network. It is also estimated from a multitude of satellites in the GPS (Global Positioning System). The 3G systems feature *assisted GPS*, where required information for measurement setup and positioning calculations are broadcasted in each cell by the base stations in the cellular network, which improves reliability and reduces time to position fix. Such measurements can also be seen as TOA measurements. The performance depends mainly on the synchronization accuracy, and

GPS typically feature an accuracy in the order of 10 m in unobstructed environments.

- **Time difference of arrival:** $y_t^{i,j} = |p_t - p^i| - |p_t - p^j| + e_t^{ij} = h_{TDOA}(p_t, p^i, p^j) + e_t^{ij}$ is a practical mobile measurement related to relative distance. The measurements are reported to the network, which performs necessary computations and it is not necessary to communicate the network synchronization nor the reference point locations to the mobile. As for TOA, the synchronization accuracy determines the performance, but also the base station locations. The observed TDOA accuracy requirement for location purposes in WCDMA is 0.5 chip Zhao (2002) which means an error of about 40 m ($\sigma_e \approx 20$m). Similarly, a TDOA accuracy requirement of 0.5 chip in cdma2000 (advanced forward link trilateration - A-FLT) means 120 m ($\sigma_e \approx 60$m) due to the lower chip rate. Satellite navigation systems feature even better accuracies. On the other hand, satellite navigation systems have a much higher chip rate, so for instance *assisted GPS* can provide TDOA measurements with $\sigma_e \approx 1$m.

- **Angle of arrival:** $y_t^i = h_{AOA}(p_t, p^i) + e_t^i$ is today mainly available as a very crude sector information (*e.g* $120°$). With the use of group antennas this will be improved to about $30°$ beam width ($\sigma_e \approx 8$ degrees), and perhaps even better. Geometrically, the spatial resolution of the intersection of two perfectly complementing AOA measurements is limited to $2D\sin(\alpha/2)$, where α is the angular resolution and D the distance between the antennas. For $\alpha = 30°$, this means 52% of D.

- **Digital map information:** $y_t = h_{MAP}(p_t, p^i) + e_t$ contains for instance RSS measurements relative the reference points either predicted or provided via dedicated measurement scans in the service area. The former is conducted in the network deployment phase using graphical information systems dedicated for network planning, like the TEMS CellPlanner Universal (http://www.ericsson.com/tems/) by Ericsson. The prediction grid is usually from 100 m down to a few meters when considering building data. Figure 14.12 illustrates such digital prediction maps. Performing actual measurements is only plausible in very limited service areas, like indoor. Most of the shadowing is included in the maps, and the remaining shadow fading component has typically Std(e_t^i) ≈ 3 dB depending on the environment and spatial map resolution. The map can also be a commercial street map for automotive terminals Gustafsson et al. (2002) as examplified in Figure 14.12. A support to GPS for sea navigation was proposed in Karlsson and Gustafsson (2003) using sonar depth measurements and a depth map.

14.4.2 Summary and Typical Performances

Table 14.8 summarizes the different measurements and their approximate accuracy. The noise standard deviation values in paranthesis indicate values used in numerical evaluations, and represent a favorable situation with essentially line-of-sight measurements. Note that the quality of the sensor information depends not only on the the noise variance, but also on the size and variation in $h(p)$. This is discussed in Section 4.2.3.

Table 14.8: *Mathematical notation of available measurements in wireless communication systems together with approximative noise standard deviations σ_e. Here y is the actual numerical value, h denotes a general nonlinear model, p_t is the sought position, p^i denotes the position of base station/antenna number i and e denotes measurement noise with a probability density function $p_e(\cdot)$.*

Measurement	Notation	Standard deviation
Received signal strength:	$y_t^i = h_{RSS}(\|p_t - p^i\|) + e_t^i$	4-12 dB (10 dB)
Time of arrival:	$y_t^i = \|p_t - p^i\| + e_t^i$	5-100 m (14 m)
	$= h_{TOA}(p_t, p^i) + e_t^i$	
Time difference of arrival:	$y_t^{i,j} = \|p_t - p^i\| - \|p_t - p^j\| + e_t^{ij}$	10-60 m (20 m)
	$= h_{TDOA}(p_t, p^i, p^j) + e_t^{ij}$	
Angle of arrival:	$y_t^i = h_{AOA}(p_t, p^i) + e_t^i$	$5^o - 10^o$ (6°)
Digital map information:	$y_t = h_{MAP}(p_t) + e_t$	(RSS map, 3 dB)

For an overview of typical wireless network measurements, see Drane et al. (1998). See also 3GPP (2003) for positioning support through specified measurements in UMTS.

Sector information (AOA) and received signal strength (RSS) provide coarse position information with accuracy in the order of hundreds of meters. Power attenuation maps refine the RSS information, how much depends on the terrain.

Timing measurements as TOA and TDOA have better resolution but are fundamentally limited by the symbol length, that depends on chip rate and speed of light. For instance, the smallest chip in WCDMA has a spatial duration of about 80 meters. Advanced synchronization can bring down the time resolution further. With an integrated satellite navigation unit, the position information is in the order of 10 meter accuracy.

14.4.3 Specific Observation Models

Let the sensor network consist of N nodes positioned at p_k, $k = 1, \ldots, N$. The notation assumes a two-dimensional position, $p_k = (p_{k,1}, p_{k,2})^T$, but can be extended to more dimensions in a straightforward way. The observed signal is in continuous time assumed to be of the form

$$y_k(t) = a_k s(t - \tau_k) + e_k(t), \quad k = 1, 2, \ldots N, \qquad (14.40)$$

14.4 Wireless Network Measurements

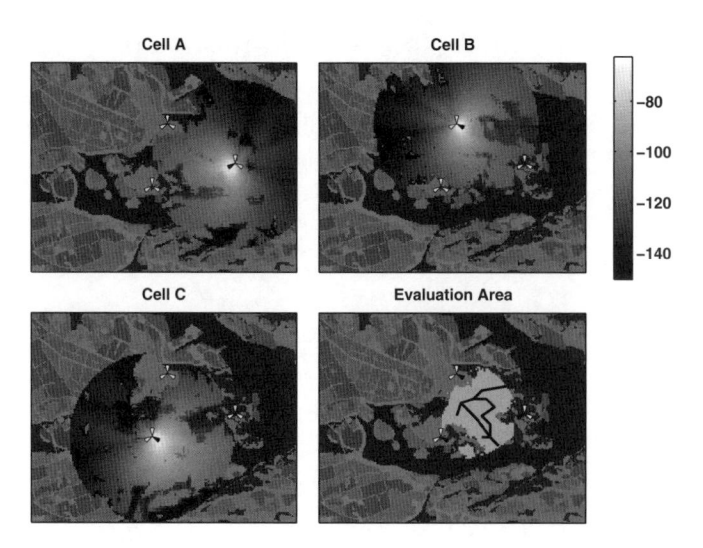

Figure 14.12: Example of RSS predictions with the spatial resolution 40 m using TEMS CellPlanner Universal. The artificial network is deployed at Djurgården, Stockholm, Sweden (true terrain data, and fictitious but realistic base station locations). Predicted power gain from three cells A, B and C is roughly directed towards the interior area between the base station. The evaluation area used in examples is the shaded area in the last plot, where all three cells can be detected. A simple and artifical road map is also depicted.

where $s(t)$ is the emitted signal from the target and $y_k(t)$ is the received signal at sensor k, which is delayed and attenuated and subject to additive noise $e_k(t)$ with variance σ_e^2. Both the attenuation $a_k = a_k(\|x - p_k\|)$ and time delay $\tau_k = \tau_k(\|x - p_k\|)$ depend on the distance $\|x - p_k\|$ between target and sensor.

The sensor-near signal processing aims at converting the continuous time signal $y_k(t)$ to a measurement of the functional form $y_k(\|x - p_k\|)$. The basic measurements are

- Linear range: *time of arrival (TOA)*.
- Linear range difference: *time difference of arrival (TDOA)* when the transmission time is unknown.
- Logarithmic range: power measurements.
- Logarithmic range difference: power measurements when the emitted power is unknown.
- Bearing measurements: *direction of arrival (DOA)*.

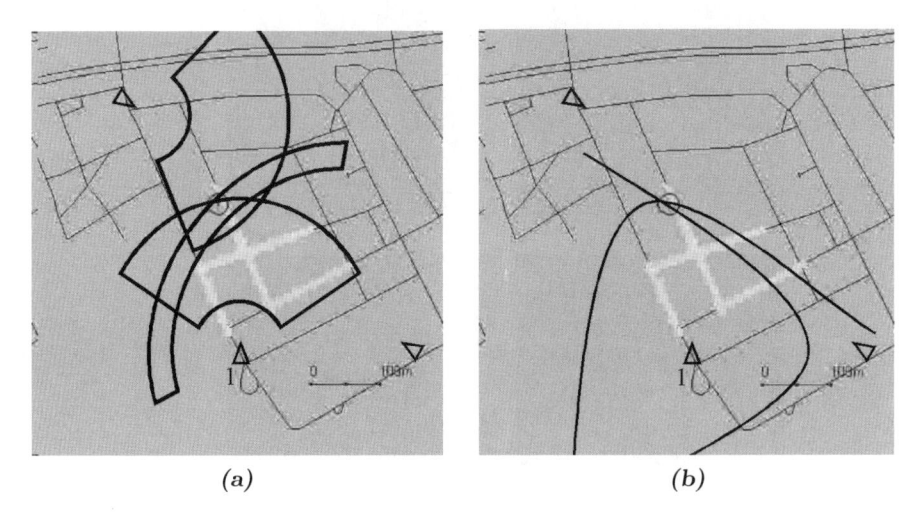

Figure 14.13: *(a) Sector antenna information from three base stations. (b) TDOA measurements from two pairs of base stations. (a)–(b) are overlayed a street map, where map matching can be used to improve tracking of automotive terminals.*

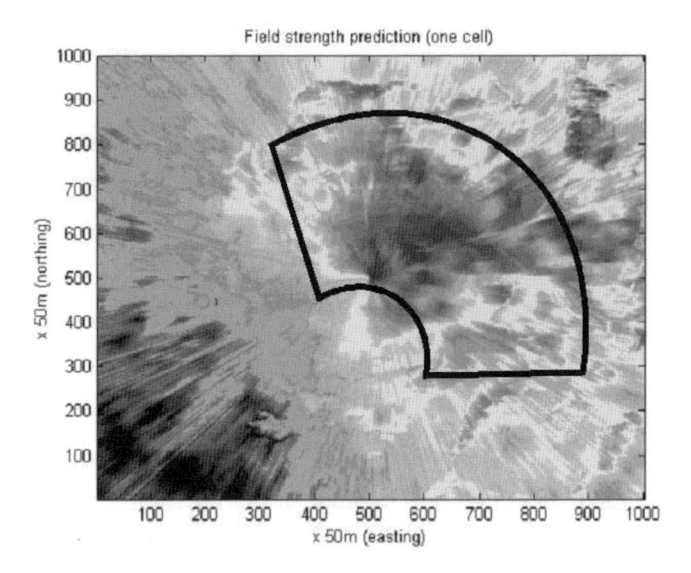

Figure 14.14: *Power attenuation map.*

14.4 Wireless Network Measurements

The goal in this section is to briefly describe the mathematical background for range, range difference and angle sensors from a network perspective. Of course, there are other single sensors providing information of this kind.

Time of Arrival Measurements

With known reference $s(t)$ and perfect synchronization, we can directly estimate τ_k (TOA) using correlation based time-delay estimation or the maximum likelihood approach. The basic idea is to maximize the correlation

$$\hat{\tau} = \arg\max_{\tau} \int_0^T s^*(u) y(t - u - \tau) du, \qquad (14.41)$$

where T is the signal duration.

The CRLB for the TOA is provided in Knapp and Carter (1976) and Patwari et al. (2003) as

$$\text{Var}(\hat{\tau}) \geq \frac{1}{8\pi^2 BT f_c^2 \mathbf{SNR}},$$

where B is the signal bandwidth, f_c its center frequency and the *signal to noise ratio* (*SNR*) is defined as

$$\mathbf{SNR} = \int_0^T |s(t)|^2 dt / \sigma_e^2. \qquad (14.42)$$

This can be used as an approximative (optimistic) value for the variance in sub-sequent estimation problems.

The time delay can be converted to a range observation (TOA measurement),

$$r_k = \tau_k v = \|x - p_k\| = \sqrt{(x_1 - p_{k,1})^2 + (x_2 - p_{k,2})^2}. \qquad (14.43)$$

Here, v is the speed of the propagation media (speed of sound, light or water), depending on if the signal is a radio, electromagnetic, acoustic or sonar wave.

Power Models

The power of the received signal over a time interval T is

$$\bar{P}_k = \int_{t-T}^t |y_k(u)|^2 du = \int_{t-T}^t |a_k s(u) + e(u)|^2 du. \qquad (14.44)$$

The letter P is in this section only used for power and not for covariance. The bar denotes here and in the sequel power in linear scale, and without bar the power is in a logarithmic scale. There are two different cases:

- The emitted signal $s(t)$ is unknown. Then, only the relative attenuation $\bar{P}_j/\bar{P}_i$ at two different sensors contains information about the target position. The signal may be low-pass or band-pass filtered before calculating the signal power to capture the power in the frequency band where the signal is expected to reside.

- The emitted signal $s(t)$ is known and the transmitter and receiver are perfectly synchronized, in which case the attenuation can be estimated. The observation model (14.40) is a continuous time linear model in a_k, and the least squares estimate from an interval T is

$$\hat{a}_k = \frac{\int_{t-T}^{t} s^*(u)y(u)du}{\int_{t-T}^{t} |s(u)|^2 du}, \qquad (14.45a)$$

$$\mathrm{Var}\left(\hat{a}_k\right) = \frac{1}{T \cdot \mathbf{SNR}}. \qquad (14.45b)$$

For Gaussian noise, the CRLB coincides with the variance above.

The observed power is related to the relative distance between two nodes

$$\bar{P}_k = h_{RSS}(\|x - p_k\|) + e_k. \qquad (14.46)$$

Radio frequency signals, acoustic signals, seismic signals, radar signals etc have in common that the emitted signal strength $\bar{P}_k^0$ at node k decays quickly with distance. For example, this is observed for radio signals in e.g. Hata (1980); Y.Okumura et al. (1968). Hence, the received signal strength $\bar{P}_k$ at node k can be modeled as

$$\bar{P}_k = \bar{P}^0 \|x - p_k\|^{-\beta}, \qquad (14.47)$$

where β denotes the so called *path loss constant*. The corresponding relation in logarithmic scale is

$$P_k = P^0 - \beta \log(\|x - p_k\|). \qquad (14.48)$$

It is a matter of taste whether a model in linear or in logarithmic scale is selected. One aspect is that the uncertainty e_k with good approximation can be modeled as Gaussian when the received signal strength measurement equation is in logarithmic scale Hata (1980); Y.Okumura et al. (1968). Equations (14.47) and (14.48) yield two alternative nonlinear functions

$$h_{RSS,lin} = \bar{P}^0 \|x - p_k\|^{-\beta}, \qquad (14.49a)$$
$$h_{RSS,log} = P^0 - \beta \log(\|x - p_k\|). \qquad (14.49b)$$

Figure 14.15 illustrates radio, acoustic and seismic waves, respectively, which propagation can be described with Equation (14.49) with good accuracy. The

14.4 Wireless Network Measurements

(a) Acoustic and seismic waves

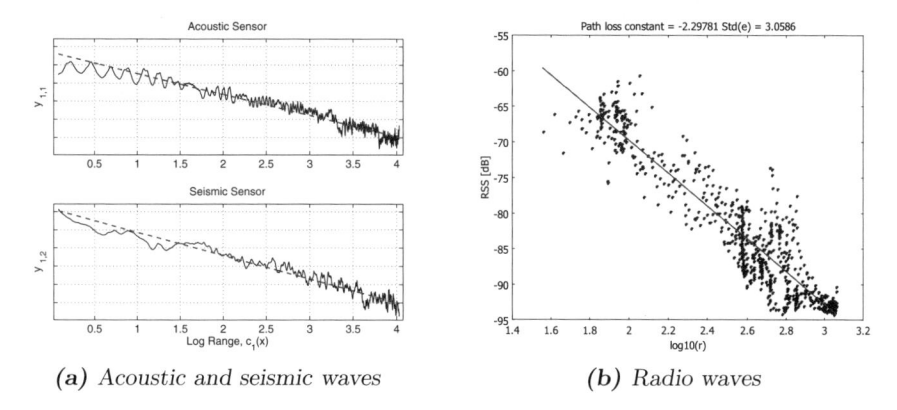

(b) Radio waves

Figure 14.15: Received sensor energy in log scale P_k versus log range $\log(\|x - p_k\|)$, together with a fitted linear relation as modeled in (14.49). (a) Acoustic and seismic waves emitted by a motorcycle. (b) Wimax radio waves.

path loss constant β is -2.3, -2.6 and -2.3, respectively. In conclusion, similar physical logarithmic relations hold for acoustic, seismic and magnetic waves.

If both P^0 and β are known, this measurement can be transformed to a range form,

$$r_k = \|x - p_k\| = \left(\frac{y_k}{P^0}\right)^{1/\beta}. \qquad (14.50)$$

However, in this case the measurement noise may not be additive. For sensor fusion, the logarithmic form (14.48) is clearly recommended.

In wireless networks, an RSS measurement is often considered to be a power attenuation measurement, because the network is aware of the transmitted power. For surveillance networks, this is not the case, so the power at the transmitter is a nuisance parameter.

Another nuisance parameter is the power path loss constant. Again, in a wireless network, this number can be calibrated for a fixed deployment. In rapidly deployed networks this is not possible, and in dynamic environments the number may vary over time.

14.5 Summary

14.5.1 Theory

Sensors for localization and motion estimation can be divided into two groups: tracking sensors and on-board navigation sensors. There are four main categories of measurements for tracking sensors receiving electromagnetic or acoustic waves:

- Range and bearing pairs $y_k = (R_k, \theta_k)^T$ as obtained from active radar, lidar and sonar. In 3D, the azimuth angle is also included as $y_k = (R_k, \theta_k, \varphi_k)^T$.

- Range rate $y_k = \dot{R}_k$ obtained from the Doppler shift, or relative range rate between two receivers when the frequency is unknown.

- Bearing $y_k = \theta_k$ or bearing and azimuth $y_k = (\theta_k, \varphi_k)^T$ as obtained by passive radar, lidar and sonar and by electro-optical and IR sensors (cameras).

- Log range $y_k = P_k - n_k \log(R_k)$ originating from an exponential path attenuation relation. Here, n_k is the path loss constant. This relation applies to a range of sensors as radio signals, seismic signals, magnetic field variations, acoustic signals and so on.

Navigation sensors include

- Inertial sensors (accelerometers and gyroscopes), magnetometer and barometer.

- The versatile wheel speed sensors also provide a kind of inertial information that can be used for odometry but also as a basis for many virtual sensors.

- Beacon based reference systems as satellite navigation systems (GPS). These return absolute position and are a perfect complement to inertial sensors.

14.5.2 Software

Simple measurement models for ranging and bearing sensors are available as examples in `exsensor`.

14.5 Summary

exsensor(fun)	Description
toa	$\|x_n - p_m\|$ 2D TOA as range measurement
tdoa1	$\|x_n - p_m\| + x_{N+1}$ 2D TDOA as biased range measurement
tdoa2	$\|x_n - p_m\| - \|x_k - p_m\|$ 2D TDOA as range differences
doa	$\arctan 2(x_n - p_m)$ 2D DOA as bearing measurement
radar	Combination of TOA and DOA above
rss1	$\theta_1 + \theta_2 10 \log_{10}(\|x_{n,1:2} - p_m\|)$ Parametric emitted power and path loss constant
rss2	$x_{n,n_x+1} + x_{n,n_x+2} 10 \log_{10}(\|x_{n,1:2} - p_m\|)$ Emitted power and path loss constant are considered as states
gps2d	x 2D position
gps3d	x 3D position
mag2d	2D magnetic field disturbance (14.10)
mag3d	3D magnetic field disturbance (14.9)
quat	$1 = \sum_{i=1}^{4} q_i^2 + e$ Quaternion constraint
*slam	as above, but with p_m as states instead of parameters

Use exsensor('list') to get an array with the current options.

A sensor model can be combined with a motion model (for instance taken from the datebase of models in exmotion) using the addsensor) method. Example:

```
m=exmotion('ctcv2d');
s=exsensor('radar',1);
ms=addsensor(m,s);
```

Several sensors can be added iteratively.

15
Filter and Model Validation

A filter is a black-box operation taking the data $y_{1:N}$, a model represented with $\{f(x,v),\ h(x,e),\ p_e(e),\ p_v(v)\}$ and an algorithm as inputs and giving for example the state estimate $\hat{x}_k$ and its covariance matrix P_k as outputs. The preceding chapters have focused on the theory and analytical design tools, but to validate a filter in practice is an engineering problem where there is no given answer. We will here try to provide systematic ways to validate a filter and its performance. We will focus on the iterative process of refining the model and filter, but exclude the tuning aspects of the filter in this discussion.

15.1 Parametric Uncertainty

Parametric uncertainty is often neglected in filter applications, and it is the purpose here to provide constructive methods to quantify how this uncertainty affects the filter outputs. The uncertain parameters are denoted as θ, and they can correspond to constants in the model, sensor position, noise variances, etc.. Typically, the user has some nominal values for these parameters, and in the filtering applications these are used as the ground truth. The filter output will here be written as an explicit function $\hat{x}_k(\theta)$, $P_k(\theta)$ of the parameters, to highlight that it depends on a nominal parameter value.

First, an example is given to motivate the problem.

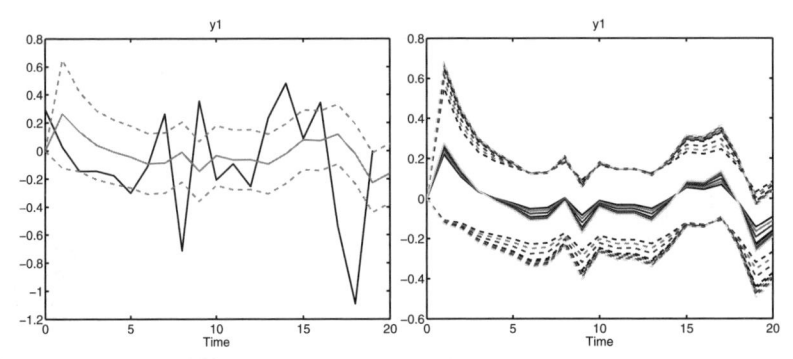

(a) Simulated signal and estimate using true system.

(b) Estimated signal using nominal systems.

Figure 15.1: Time-varying Kalman filter for predicting the next observation. The result is somewhat sensitive to the uncertain parameter a in (15.1).

---**Example 15.1: Parametric uncertainty in a linear system**---

Consider the linear system

$$x_{k+1} = ax_k + v_k, \qquad v_k \sim \mathcal{N}(0, 0.01), \qquad (15.1a)$$
$$y_k = x_k + e_k, \qquad e_k \sim \mathcal{N}(0, 1), \qquad (15.1b)$$

where the parameter a is unknown. Suppose the true value is $a = 0.9$. The code below defines the system, simulates 20 observations, applies the time-varying Kalman predictor, and illustrates the output prediction with a 90% confidence bound. The result is shown in Figure 15.1(a).

```
m=lss(0.9,[],1,[],0.01,0.1,1)
x[k+1] = 0.9 x[k] + v[k]
y[k]   = 1 x[k] + e[k]
Q = Cov(v) = 0.01
R = Cov(e) = 0.1
y=simulate(m,20);
xhat=kalman(m,y,'alg',2,'k',1);   % Time-varying one-step pred.
plot(y,xhat,'conf',90,'view','interp')
```

Suppose now the filter constant a is uncertain and only known to be in the interval $[0.75, 0.95]$. The code below examines the sensitivity with respect to a.

```
a=0.75:0.05:0.95;
for k=1:length(a)
    m.A=a(k);
    xhata{k}=kalman(m,y,'alg',2,'k',1);
end
plot(xhata{:},'conf',90,'view','interp')
```

15.1 Parametric Uncertainty

Figure 15.1(b) shows the result. It can be seen that both the estimate and the covariance (check for instance sample 8 where the estimate is constant) are affected by the value of a.

15.1.1 Deterministic Sensitivity Analysis

The partial derivatives of the filter outputs,

$$\frac{\partial \hat{x}_k(\theta)}{\partial \theta_i}, \quad \frac{\partial P_k(\theta)}{\partial \theta_i}, \quad (15.2)$$

should be small in a robust algorithm. A large gradient indicates high sensitivity for the parameter θ_i. To judge the size of the gradient, it can be normalized as

$$T_{\theta,i} = \frac{1}{N} \sum_{k=1}^{N} \left(\frac{\partial \hat{x}_k(\theta)}{\partial \theta_i}\right)^T P_k^{-1} \left(\frac{\partial \hat{x}_k(\theta)}{\partial \theta_i}\right), \quad (15.3a)$$

$$T_{P,i} = \frac{1}{N} \sum_{k=1}^{N} \frac{\left\|\frac{\partial P_k(\theta)}{\partial \theta_i}\right\|}{\|P_k\|}. \quad (15.3b)$$

Any indicator T above that is significantly different from zero indicates high sensitivity for this parameter, which can be used in one of two ways:

- If θ_i is a physical parameter in the model, for instance corresponding to calibration errors, then one should perform dedicated identification experiments to assure that its uncertainty is small enough.

- If θ_i is a tuning parameter (for instance, an entry in the state noise covariance Q), then high sensitivity indicates that this is an important tuning parameter. Contrary, if $T_{\theta,i}, T_{P,i}$ are both small, compared to other parameters, there is no need to waste energy on tuning this parameter, or it should be changed in large steps in the tuning.

A more formal analysis of parametric uncertainty is performed in the following section.

15.1.2 Stochastic Sensitivity Analysis

Suppose now that θ is a stochastic variable estimated with $\hat{\theta}$, where $\mathrm{E}(\hat{\theta}) = \theta^o$ is unknown and $\mathrm{Cov}(\hat{\theta}) = P_\theta$ is known. This is the case when gray-box modeling is used for instance. The filter thus provides $\hat{x}_k(\hat{\theta})$, $P_k(\hat{\theta})$. The bias and variance induced by the parametric uncertainty can be expressed as

$$E(x_k) = E_\theta(\hat{x}_k(\theta)), \tag{15.4a}$$
$$\text{Cov}(x_k) = E_\theta(\text{Cov}_{x_k}(x_k|\theta)) + \text{Cov}_\theta(E_{x_k}(x_k|\theta)) \tag{15.4b}$$
$$= E_\theta(P_k(\theta)) + \text{Cov}_\theta(\hat{x}_k(\theta)). \tag{15.4c}$$

The conditioning on old measurements in filtering applications is here omitted for brevity. For a given data set $y_{1:N}$, a Monte Carlo evaluation with random samples of θ can be used to evaluate the bias and excessive variance caused by the parametric uncertainty, see Section A.3.2. Take random samples θ^i from its modelled distribution, and compute

$$\hat{\bar{x}}_k = \frac{1}{N}\sum_{k=1}^{N} \hat{x}_k(\theta^i), \tag{15.5a}$$

$$\hat{P}_k = \frac{1}{N}\sum_{k=1}^{N} P_k(\theta^i) + (\hat{x}_k(\theta^i) - \hat{\bar{x}}_k)(\hat{x}_k(\theta^i) - \hat{\bar{x}}_k)^T. \tag{15.5b}$$

As an alternative to Monte Carlo sampling, the filter can be seen as a nonlinear mapping from a Gaussian vector θ to $\hat{x}_k(\theta)$, and the unscented transform can be used to find the bias and covariance with a small number of points ($N = 2n_\theta + 1$ above), see Section A.3.3.

The *mean square error (MSE)* consists of squared bias and variance from (15.5),

$$\text{MSE} = (x_k - E_\theta(\hat{x}_k(\theta)))(x_k - E_\theta(\hat{x}_k(\theta)))^T$$
$$+ E_\theta(P_k(\theta)) + \text{Cov}_\theta(\hat{x}_k(\theta)),$$

where x_k denotes the true state at time k. The goal should be that the MSE is dominated by $P_k(\theta)$, so the bias and spread of the mean contributions are negligible, that is,

$$x_k(\hat{\theta}) \approx E_\theta(\hat{x}_k(\theta)), \tag{15.6a}$$
$$P_k(\hat{\theta}) \approx E_\theta(P_k(\theta)) \gg \text{Cov}_\theta(\hat{x}_k(\theta)). \tag{15.6b}$$

These conditions should always be checked if there is a stochastic parametric uncertainty in the model.

15.2 Ground Truth Data

Access to *ground truth* data is of course a great asset in filter validation and evaluation, but it is perhaps not that obvious how to use it in a systematic way

15.2 Ground Truth Data

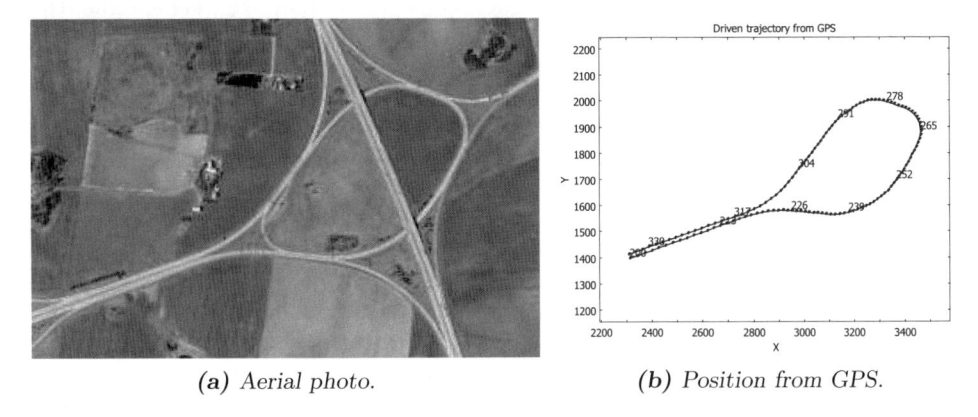

(a) Aerial photo.   **(b)** Position from GPS.

Figure 15.2: Test drive for the navigation problem explained in Section 1.2.

to improve the filter. We will in this section describe systematic approaches for validating each part of the model. Once this is done, different filter algorithms can be compared systematically.

We denote the ground truth state x_k^o, but also realize that it is by necessity not perfectly known, so the validation trajectory is $\hat{x}_k^o \approx x_k^o$ with an associated covariance matrix $P_k^o = \text{Cov}(x_k^o)$. The goal should be that the ground truth is sufficiently accurate, so that this uncertainty can be neglected compared to the final filter performance. In a way, there is a similarity betwee the uncertainty in $\hat{x}_k^o$ and the parametric unceatainty in θ discussed in Section 15.1. As a rule of thumb, P_k^o should be small compared to P_k.

15.2.1 How to get Ground Truth State

The obvious way to get ground truth is to use one accurate sensor for each state, so $y_k = \hat{x}_k^o \approx x_k^o$. This might be too expensive or inconvenient in general. Good compromises in practice include the application cases below. We will use the navigation problem in Section 1.2, to illustrate the different concepts of getting ground truth data.

Figure 15.2 illustrates the experiment for convenience. The filtering goal is both to find the true trajectory, but also to estimate the velocities and accelerations accurately.

- For target tracking, the navigation state from the target can be used as ground truth. This includes a possible application where a surveillance radar is measuring range and bearing to the vehicle in Figure 15.2.

- For navigation, an accurate differential GPS can be used in dedicated experiments.

However, a general approach is based on a combination of two or more of the following options, in order of importance:

- Feasible extra sensors. In contrast to the example above, these do not have to be high-performance. For instance, the speedometer, control inputs as accelerator and steering wheel angle, or wheel speeds in the car can be used in the fusion filter for navigation in Figure 15.2.

- There is also a possibility to use state constraints (sometimes called *non-holonomic constraints*) as *virtual measurements*. Consider, for instance, the trajectory in Figure 15.2. The final speed is about the same as the initial one ($v_N = v_1 + e_{v_N}$) and the final course is the opposite of the initial course ($\psi_N = \psi_1 + 180° + e_{\psi_N}$), since the start and end points are from the same location of the road network.

- Smoothing instead of filtering. This removes the lag in the filtered state estimates, and allows for a slower filter with more noise reduction than is appropriate in filter applications. It is well-known in tracking applications that the vehicle appears to always turn too late after causal filtering, a problem that is avoided with smoothing.

- More sophisticated filters. Validation is an off-line procedure, and for instance a particle filter with quite a lot of particles can be used to get $\hat{x}_k^o$, even though the filter to be implemented is an EKF. For instance, a dynamic model of the vehicle using known control inputs can be used at this stage.

- Experiment based tuning, where prior knowledge of the process noise can be used to hand-tune a time-varying state covariance. One can for instance visually detect onset times of maneuvers in tracked targets and momentarily increase the state noise at these times. For the driving experiment in Figure 15.2, the test driver can log his actions, for instance by pushing a button each time he maneuvers, and such binary data can later be used for selecting the process noise covariance matrix.

- More sophisticated models can be used at this stage. The only condition is that the state vector of this model should include the states in the original model.

In particular, smoothing is an important tool for getting a state estimate that can be used as a substitute for ground truth.

There is a twist here. To validate the model and filter intended for the application, we need even more advanced models and filters according to the recommendations above, which have to be validated themselves, and so on. The key is often the aiding validation sensors, which simplify the design of a robust and easy to tune ground truth filter and model.

15.2.2 All-in-One Validation

The obvious validation step is to compare the state estimates directly to the ground truth data. This can be done in the following ways:

- Plot each state separately as a function of time, where confidence intervals indicate if the performance is good.

- Plot two states jointly as a trajectory. Confidence ellipses at regular time instants indicate the accuracy.

- Objective performance measures and hypothesis tests. The following test statistic can be computed,

$$T = \sum_{k=1}^{N} \left(\hat{x}_k - \hat{x}_k^o \right)^T P_k^{-1} \left(\hat{x}_k - \hat{x}_k^o \right). \tag{15.7}$$

The test statistic above is distributed as $T \sim \chi^2_{Nn_x}$ under the hypothesis that (i) the system is linear and Gaussian, (ii) the model coincides exactly with the system, (iii) the covariance matrices in the model are correct and (iv) there is no bug in the filter implementation. That is, if the test statistic falls into an $1 - 2\alpha$ symmetric confidence interval,

$$h_{\chi^2_{Nn_x}}(\alpha) < T < h_{\chi^2_{Nn_x}}(1 - \alpha), \tag{15.8}$$

then this indicates that the filter and model are appropriate. Here, $h_{\chi^2_{Nn_x}}(\alpha)$ denotes the threshold corresponding to the probability $P(T < h_{\chi^2_{Nn_x}}(\alpha)) = \alpha$. Thus, this all-in-one validation tests everything in one step. The problem occurs if the result does not pass these tests, or even diverges, where should the user start to look? This approach *is not recommended* for initial tests, and should be saved for the last step in the validation chain.

Even in the final step, this test statistic should be used with some caution. It is easily shown that the covariance matrix P_k from the KF can be scaled a factor β by just multiplying all covariance matrices P_0, Q_k and R_k with the same constant β. The state estimate $\hat{x}_k$ is not affected by this scaling. This means that one can always tune the KF to give an arbitrary T without changing the state estimate. Basically the same conclusion should hold also for nonlinear filters. The easiest remedy is to fix the measurement covariance R by sensor modeling as described in the next section.

15.2.3 Sensor Model Validation

Consider a measurement model with additive noise,

$$y_k = h(x_k) + e_k. \tag{15.9}$$

Now, the ground truth data can be substituted for the state,

$$\hat{e}_k = y_k - h(\hat{x}_k^o). \tag{15.10}$$

The sensor model can be validated by checking the unbiasedness and whiteness of the obtained residual. Further, and more importantly, the noise model can be fitted to the residuals. For instance, a histogram plot of a one-dimensional residual $\hat{e}_k$ together with the model distribution is quite revealing. A time-invariant covariance can be estimated with

$$\hat{R} = \frac{1}{N} \sum_{k=1}^{N} \hat{e}_k \hat{e}_k^T. \tag{15.11}$$

If the noise is modeled as Gaussian, $e_k \sim \mathcal{N}(0, R_k)$, one can compute the following test statistic as an indicator

$$T = \sum_{k=1}^{N} \hat{e}_k^T R_k^{-1} \hat{e}_k \sim \chi^2_{Nn_y}. \tag{15.12}$$

If the obtained T is small, one can accept (or more formally, one cannot reject) the hypothesis that both the sensor model and the measurement noise model are correct. If T is large, either the sensor model or the measurement error distribution is incorrect, or both. This requires further investigation by studying the residuals $\hat{e}_k$ more systematically. Collecting measurements from a sensor with a constant input can be a good approach to model its noise properties, see Section 12.5.2.

There is a catch to be aware of in this context. If the covariance estimate (15.11) is plugged into (15.12), the result is

$$T = \sum_{k=1}^{N} \hat{e}_k^T \hat{R}^{-1} \hat{e}_k = \sum_{k=1}^{N} \operatorname{tr}\left(\hat{R}^{-1} \hat{e}_k \hat{e}_k^T\right) = \sum_{k=1}^{N} \operatorname{tr}\left(I_{n_y}\right) = N n_y. \tag{15.13}$$

That is, the noise distribution cannot be estimated and validated using the same data in this way. However, there are established methods in mathematical statistics for this task. These are based on measuring how well a histogram of observed errors matches the modeled distribution.

If the sensor model (15.9) is replaced with an implicit one, $h(y_k, x_k) = e_k$, nothing changes since it can be solved for the noise, $\hat{e}_k = h(y_k, \hat{x}_k^o)$. If the noise is implicit as well, $h(y_k, x_k, e_k) = 0$, one has to solve for e_k numerically for each k.

15.2.4 Motion Model Validation

The same principle as above applies also for the dynamic motion model,

$$x_{k+1} = f(x_k) + v_k. \tag{15.14}$$

15.2 Ground Truth Data

Now, the ground truth data can be substituted for the state,

$$\hat{v}_k = \hat{x}^o_{k+1} - f(\hat{x}^o_k). \tag{15.15}$$

Whiteness (and unbiasedness) of $\hat{v}_k$ is critical for model validation. A state noise covariance function can be defined by and estimated with

$$Q(\tau) = \mathrm{E}\bigl(v_k v^T_{k-\tau}\bigr), \tag{15.16a}$$

$$\hat{Q}(\tau) = \frac{1}{N} \sum_{k=\tau+1}^{N} \hat{v}_k \hat{v}^T_{k-\tau}, \tag{15.16b}$$

respectively. If $\|\hat{Q}(\tau)\| \ll \|\hat{Q}(0)\|$ for all $\tau \neq 0$, then the process noise can be considered white, and the dynamical model is most likely appropriate.

Otherwise, the same validation steps and comments as in the previous section, about the sensor model, apply also here. In particular, implicit process noise is quite common for many motion models of the form $x_{k+1} = f(x_k, v_k)$. In the case $n_v = n_x$, one can apply a numerical solver to get

$$\hat{v}_k = \mathrm{sol}_{v_k}\bigl(\hat{x}^o_{k+1} = f(\hat{x}^o_k, v_k)\bigr). \tag{15.17}$$

Usually, $n_v < n_x$ and the solution can be defined in the least squares sense using the Q-norm,

$$\hat{v}_k = \arg\min_{v_k} \bigl\|\hat{x}^o_{k+1} - f(\hat{x}^o_k, v_k)\bigr\|^2_Q. \tag{15.18}$$

In the linear case $x_{k+1} = Fx_k + Gv_k$, the solution is given by using the pseudo-inverse $G^\dagger$ of G in $\hat{v}_k = G^\dagger\bigl(\hat{x}^o_{k+1} - F\hat{x}^o_k\bigr)$.

15.2.5 Choice of Nonlinear Transformation Method

Consider for instance the motion model (15.14). In a Kalman filter approach, a nonlinear transformation has to be chosen, and as was concluded in Chapter 8, there is no single choice that is uniformly best in all applications. Ground truth data can be used to select the transformation that is most appropriate for the function $f(x_k)$ in realistic scenarios.

To recapitulate Section A.3, the nonlinear transformation methods aim at approximating the distribution of a nonlinear mapping $z = g(x)$ of a Gaussian variable x with a Gaussian distribution. For the sensor model, we have

$$x_k \sim \mathcal{N}(\mu_x, P_x),\ e_k \sim \mathcal{N}(0, R) \Rightarrow y_k = h(x_k, e_k) \sim \mathcal{N}(\hat{y}_k, S_k). \tag{15.19}$$

Given such a mapping (for instance a first order Taylor expansion, or an unscented transform), one can evaluate its correctness using the relation

$$T = \sum_{k=1}^{N} (y_k - \hat{y}_k) S_k^{-1} (y_k - \hat{y}_k) \sim \chi^2_{Nn_y}. \tag{15.20}$$

Transformation methods that give a T which is significantly larger than a given confidence bound should not be used in the measurement update. The same test should also be applied to the motion model to select the mapping for the time update.

15.3 Sensor Calibration Issues

The nominal sensor model is expressed as $y_k = h(x_k) + e_k$. However, there are at least three kinds of calibration issues to consider carefully: scaling and offsets, time alignment and sensor location. The first kind is a generic problem, and can be formulated as

$$y(t_k) = ch(x_k) + b + e_k, \qquad (15.21)$$

for an unknown scaling c and offset b. The other calibration issues are dealt with in separate sections below.

15.3.1 Time Delays

The observation time t_k does not always coincide with the sample time kT corresponding to the state $x_k = x(kT)$. First, if $kT - t_k$ are independent stochastic variables we talk about sampling jitter. Second, if $kT - t_k$ is constant, then we have a time delay. To exemplify, a sensor connected to the computer via the USB gives a time delay of about 1 ms with a small jitter. The processing delay in a GPS can be up to 100 ms long, which causes a time delay in the outputted position. A wireless sensor interface, for instance the Bluetooth standard, gives varying time delays up to the order of 100 ms. An overview of synchronization issues is provided in Skog and Händel (2009), and an approach of augmenting the state vector with the time offset is proposed in Skog and Händel (2008).

Even if *time stamps* are provided to each measurement, the clocks may differ, see Skog and Händel (2010). An external synchronization clock is always to prefer when possible to guarantee correct time stamps. The delay can then be handled with *out of sequence* updates as described in Section 7.2.5.

A third, more complicated, situation is when the time delay is state dependent. For example, this occurs for ranging sensors where the speed of a tracked object is not negligible to the speed of the media. With a sensor in the origin and p_k denoting the position of the target (a subset of the state vector), then we have $t_k - kT = c\|p_k\|$, where c is the speed of the medium and $\|p_k\|$ is the range to the target. This leads to a quite complicated problem. The problem is called *propagation delayed measurements* in Orguner and Gustafsson (2010), where more background and information on this problem are given.

Dedicated filter solutions to these problems are outside the scope of this section. Instead, we recommend to investigate the sensitivity issue. How much jitter or delay can the filter tolerate without a significant degradation in performance? This is easiest checked with a dedicated simulation study, where the sensor observation is delayed randomly. It is here important to know the sensor and its communication link to the processor so that reasonable delay values can be selected. The observations can be computed at high time resolution using the fast sampling described in Section 12.6.1.

15.3.2 Sensor Location

The sensor model in (15.21) is in this section assumed to measure something related to an object's physical position p_k, velocity or acceleration, which is often the case in sensor fusion. The sensor position l might not be exactly what was designed in the sensor model $h(p_k, l)$, and a possible translation t and rotation R error can be expressed as $Rl+t$. Once the problem is identified, it can easily be recast into a parametric uncertainty by letting θ include the unknowns in R, t. Then, the methods in Section 15.1 apply.

If the sensor origin turns out to be crucial, dedicated *calibration experiments* must be performed. This can be done using the standard NLS framework in Section 12.7. Here, ground truth data are really useful.

15.4 Summary

Given a filter $\mathcal{F}$, data $y_{1:N}$, a dynamic model $f(x_k, v_k)$, a sensor model $h(x_k, e_k)$ and noise distributions $p_v(\cdot)$, $p_e(\cdot)$. How can we validate that the output $(\hat{x}_{1:N}, P_{1:N}) = \mathcal{F}(y_{1:N}, f, h, p_v, p_e)$ gives a good description of the true state $x_{1:N}$?

1. Checklist for ground truth state:

    (a) Can additional validation sensors or external information be used in a dedicated validation experiment?

    (b) Can the filter output be improved by smoothing, a more sophisticated filter, a more sophisticated model or hand-tuning using prior knowledge of the experiment?

    The ground truth state $\hat{x}_k^o$ can be used to validate the model in the following systematic ways:

    - The sensor model residuals $\hat{e}_k = y_k - h(\hat{x}_k^o)$ should be white (independent over time and with $\hat{x}_k^o$) and their distribution should follow $p_e(\cdot)$.

    - The motion model residuals $\hat{v}_k = \hat{x}_{k+1}^o - f(\hat{x}_k^o)$ should be white (independent over time and with $\hat{x}_k^o$) and their distribution should follow $p_v(\cdot)$.

2. Checklist for validation:

    (a) The "I'm feeling lucky" alternative: apply the intended filter to the data and compare to ground truth. This unsystematic approach is not recommended. Typically, the first implementation includes a bug or is so poorly tuned that the output is rubbish, and the user has then no clue of what went wrong.

    (b) Validate the sensor model using the ground truth data. The first step is to compute the model residuals $\hat{e}_k$ from the measurements and ground truth data. These should be independent of the state for the sensor model to be correct. The sensor model and the noise covariance can be validated with the test statistic

    $$T = \sum_{k=1}^{N} \hat{e}^T R_k^{-1} \hat{e}_k \sim \chi^2_{Nn_y}.$$

    A parametric or non-parametric noise distribution can be estimated from the residuals, and later used in a PF, PMF or KFB.

15.4 Summary

(c) Validate the motion model using the ground truth data in the same way as the sensor model. The first step is to compute the model residuals $\hat{v}_k$. These should be independent for the motion model to be correct. The sensor model and the noise covariance can be validated with the test statistic

$$T = \sum_{k=1}^{N} \hat{v}^T Q_k^{-1} \hat{v}_k \sim \chi^2_{Nn_v}.$$

A parametric or non-parametric noise distribution can be estimated from the residuals, and later used in a PF, PMF or KFB.

(d) The final step, when all of $f(x_k, v_k)$, $h(x_k, e_k)$, $p_v(\cdot)$ and $p_e(\cdot)$ have been validated, is to validate the filter. This can be done with state plots with confidence bounds, or with the test statistic

$$T = \sum_{k=1}^{N} \left(\hat{x}_k - \hat{x}_k^o\right)^T P_k^{-1} \left(\hat{x}_k - \hat{x}_k^o\right).$$

As a comment, this systematic approach is particularly important when the implementation platform is not the same as the development environment. Much of the validation work can be done in MATLAB^{TM}, but still one has to validate parts of the model and code in parallel and carefully compare the results.

3. Checklist for evaluating the influence of parametric uncertainty:

   (a) Are the sensor positions critical? In such case, how accurately are these positions known?

   (b) Is there infrastructure in terms of markers, beacons or base stations whose positions are critical? In such case, how accurately are these positions known?

   (c) Are there parameters in the motion model or sensor model that are not exactly known?

The parametric uncertainties in the model can be evaluated by the sensitivity function

$$\frac{\partial \hat{x}_k(\theta)}{\partial \theta_i}, \quad \frac{\partial P_k(\theta)}{\partial \theta_i},$$

or the stochastic uncertainty

$$\mathrm{E}(x_k) = \mathrm{E}_\theta\left(\hat{x}_k(\theta)\right),$$
$$\mathrm{Cov}(x_k) = \mathrm{E}_\theta\left(P_k(\theta)\right) + \mathrm{Cov}_\theta\left(\hat{x}_k(\theta)\right).$$

If any of these indicate that the actual uncertainty can give a contribution to the filter output, dedicated *calibration experiments* must be performed, where the core algorithms are described in Section 12.7.

16

Applications

The purpose of this chapter is to present some larger application cases where the various models and algorithms in the preceding chapters have been combined to solve real world problems. A common theme is that the complexity of each application exceeds the format of an example.

16.1 Sensor Networks

Some sensor network applications using the most common RSS, TOA, TDOA and DOA measurements were described in Chapter 4. We here give an application study based on a non-standard measurement presented in Lindgren et al. (2009). The goal is to estimate the position of a sniper based on a microphone network. Each microphone registers the arrival time for explosive sound, which is called *muzzle blast* (*MB*). Microphones inside a certain area can also detect a *shockwave* (*SW*) generated by super-sonic bullets (compare with the bow wave from a boat). Figure 16.1(a) illustrates a typical microphone signal, and Figure 16.1(b) describes the geometrical relations.

According to the clock at microphone k, the *muzzle blast (MB) sound* is assumed to reach the sensor located in position p_k at the time

$$y_k^{\text{MB}} = t_0 + b_k + \frac{1}{c}\|p_k - x\| + e_k^{\text{MB}}. \tag{16.1}$$

Here, t_0 is the unknown shooting time, b_k an unknown time offset in each sensor clock, c is the speed of sound in air, y_k^{MB} is the detected time for the MB, and e_k^{MB} is the noise. The shooter position x and microphone location p_k are in $\mathbb{R}^n$, where generally $n = 3$. However, both computational and

(a) Microphone recording.

(b) Geometry of the shock wave.

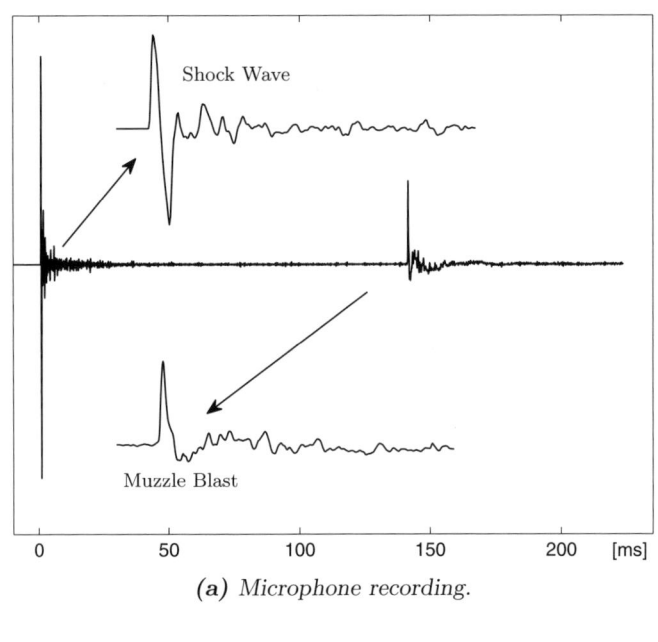

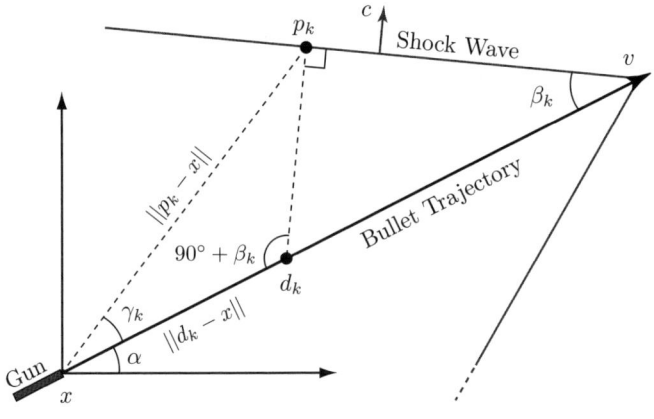

Figure 16.1: (a) Signal from a microphone placed 180 m from a firing gun. Initial bullet speed is 767 m/s. The bullet passes the microphone at a distance of 30 m. The shockwave from the supersonic bullet reaches the microphone before the muzzle blast. (b) Geometry of supersonic bullet trajectory and shock wave. Given the shooter location x, the shooting direction (aim) α, the bullet speed v, and the speed of sound c, the time it takes from firing the gun to detecting the shock wave can be calculated.

16.1 Sensor Networks

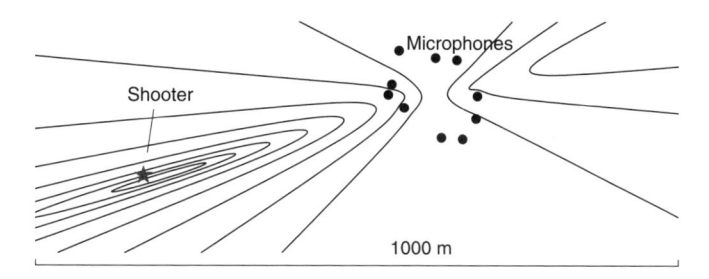

Figure 16.2: *Level curves of the muzzle blast localization least squares criterion $\sum_{k=1}^{N}\left(y_k^{\mathrm{MB}} - \hat{y}_k^{\mathrm{MB}}\right)^2$ using the observation model (16.1) and the geometry and data from a field trial. The SLS approach is used to eliminate the shooting time t_0, and the sensors are perfectly synchronized $b_k = 0$.*

numerical issues occasionally motivate a simplified plane model with $n = 2$. The notation is summarized in Table 16.1.

In a synchronized network where b_k is constant, a TDOA approach would make it possible to eliminate the shooting time t_0, see Section 4.3.4. Figure 16.2 shows the loss function of the *separable least squares (SLS)* approach as described in Section 3.6.1. Here, the shooting time t_0 is eliminated by the least squares principle, and the remaining least squares loss function is a function of the 2D position and can simply be computed on a grid. The clock offsets are in Figure 16.2 assumed to be zero, $b_k = 0$. It turns out that the criterion is quite sensitive to random clock errors, which implies that either dedicated synchronization procedures must be implemented, or other approaches should be investigated. We will below show that the shock wave can be used to remove the influence of the clock bias.

To explain the shock wave sensor model, consider Figure 16.1(b). The bullet travel time to point d_k and the wave propagation time from d_k to p_k sum up to the total time from firing to detection,

$$y_k^{\mathrm{SW}} = t_0 + b_k + \frac{1}{r}\log\frac{v_0}{v_0 - r\|d_k - x\|} + \frac{1}{c}\|d_k - p_k\| + e_k^{\mathrm{SW}}.$$

Here, d_k is implicitly given by the equations

$$\frac{\sin(90° - \beta_k - \gamma_k)}{\|d_k - x\|} = \frac{\sin(90° + \beta_k)}{\|p_k - x\|}, \quad (16.2a)$$

$$\sin\beta_k = \frac{c}{v} = \frac{c}{v_0 - r\|d_k - x\|}. \quad (16.2b)$$

This system of equations has to be solved numerically at each time.

One way to avoid accurate synchronization in the network, is to report the time differences $y_k = y_k^{\mathrm{SW}} - y_k^{\mathrm{MB}}$. The point is that both the shooting time

Table 16.1: Notation. MB, SW, and MB–SW are different models, and L/N indicates if model parameters or signals enter the model linearly (L) or nonlinearly (N).

Variable	MB	SW	MB–SW	Description
M				Number of microphones
S				Number of microphones receiving shock wave
x	N	N	N	Position of shooter
p_k	N	N	N	Position of microphone k
y_k	L	L	L	Measured detection time for microphone at position p_k
t_0	L	L		Rifle or gun firing time
c	L	N	N	Speed of sound
v_0		N	N	Initial speed of bullet
v				Time-varying speed of bullet
α		N	N	Shooting direction
b_k	L	L		Clock error for microphone k
e_k	L	L	L	Detection error at microphone k
r		N	N	Bullet speed decay rate
d_k				Point of origin for shock wave at microphone k
β				Mach angle, $\sin\beta = c/v$

and the clock bias are eliminated. The fusion center can then use $y_k^i = h(x_k) + e_k^i$ from each sensor i to estimate the parameters x, as given in Table 16.1. Note that the state vector includes the shooter's aiming angle, the bullet speed and the bullet retardation. The latter two states are useful for classification of the ammunition.

Figure 16.3 illustrates the potential of an acoustic network for sniper detection. The position of the shooter is estimated quite accurately from the supersonic shock wave and the muzzle blast. Actually, also the shooting angle and various feature of the ammunition (bullet length, speed and air drag) can be estimated accurately. Besides being an interesting application in itself, it highlights the real strength of sensor fusion: by combining multiple sensor signals with accurate physical models, surprisingly much information can be extracted. Dedicated expensive sensor systems can many times be replaced with several standard sensors (in this case cheap microphones) and sophisticated signal processing algorithms.

16.2 Kalman Filtering   431

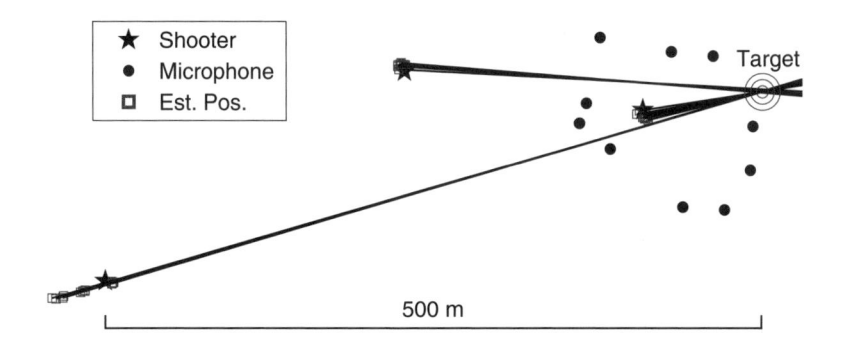

Figure 16.3: *Acoustic sensor network for estimating shooting directions. The plot illustrates estimated position and aiming angle from a field trial using three different shooter positions, different weapons and ammunition but the same shooting target. In particular the bullet path is accurately estimated.*

16.2 Kalman Filtering

16.2.1 Navigation using GPS and IMU

Section 1.2 stated that a *GPS (global positioning system)* and an *IMU (inertial measurement unit)* combination is a perfect illustration of the sensor fusion principle: sensory data from disparate sources are combined such that the resulting information is better than would be possible if these sources were used individually. This combination is today standard in most transportation systems. Navigation systems in cars have a simple integration of GPS with only a yaw rate gyro, but there are many studies on how the full integration should be performed, see Skog and Händel (2009) for a survey with a focus on car navigation systems. The literature distinguishes several levels of fusion:

- *Loose integration*, where the GPS is seen as a sensor $y_k = p_k$ of the position p_k.

- *Tight integration*, where each satellite provides a pseudo-range observation $y_k^i = \|p_k - p_k^i\|/c + t_k$, where p_k^i is the known position of satellite i, c the speed of light and t_k is the time error in the receiver.

- *Deep integration* is characterized by a feedback from the fusion filter to the GPS receiver, used to control the phase locked loops in the GPS hardware, which determines where to look for each satellite's correlation sequence. The advantage is a more robust system that can handle large accelerations, but the basic sensor model is the same as for tight integration.

However, the total reliance on GPS today comes with obvious risks. For critical navigation applications, GPS cannot be the only positioning sensor. In military applications, an independent backup sensor insensitive to GPS jamming is often used, but also in civil applications the robustness against jamming may constitute one of the main design issues in the future. For marine applications, the reports Volpe (2001) and Forum (2001) discuss the problem of intentional or unintentional GPS jamming and alternative backup systems are strongly recommended. Basically, the required redundancy can be achieved either with an existing wireless network, as mentioned above, or with a dedicated network of sensors or beacons, or with support from a geographical information system (GIS).

Loose Integration

(a) Actual road segment.

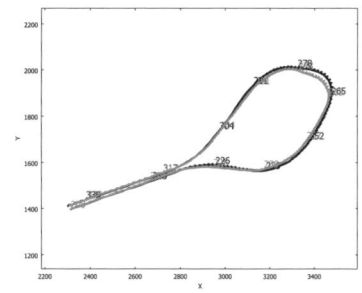

(b) Trajectory from GPS and fusion filter.

Figure 16.4: *Satellite view of a highway intersection from Figure 1.3, and position estimates by fusing GPS and IMU in a Kalman filter (labeled with some time markers in seconds).*

Figure 16.4 returns to the example in Section 1.2, and it shows the result from a Kalman filter using a coordinated turn model from Table 13.4. The result is a trajectory that improves the one from GPS slightly, but the most important advantages are:

- The velocities can be computed with little noise.

- The accelerations are estimated with offset compensation, and with less noise than the raw IMU measurements.

- In short periods without GPS coverage, the model can *dead-reckon* the position due to the offset compensation in the IMU.

The only drawback with this solution is that at least four satellites must be in line of sight most of the time to give a useful GPS position fix. This drawback is circumvented in tight integration.

The data in this section come from Vinkvist and Nilsson (2008), which also presents detailed fusion filter. A similar filter with racing applications are presented in Blomfeldt and Haverstad (2008).

Tight Integration

One remaining area where GPS/IMU integration is not used today is in rocket missions, as the one illustrated in Figure 16.5. Export restrictions for the GPS software in high-speed objects (such as missiles) prevent a direct application. Instead, dedicated fusion solutions for rockets, that would not be applicable to other similar applications, are needed.

We here investigate data from the Maxus launch illustrated in Figure 16.5(a). The mission of this sounding rocket is to get outside the air drag of the atmosphere and then to fall back freely without any gravitational or control forces. A tactical grade IMU is used for navigation. It is quite accurate, and the drift during the whole mission is only in the order of hundreds of meters in this case, as illustrated in Figure 16.5(b).

GPS was not available in this rocket launch. Instead, simulation software was used to generate the actual satellite constellation according to the time stamps from the IMU. The software applies wave propagation models to generate realistic GPS pseudo-range measurements. These are then used together with the actual IMU data in an *EKF*. Figure 16.5(c) illustrates that the position error comes down to meter accuracy.

See Törnqvist et al. (2010) for more details on this application of tight integration. Deep integration is discussed in Vinkvist and Nilsson (2008) for sounding rockets.

16.2.2   Yaw Rate Estimation in Vehicles

Accurate vehicle course information is crucial for several *advanced driver assistance systems* (*ADAS*). For instance, the *electronic stability control* (*ESC*) system mitigates over- and under-steering by braking individual wheels in order to make the vehicle follow the yaw rate reference from the steering wheel. For that reason, an accurate drift-free estimate of the actual yaw rate is needed. Standard low-cost yaw rate sensors have a substantial drift of several degrees per second, a fact that has forced manufacturers of current ESC systems to control the yaw acceleration rather than the more natural yaw rate. We here describe a sensor fusion approach, where the low-cost yaw rate gyro is supported with wheel speed information, see also Section 13.4.2.

(a) We know where it starts, but where will it land?

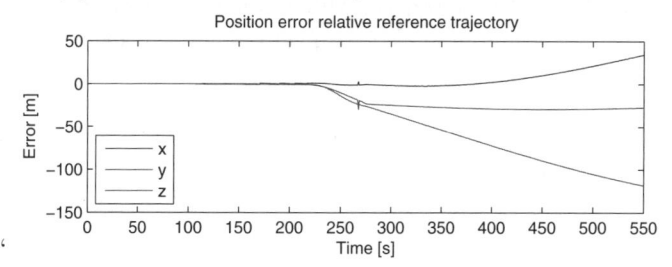

(b) Position error using dead-reckoning of IMU

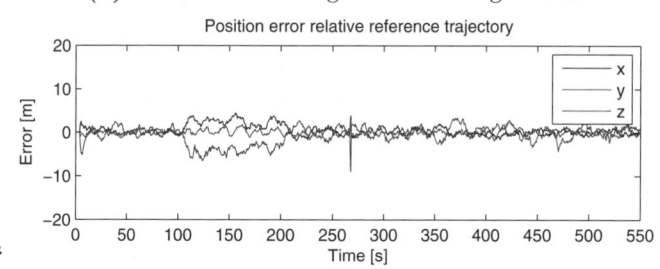

(c) Position error using robust sensor fusion of GPS and IMU

Figure 16.5: The Maxus launch in (a) is one of the 470 rocket launches at Esrange in Kiruna, Sweden. All these have been based on only IMU navigation, and the rockets have in most cases landed inside the huge area of Esrange. Even very expensive IMU systems drift over time. With robust fusion with GPS, the landing area could be restricted considerably, with a cheaper and lighter navigation system as a bonus.

16.2 Kalman Filtering

The basic sensor models are first a yaw rate sensor with time-varying offset b_k, and secondly the virtual yaw rate sensor from (14.35d), extended with a time-varying offset model for the wheel radii. The latter is simplified somewhat by assuming the nominal wheel radii in the factor corresponding to speed. The resulting model is

$$y_k^1 = \dot{\psi}_k + b_k + e_k^1, \tag{16.3a}$$

$$y_k^2 = \frac{\omega_3 r_{nom} + \omega_4 r_{nom}}{2} \frac{2}{B} \frac{\frac{\omega_3}{\omega_4} \frac{r_{k,3}}{r_{k,4}} - 1}{\frac{\omega_3}{\omega_4} \frac{r_{k,3}}{r_{k,4}} + 1} + e_k^2. \tag{16.3b}$$

If a lateral accelerometer is available, it can also be included as described in Section 13.4.2. The state vector is

$$x_k = \left(\dot{\psi}_k, \ddot{\psi}_k, b_k, \frac{r_{k,3}}{r_{k,4}}\right), \tag{16.4}$$

and the simplest possible random walk model of the three last states is applied. Figure 16.6 shows a block-diagram for the system, together with a plot illustrating how the critical yaw rate gyro offset is successfully estimated in one minute.

Figure 16.7 shows four snapshots of a test drive, consisting of four laps in a round-about. The actual starting position is assumed known, and then dead-reckoning according to (13.29) in Section 13.5 is applied. The plot that is over-layed a map (here used as *ground truth*) shows firstly dead-reckoning based on speed and measured yaw rate, and secondly the result of dead-reckoning the states of a fusion filter. Basically, the state vector (16.4) is augmented with global position X, Y, and the dynamic equation is extended with (13.29). Without offset compensation, the drift is $180°$, while the fusion filter gives virtually no drift at all.

16.2.3 Roll Angle Estimation on Motorcycles

The roll angle, illustrated in Figure 16.8, is a critical state for stability of two-wheeled vehicles. Excessive braking or acceleration when the roll angle is high may cause instability. For that reason, *advanced driver assistance systems* (*ADAS*) as *anti-locking brake systems* (*ABS*) and anti-spin system can be improved with accurate knowledge of the roll angle. There are also mechanically controlled head-lights that can direct the light beam to the correct ground area if only the roll angle was known.

Table 16.2 summarizes potential inputs, measurements and states for such a system. The sensor model for the lateral and vertical accelerometers and

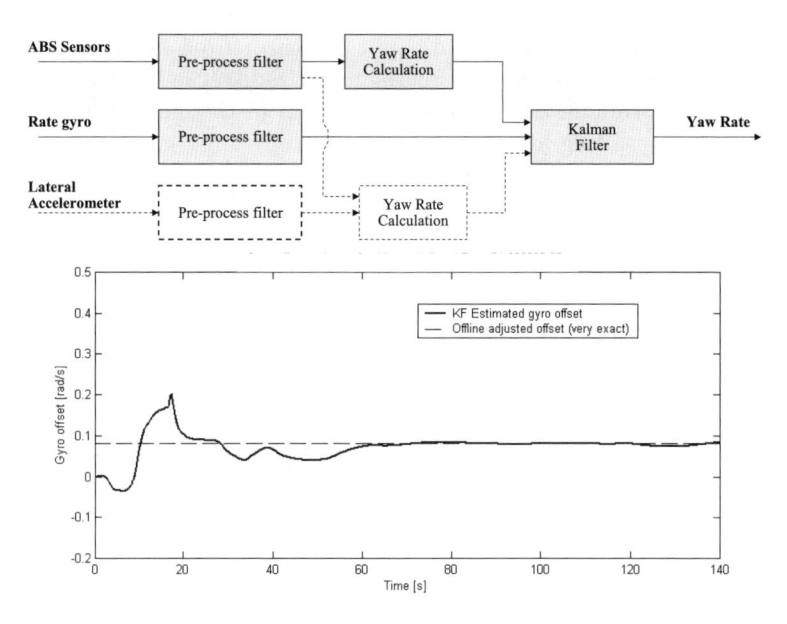

Figure 16.6: *Block diagram of a sensor fusion system for computing the yaw rate, combining information from a yaw rate gyroscope, wheel speeds and possibly also a lateral accelerometer. Lower plot shows an example of how the offset in the gyro is estimated adaptively.*

the roll gyro is

$$y = h(x) = \begin{pmatrix} ux_4 - z_y x_3 + z_y x_4^2 \tan(x_1) + g\sin(x_1) + x_6 \\ -ux_4 \tan(x_1) - z_z\left(x_2^2 + x_4^2 \tan^2(x_1)\right) + g\cos(x_1) + x_7 \\ -a_1 x_3 + a_2 x_4^2 \tan(x_1) - ux_4 J + x_6 \\ x_2 + x_8 \end{pmatrix} \tag{16.5}$$

where z_y, z_z, a_1, a_2, J are constants relating to geometry and inertias of the motorcycle. The last equation is an example of *virtual sensors*, saying that in average the roll rate is zero (the vehicle is not rolling around the x axis).

A roll gyro is of course the single best sensor. However, today a gyro is several order of magnitudes more expensive than accelerometers. For that reason, it is interesting to compare different combinations of sensors, and evaluate their price–performance ratio. Figure 16.9 shows four such combinations, compared to a reference system (*ground truth sensor*). Quite interestingly, good results are obtained using only one or two accelerometers.

16.2 Kalman Filtering

Figure 16.7: *Example of a validation test in a round-about: four snapshots at initialization, after one lap, two laps and four laps, respectively. The dark line shows the sensor fusion output, and the gray line shows the dead-reckoned position using only the yaw rate.*

Figure 16.8: *Knowing the roll angle enables new funtionalities as adaptive headlight control and improved ABS and traction control systems.*

Table 16.2: *Full state vector, measurements and inputs in roll angle estimation.*

x_1	φ	Roll angle
x_2	$\dot{\varphi}$	Roll rate
x_3	$\ddot{\varphi}$	Roll acceleration
x_4	$\dot{\psi}$	Yaw rate
x_5	$\ddot{\psi}$	Yaw acceleration
x_6	δ_{ay}	Offset in lateral accelromter
x_7	δ_{az}	Offset in vertical accelromter
x_8	$\delta_{\dot{\varphi}}$	Offset in roll gyro
y_1	a_y	Lateral acceleration in body system
y_2	a_z	Vertical acceleration in body system
y_3	$\dot{\varphi}$	Roll rate
y_4	$\dot{\varphi} + \delta_{\dot{\varphi}}$	Virtual measurement for constraint
u	v_x	Longitudinal speed

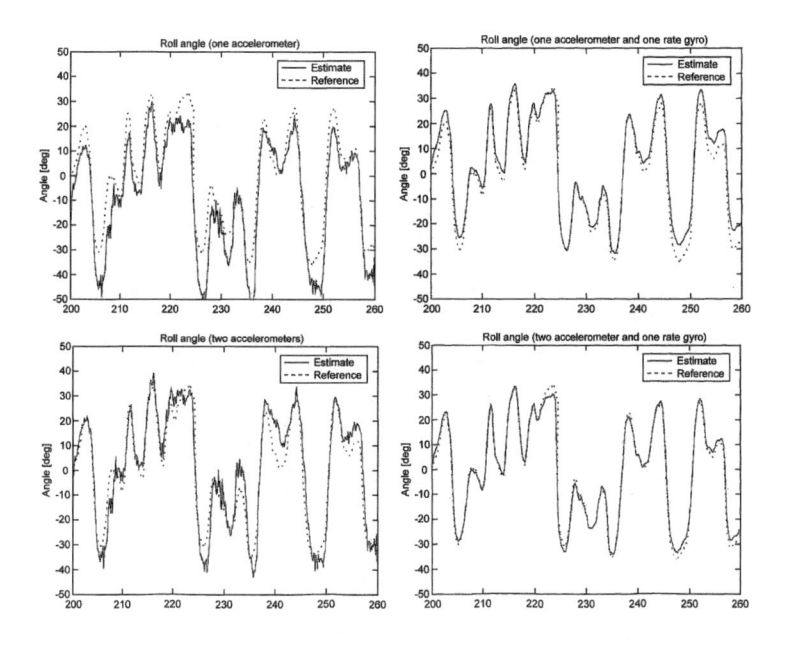

Figure 16.9: *Roll angle estimates for different sensor configurations: top left: a_z, bottom left: a_z, a_y, top right: $\dot{\varphi}, a_z$, bottom right: $\dot{\varphi}, a_z, a_x$. The states and measurements are subsets of the variables in Table 16.2 and sensor models in (16.5).*

16.3 Particle Filter Positioning Applications

This section is concerned with four positioning applications of underwater vessels, surface ships, wheeled vehicles (cars), and aircraft, respectively. Though these applications are at first glance quite different, almost the same particle filter can be used in all of them. In fact, successful applications of the PF are described in literature which are all based on the same state-space model and similar measurement equations.

16.3.1 Model Framework

The positioning applications are all based on the model Gustafsson et al. (2002)

$$x_k = (X_k, Y_k, \psi_k)^T, \tag{16.6a}$$
$$u_k = (V_k, \dot{\psi}_k)^T, \tag{16.6b}$$
$$X_{k+1} = X_k + TV_k \cos(\psi_k), \tag{16.6c}$$
$$Y_{k+1} = Y_k + TV_k \sin(\psi_k), \tag{16.6d}$$
$$\psi_{k+1} = \psi_k + T\dot{\psi}_k, \tag{16.6e}$$
$$y_k = h(x_k) + e_k. \tag{16.6f}$$

Here, X_k, Y_k denote the Cartesian position, ψ_k the course or heading, T is the sampling interval, V_k is the speed and $\dot{\psi}_k$ the yaw rate. The inertial signals V_k and $\dot{\psi}_k$ are considered as inputs to the dynamic model, and are given by on-board sensors. These are different in each of the four applications, and they will be described in more detail in the subsequent sections. The measurement relation is based on a distance measuring equipment (DME) and a geographical information system (GIS). Both the DME and the GIS are different in the four applications, but the measurement principle is the same. By comparing the measured distance to objects in the GIS, a likelihood for each particle can be computed.

It is important to make clear that neither an EKF, UKF nor KF bank is suited for such problems. The reason is that it is typically not possible to linearize the database other than in a very small neighborhood. Put in other words, the posterior is not well approximated by a Gaussian density, and it is further in many cases multi-modal.

In common for the applications is that they do not rely on satellite navigation systems, which are assumed unavailable or to provide insufficient navigation integrity. First, the inertial inputs, DME and GIS for the four applications are described. Conclusions concering the PF from these applications practice are summarized in Section 16.3.7. Different ways to augment the state vector are described for each application in Section 16.3.6. The point is that the dimension of the state vector has to be increased in order to account

Table 16.3: *Model summary for the UW application.*

Model	Description
State dynamics $f(x)$	(16.6)
State noise $p_v(v)$	Gaussian $\mathcal{N}(0,Q)$
Observation $h(x)$	Interpolation of an unstructured table $h(X_i, Y_i)$, obtained by separate mapping experiments
Measurement noise $p_e(e)$	Gaussian $\mathcal{N}(0,R)$, or Gaussian mixture with one out-lier mode
Extra states in MPF	Z, φ, θ, v (3D) and ω (3D)

for model errors and more complicated dynamics. This implies that the PF is simply not applicable, due to the high dimensional state vector.

The outline follows a bottom–up approach, starting with underwater vessels below sea level and ending with fighter aircraft in the air.

16.3.2 Underwater Positioning using a Topographic Map

The goal is to compute the position of an underwater (UW) vessel. A sonar is measuring the distance d_1 to the sea floor. The depth of the platform itself d_2 can be computed from pressure sensors, or from a sonar directed up-wards. By adding these distances, the sea depth at the position (X_k, Y_k) is measured. This can be compared to the depth in a dedicated sea chart with detailed topographical information, and the likelihood takes the combined effect of errors in the two sensors and the map into account, see Karlsson and Gustafsson (2006). Figure 16.10 provides an illustration, and Table 16.3 summarizes the application specific models.

The speed V_k and yaw rate $\dot{\psi}_k$ in (16.6) are computed using simplified dynamic motion models based on the propeller speed and the rudder angle. It is important to note that since the PF does not rely on pure dead-reckoning, such models do not have to be very accurate, see Fauske et al. (2007) for one simple linear model. An alternative is to use inertial measurement units (IMU) for measuring and computing speed and yaw rate.

Detailed seabed charts are so far proprietary military information, and most applications are also military. As an example of civilian use, oil companies are starting to use unmanned UW vessels for exploring the sea and oil platforms, and in this way building up their own maps.

16.3.3 Surface Positioning using a Sea Chart

The same principle as in the previous section can of course also be used for surface ships, which are constrained to be on the sea level ($d_2 = 0$). However,

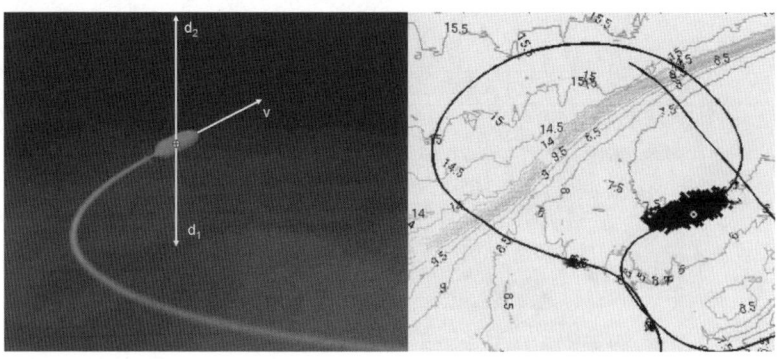

16.3 Particle Filter Positioning Applications

(a) Depth measurement principle   (b) Posterior snapshot from PF

Figure 16.10: (a) Illustration of an UW vessel measuring distance d_1 to sea bottom, and absolute depth d_2. (b) The sum $d = d_1 + d_2$ is compared to a bottom map, illustrated with the contours in the plot. The particle cloud illustrates a snapshot of the PF from a known validation trajectory in a field trial, see Karlsson (2002).

Table 16.4: Model summary for the surface ship navigation application.

Model	Description
State dynamics $f(x)$	(16.6)
State noise $p_v(v)$	Gaussian $\mathcal{N}(0, Q)$
Observation $h(x)$	Ray casting using a ray from position (X, Y) with angle θ to the closest shore in the sea chart
Measurement noise $p_e(e)$	Gaussian $\mathcal{N}(0, R)$, or an asymmetric distribution with heavy right tail
Extra states in MPF	Δv, $\Delta \psi$, $\Delta \theta$

vectorized sea charts (for instance the S-57 standard) contain a commercially available world-wide map.

The idea is to use the radar as DME and compare the detections with the shore profile, which is known from the sea chart conditioned on the position (X_k, Y_k) and course ψ_k (indeed the ship orientation, but more on this later), see Karlsson and Gustafsson (2006). The likelihood function models the radar error, but must also take clutter (false detections) and other ships into account. Table 16.4 summarizes the application specific models.

Figure 16.11(a) illustrates the measurements provided by the radar, while Figure 16.11(b) shows the radar detections from one complete revolution overlayed on the sea chart. The inertial data can be computed from propeller speed

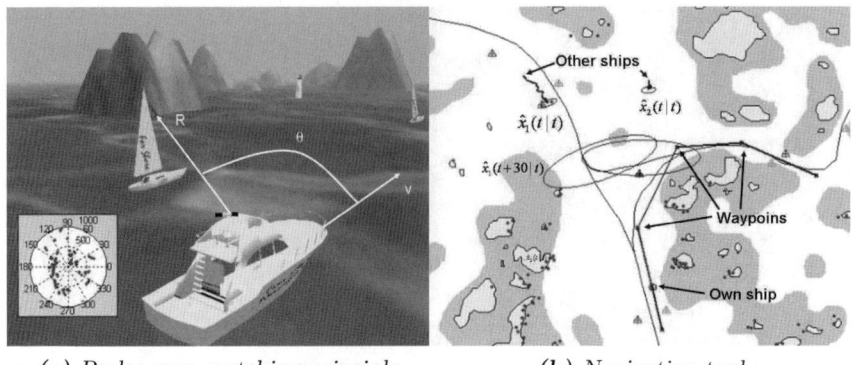

(a) Radar map matching principle   **(b)** Navigation tool

Figure 16.11: (a) The rotating radar returns detections of range R at body angle θ. The result of one radar revolution is conventionally displayed in polar coordinates as illustrated. (b) Comparing the (R, θ) detections to a sea chart, the position and course are estimated by the PF. When correctly estimated, the radar overlay principle can be used for visual validation as also illustrated in the sea chart. The PF has to distinguish radar reflections from shore with clutter and other ships, see Dahlin and Mahl (2007). The latter can be used for conventional target tracking algorithms, and collision avoidance algorithms, as also illustrated to the right, see Rönnebjerg (2005).

and rudder angle using simplified dynamical models as above.

American and European maritime authorities have recently published reports highlighting the need for a backup and support system to satellite navigation to increase integrity. The reason is accidents and incidents caused by technical problems with the satellite navigation system, and the risk of accidental or deliberate jamming. The LORAN standard offers one such supporting technique based on triangulation to radio beacons, see Lo et al. (2007). The PF solution here is a promising candidate, since it is in contrast to LORAN not sensitive to jamming nor does it require any infrastructure.

16.3.4 Vehicle Positioning using a Road Map

The goal here is to position a car relative to a road map by comparing the driven trajectory to the road network. The speed V_k and yaw rate $\dot{\psi}_k$ in (16.6) are computed from the angular velocities of the non-driven wheels on one axle, using rather simple geometrical relations. Dead-reckoning (16.6) provides a profile that is to be fitted to the road network. Table 16.5 summarizes the application specific models.

16.3 Particle Filter Positioning Applications

Table 16.5: *Model summary for the car navigation application (one option, there are other ones for handling the road constraints).*

Model	Description
State dynamics $f(x)$	(16.6)
State noise $p_v(v)$	Gaussian $\mathcal{N}(0, Q)$
Observation $0 = h(x) + e$	Binary road map, $h(x) = 1$ for being on the road, and $h(x) = 0$ otherwise
Measurement noise $p_e(e)$	Binary, with high probability for $e = -1$ (on-road) and low probability $e = 0$ (off-road)
Extra states in MPF	Gyro offset b_k and wheel radii offset $r_{k,3}/r_{k,4}$

The measurement relation is in its simplest form a binary likelihood which is zero for all positions outside the roads, and a non-zero constant otherwise. In this case, the DME is basically the prior that the vehicle is located on a road, and not a conventional physical sensor. See Forssell et al. (2002); Gustafsson et al. (2002) for more details, and Figure 16.12 for an illustration. More sophisticated applications use vibrations in wheel speeds and vehicle body as a DME. When a rough surface is detected, this DME can increase the likelihood for being outside the road. Likewise, if a forward-looking camera is present in the vehicle, this can be used to compute the likelihood that the front view resembles a road, or if it is rather a non-mapped parking area or smaller private road.

The system is suitable as a support to satellite navigation in urban environments, in parking garages or tunnels or whenever satellite signals are likely to be obstructed. It is also a stand-alone solution to the navigation problem. Road databases covering complete continents are available from two main vendors (NAVTEQ and TeleAtlas).

16.3.5 Aircraft Positioning using a Topographic Map

The principal approach here is quite similar to the underwater positioning application, and extends the one-dimensional example in Section 9.9 to two dimensions. Table 16.6 summarizes the application specific models.

A high-end IMU is used in an inertial navigation system (INS) which dead-reckons the sensor data to speed V_k and yaw rate $\dot{\psi}_k$ in (16.6) with quite high accuracy. Still, absolute position support is needed to prevent long-term drifts.

The DME is a wide-lobe downward looking radar that measures the distance to the ground. The absolute altitude is computed using the INS and a supporting barometric pressure sensor. Figure 16.13 shows one example just before convergence to a unimodal filtering density.

(a) Example of multi-modal posterior
(b) Snapshot from PF-based navigator

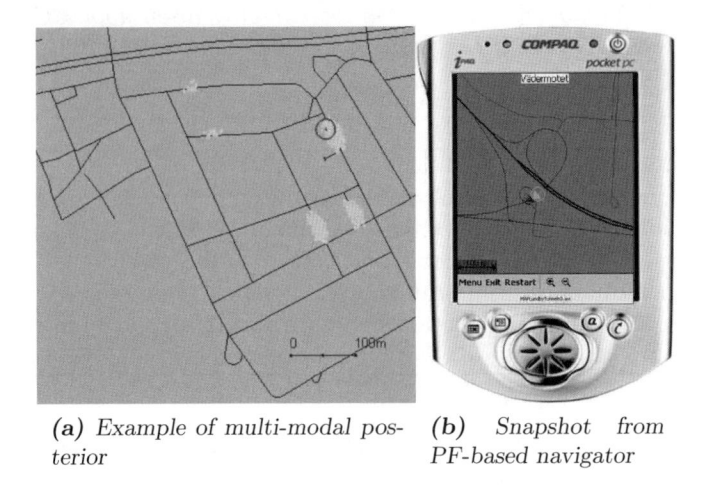

Figure 16.12: (a) Example of multimodal posterior represented by a number of distinct particle clouds from NIRA Dynamics navigation system. This is caused by the regular road pattern and will be resolved after a sufficiently long sequence of turns. (b) PF in an embedded navigation solution runs in real-time on a pocket PC with a serial interface to the vehicle CAN data bus, see Hall (2001).

Commercial databases of topographic information are available on land (but not below sea level), with a resolution of 50–200 meters.

16.3.6 Marginalized Particle Filter Applications

This section continues the applications in Section 16.3 with extended motion models where the marginalized particle filter (MPF) has been applied.

Underwater Positioning

Navigating an unmanned or manned UW vessel requires knowledge of the full three-dimensional position and orientation, not only the projection in a horizontal plane. That is, at least six states are needed. For control, also the velocity and angular velocities are needed, which directly implies at least a twelve dimensional state vector. The PF cannot be assumed to perform well in such cases, and MPF is a promising approach Karlsson and Gustafsson (2006).

Table 16.6: Model summary for the aircraft navigation application.

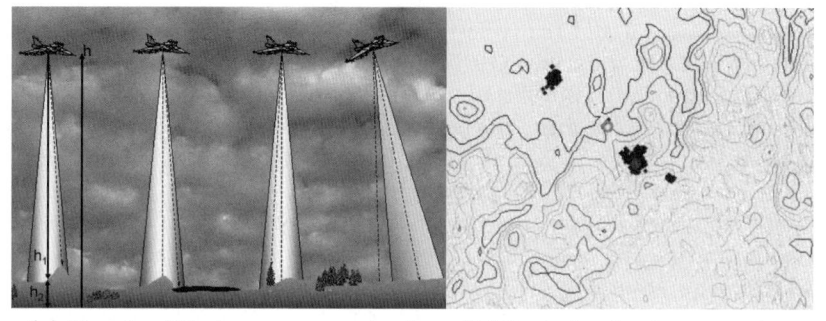

Model	Description
State dynamics $f(x)$	(16.6) or full navigation model
State noise $p_v(v)$	Gaussian $\mathcal{N}(0,Q)$
Observation $h(x)$	Interpolated terrain altitude using a database with a 50 m grid resolution.
Measurement noise $p_e(e)$	Gaussian mixture $\alpha_1 \mathcal{N}(0, R_1) + \alpha_2 \mathcal{N}(\mu_2, R_2)$ (the second one is the 'tree-top distribution')
Extra states in MPF	Full navigation state except for the PF states X, Y, ψ

(a) Terrain altitude measurement principle

(b) Posterior density approximation from PF

Figure 16.13: (a) Illustration of an aircraft measuring distance h_1 to ground. The on-board baro-altitude supported INS system provides absolute altitude over sea level h, and the difference $h_2 = h - h_1$ is compared to a topographical map. (b) Snapshot of the PF particle cloud, just after the aircraft has left the sea in the upper left corner. There are three distinct modes, where the one corresponding to the correct position dominates.

Surface Positioning

There are two bottlenecks in the surface positioning PF that can be mitigated using the MPF. Both relates to the inertial measurements. First, the speed sensed by the log is the speed in water, not the speed over ground. Hence, the local water current, Δv, is a parameter to include in the state vector. Second, the radar is strap-down and measures relative to body orientation, which is not the same as the course ψ_k. The difference is the so called crab angle, $\Delta \psi$, which depends on currents and wind. This can also be included in the state vector. Further, there is in the system described in Dahlin and Mahl (2007) an unknown and time-varying offset in the reported radar angle, $\Delta\theta$, which has to be compensated for. All these parameters enter linearly in the model and can with great computational benefits be marginalized.

Vehicle Positioning

The bottleneck of the first generation of vehicle positioning PF is the assumption that the vehicle must be located on a road. As previously hinted, one could use a small probability in the likelihood function for being off-road, but there is no real benefit for this without an accurate dead-reckoning ability, so re-occurrence on the road network can be predicted with high reliability.

The speed and yaw rate computed from the wheel angular velocity are limited by the insufficient knowledge of wheel radii. However, the deviation between actual and real wheel radii on the two wheels on one axle can be included in the state vector. Similarly, with a yaw rate sensor available (standard component in electronic stability programs (ESP) and navigation systems), the yaw rate drift can be included in the state vector. These two offsets are the same as b_k and $r_{k,3}/r_{k,4}$ in (16.4) for the yaw rate estimation application in Section 16.2.2. Note that the sensor model (16.3) can quite accurately be linearized in these two offsets and an approximative marginalization procedure can be applied. From an observability point of view, these parameters are accurately estimated when the vehicle is on the road. Once the offsets are learned, improved dead-reckoning can be achieved during off-road driving. Tests in demonstrator vehicles have shown that the exit point from parking garages and parking areas are well estimated, and that shorter unmapped roads are not a problem, see Figure 16.14.

Aircraft Positioning

The primary role of the terrain based navigation (TERNAV) module is to support the INS with absolute position information. The INS consists of an extended Kalman filter based on a state vector with over 20 motion states and sensor bias parameters. The current bottleneck is the interface between TERNAV and INS. The reason is that TERNAV outputs a possibly multimodal position density, while the INS EKF expects a Gaussian observation. The natural idea is to integrate both TERNAV and INS into one filter. This gives a high-dimensional state vector, where one measurement (radar altitude) is very nonlinear. The MPF handles this elegantly, by essentially keeping the EKF from the existing INS and using the PF only for the radar altitude measurement.

The altitude radar gives a measurement outlier when the radar pulse is reflected in trees. Tests have validated that a Gaussian mixture where one mode has a positive mean models the real measurement error quite well. This

16.3 Particle Filter Positioning Applications

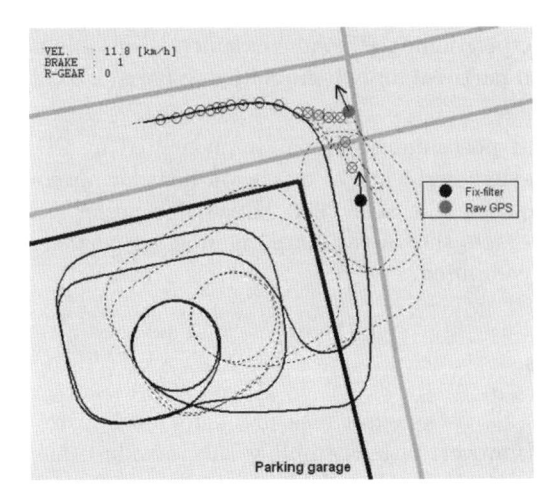

Figure 16.14: Navigation of a car in a parking garage. Results for MPF when relative wheel radii and gyro offset are added to the state vector. The two trajectories correspond to the map-aided system and an EKF with the same state vector, but where GPS is used as position sensor. Since the GPS gets several drop-outs before the parking garage, the dead-reckoning trajectory is incorrect, see Kronander (2003).

Gaussian mixture distribution can be used in the likelihood computation, but such a distribution is in this case logically modeled by a binary Markov parameter, which is one in positions over forest and zero otherwise. In this way, the positive correlation between outliers is modeled, and a prior from ground type information in the GIS can be incorporated. This example motivates the inclusion of discrete states in the model framework. See Schön et al. (2005) and Nordlund and Gustafsson (2009) for the details.

16.3.7 Conclusions

This section summarizes practical experience from the applications in Sections 16.3, in particular Section 16.3.6, with respect to the theoretical survey in Sections 9.2 and 9.8.

Real-Time Issues

The PF has been applied to real data and implemented on hardware targeted for the application platforms. The sampling rate has been chosen in the order 1–2 Hz, and there is no problem to achieve real-time performance in any of the applications. Some remarkable cases:

- The vehicle positioning PF was implemented on a hand-held computer using 15000 particles already in 2001, see Forssell et al. (2002).

- The aircraft positioning PF was implemented in ADA and shown to satisfy real-time performance on the on-board computer in the Swedish fighter Gripen in the year 2000. Real-time performance was reached, despite the facts that a very large number of particles were used on a rather old computer.

Sampling Rates

The DME can in all cases deliver measurements much faster than the chosen sampling rate. However, faster sampling will introduce an unwanted correlation in the observations. This is due to the fact that the databases are quantized, so the platform should make a significant move between two measurement updates.

Implementation

Implementing and debugging the PF has not been a major issue. On the contrary, students and non-experts have faced less problems with the PF than for similar projects involving the EKF. In many cases, they obtained deep intuition for including non-trivial but *ad-hoc* modifications. There are today several hardware solutions reported in literature, where the parallel structure of the PF algorithms can be utilized efficiently. For instance, an FPGA implementation is reported in Athalye (2007), and on a general purpose graphics processing unit (GPGPU) in Hendeby et al. (2007a). Analog hardware can further be used to speed up function evaluations Velmurugan et al. (2006).

Dithering

Both the process noise and measurement noise distributions need some dithering (increased covariance). Dithering the process noise is a well-known method to mitigate the sample depletion problem, see Gordon et al. (1993). Dithering the measurement noise is a good way to mitigate the effects of outliers and to robustify the PF in general. One simple and still very effective method to mitigate sample depletion is to introduce a lower bound on the likelihood. This lower bound was first introduced more or less *ad hoc*. However, recently this algorithm modification has been justified more rigorously. In proving that the particle filter converges for unbounded functions, like the state x_k itself, it is sufficient to have a lower bound on the likelihood, see Hu et al. (2008) for details.

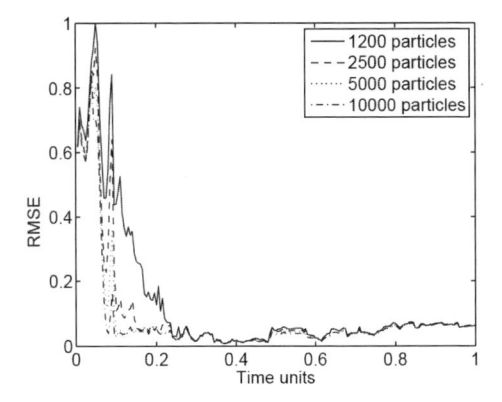

Figure 16.15: RMSE performance for aircraft terrain navigation as a function of the number of particles.

Number of Particles

The number of particles is chosen quite large to achieve good transient behaviour in the start up phase and to increase robustness. However, it has been concluded that in the normal operational mode the number of particles can be decreased substantially (typically a factor of ten). Figure 16.15 shows experimental results for the terrain navigation application. The transient improves when going from $N = 1200$ to $N = 2500$, but using more particles gives no noticable improvement after convergence.

A real-time implementation should be designed for the worst case. However, using an adaptive sampling interval T and number of particles N is one option. The idea is to use a longer sampling interval and more particles initially, and when the PF has converged to a few distinct modes, T and N can be decreased in such a way that the complexity N/T is constant.

Choosing the Proposal Density

The standard sampling importance resampling (SIR) PF works fine for an initial design. However, the maps contain rather detailed information about position, and can in the limit be considered as state constraints. In such high signal-to-noise applications, the standard proposal density used in the SIR PF is not particularly efficient. An alternative, that typically improves the performance, is to use the information available in the next measurement already in the state prediction step. Note that the proposal in its most general form includes the next observation. Consider for instance positioning based on road maps. In standard SIR PF, the next positions are randomized around the predicted position according to the state noise, which is required to obtain diversity. Almost all of these new particles are outside the road network, and will not survive the resampling step. Obviously this is a waste of particles. By looking in the map how the roads are located locally around the predicted position, a much more clever process noise can be computed, and the particles explore the road network much more efficiently.

Divergence Monitoring

Divergence monitoring is fundamental for real-time implementations to achieve the required level of integrity. After divergence, the particles do not reflect the true state distribution and there is no mechanism that automatically stabilizes the particle filter. Hence, divergence monitoring has to be performed in parallel with the actual PF code, and when divergence is detected, the PF is re-initialized.

One indicator of particle depletion is the effective number of samples N_{eff}, used in the PF. This number monitors the amount of particles that significantly contribute to the posterior, and it is computed from the normalized weights. However, the un-normalized likelihoods are a more logical choice for monitoring. Standard hypothesis tests can be applied for testing whether the particle predictions represent the likelihood distribution.

Another approach is to use parallel particle filters interleaved in time. The requirement is that the sensors are faster than the chosen sampling rate in the PF. The PF's then use different time delays in the sensor observations.

The re-initialization procedure issued when divergence is detected is quite application dependent. The general idea is to use a very diffuse prior, or to infer external information. For the vehicle positioning application in Forssell et al. (2002), a cellular phone operator took part in the demonstrator, and cell information was used as a new prior for the PF in case of occasional divergence.

Performance Bounds

For all four GPS-free applications the positioning performance is in the order of ten meter root mean square error (RMSE), which is comparable to GPS performance. Further, the performance of the PF has been shown to be close to the Cramér-Rao lower bound (CRLB) for a variety of examined trajectories. In Figure 16.16 two examples of performance evaluations in terms of the RMSE are depicted. Figure 16.16(a) shows the position RMSE and CRLB for the UW application and Figure 16.16(b) illustrates the horizontal position error for the aircraft application.

Particle Filter in Embedded Systems

The primary application is to output position information to the operator. However, in all cases there have been decision and control applications built on the position information, which indicates that the PF is a powerful software component in embedded systems:

16.3 Particle Filter Positioning Applications

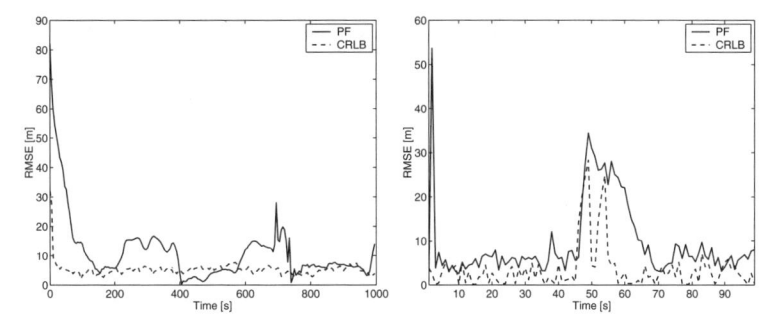

Figure 16.16: *The position RMSE for the UW (left) and surface (right) applications respectively, compared to the CRLB.*

- UW positioning: Here, the entire mission relies on the position, so path planning and trajectory control are based on the output from the PF. Note that there is hardly any alternative below sea level, where no satellites are reachable, and deploying infrastructure (sonar buoys) is quite expensive.

- Surface positioning: Differentiating radar detections from shore, clutter and other ships is an essential association task in the PF. It is a natural extension to integrate a collision avoidance system in such an application, as illustrated in a sea chart snapshot in Figure 16.11.

- Vehicle positioning: The PF position was also used in a complete voice controlled navigation system with dynamic route optimization, see Figure 16.12.

- Aircraft navigation: The position from the PF is primarily used as a supporting sensor in the INS, whose position is a refined version of the PF output.

Marginalized Particle Filtering

Finally, the marginalized particle filter offers a scalable extension of the PF in all applications surveyed here and many others. MPF is applicable for instance in the following localization, navigation and tracking problems:

- Three-dimensional position spaces.

- Motion models with velocity and acceleration states.

- Augmenting the state vector with unknown nuisance parameters as sensor offsets and drifts.

State of the art is the FastSLAM algorithm, see Montemerlo et al. (2002), that applies MPF to the Simultaneous Localization and Mapping (SLAM) problem. FastSLAM has been applied to applications where thousands of two-dimensional landmark features are marginalized out from a three dimensional motion state. Further, in Karlsson et al. (2008) a double marginalization process was employed to handle hundreds of landmark features and a 24-dimensional state vector for three-dimensional navigation of an unmanned aerial vehicle in an unknown environment.

Appendices

APPENDIX

A

Statistics Theory

This chapter summarizes some properties of selected statistical distributions. The focus is on properties that are easy to define and tabulate, and do not require any deeper understanding of the applications. These properties are then applied to estimation problems in Chapter C.

This chapter contains:

- Tables for some selected distributions in the exponential family.

- A list of conjugate priors for likelihood functions in the exponential family.

- Methods to approximate the distribution of nonlinear transformations of stochastic variables.

A.1   Selected Distributions

The tables in this section list properties of some members in the *exponential family*, where the PDF can be written as:

$$p(y|\theta) = a(\theta)b(y)e^{-0.5\,\mathrm{tr}(g^T(\theta)h(y))}, \quad y \in \Omega_y. \tag{A.1}$$

Besides the PDF, Tables A.1–A.4 give the mean

$$\mathrm{E}(y) = \int_{y \in \Omega_y} y p(y|\theta) dy, \tag{A.2}$$

455

Table A.1: Scalar normal distribution $\mathcal{N}(\mu, \sigma^2)$.

$$
\begin{aligned}
p(y) &= \frac{1}{\sqrt{2\pi\sigma^2}} e^{-\frac{(y-\mu)^2}{2\sigma^2}} \\
\mathrm{E}(y) &= \mu \\
\mathrm{Var}(y) &= \sigma^2 \\
\mathrm{E}(y^n) &= 1 \cdot 3 \cdot 5 \ldots (n-1)\sigma^n \\
Q(y) &\approx 1 - \frac{1}{\sqrt{2\pi}(y-\mu)/\sigma} e^{-\frac{(y-\mu)^2}{2\sigma^2}} \\
a(\theta) &= \frac{1}{\sigma} e^{-\frac{\mu^2}{2\sigma^2}} \\
g(\theta) &= \frac{1}{\sigma^2}(1, -2\mu)^T \\
b(y) &= \frac{1}{\sqrt{2\pi}} \\
h(y) &= (y, y^2)^T
\end{aligned}
$$

Table A.2: Exponential distribution $\mathrm{Exp}(\mu)$

$$
\begin{aligned}
p(y) &= \tfrac{1}{\mu} e^{-\frac{y}{\mu}},\ y \geq 0 \\
a(\theta) &= \tfrac{1}{\mu},\quad g(\theta) = \tfrac{-2}{\mu},\quad h(y) = y,\quad b(y) = 1, \\
\mathrm{E}(y) &= \mu \\
\mathrm{Var}(y) &= \mu^2 \\
Q(y) &= 1 - e^{-\frac{y}{\mu}}
\end{aligned}
$$

variance (or covariance in the multivariate case)

$$\mathrm{Cov}(y) = \int_{y \in \Omega_y} (y - \mathrm{E}(y))(y - \mathrm{E}(y))^T p(y|\theta) dy, \tag{A.3}$$

and for univariate distributions also the cumulative distribution

$$Q(y) = \int_{-\infty}^{y} p(z|\theta) dz. \tag{A.4}$$

Table A.3: Chi-square distribution χ_ν^2

$$
\begin{aligned}
p(y) &= \tfrac{1}{2^{\nu/2}\Gamma(\nu/2)} y^{-\nu/2-1} e^{-y/2},\ y > 0 \\
\Gamma(u) &= \int_0^\infty t^{u-1} e^{-t} dt \\
a(\theta) &= \tfrac{1}{2^{\nu/2}\Gamma(\nu/2)},\quad g(\theta) = \nu + 2,\quad h(y) = \log(y),\quad b(y) = e^{-y/2}, \\
\mathrm{E}(y) &= \nu \\
\mathrm{Var}(y) &= 2\nu \\
Q(y) &= 1 - e^{-y/2} \sum_{k=0}^{\nu/2-1} \tfrac{(y/2)^k}{k!},\ \nu \geq 2 \\
\text{Comments:} &\quad \chi_2^2 = \mathrm{Exp}(2).\ \text{If } y = \sum_{i=1}^\nu y_i^2,\ y_i \in \mathcal{N}(0,1),\ \text{then } y \in \chi_\nu^2 \\
&\quad \Gamma(n) = (n-1)!,\ \Gamma(u) = (u-1)\Gamma(u-1),\ \Gamma(0.5) = \sqrt{\pi}
\end{aligned}
$$

Table A.4: Non-central chi-square distribution $\chi^2_\nu(\lambda)$

$p(y) =$	$\frac{1}{2}\left(\frac{y}{\lambda}\right)^{\frac{\nu-2}{4}} e^{-(y+\lambda)/2} I_{\nu/2-1}(\sqrt{\lambda y}), y > 0$
$I_r(u) =$	$\sum_{k=0}^{\infty} \frac{(u/2)^{2k+r}}{k!\Gamma(r+k+1)}$ Bessel function
$E(y) =$	$\nu + \lambda$
$\text{Var}(y) =$	$2\nu + 4\lambda$
Comments:	If $y = \sum_{i=1}^{\nu} y_i^2, y_i \in \mathcal{N}(\mu_i, 1)$, then $y \in \chi^2_\nu(\sum_i \mu_i^2)$

Table A.5: Gamma distribution $\text{GAMMA}(k, \lambda)$

$p(y) =$	$\frac{1}{\Gamma(k)\lambda^k} y^{\lambda-1} e^{-\frac{y}{\lambda}}, y > 0$
$a(\theta) =$	$\frac{1}{\Gamma(k)\lambda^k}$, $g(\theta) = \left(\frac{2}{\lambda}, -2\lambda\right)^T$ $h(y) = (y, , \log(y))^T$, $b(y) = \frac{1}{y}$,
$E(y) =$	$k\lambda$
$\text{Var}(y) =$	$k\lambda^2$
$\frac{E(y^3)}{E(y^2)^{3/2}} =$	$\frac{2}{\sqrt{k}}$
$\frac{E(y^4)}{E(y^2)^2} =$	$\frac{6}{k}$
$Q(y) =$	$\frac{\gamma(k, x/\lambda)}{\Gamma(k)}$
Comment 1:	$k > 0$ shape, $\lambda > 0$ scale,
Comment 2:	Γ =gamma and γ =gammainc in MATLAB™
Comment 3:	If $y \sim \text{GAMMA}(k, \lambda)$, then $cy \sim \text{GAMMA}(k, c\lambda)$
Comment 4:	If $y_i \sim \text{GAMMA}(k_i, \lambda)$, then $\sum_i y_i \sim \text{GAMMA}(\sum_i k_i, \lambda)$
Comment 5:	$\text{GAMMA}(1,1) = \text{EXP}(1)$, $\text{GAMMA}(1, \lambda) = \text{EXP}(\lambda)$

A.2 Conjugate Priors

Table A.6 lists combinations of likelihoods and *conjugate priors* from the exponential family, starting with the notational and symbolical definitions of the *exponential family*, respectively. The idea is that the posterior gets the same distribution as the prior distribution when applying Bayes law as

$$p(x|y) = \frac{p(y|x)p(x)}{p(y)}. \tag{A.5}$$

Only the scale factor $p(y)$ is indetermined. See also Section C.8, and the text book Gelman et al. (2004).

A.3 Nonlinear Transformations

This section summarizes different methods to approximate the distribution of a nonlinear mapping $z = g(x)$ of a Gaussian variable x with a Gaussian distribution,

$$x \sim \mathcal{N}(\mu_x, P_x) \to z \sim \mathcal{N}(\mu_z, P_z). \tag{A.6}$$

Table A.6: *Examples of members in the exponential family: Normal inverse Gamma (NIG), Gamma (GAMMA), Poisson (PO), Bernoulli (BER), Pareto, Beta, multi-nomial (MUNOM), and Dirichlet (DIR). $\bar{e}_k$ denotes the i'th unit vector.*

Likelihood $p(y_k\|\theta)$	θ	Prior $p(\theta\|y_{1:k-1})$	Posterior $p(\theta\|y_{1:k})$
$a(x)b(y)e^{-0.5\operatorname{tr}(g^T(x)h(y))}$	x	$a(x)^\nu e^{-0.5 g^T(x)v}$	$a(x)^{\nu+1} e^{-0.5 g^T(x)(v+h(y))}$
$\sim \mathrm{EF}(1, h(y_k))$	x	$\mathrm{EF}(\nu_{k-1}, v_{k-1})$	$\mathrm{EF}(\nu_{k-1}+1, v_{k-1}+h(y_k))$
$\mathcal{N}(y_k; x, R)$	x	$\mathcal{N}(\nu_{k-1}, v_{k-1})$	$\mathcal{N}(\nu_{k-1}+1, v_{k-1}+y_k)$
$\mathcal{N}(y_k; H_k x, R)$	x	$\mathcal{N}(\nu_{k-1}, v_{k-1})$	$\mathcal{N}\left(\nu_{k-1}+1, v_{k-1}+\begin{pmatrix} H_k^T R_k^{-1} y_k \\ H_k^T R_k^{-1} H_k \end{pmatrix} \begin{pmatrix} y_k y_k^T & y_k \\ y_k^T & 1 \end{pmatrix}\right)$
$\mathcal{N}(y_k; x, rI)$	x, r	$\mathrm{NIG}(\nu_{k-1}, v_{k-1})$	$\mathrm{NIG}\left(\nu_{k-1}+1, v_{k-1}+\begin{pmatrix} y_k y_k^T & y_k \\ y_k^T & 1 \end{pmatrix}\right)$
$\mathrm{EXP}(y_k; x)$	x	$\mathrm{GAMMA}(a_{k-1}, b_{k-1})$	$\mathrm{GAMMA}(a_{k-1}+1, b_{k-1}+y_k)$
$\mathrm{GAMMA}(y_k; a, b)$	b	$\mathrm{GAMMA}(a_{k-1}, b_{k-1})$	$\mathrm{GAMMA}(a_{k-1}+a, b_{k-1}+y_k)$
$\mathrm{GAMMA}(y_k; a, b)$	a, b	$F(p,q,r,s) = \frac{1}{Z} \frac{p^{a-1} e^{-bq}}{\Gamma(a) b^{-as}}$	$F(py_k, q+y_k, r+1, s+1)$
$\mathcal{U}(y_k; 0, x)$	x	$\mathrm{PARETO}(x_{k-1}^o, k_{k-1})$	$\mathrm{PARETO}(\max(x_{k-1}^o, y_k), k_{k-1}+1)$
$\mathrm{BER}(y_k; p)$	p	$\mathrm{BETA}(\alpha_{k-1}, \beta_{k-1})$	$\mathrm{BETA}(\alpha_{k-1}+y_k, \beta_{k-1}+1-y_k)$
$\mathrm{PO}(y_k; \lambda)$	λ	$\mathrm{GAMMA}(\alpha_{k-1}, \beta_{k-1})$	$\mathrm{GAMMA}(\alpha_{k-1}+y_k, \beta_{k-1}+1)$
$\mathrm{MUNOM}(y_k; p)$	p	$\mathrm{DIR}(\alpha_{k-1})$	$\mathrm{DIR}(\alpha_{k-1}+\bar{e}_{y_k})$

A.3 Nonlinear Transformations

The following subsections describe different approaches to compute μ_z and P_z. The underlying idea is that it is often easier to approximate a distribution than a general nonlinear function, so the analytic distribution of $g(x)$ is never even considered here.

A.3.1 Taylor Expansion

Consider a general nonlinear transformation and its second order Taylor expansion

$$z = g(x) = g(\mu_x) + g'(\mu_x)(x - \mu_x) + \underbrace{\left[\tfrac{1}{2}(x - \mu_x)^T g_i''(\xi)(x - \mu_x)\right]_i}_{r\left(x;\mu_x,g''(\xi)\right)}, \quad (A.7)$$

where n_x is the dimension of the vector x, $x \in \mathbb{R}^{n_x}$, and $z \in \mathbb{R}^{n_z}$. The notation $[v_i]_i$ is used to denote a vector in which element i is v_i. Analogously, the notation $[m_{ij}]_{ij}$ will be used to denote the matrix where the (i, j) element is m_{ij}. Theorem A.1 gives the theoretical mean and covariance of (A.7) when ξ is substituted with μ_x.

Theorem A.1 (First moments of the Taylor expansion)
Consider the mapping

$$z = g(\mu_x) + g'(\mu_x)(x - \mu_x) + \left[\tfrac{1}{2}(x - \mu_x)^T g_i''(\mu_x)(x - \mu_x)\right]_i, \quad (A.8a)$$

from $\mathbb{R}^{n_x}$ to $\mathbb{R}^{n_z}$. Let $\mathrm{E}(x) = \mu_x$ and $\mathrm{Cov}(x) = P$, then the first moment of z is given by

$$\mu_z = g(\mu_x) + \tfrac{1}{2}[\mathrm{tr}(g_i''(\mu_x)P)]_i. \quad (A.8b)$$

Further, let $x \sim \mathcal{N}(\mu_x, P)$, then the second moment of z is given by

$$P_z = g'(\mu_x) P \big(g'(\mu_x)\big)^T + \tfrac{1}{2}\left[\mathrm{tr}(g_i''(\mu_x) P g_j''(\mu_x) P)\right]_{ij}, \quad (A.8c)$$

with $i, j = 1, \ldots, n_z$.

The result is in for instance Bar-Shalom et al. (2001) given without proof, so we include it here for completeness and for preparing for Theorem A.2.

Proof: Suppose without loss of generality that $\hat{x} = \mathrm{E}[x] = 0$ and $\mathrm{Cov}(x) = P$. Further, to simplify notation, let $G = g''(\mu_x)$. Then, one direct way to express the expected value of the rest term is to use the trace linearity property and $\mathrm{tr}(AB) = \mathrm{tr}(BA)$,

$$\mathrm{E}[x^T G x] = \mathrm{E}[\mathrm{tr}(G x x^T)] = \mathrm{tr}(G \mathrm{E}[x x^T]) = \mathrm{tr}(GP). \quad (A.9)$$

The variance is more complicated to compute, and the Gaussian assumption is needed. Below, a derivation of both mean and variance is provided.

First, let $q = G^{1/2}x$, where $G = G^{T/2}G^{1/2}$, so that $\text{Cov}(q) = G^{1/2}PG^{T/2}$. Then,
$$x^T G x = q^T q. \tag{A.10}$$
The SVD $G^{1/2}PG^{T/2} = U\Sigma U^T$ gives a second transformation $v = U^T q$ which does not change the eigenvalues of the covariance, and in particular its trace is the same, since $\text{tr}(U\Sigma U^T) = \text{tr}(\Sigma)$. Thus (using $\text{E}(x^4) = 3\bigl(\text{E}(x^2)\bigr)^2 = 3\sigma^4$ for scalar zero mean Gaussian variables),

$$\text{E}[v^T v] = \sum_{i=1}^{n_x} \sigma_i^2, \tag{A.11a}$$

$$\text{E}[(v^T v)^2] = \sum_{i=1}^{n_x} 3\sigma_i^4 + \sum_{i \neq j} \sigma_i^2 \sigma_j^2, \tag{A.11b}$$

$$\text{Var}[v^T v] = \text{E}[(v^T v)^2] - \bigl(\text{E}[(v^T v)]\bigr)^2 = 2\sum_{i=1}^{n_x} \sigma_i^4. \tag{A.11c}$$

Now, the sum of the diagonal elements of a matrix Σ^2 can be expressed as the trace of the square of the matrix in the SVD, so

$$\text{Var}[v^T v] = 2\,\text{tr}(G^{1/2}PG^{T/2}G^{1/2}PG^{T/2}) = 2\,\text{tr}(GPGP). \tag{A.12}$$

Further, if the function g is vector-valued, the covariance between different rows can be derived in a similar way. Let $G_i = g_i''(\hat{x})$ be the Hessian of the i'th row of $g(x)$. Then the result is

$$\text{E}[x^T G_i x x^T G_j x] - \text{E}[x^T G_i x]\text{E}[x^T G_j x] = 2\,\text{tr}(G_i P G_j P). \tag{A.13}$$

In summary, the rest term for a vector valued function $g(z)$ has mean and covariance given by

$$\text{E}\bigl((x-\hat{x})^T g''(\mu_x)(x-\hat{x})\bigr) = [\text{tr}(g_i''(\mu_x)P)]_i, \tag{A.14a}$$
$$\text{Cov}\bigl((x-\hat{x})^T g''(\mu_x)(x-\hat{x})\bigr) = [2\,\text{tr}(g_i''(\mu_x)Pg_j''(\mu_x)P)]_{ij}. \tag{A.14b}$$

This concludes the proof. $\square$

The following remarks are important:

- For quadratic functions $g_i(x) = a_i + B_i x + x^T C_i x$, the Hessian $g_i''(x) = C_i$ is independent of x. That is, Theorem A.1 gives the correct first and second order moments.

- For polynomial functions $g(x)$, the principle of moment matching can be applied to compute μ_x and P_x analytically, see S.Saha et al. (2009). For a Gaussian x, all moments of z can be expressed as polynomial functions of μ_x and P_x.

A.3 Nonlinear Transformations

- The mean and covariance can also be derived from a linear regression formulation, where the sigma points become the regressors Lefebvre et al. (2002).

- The moment integrals

$$\mu_z = \int g(x)p(x)\,dx, \qquad (A.15a)$$

$$P_z = \int g(x)g^T(x)p(x)\,dx, \qquad (A.15b)$$

where $p(x)$ is the probability density function of x, can be approximated with numerical integration techniques. The Gauss-Hermite quadrature rule is examined in Arasaratnam et al. (2007), and the cubature rule is investigated in Arasaratnam et al. (2009). The latter reference gives a nice link to the unscented transform which we will come back to.

To summarize the theorem, the first order Taylor approximation (TT1) is

$$\text{TT1}: \ x \sim \mathcal{N}(\mu_x, P) \to z \sim \mathcal{N}\big(g(\mu_x), g'(\mu_x) P \big(g'(\mu_x)\big)^T\big), \qquad (A.16)$$

and the second order Taylor approximation (TT2) is

$$\text{TT2}: \ x \sim \mathcal{N}(\mu_x, P) \to z \sim \mathcal{N}\Big(g(\mu_x) + \tfrac{1}{2}[\text{tr}(g_i''(\mu_x)P)]_i,$$
$$g'(\mu_x) P \big(g'(\mu_x)\big)^T + \tfrac{1}{2}\Big[\text{tr}(Pg_i''(\mu_x)Pg_j''(\mu_x))\Big]_{ij}\Big). \qquad (A.17)$$

It is a trivial fact that the gradient and Hessian in TT1 and TT2, respectively, can both be computed using numerical methods. It is worth stressing that both $g_i'(x)$ and $g_i''(x)$ are in all illustrations computed using numerical methods. That is, only function evaluations of the nonlinear function $g(x)$ are assumed to be available.

A.3.2 Monte Carlo Transformation

The *Monte Carlo Transformation* (MCT) provides a general framework to compute an accurate approximation, which asymptotically should be the best possible one. The method is straightforward. First, generate a number N of random points $x^{(i)}$, let these pass the nonlinear function, and then estimate the mean and covariance as follows:

$$x^{(i)} \sim \mathcal{N}(\mu_x, P), \quad i = 1, \ldots, N, \tag{A.18a}$$

$$z^{(i)} = g(x^{(i)}), \tag{A.18b}$$

$$\mu_z = \frac{1}{N} \sum_{i=1}^{N} z^{(i)}, \tag{A.18c}$$

$$P_z = \frac{1}{N-1} \sum_{i=1}^{N} \left(z^{(i)} - \mu_z\right)\left(z^{(i)} - \mu_z\right)^T. \tag{A.18d}$$

A.3.3 Unscented Transform

The *unscented transform* (UT) is in a sense similar to the MCT approach in that it selects a number of points $x^{(i)}$, maps these to $z^{(i)} = g(x^{(i)})$, and then estimates the mean and covariance in the standard way. The difference lies in how the points $x^{(i)}$ are selected.

First define, u_i and σ_i from the *singular value decomposition* (SVD) of the covariance matrix P,

$$P = U \Sigma U^T = \sum_{i=1}^{n_x} \sigma_i^2 u_i u_i^T,$$

where $u_i = U_{:,i}$ is the i'th column of U and $\sigma_i^2 = \Sigma_{i,i}$ is the i'th diagonal element of Σ. Then, let

$$x^{(0)} = \mu_x, \qquad x^{(\pm i)} = \mu_x \pm \sqrt{n_x + \lambda} \sigma_i u_i, \tag{A.19a}$$

$$\omega^{(0)} = \frac{\lambda}{n_x + \lambda}, \qquad \omega^{(\pm i)} = \frac{1}{2(n_x + \lambda)}, \tag{A.19b}$$

where $i = 1, \ldots, n_x$. Let $z^{(i)} = g(x^{(i)})$, and apply

$$\mu_z = \sum_{i=-n_x}^{n_x} \omega^{(i)} z^{(i)}, \tag{A.20a}$$

$$P_z = \sum_{i=-n_x}^{n_x} \omega^{(i)} (z^{(i)} - \mu_z)(z^{(i)} - \mu_z)^T \tag{A.20b}$$

$$+ (1 - \alpha^2 + \beta)(z^{(0)} - \mu_z)(z^{(0)} - \mu_z)^T, \tag{A.20c}$$

where $\omega^{(0)} + (1 - \alpha^2 + \beta)$ is often denoted $\omega_c^{(0)}$ and used to make the notation more compact for the covariance matrix expression.

The design parameters of UT have here the same notation as in UKF literature (e.g., Wan and van der Merwe):

A.3 Nonlinear Transformations

Table A.7: Different versions of the UT (counting the CT as a UT version given appropriate parameter choice) in (A.19) using the definition $\lambda = \alpha^2(n_x + \kappa) - n_x$.

Parameter	UT1		UT2	CT	DFT
α	$\sqrt{3/n_x}$		10^{-3}	1	–
β	$3/n_x - 1$		2	0	–
κ	0		0	0	–
λ	$3 - n_x$		$10^{-6}n_x - n_x$	0	0
$\sqrt{n_x + \lambda}$	$\sqrt{3}$		$10^{-3}\sqrt{n_x}$	$\sqrt{n_x}$	$\frac{1}{a}\sqrt{n_x}$
$\omega^{(0)}$	$1 - n_x/3$		-10^6	0	0

- λ is defined by $\lambda = \alpha^2(n_x + \kappa) - n_x$.

- α controls the spread of the sigma points and is suggested to be approximately 10^{-3}.

- β compensates for the distribution, and should be chosen to $\beta = 2$ for Gaussian distributions.

- κ is usually chosen to zero.

Note that $n_x + \lambda = \alpha^2 n_x$ when $\kappa = 0$, and that for $n_x + \lambda \to 0^+$ the central weight $\omega^{(0)} \to -\infty$. Furthermore, $\sum_i \omega^{(i)} = 1$. We will consider the two versions of UT in Table A.7, corresponding to the original one in Julier et al. (1995) and an improved one in Wan and van der Merwe.

The *cubature transform* (CT), that is used in the *cubature Kalman filter* (CKF, see Arasaratnam et al. (2009)), is derived using different principles than the UT. However, it still fits the UT framework for a particular parameter tuning. The CT parameters are given in Table A.7 for comparison.

The derivative-free EKF (*DF-EKF*) in Quine (2006) avoids the center sigma points just as the CT, but includes also an arbitrary scaling factor a to the other sigma points. For the case $a = 1$, the method coincides with CKF. Here the transformation used is denoted *derivative-free transform* (DFT).

In summary, TT1 is a computationally cheap approximation, TT2 recovers the first two moments if the gradient and Hessian in μ_x are available (for functions $g(x)$ quadratic in x, TT2 is completely correct, otherwise it is often a good approximation), the MC approach is always asymptotically correct. The UT is a fairly good compromise between TT2 and MC, that improves computational complexity to MC and the need for prior knowledge to TT2.

The unscented transform may have a negative weight for the center point $z^{(0)}$. This might cause problems when implementing the UKF, for instance using the square root form. On the other hand, the cubature filter described in Arasaratnam et al. (2009) has a similar set of sigma points. The points all have positive weights, and the central point is left out.

A.3.4 Analytical Comparison of TT2 and UT

In the following theorem, the relation between TT2 and UT will be analyzed, and expressions for the resulting mean and covariance are given and interpreted in the limit as the sigma points in the UT approach the center point.

Theorem A.2 (Asymptotic property of UT)
Consider a mapping $z = g(x)$ from $\mathcal{R}^{n_x}$ to $\mathcal{R}^{n_z}$ of the stochastic variable x with mean μ_x and covariance P_x. The UT yields mean μ_z and covariance P_z asymptotically as the sigma points in UT2 tend to the mean given by

$$\mu_z^{\text{UT}} = g(\mu_x) + \tfrac{1}{2}\big[\operatorname{tr}(g_i'' P)\big]_i, \tag{A.21a}$$

$$P_z^{\text{UT}} = g'(\mu_x) P \big(g'(\mu_x)\big)^T + \tag{A.21b}$$

$$\tfrac{(\beta-\alpha^2)}{4}\Big[\operatorname{tr}\big(P g_i''(\mu_x)\big) \operatorname{tr}\big(P g_j''(\mu_x)\big)\Big]_{ij} \tag{A.21c}$$

For $n_x = 1$, equality $P_z^{\text{TT2}} = P_z^{\text{UT}}$ holds if $\beta - \alpha^2 = 2$.

Proof: Reorganizing the terms in (A.20) gives

$$\mu_z = z^{(0)} + \frac{1-\omega^{(0)}}{2n_x}\sum_{i=1}^{n_x}(z^{(i)} - 2z^{(0)} + z^{(-i)}) \tag{A.22a}$$

$$\Sigma_z = \big(\omega^{(0)} + (1-\alpha^2+\beta)\big)(z^{(0)} - \mu_z)(\cdot)^T$$
$$+ \sum_{i \neq 0} \tfrac{1-\omega^{(0)}}{2n_x}(z^{(i)} - \mu_z)(\cdot)^T$$
$$= (1-\alpha^2+\beta)\tfrac{(1-\omega^{(0)})^2}{4n_x^2}\Big(\sum_{i=1}^{n_x}(z^{(i)} - 2z^{(0)} + z^{(-i)})\Big)(\cdot)^T$$
$$- \tfrac{(1-\omega^{(0)})^2}{4n_x^2}\Big(\sum_{i=1}^{n_x}(z^{(i)} - 2z^{(0)} + z^{(-i)})\Big)(\cdot)^T$$
$$+ \tfrac{1-\omega^{(0)}}{2n_x}\sum_{i=-n_x}^{n_x}(z^{(i)} - z^{(0)})(\cdot)^T. \tag{A.22b}$$

With the sigma points in (A.19), differences can be constructed that in the limit as $n_x + \lambda \to 0^+$ (i.e., $\alpha \to 0^+$ with $\kappa = 0$), yields the derivatives:

$$\frac{z^{(i)} - z^{(0)}}{\sigma_i \sqrt{n_x + \lambda}} \to g'(\mu_x) u_i \tag{A.23}$$

$$\frac{z^{(i)} - 2z^{(0)} + z^{(-i)}}{\sigma_i^2 (n_x + \lambda)} \to \big[u_i^T g_k''(\mu_x) u_i\big]_k. \tag{A.24}$$

Note that $n_x + \lambda = n_x/(1-\omega^{(0)})$.

A.3 Nonlinear Transformations

Using this, the limit case of (A.22) can be evaluated,

$$\mu_z \to g(\mu_x) + \tfrac{1}{2}\bigl[\operatorname{tr}(g_i''(\mu_x)P)\bigr]_i \qquad \text{(A.25a)}$$

and

$$\begin{aligned}\Sigma_z \to{}& g'(\mu_x)P\bigl(g'(\mu_x)\bigr)^T \\ & + \tfrac{(\beta-\alpha^2)}{4}\bigl[\operatorname{tr}\bigl(Pg_i''(\mu_x)\bigr)\operatorname{tr}\bigl(Pg_j''(\mu_x)\bigr)\bigr]_{ij}.\end{aligned} \qquad \text{(A.25b)}$$

□

By comparing (A.25) and (A.8) for a scalar $z = g(x)$, both TT2 and UT asymptotically gives the same result. In general, the covariances of TT2 and UT differ since

$$\begin{aligned}P_z^{\text{TT2}} - P_z^{UT} ={}& \tfrac{1}{2}\bigl[\operatorname{tr}(Pg_i''(\mu_x)Pg_j''(\mu_x))\bigr]_{ij} \\ & - \tfrac{(\beta-\alpha^2)}{4}\bigl[\operatorname{tr}\bigl(Pg_i''(\mu_x)\bigr)\operatorname{tr}\bigl(Pg_j''(\mu_x)\bigr)\bigr]_{ij} \end{aligned}\qquad \text{(A.26)}$$

Note that $\operatorname{tr}(AB) = \operatorname{tr}(A)\operatorname{tr}(B)$ only if both A and B are scalar, in general. Even the diagonal elements my differ, consider for instance the example $\operatorname{tr}(I_2)\operatorname{tr}(I_2) = 4 \neq 2 = \operatorname{tr}(I_2 I_2)$. One explanation for this discrepancy is that the UT cannot express the mixed second order derivatives needed for the TT2 compensation term without increasing the number of sigma points. The quality of this approximation depends on the transformation and must be analyzed for the case at hand. See Section A.3.5 for some illustrative examples.

A.3.5 Numerical Comparisons

We here provide some examples where the following methods are compared:

TT1 First order Taylor expansion leading to Gauss' approximation formula.

TT2 Second order Taylor expansion, which compensates the mean and covariance with the quadratic second order term.

UT The unscented transformations UT1 and UT2. UT2 will be the default one in the sequel if the number is not indicated.

MCT The Monte Carlo transformation approach, which in the limit should compute correct moments.

Tables A.8 and A.9 summarize the results.

─── **Example A.1: Sum of squares** ───────────────────

The following mapping has a well-known distribution

$$z = g(x) = x^T x, \quad x \in \mathcal{N}(0, I_n) \Rightarrow z \in \chi_n^2. \qquad \text{(A.27)}$$

This distribution has mean n and variance $2n$. For the Taylor expansion, we have
$$g'(\mu) = 0_{1,n}, \quad g''(\mu) = 2I_n.$$
It follows that

$$\mu_z^{\text{TT1}} = 0, \qquad P_z^{\text{TT1}} = 0,$$
$$\mu_z^{\text{TT2}} = n, \qquad P_z^{\text{TT2}} = \frac{1}{2} \cdot 4n = 2n,$$
$$\mu_z^{\text{UT1}} = n, \qquad P_z^{\text{UT1}} = (3-n)n,$$
$$\mu_z^{\text{UT2}} = n, \qquad P_z^{\text{UT2}} = \frac{1}{2} \cdot 2n \cdot 2n = 2n^2,$$
$$\mu_z^{\text{CT}} = n, \qquad P_z^{\text{CT}} = 2n \cdot \frac{1}{2n} \cdot n.$$

That is, TT1 fails completely and TT2 works perfectly. UT gives correct mean. The standard version of UT gives negative variance, while the modified one overestimates the variance, and CT gives zero variance.

Table A.8: Nonlinear approximations of $x^T x$ for $x \in \text{N}(0_n, I_n)$. Theoretical distribution is χ_n^2 with mean n and variance $2n$. The mean and variance are below summarized as a Gaussian distribution.

n	1	2	3	4	5	n
TT1	$\mathcal{N}(0,0)$	$\mathcal{N}(0,0)$	$\mathcal{N}(0,0)$	$\mathcal{N}(0,0)$	$\mathcal{N}(0,0)$	$\mathcal{N}(0,0)$
TT2	$\mathcal{N}(1,2)$	$\mathcal{N}(2,4)$	$\mathcal{N}(3,6)$	$\mathcal{N}(4,8)$	$\mathcal{N}(5,10)$	$\mathcal{N}(n,2n)$
UT1	$\mathcal{N}(1,2)$	$\mathcal{N}(2,2)$	$\mathcal{N}(3,0)$	$\mathcal{N}(4,-4)$	$\mathcal{N}(5,-10)$	$\mathcal{N}(n,(3-n)n)$
UT2	$\mathcal{N}(1,2)$	$\mathcal{N}(2,8)$	$\mathcal{N}(3,18)$	$\mathcal{N}(4,32)$	$\mathcal{N}(5,50)$	$\mathcal{N}(n,2n^2)$
CT	$\mathcal{N}(1,1)$	$\mathcal{N}(2,2)$	$\mathcal{N}(3,3)$	$\mathcal{N}(4,4)$	$\mathcal{N}(5,5)$	$\mathcal{N}(n,n)$

Example A.2: Radar measurements

Consider the mapping of range and bearing to Cartesian coordinates

$$z = g(x) = \begin{pmatrix} x_1 \cos x_2 \\ x_1 \sin x_2 \end{pmatrix}. \tag{A.28}$$

For the first case in Table A.9, $\mu_x = (3,0)^T$, the Taylor expansion has

$$g'(\mu_x) = \begin{pmatrix} 1 & 0 \\ 0 & 3 \end{pmatrix}, \quad g_1''(\mu_x) = \begin{pmatrix} 0 & 0 \\ 0 & 0 \end{pmatrix}, \quad g_2''(\mu_x) = \begin{pmatrix} 0 & 1 \\ 1 & 0 \end{pmatrix}.$$

Note that all higher order derivatives have the unit norm, $\|g^{(n)}(x)\| = 1$ for all $n \geq 2$, so the second order Taylor expansion cannot be regarded as an

A.3 Nonlinear Transformations

accurate approximation. This is particularly the case when the angular error is large, as it is designed to be here. It follows from (A.17) and Theorem A.2 that

$$\mu_z^{\text{UT}} = \mu_z^{\text{TT2}} = \begin{pmatrix} 0 \\ 0 \end{pmatrix},$$

and that the covariance approximations differ.

Table A.9: Nonlinear approximations of the radar observations to Cartesian position mapping $(x_1 \cos x_2, x_1 \sin x_2)^T$. The mean and variance are below summarized as a Gaussian distribution. The number of Monte Carlo simulations is 10 000.

Method	x		
	$\mathcal{N}((\begin{smallmatrix}3.0\\0.0\end{smallmatrix}),(\begin{smallmatrix}1.0&0.0\\0.0&1.0\end{smallmatrix}))$	$\mathcal{N}((\begin{smallmatrix}3.0\\0.5\end{smallmatrix}),(\begin{smallmatrix}1.0&0.0\\0.0&1.0\end{smallmatrix}))$	$\mathcal{N}((\begin{smallmatrix}3.0\\0.8\end{smallmatrix}),(\begin{smallmatrix}1.0&0.0\\0.0&1.0\end{smallmatrix}))$
TT1	$\mathcal{N}((\begin{smallmatrix}3.0\\0.0\end{smallmatrix}),(\begin{smallmatrix}1.0&0.0\\0.0&9.0\end{smallmatrix}))$	$\mathcal{N}((\begin{smallmatrix}2.6\\1.5\end{smallmatrix}),(\begin{smallmatrix}3.0&-3.5\\-3.5&7.0\end{smallmatrix}))$	$\mathcal{N}((\begin{smallmatrix}2.1\\2.1\end{smallmatrix}),(\begin{smallmatrix}5.0&-4.0\\-4.0&5.0\end{smallmatrix}))$
TT2	$\mathcal{N}((\begin{smallmatrix}2.0\\-0.0\end{smallmatrix}),(\begin{smallmatrix}3.0&0.0\\0.0&10.0\end{smallmatrix}))$	$\mathcal{N}((\begin{smallmatrix}-1.4\\0.5\end{smallmatrix}),(\begin{smallmatrix}27.0&2.5\\2.5&9.0\end{smallmatrix}))$	$\mathcal{N}((\begin{smallmatrix}2.1\\2.1\end{smallmatrix}),(\begin{smallmatrix}9.0&0.0\\0.0&13.0\end{smallmatrix}))$
UT1	$\mathcal{N}((\begin{smallmatrix}1.8\\0.0\end{smallmatrix}),(\begin{smallmatrix}3.7&0.0\\0.0&2.9\end{smallmatrix}))$	$\mathcal{N}((\begin{smallmatrix}1.6\\0.9\end{smallmatrix}),(\begin{smallmatrix}3.5&0.3\\0.3&3.1\end{smallmatrix}))$	$\mathcal{N}((\begin{smallmatrix}1.3\\1.3\end{smallmatrix}),(\begin{smallmatrix}3.3&0.4\\0.4&3.3\end{smallmatrix}))$
UT2	$\mathcal{N}((\begin{smallmatrix}1.5\\0.0\end{smallmatrix}),(\begin{smallmatrix}5.5&0.0\\0.0&9.0\end{smallmatrix}))$	$\mathcal{N}((\begin{smallmatrix}1.3\\0.8\end{smallmatrix}),(\begin{smallmatrix}6.4&-1.5\\-1.5&8.1\end{smallmatrix}))$	$\mathcal{N}(\begin{smallmatrix}1.1\\1.1\end{smallmatrix}),(\begin{smallmatrix}7.2&-1.7\\-1.7&7.2\end{smallmatrix}))$
CT	$\mathcal{N}((\begin{smallmatrix}1.73\\0.0\end{smallmatrix}),(\begin{smallmatrix}2.6&0.0\\0.0&4.39\end{smallmatrix}))$	$\mathcal{N}((\begin{smallmatrix}1.52\\0.83\end{smallmatrix}),(\begin{smallmatrix}3.01&-0.75\\-0.75&3.98\end{smallmatrix}))$	$\mathcal{N}(\begin{smallmatrix}1.21\\1.24\end{smallmatrix}),(\begin{smallmatrix}3.52&-0.893\\-0.893&3.47\end{smallmatrix}))$
MCT	$\mathcal{N}((\begin{smallmatrix}1.8\\0.0\end{smallmatrix}),(\begin{smallmatrix}2.5&0.0\\0.0&4.4\end{smallmatrix}))$	$\mathcal{N}((\begin{smallmatrix}1.6\\0.9\end{smallmatrix}),(\begin{smallmatrix}2.9&-0.8\\-0.8&3.9\end{smallmatrix}))$	$\mathcal{N}((\begin{smallmatrix}1.3\\1.3\end{smallmatrix}),(\begin{smallmatrix}3.4&-1.0\\-1.0&3.4\end{smallmatrix}))$

—— **Example A.3: TOA, DOA and RSS measurements** ——

The basic measurements in sensor networks Gustafsson and Gunnarsson (2005) are *time of arrival* (TOA), direction of arrival (DOA) and *received signal strength* (RSS). These all relate to the position in a nonlinear way. Range measurements in two ($n = 2$) and three ($n = 3$) dimensions, respectively, are given by

$$g_{\text{TOA}}(x) = \|x\| = \sqrt{\sum_{i=1}^{n} x_i^2}.$$

Received signal strength in two dimensions in dB scale (where the measurement noise can be seen as additive and Gaussian Gustafsson and Gunnarsson (2005)) is given by

$$g_{\text{RSS}}(x) = c_1 - c_2 \cdot 10 \log_{10}(\|x\|^2).$$

Finally, direction of arrival is expressed as

$$g_{\text{DOA}}(x) = \arctan 2(x_1, x_2),$$

where arctan2 is the four quadrant arc-tangent function. The resulting approximation depends a lot of the assumed Gaussian distribution of position. We choose a distribution which is typical in single sensor tracking applications, where the prior distribution before the measurement update is uncertain in the direction tangential to the measurement information. The results are summarized in Table A.10.

The conclusion from the example, and many similar tests with other prior distributions of the position, is that TT1 is inferior and that the UT, and in particular the tuning provided by the CT, is to preferred to TT2. However, all of TT1, TT2, and UT can be arbitrarily bad compared to the MCT.

It should be remarked, though, that all the cases in Examples A.2 and A.3 are deliberately designed to excite higher order terms in the Taylor expansion. As the range to the target increases, the higher order terms will decrease with a rate that for the higher order terms is faster than for the lower order terms. Based on the preceding analysis, one may expect that the TT2 will converge faster than TT1 and UT.

A.3 Nonlinear Transformations

Table A.10: *Numerical comparison of approximate transformations for nonlinear measurement models in sensor network applications.*

TOA 2D $g(x) = \|x\|$	$\mathcal{N}([3;0],[1,0;0,10])$
TT1	$\mathcal{N}(3,1)$
TT2	$\mathcal{N}(4.67, 6.56)$
UT1	$\mathcal{N}(4.08, 3.34)$
UT2	$\mathcal{N}(4.67, 6.56)$
CT	$\mathcal{N}(4.19, 2, 42)$
MCT	$\mathcal{N}(4.25, 2.4)$

TOA 3D $g(x) = \|x\|$	$\mathcal{N}([3;0;0],[1,0,0;0,10,0;0,0,10])$
TT1	$\mathcal{N}(3,1)$
TT2	$\mathcal{N}(6.33, 12.1)$
UT1	$\mathcal{N}(5.16, 3.34)$
UT2	$\mathcal{N}(6.33, 23.2)$
CT	$\mathcal{N}(5.16, 3.34)$
MCT	$\mathcal{N}(5.17, 2.95)$

DOA $g(x) = \arctan(x_2, x_1)$	$\mathcal{N}([3;0],[10,0;0,1])$
TT1	$\mathcal{N}(0, 0.111)$
TT2	$\mathcal{N}(0, 0.235)$
UT1	$\mathcal{N}(0.524, 1.46)$
UT2	$\mathcal{N}(0, 0.111)$
CT	$\mathcal{N}(0.785, 1.95)$
MCT	$\mathcal{N}(-0.004, 1.38)$

RSS $g(x) = 10 - 20\log_{10}(\|x\|^2)$	$\mathcal{N}([3;0],[10,0;0,1])$
TT1	$\mathcal{N}(-18, 12.1)$
TT2	$\mathcal{N}(-21.1, 36.5)$
UT1	$\mathcal{N}(-19.9, 25.5)$
UT2	$\mathcal{N}(-21.1, 31.6)$
CT	$\mathcal{N}(-20.1, 21.3)$
MCT	$\mathcal{N}(-20.1, 16.4)$

B

Sampling Theory

This appendix surveys classical sampling methods:

- Methods for generation of random numbers are overviewed in Sections B.1, B.2 and B.3.
- Resampling strategies are given in Section B.4.
- Stochastic integration is described in Section B.5.
- Monte Carlo Markov Chain methods for generating random numbers are given in Sections B.6 and B.7.

B.1 Generating Samples from Uniform Distribution

Generation of uniform random numbers is instrumental in all methods in the chapter, and we assume that a good uniform random number generator is available.

For instance, the MATLAB™ uniform random number generator **rand** simulates $u \sim \mathcal{U}(0, 1)$, and it has the properties that the values are in the interval $u \in [2^{-53}, 1 - 2^{-53}]$ and there are $2^{19937} - 1$ values before repetition.

Consider now the problem of generating random number from another distribution $\pi(x)$. If the inverse of its probability distribution function $\Pi(x) = \int_{-\infty}^{s} \pi(s)ds$ exists, then random numbers can be generated by transforming uniform random numbers $u^{(i)}$ using $x^{(i)} = \Pi^{-1}(u^{(i)})$. Otherwise, one of the algorithms outlined below has to be used.

B.2 Accept-Reject Sampling

The most basic algorithm to generate random numbers from a distribution $\pi(x)$ is the *accept-reject sampling* algorithm, see Algorithm B.1. This algorithm is sometimes called just *rejection sampling*. The only thing required is a proposal density $q(x)$, from which random numbers can easily be generated and whose support covers at least the support of $\pi(x)$.

Algorithm B.1 Accept-Reject Sampling

Choose a proposal density $q(x)$ such that $\pi(x) > 0 \Rightarrow q(x) > 0$. Choose M such that $Mq(x) > \pi(x)$ for all x.
Repeat

1. Sample $x \sim q(x)$.

2. Sample $u \sim \mathcal{U}(0,1)$.

3. Accept the sample x if $u < \frac{\pi(x)}{Mq(x)}$, otherwise repeat from 1.

To prove that the algorithm works, the probability distribution function of the random number it generates is given by

$$\Pi(t) = \Pr(x \le t | x \text{ is accepted}) = \frac{\Pr(x \le t \text{ and } x \text{ is accepted})}{\Pr(x \text{ is accepted})}$$

$$= \frac{\int_{-\infty}^{t} \frac{\pi(x)}{Mq(x)} q(x)\, dx}{\int_{\mathbb{R}} \frac{\pi(x)}{Mq(x)} q(x)\, dx} = \frac{\frac{1}{M}\int_{-\infty}^{t} \pi(x)\, dx}{\frac{1}{M}} = \int_{-\infty}^{t} \pi(x)\, dx$$

which confirms that $\Pi(t)$ is the probability distribution function of $\pi(x)$. Here, only Bayes rule $P(B|A) = \frac{P(B,A)}{P(A)}$ and basic definitions have been used.

In a Bayesian framework, we often want to sample from the posterior

$$\pi(x) = p(x|y) \propto p(y|x)p(x),$$

which is easily expressed in terms of prior and likelihood. However, the scaling factor $p(y) = \int p(y|x)p(x)\, dx$ may be difficult to compute. If an upper bound C on the likelihood is known, the prior can be used as a proposal density $q(x) = p(x)$, since

$$\frac{\pi(x)}{Mq(x)} = \frac{p(y|x)p(x)}{Mq(x)p(y)} = \frac{p(y|x)}{Mp(y)} \le \frac{C}{Mp(y)} \le 1$$

if $M > C/p(y)$.

The following example makes use of all of the aforementioned strategies: uniform number generation, an analytic method and accept-reject.

B.2 Accept-Reject Sampling

---- **Example B.1: Accept-reject** ----

Consider a PDF defined as a squared sinc function,

$$\pi(x) = c\,\text{sinc}^{2k}(\pi x/(2k\Delta)) = c\frac{\sin^{2k}\left(\frac{\pi x}{2k\Delta}\right)}{\left(\frac{\pi x}{2k\Delta}\right)^{2k}}, \tag{B.1}$$

$$c = \int \text{sinc}^{2k}(\pi x/(2k\Delta))\,\mathrm{d}x. \tag{B.2}$$

This distribution is characterized by its unique property of having a compact support in the frequency domain. Here, k is the order and Δ a scale parameter in the distribution. It is virtually impossible to analytically compute the inverse cdf, so numerical methods are needed.

The tails are quite slowly decaying, so the Cauchy distribution

$$q(x) = \frac{1}{\pi}\frac{1}{1+x^2}, \tag{B.3}$$

is one candidate as proposal distribution. Here, the inverse cdf is readily available, see below. It turns out that the Cauchy distribution with $M = 2$ satisfies the necessary conditions for all k in Algorithm B.1, and the following algorithm can be formulated.

1. Generate a standard uniform sample u.

2. Generate a proposal sample from the standard Cauchy distribution, using the transformation $x = \tan(\pi(u - 0.5))$.

3. Generate an acceptance probability p from the standard uniform distribution.

4. Accept the proposal sample x if

$$p < \frac{c_k \text{sinc}^{2k}\left(\frac{\pi x}{2k}\right)}{M\frac{1}{\pi}\frac{1}{1+x^2}}. \tag{B.4}$$

The acceptance rate is slightly larger than 50%.

Figure B.1 is generated using

```
X=bldist(1,1);
x=rand(X,10000);
plot(X,empdist(x),cauchy(0,1))
```

where the `rand` method of the `bldist` implements the accept-reject method above.

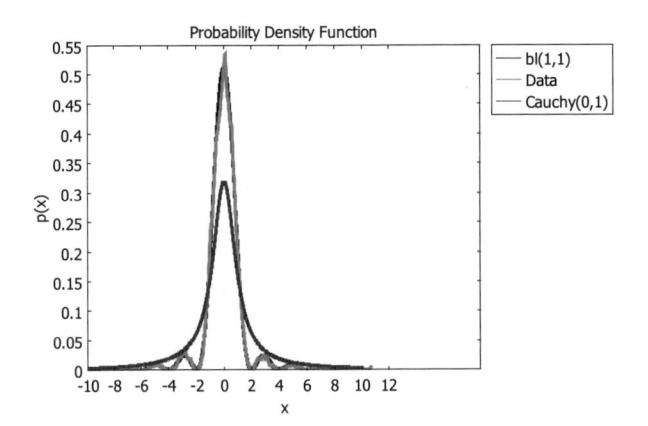

Figure B.1: True PDF and smoothed histogram approximation of the random numbers obtained from accept-reject sampling.

───── Example B.2: Generating normal distributed values ─────

The first versions of MATLAB^{TM} used randn(1,1)=sum(rand(12,1))-6, motivated by the central limit theorem. However, such simple algorithms have obvious short-comings. For instance, Gaussian variables larger than 6 are never obtained.

Prior to version 5, an accept-reject algorithm of two-dimensional normal vectors was used. The point is that the two-dimensional Gaussian probability distribution has an analytic inverse, which can be derived using a change to polar coordinates.

The following lines of code generate two-dimensional uniform samples until such a sample is inside the unit disc, after which a two-dimensional Gaussian variable can be computed analytically.

```
r=2;
while r>1
  U=2*rand(2,1)-1;
  r=U'*U;
end
X=sqrt(-2*log(r)/r)*U;
```

The acceptance ratio is equal to the ratio of the areas of the unit circle and the unit rectangle, which is $(\pi r^2)/(2r)^2 = \pi 0.5^2/1^2 \approx 0.79$.

Modern versions of MATLAB^{TM} use the Ziggurat algorithm developed by Marsaglia and Knuth. The accept-reject idea is as follows:

1. Compute once in each session a rectangle approximation of PDF $\pi(x)$,

where the grid points are computed from x_n as

$$x_k : (\pi(x_k) - \pi(x_{k+1}))x_k = c, \quad k = 1, 2, \ldots, n-1, \quad \text{(B.5)}$$

$$\sigma_k = \frac{x_{k+1}}{x_k}. \quad \text{(B.6)}$$

2. Given these vectors x_k and σ_k, the code for acceptance (97% accepted) looks as follows ($n = 128$):

```
k=ceil(128*rand);
U=2*rand-1;
if abs(U)<sigma(k)
   X=U*x(k);
   return
end
```

B.3 Bootstrap

Given a sequence of i.i.d. random variables $x^{(i)}$ from $\pi(x)$, the bootstrap idea is to generate a new sequence by randomly resampling with replacement N samples from this set. The idea is motivated by the fact that

$$I = \int g(x)p(x)dx \approx \frac{1}{N}\sum_i g(x^{(i)}) \approx \sum_i \frac{N(i)}{N} g(x^{(i)}), \quad \text{(B.7)}$$

where $N(i)$ is the number of replicas of $x^{(i)}$ after the resampling, where $\sum_i N(i) = N$.

The point is that many data sets of N samples can be generated from just one, and one can get an idea of variability in the estimates that are non-linear transformation of the samples.

B.4 Resampling

Resampling is needed in the bootstrap, and is a crucial step in the particle filter, see Section 9.7.1. The following options are the most common in particle filter literature, see Hol et al. (2006) for more details:

1. **Multinomial resampling**
   Generate N ordered uniform random numbers

   $$u_k = u_{k+1}\tilde{u}_k^{\frac{1}{k}}, \quad u_N = \tilde{u}_N^{\frac{1}{N}}, \text{ with } \tilde{u}_k \sim \mathcal{U}(0,1)$$

and use them to select x_k^* according to the multinomial distribution. That is,

$$x_k^* = x(F^{-1}(u_k))$$
$$= x_i \text{ with } i \text{ s.t. } u_k \in \left[\sum_{s=1}^{i-1} w_s, \sum_{s=1}^{i} w_s\right),$$

where F^{-1} denotes the generalised inverse of the cumulative probability distribution of the normalised particle weights.

2. **Stratified resampling**
   Generate N ordered random numbers

   $$u_k = \frac{(k-1) + \tilde{u}_k}{N}, \text{ with } \tilde{u}_k \sim \mathcal{U}(0,1)$$

   and use them to select x_k^* according to the multinomial distribution.

3. **Systematic resampling**
   Generate N ordered numbers

   $$u_k = \frac{(k-1) + \tilde{u}}{N}, \text{ with } \tilde{u} \sim \mathcal{U}(0,1)$$

   and use them to select x_k^* according to the multinomial distribution.

4. **Residual resampling**
   Allocate $n_i' = \lfloor Nw_i \rfloor$ copies of particle x_i to the new distribution. Additionally, resample $m = N - \sum n_i'$ particles from $\{x_i\}$ by making n_i'' copies of particle x_i where the probability for selecting x_i is proportional to $w_i' = Nw_i - n_i'$ using one of the resampling schemes mentioned earlier.

All these algorithms are unbiased and can be implemented in $\mathcal{O}(N)$ as the random numbers are ordered, but have different computational complexities.

B.5 Stochastic Integration by Importance Sampling

Monte Carlo integration is the most basic application of *stochastic integration*. The expected value of any function $f(x)$ is given by

$$I = \int_{\mathbb{R}^n} f(x)\pi(x)\,dx, \tag{B.8}$$

and approximated with

$$\hat{I}_N = \frac{1}{N}\sum_{i=1}^{N} f(x^{(i)}), \tag{B.9}$$

B.5 Stochastic Integration by Importance Sampling

where $x^{(i)}$ are (Monte Carlo) samples from $\pi(x)$. This works fine as long as it is possible to easily generate samples from $\pi(x)$.

Importance sampling substitutes $\pi(x)$ with a *proposal distribution* $q(x)$ which is easier to generate samples from. However, the only general assumption on the *importance function* $q(x)$ is that its support covers the support of $\pi(x)$, i.e., that $\pi(x) > 0 \Rightarrow q(x) > 0$ for all $x \in \mathbb{R}^n$.

The problem in consideration with importance sampling is, however, not to draw random numbers. For this purpose, the sampling importance resampling algorithm that follows can be used. Instead, the focus in importance sampling is to approximate integrals of the form

$$I = \int_{\mathbb{R}^n} f(x)\pi(x)\,dx = \int_{\mathbb{R}^n} f(x)\frac{\pi(x)}{q(x)}q(x)\,dx. \tag{B.10}$$

That is, importance sampling is an alternative to the Monte Carlo approach in the case sampling from $\pi(x)$ is not possible.

A Monte Carlo estimate is computed by generating $N \gg 1$ independent samples $x^{(i)}$ from $q(x)$, and forming the weighted sum

$$\hat{I}_N = \frac{1}{N}\sum_{i=1}^{N} f(x^{(i)}) \underbrace{\frac{\pi(x^{(i)})}{q(x^{(i)})}}_{w(x^{(i)})}, \tag{B.11}$$

where $w(x^{(i)})$ denotes the *importance weight*. Monte Carlo sampling can be seen as a special case of importance sampling, when $q(x) = \pi(x)$. The normalization factor of $\pi(x)$ may be unknown, in which case the following algorithm applies

$$\hat{I}_N = \frac{\sum_{i=1}^{N} f(x^{(i)})w(x^{(i)})}{\sum_{j=1}^{N} w(x^{(j)})}, \tag{B.12}$$

$$w(x^{(i)}) \propto \frac{\pi(x^{(i)})}{q(x^{(i)})}. \tag{B.13}$$

One such case occurs in the Bayesian framework, where

$$\pi(x) = p(x|y) = \frac{p(y|x)p(x)}{p(y)} \propto p(y|x)p(x).$$

The prior $p(x)$ can thus be used as importance distribution, and the unnormalized weights are given by $w(x) = p(y|x)$. This case is referred to as *Bayesian importance sampling*.

Theorem B.1
When (B.10) exists and is finite, the estimate (B.12) converges almost everywhere

$$\Pr\left(\lim_{N\to\infty} \hat{I}_N = I\right) = 1.$$

Additionally, if $\mathrm{E}w(x) < \infty$ and $\mathrm{E}f^2(x)w(x) < \infty$,

$$\lim_{N \to \infty} \sqrt{N}(\hat{I}_N - I) \sim \mathcal{N}(0, \sigma^2) \qquad (B.14)$$

where $\sigma^2 = \mathrm{E}(f(x) - I)^2 w(x)$. All expectations above are performed w.r.t. the density $\pi(x)$.

With a bounded likelihood and some weak regularity conditions on $\mathrm{E}f(x)$ and $\mathrm{E}f^2(x)$, the Bayesian importance sampling procedure will thus yield asymptotically consistent estimates of (B.10).

---**Example B.3: Importance sampling**---

Consider the problem of determining the mean ($I = \mathrm{E}(x)$) of a Gaussian mixture $\pi(x) = 0.5\mathcal{N}(x; 0, 1) + 0.5\mathcal{N}(x; 3, 0.5)$ constrained to the interval $[-5, 5]$. A feasible importance distribution $q(x)$ is a uniform one $\mathcal{U}(-5, 5)$. The numeric mean using 10000 samples is found to be 1.487, which is by (B.14) not significantly different to the mean 1.5 of the untruncated Gaussian mixture. Figure B.2 shows a histogram of importance weights.

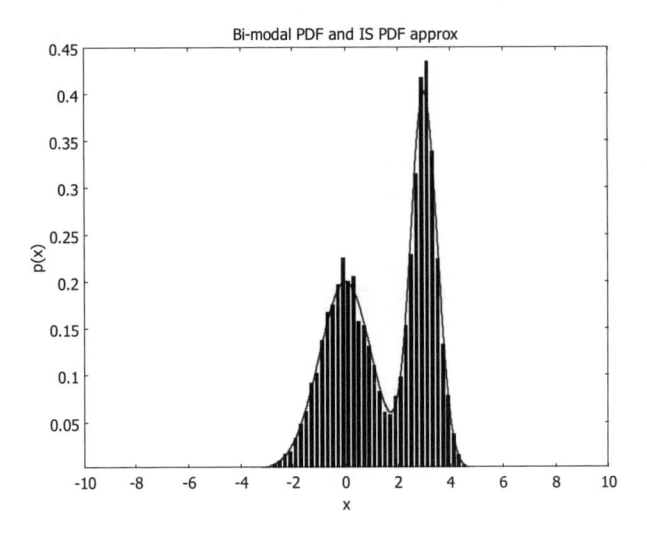

Figure B.2: *Histogram of samples generated with importance sampling and true distribution density function.*

The *sampling importance resampling* (SIR) method in Algorithm B.2 is a procedure for generating an approximately independent draw from $\pi(x)$ using its weighted approximation. The independent draw from $\pi(x)$ is obtained by

Algorithm B.2 Sampling Importance Resampling, SIR

1. Generate M independent samples $\{x^{(i)}\}_{i=1}^{M}$ with common distribution $q(x)$.

2. Compute the weight $w_i \propto \pi(x^{(i)})/q(x^{(i)})$ for each $x^{(i)}$.

3. Normalize the weights $w_i := w_i / \sum_{j=1}^{M} w_j$.

4. Resample with replacement N times from the discrete set $\{x^{(i)}\}_{i=1}^{M}$ where $\Pr(\text{resampling } x^{(i)}) = w_i$.

---

inserting a resampling step in the spirit of Effron's Bootstrap (Politis, 1998) after the weight calculations.

Here, M should be chosen much greater than N, and as a rule of thumb one can use at least $M = 10N$. As another rule of thumb, the importance function should be chosen similar to the density $\pi(x)$ but with heavier tails.

Note that the SIR algorithm only asymptotically generate samples from $\pi(x)$. The rejection sampling algorithm, on the other hand, generates samples from $\pi(x)$ without approximations.

B.6 Markov Chain Monte Carlo

Markov chain Monte Carlo (MCMC) is a combination of

- Monte Carlo methods.
- Accept-reject sampling.
- Importance sampling.

The main difference is that the proposal distribution depends on the previous sample $x^{(i+1)} \sim q(z|x^{(i)})$. The samples thus form a Markov Chain. The idea is that the sequence of samples will converge to $\pi(x)$. In the Metropolis–Hastings algorithm, a candidate sample z is drawn from the proposal $q(z|x)$ and accepted with a certain *acceptance probability*. If the candidate is accepted the chain moves to the new position, while a rejection of the candidate leaves the chain at the current value.

An interpretation of (B.15) is that all candidates that yield an increase of $\pi(\cdot)$, and are not too unlikely to return from, are accepted.

One very important feature of the Metropolis–Hastings algorithm is that the distributions $\pi(x)$ only need to be known up to a normalizing constant. The normalizing factor of $\pi(x)$ cancels in the expression for the acceptance probability (B.15), which thus may be evaluated even if $\int \pi(x)\,dx$ is unknown.

Algorithm B.3 Metropolis–Hastings

Choose a proposal density $q(z|x)$ and initial value x^0. Iterate for $i = 0, 1, 2, \ldots, N$:

1. Sample $z \sim q(z|x^{(i)})$.

2. Sample $u \sim \mathcal{U}(0,1)$.

3. Compute the acceptance probability

$$\alpha(x,z) = \min\left(1, \frac{\pi(z)q(x|z)}{\pi(x)q(z|x)}\right). \tag{B.15}$$

4. If $u \leq \alpha(x^{(i)}, z)$, then let $x^{(i+1)} = z$, otherwise let $x^{(i+1)} = x^{(i)}$.

Consider the general case of Bayesian estimation where

$$\pi(x) = p(x|y) = \frac{p(y|x)p(x)}{p(y)}.$$

After generating a candidate sample from a suitable $q(z|x)$, this sample is accepted with probability

$$\alpha(x,z) = \min\left(1, \frac{p(y|z)p(z)q(x|z)}{p(y|x)p(x)q(z|x)}\right).$$

Using the prior as proposal distribution $q(z|x) = p(z)$, the acceptance probability simplifies to

$$\alpha(x,z) = \min\left(1, \frac{p(y|z)}{p(y|x)}\right).$$

That is, the candidate is always accepted if it increases the likelihood.

B.7 Gibbs Sampling

For high-dimensional PDF's $\pi(x)$, the marginals may be much simpler to generate samples from than the multivariate distribution. The idea in Gibbs sampling is to sample from one marginal at the time. The *Gibbs sampling* algorithm by Geman and Geman (1984) is the most commonly applied MCMC algorithm.

Example B.4: Gibbs sampling

Consider the problem of generating samples from the *Laplace distribution* $L(x, \sigma^2)$. The following relations for the marginals of X and σ^2 can be used

B.7 Gibbs Sampling

Algorithm B.4 Gibbs sampling

Choose a proposal density $q(z|x)$ and initial value x^0. Iterate for $i = 0, 1, 2, \ldots N$:

1. Sample from the marginal distributions
   - $x_1^{(i)} \sim \pi(x_1 | x_{\neg 1}^{(i)})$
   - $x_2^{(i)} \sim \pi(x_2 | x_{\neg 2}^{(i)})$
   - $\vdots$
   - $x_n^{(i)} \sim \pi(x_n | x_{\neg n}^{(i)})$

Above, $x_{\neg k}^{(i)} \triangleq [x_1^{(i)}, \ldots, x_{k-1}^{(i)}, x_{k+1}^{i-1}, \ldots, x_n^{i-1}]^T$.

in a Gibbs sampler:

$$x|\sigma^2 \sim \mathcal{N}(0, \sigma^2), \tag{B.16a}$$
$$\sigma^2|x \sim \exp(1). \tag{B.16b}$$

Figure B.3 shows a histogram of the samples $x^{(i)}$ together with the PDF.

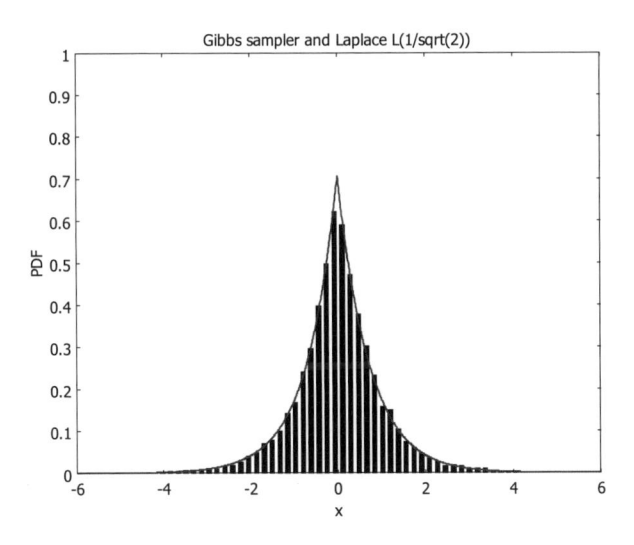

Figure B.3

C

Estimation Theory

This appendix briefly defines the main concepts and results in estimation theory. It is in no way complete. For a thorough treatment of estimation theory, see Lehmann (1991a) for a statistical perspective and Kay (1993) for a signal processing perspective.

C.1 Basic Concepts

Consider an observation y that depends on a parameter θ through a model in the form of a conditional *probability density function* (*PDF*) $p(y|\theta)$, which is referred to as the *likelihood*. Both y and θ are possibly vectors. We do not generally differ in notation between a stochastic variable and an observation of it. That is, $p(y|\theta)$ denotes both a likelihood given a value of the parameter vector, as is done in Fisher statistics, and a conditional density function in a Bayesian setting, where θ is a stochastic variable. The context decides the interpretation.

We are looking for a good estimator, which is a mapping α from the observations to an estimate $\hat{\theta} = \alpha(y)$. When several observations are available, each one is labeled y_k, $k = 1, 2, \ldots, N$. The vector of all observations is then denoted $\mathbf{y}$, or $y_{1:N}$ when subsets of the observations are needed.

Denote the true parameter vector θ^o. Define the *mean square error* (*MSE*) as

$$\mathrm{MSE}(\hat{\theta}) = \mathrm{E}\big((\theta^o - \hat{\theta})^2\big) = \underbrace{\mathrm{E}\big((\hat{\theta} - \mathrm{E}(\hat{\theta}))^2\big)}_{\mathrm{Var}(\hat{\theta})} + \underbrace{\big(\theta^o - \mathrm{E}(\hat{\theta})\big)^2}_{\mathrm{Bias}^2(\hat{\theta})} \qquad (\mathrm{C.1})$$

The estimator $\hat{\theta} = \alpha(y)$ is characterized by the following properties:

- It is *unbiased* if $\text{Bias}(\hat{\theta}) = 0$.
- It is *minimum variance* (MV) if $\hat{\theta}(y) = \arg\min_\alpha \text{Var}(\alpha(y))$. Without constraints on the estimator, the MV property is useless since for instance $\alpha(y) = 0$ gives $\text{Var}(\alpha(y)) = 0$.
- The *minimum variance unbiased* (MVU) principle constrains the estimator to the class of unbiased ones.
- The *best linear unbiased estimator* $(BLUE)$ constrains the estimator to be unbiased and the linear functions $\hat{\theta} = \alpha(y) = Ly$ that minimimes the MSE.
- It is *consistent*, if $\text{Bias}(\hat{\theta}) \to 0$ and $\text{Var}(\hat{\theta}) \to 0$ as $N \to \infty$.
- It is *efficient*, if $\text{Bias}(\hat{\theta}) = 0$ and $\text{Var}(\hat{\theta}) \to \text{CRLB}$ as $N \to \infty$, where CRLB denotes the *Cramér-Rao lower bound* discussed in Section C.2.

C.2 Cramér-Rao Lower Bound

The general estimation problem is to estimate a function $\eta = g(\theta)$ given observations y from $p(y|\theta)$ using an estimator $\alpha(y)$. If the estimator is unbiased, the variance of any such estimator is, for a certain regularity condition discussed below, bounded by the *Cramér-Rao lower bound* $(CRLB)$

$$\text{Cov}(\hat{\eta}) \geq \left(\frac{d\eta(\theta)}{d\theta}\right)\left(-\text{E}\left(\frac{d^2 \log p(y|\theta)}{d\theta^2}\right)\right)^{-1}\left(\frac{d\eta(\theta)}{d\theta}\right)^T \quad (C.2)$$

C.2.1 Regularity Conditions

The CRLB applies if certain *regularity conditions* of the likelihood function is assumed:

1. The identity
$$\text{E}\left[\frac{d \log p(y|\theta)}{d\theta}\right] = 0 \quad (C.3)$$
holds.

2. The set of admissible values of θ is open.

3. The *support* (range of values of θ where $p(y|\theta)$ is non-zero) of $p(y|\theta)$ is independent of θ.

4. The partial derivative $\frac{dp(y|\theta)}{d\theta}$ exists and is finite.

5. The *Fisher information* is positive definite,

$$\mathcal{I}(\theta) = \mathrm{E}\left(\frac{d\log p(y|\theta)}{d\theta}\right)\left(\frac{d\log p(y|\theta)}{d\theta}\right)^T \geq 0. \tag{C.4}$$

The first condition is the core property of the likelihood used in the derivation. The conditions guarantee that the likelihood is sufficiently smooth, so the order of integration and differentiation can be interchanged

$$\mathrm{E}\left[\frac{d\log p(y|\theta)}{d\theta}\right] = \int \frac{d\log p(y|\theta)}{d\theta} p(y|\theta) dy \tag{C.5a}$$

$$= \int \frac{dp(y|\theta)}{d\theta} dy = \frac{d}{d\theta} \underbrace{\int p(y|\theta) dy}_{1} = 0. \tag{C.5b}$$

For instance, the uniform distribution is ruled out for CRLB analysis, since it does not have a smooth likelihood with respect to the interval parameters.

C.2.2 Derivation

The derivation is done for a scalar θ and η, but is easily extended to the vector case. The assumption of an unbiased estimator implies

$$\int \alpha(y) p(y|\theta) dy = g(\theta). \tag{C.6}$$

Differentiating this gives

$$\int \alpha(y) \frac{dp(y|\theta)}{d\theta} dy = \int \alpha(y) \frac{d\log p(y|\theta)}{d\theta} p(y|\theta) dy = \frac{dg(\theta)}{d\theta}. \tag{C.7}$$

The regularity condition 1 implies

$$\int (\alpha(y) - \eta) \frac{d\log p(y|\theta)}{d\theta} p(y|\theta) dy = \frac{dg(\theta)}{d\theta}. \tag{C.8}$$

Cauchy-Schwartz inequality now gives

$$\int (\alpha(y) - \eta)^2 p(y|\theta) dy \int \left(\frac{d\log p(y|\theta)}{d\theta}\right)^2 p(y|\theta) dy \geq \left(\frac{dg(\theta)}{d\theta}\right)^2, \tag{C.9}$$

from which we get

$$\mathrm{Var}(\hat{\eta}) = \mathrm{Var}(\alpha(y)) \geq \frac{\left(\frac{dg(\theta)}{d\theta}\right)^2}{\int \left(\frac{d\log p(y|\theta)}{d\theta}\right)^2 p(y|\theta) dy}. \tag{C.10}$$

The denominator is the Fisher information. It can also be expressed as

$$\mathcal{I}(\theta) = \int \left(\frac{d\log p(y|\theta)}{d\theta}\right)^2 p(y|\theta) dy \tag{C.11a}$$

$$= \mathrm{E}\left(\frac{d\log p(y|\theta)}{d\theta}\right)^2 \tag{C.11b}$$

$$= -\mathrm{E}\left(\frac{d^2 \log p(y|\theta)}{d\theta^2}\right). \tag{C.11c}$$

C.2.3 Special Cases

For the special case where $g(\theta) = \theta$, the CRLB simplifies to

$$\mathrm{Cov}(\hat{\theta}) \geq \mathcal{I}^{-1}(\theta). \tag{C.12}$$

Information is additive, so using the data sets $y_{1:N}$, each having the information $\mathcal{I}_k(\theta)$, we get

$$\mathrm{Cov}(\hat{\theta}) \geq \left(\sum_{k=1}^{N} \mathcal{I}_k(\theta)\right)^{-1}. \tag{C.13}$$

In case the estimator is biased, so

$$\mathrm{E}(\hat{\theta}) = \theta + b(\theta), \tag{C.14}$$

then the following bound applies, see Van Trees (1971),

$$\mathrm{Cov}(\hat{\theta}) \geq (I + b'(\theta)) \mathcal{I}^{-1}(\theta) (I + b'(\theta))^T \tag{C.15}$$

The bias can in some cases be cleverly chosen such that the lower bound actually decreases compared to an unbiased estimator. Take for instance the estimator $\hat{\theta} = 0$ which has bias $b(\theta) = -\theta$ and the right hand side of (C.15) becomes zero. A more fair comparison is based on the *MSE (mean square error)*

$$\mathrm{E}\left(\|\hat{\theta} - \theta\|^2\right) = \|b(\theta)\|^2 + \mathrm{tr}\left(\mathrm{Cov}(\theta)\right) \tag{C.16}$$

$$\geq \|b(\theta)\|^2 + \mathrm{tr}\left((I + b'(\theta)) \mathcal{I}^{-1}(\theta) (I + b'(\theta))^T\right). \tag{C.17}$$

Still, the bias can be an instrument to decrease the MSE, see Eldar (2006).

C.2.4 Examples

The first example is a standard problem with unknown mean in Gaussian noise.

C.2 Cramér-Rao Lower Bound

―― **Example C.1: Unknown mean in Gaussian noise** ――

Consider an unknown mean in Gaussian noise,

$$y_k = \theta + e_k, \quad e_k \in \mathcal{N}(0, \sigma^2). \tag{C.18}$$

The second derivative of the log likelihood function does not include any stochastic terms, so the CRLB follows as

$$p(y_{1:N}|\theta) = \frac{1}{(2\pi\sigma^2)^{N/2}} e^{\frac{-1}{2\sigma^2}\sum(y_k-\theta)^2},$$

$$\frac{d^2 \log p(y_{1:N}|\theta)}{d\theta^2} = \frac{N}{\sigma^2},$$

$$\operatorname{Var}(\hat{\theta}) \geq \frac{\sigma^2}{N}.$$

The sample average as a mean estimator is unbiased with variance,

$$\hat{\theta} = \frac{1}{N}\sum_{k=1}^{N} y_k,$$

$$\operatorname{E}(\hat{\theta}) = \theta,$$

$$\operatorname{Var}(\hat{\theta}) = \frac{\sigma^2}{N}.$$

That is, the sample average is the minimum variance estimator and thus a consistent estimator. Further, the variance is equal to the CRLB, so it is also efficient.

This example can be generalized to a general multi-variable Gaussian signal.

―― **Example C.2: General Gaussian distribution** ――

Consider an arbitrarily parametrized Gaussian distribution,

$$y \in \mathcal{N}(\mu(\theta), P(\theta)),$$

$$p(y|\theta) = (2\pi)^{-N/2}|P(\theta)|^{-1/2} e^{-\frac{1}{2}(y-\mu(\theta))^T P^{-1}(\theta)(y-\mu(\theta))}.$$

It can be shown Kay (1993) that

$$\mathcal{I}(\theta) = \left(\frac{d\mu(\theta)}{d\theta}\right)^T P^{-1}\left(\frac{d\mu(\theta)}{d\theta}\right) + \frac{1}{2}\operatorname{tr}\left[\left(P^{-1}(\theta)\frac{dP(\theta)}{d\theta}\right)^2\right]$$

In the special case of an unknown mean, this simplifies to

$$\mu(\theta) = \theta, \quad P(\theta) = P, \quad \Rightarrow \mathcal{I}(\theta) = P^{-1}.$$

Typically, the CRLB decays as $1/N$ as in the examples above. However, there are examples where it may decay much faster.

---**Example C.3: Estimating the frequency of a sinusoid**---

To estimate the frequency of a sinusoid with known amplitude and phase,

$$y_k = A\sin(2\pi f k) + e_k, \tag{C.19a}$$

$$e_k \in \mathcal{N}(0, \sigma^2), \tag{C.19b}$$

it can be shown Kay (1993) that

$$\text{Var}(\hat{f}) \geq \frac{\sigma^2}{A^2}\mathcal{O}\left(N^{-3}\right). \tag{C.19c}$$

For a Gaussian stationary process, the information matrix can be expressed in terms of the spectrum.

---**Example C.4: Spectral information**---

Consider a stationary process y_k, which by *Wold's decomposition theorem* can be expressed as $y_k = h_k(\theta) \star e_k$ for some linear filter $h_k(\theta)$, where θ are the unknown parameters in h_k and e_k is a white noise process. The "*super formula*" gives the *signal spectrum* $\Phi_{yy}(f;\theta) = |H(f;\theta)|^2 \sigma_e^2$. The information matrix has components

$$[\mathcal{I}(\theta)]_{ij} = \int \frac{d\log\Phi_{yy}(f;\theta)}{d\theta_i}\frac{d\log\Phi_{yy}(f;\theta)}{d\theta_j}df. \tag{C.20}$$

C.3 Sufficient Statistics

A sufficient statistics can be seen as a summary of the information in data, where redundant and uninteresting information has been removed. One application is as a constructive way to derive MVU estimators.

The key step is the *Neyman-Fisher factorization* of the PDF:

$$p(y|\theta) = g(T(y), \theta)h(y). \tag{C.21}$$

Here, $T(y)$ is the sufficient statistics, and the function $g(T(y), \theta)$ relates the sufficient statistics to the parameter θ, while $h(y)$ is a θ independent normalization factor of the PDF.

Compare this definition with the definition of the exponential family (A.1). It should be clear that the $h(y)$ in (C.21) corresponds to the $h(y)$ in the exponent of (A.1), which shows that a sufficient statistic is always available for all members in the exponential family.

C.3 Sufficient Statistics

Example C.5: Sufficient statistics for a Gaussian signal

For a sequence of independent Gaussian variables with mean μ

$$y_k = \mu + e_k \in \mathcal{N}(\mu, \sigma^2), \quad k = 1, 2, \ldots, N, \tag{C.22a}$$

the likelihood function can be rewritten as

$$p(y|\mu) = (2\pi\sigma^2)^{-N/2} e^{-\frac{1}{2\sigma^2}\sum(y_k-\mu)^2} \tag{C.22b}$$

$$= \underbrace{e^{-\frac{1}{2\sigma^2}\sum\mu^2 + 2\mu\sum y_k}}_{g(T(y);\mu)} \underbrace{(2\pi\sigma^2)^{-N/2} e^{-\frac{1}{2\sigma^2}\sum y_k^2}}_{h(y)}. \tag{C.22c}$$

For the problem with unknown mean, $T(y_{1:N}) = \sum y_k$. However, the sufficient statistics depends on what is unknown:

- Unknown mean: $\theta = \mu \Rightarrow T(y_{1:N}) = \sum y_k$ (the case above).
- Unknown variance: $\theta = \sigma^2 \Rightarrow T(y_{1:N}) = \sum(y_k - \mu)^2$.
- Unknown mean and variance: $\theta = (\mu, \sigma^2)^T \Rightarrow T(y_{1:N}) = (\sum y_k, \sum y_k^2)^T$.

A *complete sufficient statistics* is defined by

$$\int v(T)g(T,\theta)\,dT = 0 \Rightarrow v(T) = 0. \tag{C.23}$$

The MVU estimator can now be constructed as an unbiased function of $T(y)$, and the formal conditions for an estimator $\hat{\theta}(y)$ to be MVU are

1. The sufficient statistics $T(y)$ is complete.
2. The parameter can be expressed as the expected value of (a function of) the sufficient statistics in one of the following ways:
   (a) $E(T(y)) = \theta \Rightarrow \hat{\theta}(y) = T(y)$.
   (b) $E(\alpha(T(y))) = \theta \Rightarrow \hat{\theta}(y) = \alpha(T(y))$ for some function α.

Here the first condition is a special case of the second more general one.

Example C.6: Sufficient statistics for a Gaussian signal, continued

The MVU estimator for $y_k \in \mathcal{N}(\mu, \sigma^2)$ in Example C.5 is

- $\theta = \mu \Rightarrow \hat{\mu} = \frac{1}{N}\sum y_k$.
- $\theta = \sigma^2 \Rightarrow \widehat{\sigma^2} = \frac{1}{N}\sum(y_k - \mu)^2$.
- $\theta = (\mu, \sigma^2)^T \Rightarrow \widehat{(\mu, \sigma^2)} = \left(\frac{1}{N}\sum y_k, \frac{1}{N}\sum y_k^2 - \left(\frac{1}{N}\sum y_k\right)\right)$.

It can be shown that the MVU conditions are satisfied for these cases.

C.4 Rao-Blackman-Lehmann-Scheffe's Theorem

The *Rao-Blackman-Lehmann-Scheffe's theorem* provides another way to derive the MVU estimator from the sufficient statistics using the following two steps:

1. Take any unbiased estimator $\alpha(y)$.

2. Let $\hat{\theta} = \mathrm{E}(\alpha(y)|T(y))$

The theorem says that $\hat{\theta}$ constructed in this way is the MVU estimator.

---**Example C.7: Sufficient statistics for a Gaussian signal, continued**

For the Gaussian signal in Example C.5, let the preliminary estimator for the mean simply be the first sample $\bar{\mu} = y_1$. Then the MVU estimate may be derived as

$$\hat{\mu} = \mathrm{E}(y_1|T(y_{1:N})) = \mathrm{E}(y_1|\sum y_k) = \frac{1}{N}\sum y_k. \quad (C.24)$$

The following is a standard example to illustrate that BLUE is not the same as MVE.

---**Example C.8: Mean of uniform distribution**---

Let $y_k = e_k \in \mathcal{U}(0,b)$ be samples from the uniform distribution, where $\theta = \mathrm{E}(y_k) = b/2$. $T(y_{1:N}) = \max(y_k)$ is complete sufficient statistics, so

$$\hat{\theta} = \frac{N+1}{2N}\max y_k, \quad (C.25)$$

it is unbiased and thus MVU. Note that the BLUE $\hat{\theta} = \frac{1}{N}\sum y_k$ is not MVU for this problem.

C.5 Maximum Likelihood Estimation

The *maximum likelihood estimator* (MLE) is defined as

$$\hat{\theta}^{ML} = \arg\max_{\theta} p(y|\theta). \quad (C.26)$$

One important property of the MLE is that it is asymptotically Gaussian distributed as

$$\sqrt{N}\hat{\theta}^{ML} \sim \mathrm{As}\mathcal{N}(\theta^o, \mathcal{I}^{-1}(\theta^o)), \quad (C.27)$$

if (necessary conditions)

- The likelihood function $p(y|\theta)$ is twice differentiable.

- The likelihood function satisfies the regularity conditions.

- The information matrix is non-zero at the true parameter $\mathcal{I}(\theta^o) \neq 0$.

The result (C.27) implies that the ML estimator asymptotically satisfies the CRLB and is asymptotically an efficient estimator. For finite data sets, the ML estimator might not even be unbiased.

The ML is by definition invariant under nonlinear transformations, so

$$\eta = g(\theta) \Rightarrow \hat{\eta}^{ML} = g(\hat{\theta}^{ML}). \tag{C.28}$$

C.6 The Method of Moments

The key idea in the *method of moments* is to estimate the first p moments of data, and match these to the analytical moments of the parametric distribution $p(y|\theta)$:

$$\mu_i = \mathrm{E}(y_k^i) = g_i(\theta), \quad i = 1, 2, \ldots, p \tag{C.29a}$$

$$\hat{\mu}_i = \frac{1}{N} \sum_{k=1}^{N} y_k^i, \quad i = 1, 2, \ldots, p \tag{C.29b}$$

$$\mu = g(\theta), \tag{C.29c}$$

$$\hat{\theta} = g^{-1}(\hat{\mu}). \tag{C.29d}$$

The method of moments is generally not efficient and thus inferior to the ML method. However, it is in many cases easier to derive and implement. For Gaussian mixtures, the MLE does not lead to analytical solutions so numerical algorithms have to be applied directly to the definitions, where the whole data vector has to be used. Using the method of moments, closed expressions can be derived as functions of reduced data statistics.

The area of *higher order statistics* relates to problems where the method of moments with $p > 2$ is used.

─── **Example C.9: Gaussian mixture with known variances** ───

Consider a Gaussian mixture

$$p(y|\alpha) = \alpha \mathcal{N}(y; 0, \sigma_1^2) + (1-\alpha)\mathcal{N}(y; 0, \sigma_2^2). \tag{C.30a}$$

We have
$$\mu_1 = \mathrm{E}(y) = 0, \qquad \text{(C.30b)}$$
$$\mu_2 = \mathrm{E}(y^2) = \alpha\sigma_1^2 + (1-\alpha)\sigma_2^2, \qquad \text{(C.30c)}$$
$$\hat{\mu}_2 = \frac{1}{N}\sum_{k=1}^{N} y_k^2, \qquad \text{(C.30d)}$$
$$\hat{\alpha} = \frac{\hat{\mu}_2 - \sigma_2^2}{\sigma_1^2 - \sigma_2^2} \qquad \text{(C.30e)}$$

The estimate is consistent, but for a finite N it is not necessarily a feasible solution satisfying $\hat{\alpha} \in [0,1]$.

Example C.10: Gaussian mixture with unknown variances

Consider now the case
$$p(y|\theta) = \alpha\mathcal{N}(y;0,\sigma_1^2) + (1-\alpha)\mathcal{N}(y;0,\sigma_2^2), \qquad \text{(C.31a)}$$
$$\theta = (\alpha,\sigma_1^2,\sigma_2^2)^T. \qquad \text{(C.31b)}$$

There are three unknowns in θ, and since all odd moments are zero at least the following even moments are needed:
$$\mu_2 = \mathrm{E}(y^2) = \alpha\sigma_1^2 + (1-\alpha)\sigma_2^2, \qquad \text{(C.31c)}$$
$$\mu_4 = \mathrm{E}(y^4) = 3\alpha\sigma_1^4 + 3(1-\alpha)\sigma_2^4 + \alpha(1-\alpha)\sigma_1^2\sigma_2^2, \qquad \text{(C.31d)}$$
$$\mu_6 = \mathrm{E}(y^6) = 15\alpha\sigma_1^6 + 15(1-\alpha)\sigma_2^6 + 3\alpha(1-\alpha)\sigma_1^4\sigma_2^2 + 3\alpha(1-\alpha)\sigma_1^2\sigma_2^4. \qquad \text{(C.31e)}$$

This yields a nonlinear equation system in the three statistics $\hat{\mu}_2, \hat{\mu}_4, \hat{\mu}_6$.

C.7 Bayesian Methods

The main difference in the *Bayesian approach*, compared to the classical Fisher statistics, is that the parameter vector is considered to be a stochastic variable rather than an unknown parameter. The correspondence to the likelihood function is the *a posteriori* distribution, or simply *posterior*, $p(\theta|y)$. Using *Bayes' theorem* $P(A|B) = P(A,B)/P(B)$ twice we get $P(A|B) = P(A)P(B|A)/P(B)$ and this gives

$$p(\theta|y) = \frac{p(y|\theta)p(\theta)}{p(y)} = \frac{p(y|\theta)p(\theta)}{\int p(y|\theta)p(\theta)d\theta}. \qquad \text{(C.32)}$$

Here,

C.7 Bayesian Methods

- $p(\theta)$ is the *prior*.

- $p(y|\theta)$ is the likelihood function.

- $p(y)$ is for estimation purposes a normalization constant that is needed to obtain a proper distributions in θ that integrates to one.

The *maximum a posteriori* (*MAP*) estimate is defined as

$$\hat{\theta} = \arg\max_{\theta} p(\theta|y). \tag{C.33}$$

The *Bayes estimator* is defined as

$$\hat{\theta}^B = \mathrm{E}(\theta|y), \tag{C.34}$$

which minimizes the *Bayesian mean square error* (*BMSE*),

$$\mathrm{BMSE} = \mathrm{E}\big((\hat{\theta}^B - \theta^o)^2|y\big). \tag{C.35}$$

A rather naive observation is that BMSE coincides with ML if the prior is constant $p(\theta) = c$. Such a prior is usually referred to as a *noninformative prior*. However, the conceptual difference between Fisherian and Bayesian statistics is much more involved, and there are almost religous discussions in literature of pros and cons with each method. First, an *improper prior* is not a proper PDF that integrates to one. The pragmatic Bayesian do not care about this, if the posterior becomes a proper PDF. In such cases,

$$p(\theta) = c_1 \Rightarrow p(\theta|y) = c_2 p(y|\theta), \hat{\theta}^{ML} = \hat{\theta}^B. \tag{C.36}$$

There is, however, a main difference in the intepretation. For instance, a linear Gaussian model leads to the following two intepretations of the same numbers

$$\hat{\theta}^{ML} \sim \mathcal{N}(\theta^o, P), \tag{C.37}$$

$$\theta^o \sim \mathcal{N}(\hat{\theta}^B, P). \tag{C.38}$$

In the Fisherian approach, the data are stochastic, and since $\hat{\theta}^{ML}$ is a function of data it also becomes stochastic. In the Bayesian approach, the parameter is stochastic, while the data are conditioned on and thus known.

An information approach to unify the Fisherian and Bayesian approaches is to interpret the prior as a piece of information and use the fusion formula to merge prior information with data information. The prior can thus be seen as a fictitious measurement y_0 with mean $\mathrm{E}(y_0) = \theta_0$ and $\mathrm{Cov}(y_0) = P_0$.

C.7.1 Marginalization

Assume the parameter vector is partitioned as

$$\theta = \begin{pmatrix} \theta_1 \\ \theta_2 \end{pmatrix}, \tag{C.39}$$

where θ_1 is the unknown quantity of interest, while θ_2 are *nuisance parameters*, which is an unknown quantity of no interest. Using Bayes' formula and marginalization, we get

$$p(\theta_1|y) = \int p(\theta_1,\theta_2|y)d\theta_2 = \int \frac{p(y|\theta_1,\theta_2)p(\theta_1,\theta_2)}{p(y)}d\theta_2 \tag{C.40a}$$

$$= \frac{p(\theta_1)}{p(y)} \int p(y|\theta_1,\theta_2)p(\theta_2|\theta_1)d\theta_2. \tag{C.40b}$$

Note that marginalization and the *law of total probability* refer to the same formula, but with somewhat different approaches.

---**Example C.11: Gaussian signal**---

Consider the signal model $y_k = \mu + e_k$, $k = 1, 2, \ldots, N$. In a completely deterministic view, there are N equations and $N+1$ unknowns, where $\theta = (\mu, e_1, \ldots, e_N)^T$. The likelihood function is thus

$$p(y_k|\theta) = \delta(y_k - \mu - e_k), \tag{C.41}$$

$$p(y_{1:N}|\theta) = \prod_{k=1}^{N} \delta(y_k - \mu - e_k), \tag{C.42}$$

where $\delta(x)$ is the Dirac distribution. Using the marginalization formula,

$$p(y_k|\mu) = \int \delta(y_k - \mu - e_k)p(e_k)de_k = \mathcal{N}(y_k; \mu, \sigma^2), \tag{C.43}$$

$$p(y_{1:N}|\mu) = \prod_{k=1}^{N} \int \delta(y_k - \mu - e_k)p(e_k)de_k = \prod_{k=1}^{N} \mathcal{N}(y_k; \mu, \sigma^2). \tag{C.44}$$

Here, $\mathcal{N}(x; \mu, P)$ is the Gaussian PDF with mean μ and covariance P evaluated at x. One can here continue and consider σ^2 as an unknown nuisance parameter, and marginalize the result over a prior of σ^2. See Section 5.B.4 in Gustafsson (2001) for details and derivations.

Note that the Bayesian approach is needed for marginalization, since a prior is needed for the unknowns. In one way, even the maximum likelihood approach is semi-Bayesian, since the noise is considered as stochastic variables with a certain prior.

C.7.2 Bayesian Risk

The BMSE in (C.35) is based on a quadratic cost of the parameter estimation error. More generally, a cost function $C(\theta - \hat{\theta})$ can be introduced. For a scalar parameter, the BMSE is $C(\theta - \hat{\theta}) = (\theta - \hat{\theta})^2$. A more robust cost (the *Huber's cost*) is obtained by switching from quadratic to a linear cost,

$$C(\theta - \hat{\theta}) = \begin{cases} (\theta - \hat{\theta})^2, & |\theta - \hat{\theta}| \leq A, \\ A^2 - A + |\theta - \hat{\theta}|, & |\theta - \hat{\theta}| > A. \end{cases} \quad (C.45)$$

The *hit or miss cost* is defined as

$$C(\theta - \hat{\theta}) = \begin{cases} 0, & |\theta - \hat{\theta}| \leq \delta, \\ 1, & |\theta - \hat{\theta}| > \delta. \end{cases} \quad (C.46)$$

The *Bayesian risk* R is the expected value of the cost

$$R = \mathrm{E}(C(\theta - \hat{\theta})) \sim \int C(\theta - \hat{\theta}) p(\theta|y) d\theta \quad (C.47)$$

C.8 Recursive Bayesian Estimation

Consider an observation model in terms of a likelihood from the exponential family:

$$p(y_k|x) = a(x)b(y)e^{-0.5\,\mathrm{tr}(g^T(x)h(y))}, \quad x \in \Omega_x. \quad (C.48)$$

Compare with (A.1) in Appendix A. A finite-dimensional recursive estimator exists for the *conjugate prior*

$$p(x|y_{1:k-1}) = a(x)^{\nu_{k-1}} e^{-0.5 g^T(x) v_{k-1}} \quad (C.49)$$
$$= \mathrm{EF}(\nu_{k-1}, v_{k-1}), \quad x \in \Omega_x, \quad (C.50)$$

in which case the posterior is given by

$$p(x|y_{1:k}) = a(x)^{\nu_k} e^{-0.5 g^T(x) v_k} \quad (C.51)$$
$$= a(x)^{\nu_{k-1}+1} e^{-0.5 g^T(x)(v_{k-1}+h(y))}, \quad x \in \Omega_x, \quad (C.52)$$

or, in symbolic form

$$p(x|y_{1:k}) = \mathrm{EF}(\nu_k, v_k) \quad (C.53)$$
$$= \mathrm{EF}(\nu_{k-1}+1, v_{k-1}+h(y)), \quad x \in \Omega_x. \quad (C.54)$$

Different combinations of likelihoods and conjugate priors are summarized in Table A.6. One of these instances is exemplified below.

Estimation Theory

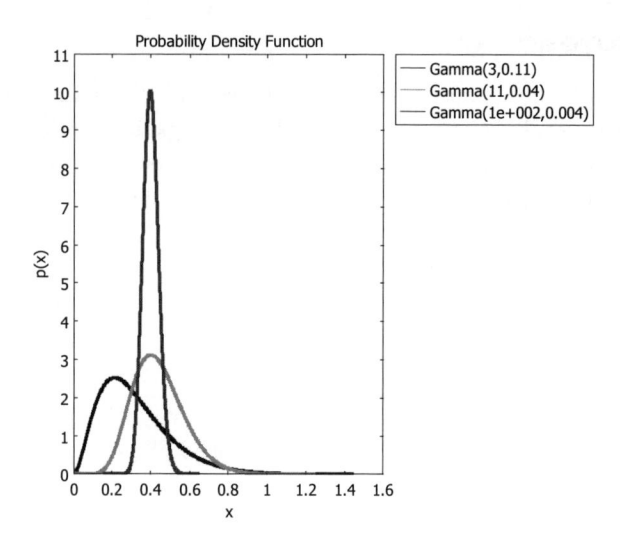

Figure C.1: *Estimating the shape parameter x in the exponential distribution using the* GAMMA *conjugate prior*

Example C.12: Recursive estimation of exponential rate

Consider a vector y of stochastic variables from the EXP$(1/x)$ distribution. The posterior of the inverse shape parameter x is given by GAMMA$(\nu_0+n, \upsilon_0+\sum_k y_k)$. The following code illustrates the principle.

```
lh=expdist;
x0=0.4;
y=rand(expdist(1/x0),100);
p0=gammadist(1,1);
p2=posterior(p0,lh,y(1:2));
p10=posterior(p0,lh,y(1:10));
p100=posterior(p0,lh,y(1:100));
plot(p2,p10,p100)
```

The resulting graph is shown in Figure C.1.

D
Detection Theory

This appendix briefly defines the main concepts and results in detection theory. It is in no way complete. For a thorough treatment of detection theory, see Lehmann (1991b) for a statistical perspective and Kay (1998) for a signal processing perspective.

D.1 Notation

Given observations y_k with known distribution under H_0 and H_1, respectively. A general detection problem can be expressed using a *test statistic* $T(y)$ as

$$\text{Decide } H_0 \text{ if } T(y) \leq \gamma$$
$$\text{Decide } H_1 \text{ if } T(y) > \gamma$$

Basic design parameters:

- The *probability of false alarm* $P_{FA} = P(H_1|H_0) = \int_{y:T(y)>\gamma} p(y|H_0)\,dy$. This is also called *type I error*.

- The *probability of detection* $P_D = P(H_1|H_1) = \int_{y:T(y)>\gamma} p(y|H_1)\,dy$, also called the *power of the test* with notation β. The complementary probability $1 - P_D$ is called *type II error*.

Standard plots in detection theory include:

- The *receiver operating characteristics* (ROC) plots P_D versus P_{FA} with γ being the parameter. Just guessing H_1 with probability P_{FA} gives

$P_D = P_{FA}$. All clever tests give a curve above this straight line. The Neyman-Pearson test defined in the next section provides in some cases an upper bound curve for all tests.

- The *detection performance* P_D versus SNR for a given P_{FA}. Again this performance is maximized for the Neyman-Pearson test.

- The *bit-error rate* (*BER*). In communication, both hypotheses are equal, and the design is to get $P_D = P_{FA}$. A BER-plot shows $P_D = P_{FA}$ versus SNR.

D.2 The Likelihood Ratio Test

The only fundamental result in detection theory is the *Neyman-Pearson's theorem*, which says that the *likelihood ratio*,

$$T(y) = \frac{p(y|H_1)}{p(y|H_0)}, \tag{D.1}$$

is the optimal test statistic for a *simple hypothesis test* of this kind, in the meaning that it maximizes P_D for a given $P_{FA} = Q(\gamma)$. The test defined by the likelihood ratio is called *likelihood ratio test* (*LRT*).

For the proof, rewrite the LRT as deciding if $\mathcal{T} = 0$ or $\mathcal{T} = 1$ using

$$\mathcal{T} = \begin{cases} 1, & \text{if } p(x|H_1) > \gamma p(x|H_0) \\ 0, & \text{if } p(x|H_1) < \gamma p(x|H_0), \end{cases} \tag{D.2a}$$

Let $0 \leq \mathcal{T}' \leq 1$ denote any test with less or equal P'_{FA}. The key inequality

$$\int_{-\infty}^{\infty} (\mathcal{T} - \mathcal{T}')\big(p(x|H_1) - \gamma p(x|H_0)\big)\, dx \geq 0, \tag{D.2b}$$

is proven by noting that both factors in the integrand always have the same sign. The result now follows by re-arranging the integral into

$$(P_D - P'_D) \geq \gamma (P_{FA} - P'_{FA}) \geq 0 \tag{D.2c}$$

where the last inequality follows from the false alarm assumption. This shows $\mathcal{T}$ optimize the power! One has to be a bit more careful if $p(x|H_1) = \gamma p(x|H_0)$ has a non-zero probability.

The ROC curve has the following properties:

- The LRT maximizes any performance criterion that values low false alarm probabilities and/or high detection probabilities.

- The curve is concave.

- The slope of the LRT ROC curve is γ.

To prove the last property,

$$E(T(x)^n|H_1) = \int \left[\frac{p(x|H_1)}{p(x|H_0)}\right]^n p(x|H_1)dx = \quad \text{(D.3a)}$$

$$\int \left[\frac{p(x|H_1)}{p(x|H_0)}\right]^{n+1} p(x|H_0)dx = E(T(x)^{n+1}|H_0) \quad \text{(D.3b)}$$

which can be written

$$\int t^n g(t|H_1)dt = \int t^{n+1} g(t|H_0)dt, \ \forall n, \quad \text{(D.3c)}$$

where $g(t|H_i)$ is the PDF for the likelihood ratio under H_i. This means that $g(t|H_1) = tg(t|H_0)$. Now

$$P_D = \int_\gamma^\infty g(t|H_1)dt, \quad \text{(D.3d)}$$

$$P_{FA} = \int_\gamma^\infty g(t|H_0)dt, \quad \text{(D.3e)}$$

and

$$\frac{dP_D}{dP_{FA}} = \frac{\frac{dP_D}{d\gamma}}{\frac{dP_{FA}}{d\gamma}} = \frac{g(\gamma|H_1)}{g(\gamma|H_0)} = \gamma. \quad \text{(D.3f)}$$

Observe that $\gamma \geq 0$.

D.3 Detection of Known Mean in Gaussian Noise

The simplest possible problem is to decide whether there is a known mean A in an observed signal or not:

$$H_0 : y_k = e_k,$$
$$H_1 : y_k = A + e_k.$$

The more general problem of detecting a given signal

$$H_0 : y_k = e_k,$$
$$H_1 : y_k = s_k + e_k.$$

can be rewritten to the previous case with an unknown mean as the following examples illustrate.

Example D.1: Communication

The signal $s(t) = 2A\sin(\omega t)$, $A > 0$, is transmitted on the carrier frequency ω and $y(t) = s(t) + e(t)$ is received. Assume that ωT is a multiple of 2π, and let

$$\bar{y} = \frac{1}{T}\int_0^T y(t)\sin(\omega t) = A + \frac{1}{T}\int_0^T e(t)\sin(\omega t) = A + \bar{e}.$$

If $e(t)$ is a Gaussian process, $\bar{e} \sim \mathcal{N}(0, \frac{\sigma^2}{T})$. Otherwise, the distribution is asymptotically Gaussian by the central limit theorem.

Example D.2: Matched filter

Detection of a general known signal s_k observed with Gaussian noise as $y_k = s_k + e_k$ is done using a matched filter defined as

$$\bar{y} = \frac{1}{N}\sum_{k=0}^{N-1} y_k s_k = A + \bar{e}$$

where $A = \frac{1}{N}\sum_{k=0}^{N-1} s_k^2$ and $\bar{e} \sim \mathcal{N}(0, \frac{\sigma^2}{N})$.

The Neyman-Pearson theorem for the general problem gives

$$T(y_{1:N}) = \frac{\frac{1}{(2\pi\sigma^2)^{N/2}} e^{-\frac{\sum_{k=0}^{N-1}(y_k - A)^2}{2\sigma^2}}}{\frac{1}{(2\pi\sigma^2)^{N/2}} e^{-\frac{\sum_{k=0}^{N-1}(y_k)^2}{2\sigma^2}}} = e^{-\frac{NA^2 - \sum_{k=0}^{N-1} 2Ay_k}{2\sigma^2}} > \gamma.$$

Taking logarithm and simplifying give

$$\bar{T}(y_{1:N}) = \frac{1}{N}\sum_{k=0}^{N-1} y_k > \frac{\sigma^2}{NA}\log(\gamma) + \frac{A}{2} = \bar{\gamma}.$$

Now

$$\bar{T}(y_{1:N}) \sim \begin{cases} \mathcal{N}\left(0, \frac{\sigma^2}{N}\right) & \text{under } H_0 \\ \mathcal{N}\left(A, \frac{\sigma^2}{N}\right) & \text{under } H_1 \end{cases}$$

D.3 Detection of Known Mean in Gaussian Noise

so

$$P_{FA} = P(\bar{T}(y_{1:N}) > \bar{\gamma}|H_0) = 1 - Q\left(\frac{\bar{\gamma}}{\sqrt{\sigma^2/N}}\right)$$

$$\bar{\gamma} = \sqrt{\frac{\sigma^2}{N}} Q^{-1}(1 - P_{FA})$$

$$P_D = P(\bar{T}(y_{1:N}) > \bar{\gamma}|H_1) = 1 - Q\left(\frac{\bar{\gamma} - A}{\sqrt{\sigma^2/N}}\right)$$

$$= 1 - Q\left(Q^{-1}(1 - P_{FA}) - \sqrt{\frac{NA^2}{\sigma^2}}\right)$$

The last expression shows that P_D is a function of the desired P_{FA} and SNR only.

Example D.3: Matlab code for ROC curve

Typical MATLAB™ code to generate a ROC curve for the general detection problem is as follows:

```
lambda=[-0.1:0.1:4];
MC=1000;
K=5;
A=1;
sigma=1;
y=sigma*randn(N,MC);
T1=mean(y);
T2=max(y);
for k=1:length(lambda)
    PD1(k)=mean(T1+A>lambda(k));
    PD2(k)=mean(T2+A>lambda(k));
    PFA1(k)=mean(T1>lambda(k));
    PFA2(k)=mean(T2>lambda(k));
end
```

Simple line search to design λ and calculate P_D given P_{FA}:

```
PFAdesigk=0.1;
ind=find(PFA1<PFAdesign); lambda1=lambda(ind(1)), PD1desigk=PD1(ind(1))
ind=find(PFA2<PFAdesign); lambda2=lambda(ind(1)), PD2desigk=PD2(ind(1))
```

Here, `interp1` is one alternative to interpolate in the plot. An example is shown in Figure D.1.

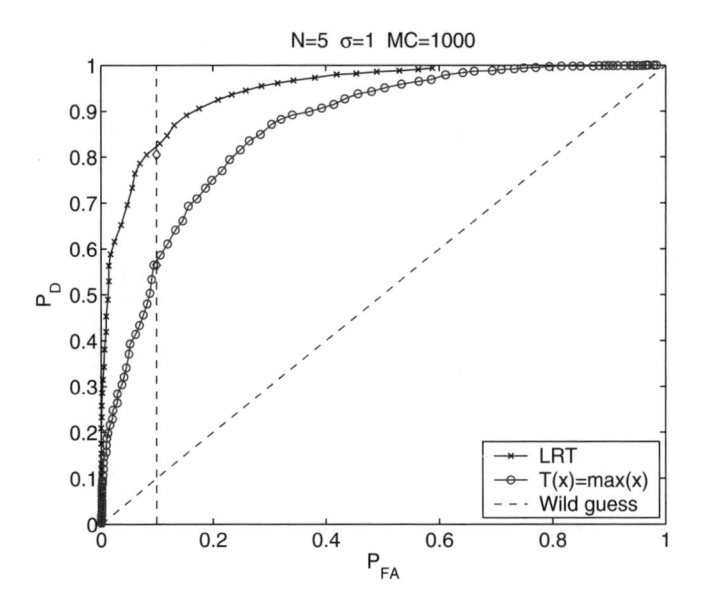

Figure D.1: ROC plot for a simple hypothesis test, illustrating a random guess, a logical ad-hoc test, and the optimal test based on the likelihood ratio.

D.4 Eliminating Unknown Parameters

D.4.1 GLRT

Consider a test between two alternative PDF's with different parameter vectors:

$$H_0 : y \sim p_0(y|\theta_0), \tag{D.4a}$$
$$H_1 : y \sim p_1(y|\theta_1). \tag{D.4b}$$

The general principle in the *generalized likelihood ratio test (GLRT)* is to plug in the maximum likelihood estimate of θ_i under H_i, and use the test statistic

$$T(y) = \frac{\max_{\theta_1} p_1(y|\theta_1)}{\max_{\theta_0} p_0(y|\theta_0)} = \frac{p_1(y|\hat{\theta}_1^{ML})}{p_0(y|\hat{\theta}_0^{ML})}. \tag{D.5}$$

A more common special case is to test a significant deviation from a nominal value of the parameter vector,

$$H_0 : \theta = \theta_0,$$
$$H_1 : \theta \neq \theta_0.$$

D.4 Eliminating Unknown Parameters

The GLRT becomes

$$T(y) = \frac{\max_{\theta_1} p(y|\theta_1)}{p(y|\theta_0)} = \frac{p(y|\hat{\theta}_1^{ML})}{p(y|\theta_0)}$$

There is a useful general result available for the asymptotic distribution of this test statistic:

$$2\log T(y) \sim \begin{cases} \chi_n^2 & \text{under } H_0, \\ \chi_n^2(\lambda) & \text{under } H_1, \end{cases} \quad \text{(D.6a)}$$

$$\lambda = (\theta_1 - \theta_0)^T \mathcal{I}(\theta_0)(\theta_1 - \theta_0). \quad \text{(D.6b)}$$

where $\mathcal{I}(\theta)$ is Fisher's information matrix.

The following example illustrates that the log GLRT is χ^2 distributed even non-asymptotically for a linear Gaussian problem.

---**Example D.4: Detection of known mean in Gaussian noise**---

Consider

$$H_0 : y_k = e_k$$
$$H_1 : y_k = A + e_k$$

with unknown A and Gaussian noise with known variance σ^2. Then

$$2\log T(y_{1:N}) = \frac{N\hat{A}^2}{\sigma^2}$$

$$\hat{A} = \frac{1}{N}\sum_{k=0}^{N-1} y_k \in \begin{cases} \mathcal{N}\left(0, \frac{\sigma^2}{N}\right) & \text{under } H_0 \\ \mathcal{N}\left(A, \frac{\sigma^2}{N}\right) & \text{under } H_1 \end{cases}$$

$$\Rightarrow 2\log T(y_{1:N}) \sim \begin{cases} \chi_1^2 & \text{under } H_0 \\ \chi_1^2\left(\frac{NA^2}{\sigma^2}\right) & \text{under } H_1. \end{cases}$$

D.4.2 MLRT

Consider again the general detection problem (D.4)

$$H_0 : y \sim p_0(y|\theta_0)$$
$$H_1 : y \sim p_1(y|\theta_1)$$

The general principle in the *marginalized likelihood ratio test* (*MLRT*) is to use a prior on θ and integrate out its influence in the LRT,

$$T(y) = \frac{\int_{\theta_1} p_1(y|\theta_1)p(\theta_1)d\theta_1}{\int_{\theta_0} p_0(y|\theta_0)p(\theta_0)d\theta_0}$$

This is called the *Bayes factor* in the *Bayesian approach*.

D.5 Nuisance Parameters

Consider now a test where there are two kind of parameters:

$$H_0 : \theta = (\theta_0, \eta), \qquad \text{(D.7)}$$
$$H_1 : \theta \neq (\theta_0, \eta). \qquad \text{(D.8)}$$

Here η is an unknown *nuisance parameter* not relevant for the test. It might for instance contain the variance of the data.

D.5.1 Mixed MLRT/GLRT Approach

The general principle using the Bayesian approach is to marginalize nuisance parameters, estimate the other parameters, which leads to a mixed marginalization/estimation approach. The LRT becomes

$$T(y) = \frac{\max_\theta \int_\eta p(y|\theta, \eta) d\eta}{\int_\eta p(y|\theta_0, \eta) d\eta} = \frac{\int_\eta p(y|\hat{\theta}^{ML}, \eta) d\eta}{\int_\eta p(y|\theta_0, \eta) d\eta}.$$

D.5.2 GLRT

The GLRT principle implies plugging in the ML estimate of nuisance parameters:

$$T(y) = \frac{p(y|\hat{\theta}^{ML}, \hat{\eta}^{ML})}{p(y|\theta_0, \hat{\eta}^{ML})}.$$

The general result for the asymptotic distribution can here be expressed as

$$2 \log T(y) \sim \begin{cases} \chi^2_{\dim(\theta)} & \text{under } H_0, \\ \chi^2_{\dim(\theta)}(\lambda) & \text{under } H_1, \end{cases}$$

$$\lambda = (\theta_1 - \theta_0)^T \underbrace{\left(\mathcal{I}_{\theta,\theta} - \mathcal{I}_{\theta,\eta} \mathcal{I}_{\eta,\eta}^{-1} \mathcal{I}_{\eta,\theta}\right)}_{[\mathcal{I}^{-1}]_{\theta,\theta}} (\theta_1 - \theta_0),$$

where $\mathcal{I}(\theta)$ is Fisher's information matrix partitioned as

$$\mathcal{I} = \begin{pmatrix} \mathcal{I}_{\theta,\theta} & \mathcal{I}_{\theta,\eta} \\ \mathcal{I}_{\eta,\theta} & \mathcal{I}_{\eta,\eta} \end{pmatrix},$$

evaluated at the true (θ_0, η), and θ_1 denotes the true value under H_1.

Note that the underbraced expression $[\mathcal{I}^{-1}]_{\theta,\theta}$ is the Schur complement to compute the upper left corner of the matrix inverse.

D.5.3 Wald Test

Use

$$T(y) = (\hat{\theta} - \theta_0)^T [\mathcal{I}^{-1}(\widehat{\theta, \eta}|H_1)]_{\theta,\theta} (\hat{\theta} - \theta_0)$$

where $\widehat{\theta, \eta}|H_1$ is the ML estimate under H_1.

D.5.4 Rao Test

Use

$$T(y) = \left.\frac{d\log(p(y|\theta,\eta))}{d\theta}\right|^T_{\theta_0,\hat{\eta}|H_0} [\mathcal{I}^{-1}(\theta_0,\hat{\eta}|H_0)]_{\theta,\theta} \left.\frac{d\log(p(y|\theta,\eta))}{d\theta}\right|_{\theta_0,\hat{\eta}|H_0}$$

where $\hat{\eta}|H_0$ is the ML estimate under H_0.

Note that θ does not have to be estimated here, as it has to in the Wald test.

D.5.5 Locally Most Powerful Test

The *locally most powerful* (*LMP*) test is a special case of the Rao test for the case of scalar parameter, no nuisance parameter and one-sided test:

$$H_0 : \theta = \theta_0,$$
$$H_1 : \theta > \theta_0.$$

Take the test statistic

$$T(y) = \frac{\left.\frac{d\log(p(y|\theta))}{d\theta}\right|_{\theta_0}}{\sqrt{\mathcal{I}(\theta_0)}} \sim \begin{cases} \mathcal{N}(0,1) & \text{under } H_0, \\ \mathcal{N}(\sqrt{\mathcal{I}(\theta_0)}(\theta_1-\theta_0),1) & \text{under } H_1, \end{cases}$$

for small changes $\theta_1 - \theta_0 > 0$. The derivation is based on local analysis by Taylor expansion.

D.6 Bayesian Extensions

D.6.1 Priors

Assume priors $p(H_0)$ and $p(H_1)$, respectively. Using a Bayesian decision rule, compare the *a posteriori* distribution of each hypothesis,

$$p(H_1|y) \underset{H_0}{\overset{H_1}{\gtrless}} p(H_0|y) \Leftrightarrow$$

$$\frac{p(y|H_1)p(H_1)}{p(y)} \underset{H_0}{\overset{H_1}{\gtrless}} \frac{p(y|H_0)p(H_0)}{p(y)}$$

$$\frac{p(y|H_1)}{p(y|H_0)} \underset{H_0}{\overset{H_1}{\gtrless}} \frac{p(H_0)}{p(H_1)} = \lambda$$

which shows that Bayes' rule provides a way to design the threshold in the LRT test.

D.6.2 Multiple Tests

The use of prior enables multiple hypotheses tests:

$$\hat{k} = \arg\max_k p(H_k|y) = \arg\max_k p(y|H_k)p(H_k)$$

D.6.3 Risk

Define a cost to each decision

$$C_{ij} = P(H_i|H_j),$$

that is, C_{10} is the false alarm cost and C_{01} is the missed detection cost. Then

$$\frac{p(y|H_1)}{p(y|H_0)} \underset{H_0}{\overset{H_1}{\gtrless}} \frac{(C_{10} - C_{00})p(H_0)}{(C_{01} - C_{11})p(H_1)} = \lambda$$

minimizes the Bayes risk or expected cost

$$\mathcal{R} = \mathrm{E}(C) = \sum_{i=0}^{1}\sum_{j=0}^{1} C_{ij}P(H_i|H_j)p(H_j).$$

D.7 Linear Model

Consider a linear Gaussian model

$$\mathbf{y} = \mathbf{H}x + \mathbf{e}, \quad \mathbf{e} \sim \mathcal{N}(0, \mathbf{R}).$$

This gives the likelihood

$$p(\mathbf{y}|x) = \frac{1}{(2\pi)^{N/2}\sqrt{\det(\mathbf{R})}} e^{-\frac{1}{2}(\mathbf{y}-\mathbf{H}x)^T \mathbf{R}^{-1}(\mathbf{y}-\mathbf{H}x)}.$$

D.7.1 Known Parameter

The alternative parameter x_1 is here specified, so the test is

$$H_0 : x = 0,$$
$$H_1 : x = x_1.$$

Neyman-Pearson gives log LRT

$$T_1(\mathbf{y}) = \log\left(\frac{p(\mathbf{y}|x_1)}{P(\mathbf{y}|x=0)}\right) = -0.5(\mathbf{y} - \mathbf{H}x_1)^T \mathbf{R}^{-1}(\mathbf{y} - \mathbf{H}x_1) + 0.5\mathbf{y}^T\mathbf{R}^{-1}\mathbf{y}$$

$$= \mathbf{y}^T\mathbf{R}^{-1}\mathbf{H}x_1 - 0.5x_1^T\mathbf{H}^T\mathbf{R}^{-1}\mathbf{H}x_1$$

D.7 Linear Model

The last term is independent of data, so the following test statistic can be used instead

$$T_2(\mathbf{y}) = \mathbf{y}^T \mathbf{R}^{-1} \mathbf{H} x_1 \sim \mathcal{N}(x^T \mathbf{H}^T \mathbf{R}^{-1} \mathbf{H} x_1, x_1^T \mathbf{H}^T \mathbf{R}^{-1} \mathbf{H} x_1),$$

To get a standard Gaussian distribution with variance one, a scaled test statistic is

$$T_3(\mathbf{y}) = \frac{\mathbf{y}^T \mathbf{R}^{-1} \mathbf{H} x_1}{\sqrt{x_1^T \mathbf{H}^T \mathbf{R}^{-1} \mathbf{H} x_1}} \sim \mathcal{N}\left(\frac{x^T \mathbf{H}^T \mathbf{R}^{-1} \mathbf{H} x_1}{\sqrt{x_1^T \mathbf{H}^T \mathbf{R}^{-1} \mathbf{H} x_1}}, 1\right),$$

$$= \begin{cases} \mathcal{N}(0, 1) & H_0 \\ \mathcal{N}\left(\sqrt{x_1^T \mathbf{H}^T \mathbf{R}^{-1} \mathbf{H} x_1}, 1\right), & H_1. \end{cases}$$

This implies that the ROC curve can be generated using the following formula:

$$P_D = 1 - Q\left(Q^{-1}(1 - P_{FA}) - \sqrt{x_1^T \mathbf{H}^T \mathbf{R}^{-1} \mathbf{H} x_1}\right).$$

D.7.2 Unknown Parameter

The alternative parameter is now unspecified, so the test is

$$H_0 : x = 0,$$
$$H_1 : x \neq 0.$$

The ML estimate of x under H_1 is

$$\hat{x} = (\mathbf{H}^T \mathbf{H})^{-1} \mathbf{H}^T \mathbf{y},$$

which gives the log GLRT

$$T(\mathbf{y}) = 2 \log\left(\frac{p(\mathbf{y}|\hat{x})}{P(\mathbf{y}|x = 0)}\right)$$
$$= -(\mathbf{y} - \mathbf{H}\hat{x})^T \mathbf{R}^{-1} (\mathbf{y} - \mathbf{H}\hat{x}) + \mathbf{y}^T \mathbf{R}^{-1} \mathbf{y}$$
$$= 2\mathbf{y}^T \mathbf{R}^{-1} \mathbf{H}\hat{x} - \hat{x}^T \mathbf{H}^T \mathbf{R}^{-1} \mathbf{H}\hat{x}$$
$$= \cdots = \mathbf{y}^T \mathbf{R}^{-1} \mathbf{H} (\mathbf{H}^T \mathbf{R}^{-1} \mathbf{H})^{-1} \mathbf{H}^T \mathbf{R}^{-1} \mathbf{y}$$
$$\sim \begin{cases} \chi^2_{n_x}, & \text{under } H_0, \\ \chi^2_{n_x}(x^T \mathbf{H}^T \mathbf{R}^{-1} \mathbf{H} x), & \text{under } H_1. \end{cases}$$

The ROC curve can now be generated using

$$P_D(x) = 1 - Q_{\chi^2_{n_x}(x^T \mathbf{H}^T \mathbf{R}^{-1} \mathbf{H} x)}\left(Q^{-1}_{\chi^2_{n_x}}(1 - P_{FA})\right).$$

E

Least Squares Theory

In this chapter, different aspects of the least squares solution for the linear regression model

$$y_k = H_k x + e_k \qquad (E.1)$$

will be highlighted. First a sequential version of weighted least squares is derived together with the LS over a sliding window (WLS). Then it is shown how certain off-line expressions that will naturally appear in estimation and detection applications can be updated sequentially. Some asymptotic approximations are derived. Finally, some marginal densities are derived.

E.1 Derivation of Least Squares Algorithms

We now derive a sequential implementation of the weighted LS algorithms. It bears much in common with the *recursive least squares* (*RLS*) algorithm, but there is no forgetting factor here. In fact, it coincides with the RLS algorithm when the forgetting factor is one

E.1.1 Weighted Least Squares in Sequential Form

We begin by deriving a sequential solution to the LS estimator. The weighted multi-variable recursive least squares algorithm, without forgetting factor (see Ljung and Söderström (1983)), minimizes the loss function

$$V_N(x) = \sum_{k=1}^{N}(y_k - H_k x)^T R_k^{-1}(y_k - H_k x) \qquad (E.2)$$

with respect to x. Here R_k should be interpreted as the measurement covariance matrix, which in this appendix can be a matrix corresponding to a vector valued measurement. Differentiation of (E.2) gives

$$V'_N(x) = 2 \sum_{k=1}^{N} (-H_k^T R_k^{-1} y_k + H_k^T R_k^{-1} H_k x). \tag{E.3}$$

Setting (E.3) equal to zero gives the least squares estimate:

$$\hat{x}_N = \underbrace{\left(\sum_{k=1}^{N} H_k^T R_k^{-1} H_k \right)^{-1}}_{\mathcal{I}_N} \underbrace{\sum_{k=1}^{N} H_k^T R_k^{-1} y_k}_{\iota_N} \tag{E.4}$$

$$= \mathcal{I}_N^{-1} \iota_N.$$

Now, the LS estimate can at any time k be evaluated as

$$\begin{aligned}
\hat{x}_k &= \mathcal{I}_k^{-1} \iota_k \\
&= \mathcal{I}_k^{-1} (\iota_{k-1} + H_k^T R_k^{-1} y_k) \\
&= \mathcal{I}_k^{-1} (\mathcal{I}_{k-1} \hat{x}_{k-1} + H_k^T R_k^{-1} y_k) \\
&= \mathcal{I}_k^{-1} (\mathcal{I}_k \hat{x}_{k-1} - H_k^T R_k^{-1} H_k \hat{x}_{k-1} + H_k^T R_k^{-1} y_k) \\
&= \hat{x}_{k-1} + \mathcal{I}_k^{-1} H_k^T R_k^{-1} (y_k - H_k \hat{x}_{k-1}).
\end{aligned} \tag{E.5}$$

This is the recursive update for the LS estimate, together with the obvious update of $\mathcal{I}_k$:

$$\mathcal{I}_k = \mathcal{I}_{k-1} + H_k^T R_k^{-1} H_k. \tag{E.6}$$

It is convenient to introduce $P_k = \mathcal{I}_k^{-1}$. The reasons are to avoid the inversion of $\mathcal{I}_k$, and so that P_k can be interpreted as the covariance matrix of $\hat{x}$. The *matrix inversion lemma*

$$[A + BCD]^{-1} = A^{-1} - A^{-1} B [DA^{-1} B + C^{-1}]^{-1} DA^{-1} \tag{E.7}$$

applied to (E.6) gives

$$\begin{aligned}
P_k &= [P_{k-1}^{-1} + H_k^T R_k^{-1} H_k]^{-1} \\
&= P_{k-1} - P_{k-1} H_k^T [H_k P_{k-1} H_k^T + R_k]^{-1} H_k P_{k-1}.
\end{aligned} \tag{E.8}$$

With this update formula for P_k we can rewrite $\mathcal{I}_k^{-1} H_k^T$ in (E.5) in the following manner:

$$\begin{aligned}
\mathcal{I}_k^{-1} H_k^T &= P_{k-1} H_k^T - P_{k-1} H_k^T [H_k P_{k-1} H_k^T + R_k]^{-1} H_k P_{k-1} H_k^T \\
&= P_{k-1} H_k^T [H_k P_{k-1} H_k^T + R_k]^{-1} R_k.
\end{aligned} \tag{E.9}$$

E.1 Derivation of Least Squares Algorithms

Algorithm E.1 Weighted LS in recursive form

Consider the linear regression (E.1) and the loss function (E.2). Assume that the LS estimate at time $k-1$ is $\hat{x}_{k-1}$ with covariance matrix P_{k-1}. Then a new measurement gives the update formulas in covariance form

$$\hat{x}_k = \hat{x}_{k-1} + P_{k-1}H_k^T[H_k P_{k-1}H_k^T + R_k]^{-1}(y_k - H_k\hat{x}_{k-1}), \quad \text{(E.10a)}$$

$$P_k = P_{k-1} - P_{k-1}H_k^T[H_k P_{k-1}H_k^T + R_k]^{-1}H_k P_{k-1}, \quad \text{(E.10b)}$$

and information form

$$\iota_k = \iota_{k-1} + H_k^T R_k^{-1} y_k, \quad \text{(E.11a)}$$

$$\mathcal{I}_k = \mathcal{I}_{k-1} + H_k^T R_k^{-1} H_k, \quad \text{(E.11b)}$$

$$\hat{x}_k = \mathcal{I}_k^{-1} \iota_k. \quad \text{(E.11c)}$$

Here the initial conditions are given by the prior, $x \sim \mathcal{N}(x_0, P_0) = \mathcal{N}(\mathcal{I}_0^{-1}\iota_0, \mathcal{I}_0^{-1})$. The *a posteriori* distribution of the parameter vector is

$$x_k \in \mathcal{N}(\hat{x}_k, P_k).$$

We summarize the recursive implementation of the LS algorithm: The interpretation of P_k as a covariance matrix, and the distribution of x_k follows from the Kalman filter. Note that the matrix that has to be inverted above is of lower dimension than $\mathcal{I}_k$ if dim $y <$ dim x. In a sensor fusion case where dim $y >$ dim x, then the information form is to prefer.

E.1.2 Windowed Least Squares

In some algorithms, the parameter distribution given the measurements in a sliding window is needed. This can be derived in the following way. The trick is to re-use old measurements with negative variance.

Lemma E.1 (Windowed LS (WLS))
The mean and covariance at time t for the parameter vector in (E.1), given the measurements $y_{k-L+1}^t = \{y_{k-L+1}, y_{k-L+2}, \ldots, y_k\}$, are computed by applying the RLS scheme in Algorithm E.1 to the linear regression

$$\begin{pmatrix} y_k \\ y_{k-L} \end{pmatrix} = \begin{pmatrix} H_k \\ H_{k-L} \end{pmatrix} x + \bar{e}_k,$$

where the (artificial) covariance matrix of $\bar{e}_k$ is

$$\begin{pmatrix} R_k & 0 \\ 0 & -R_{k-L} \end{pmatrix}.$$

Proof: This result follows directly by rewriting the loss function being minimized:

$$V_N(x) = \sum_{k=1}^{N} \left(\begin{pmatrix} y_k \\ y_{k-L} \end{pmatrix} - \begin{pmatrix} H_k \\ H_{k-L} \end{pmatrix} x\right)^T \begin{pmatrix} R_k & 0 \\ 0 & -R_{k-L} \end{pmatrix}^{-1}$$
$$\times \left(\begin{pmatrix} y_k \\ y_{k-L} \end{pmatrix} - \begin{pmatrix} H_k \\ H_{k-L} \end{pmatrix} x\right)$$
$$= \sum_{k=1}^{N} (y_k - H_k x)^T R_k^{-1} (y_k - H_k x)$$
$$- \sum_{k=1}^{N} (y_{k-L} - H_{k-L} x)^T R_{k-L}^{-1} (y_{k-L} - H_{k-L} x)$$
$$= \sum_{k=N-L+1}^{N} (y_k - H_k x)^T R_k^{-1} (y_k - H_k x).$$

Thus, the influence of the k first measurements cancels, and a minimization gives the distribution for x conditioned on the measurements y_{k-L+1}^t. ∎

E.2 Matrix Notation and QR Factorizations

The matrix notation
$$\mathbf{y} = \mathbf{H}x + \mathbf{e} \tag{E.12}$$
will be convenient for collecting N measurements in (E.1) into one equation. Here

$$\mathbf{y} = \begin{pmatrix} y_1 \\ y_2 \\ \vdots \\ y_N \end{pmatrix}, \quad \mathbf{H} = \begin{pmatrix} H_1 \\ H_2 \\ \vdots \\ H_N \end{pmatrix}, \quad \mathbf{e} = \begin{pmatrix} e_1 \\ e_2 \\ \vdots \\ e_N \end{pmatrix},$$
$$\mathbf{R} = \text{diag}(R_1, R_2, \ldots, R_N).$$

Thus, $\mathbf{R}$ is a block diagonal matrix. The off-line version of RLS in Algorithm E.1 can then be written

$$\iota_N = \mathbf{H}^T \mathbf{R}^{-1} \mathbf{y}$$
$$\mathcal{I}_N = \mathbf{H}^T \mathbf{R}^{-1} \mathbf{H}.$$

The inversion of $\mathbf{H}^T \mathbf{R}^{-1} \mathbf{H}$ may be numerically ill-conditioned, and the problem occurs already when the matrix product is formed. One remedy is to apply the *QR factorization* to the regression matrix. Assume for simplicity $\mathbf{R} = I$. The QR factorization yields

$$\mathbf{H} = Q \begin{pmatrix} R \\ 0 \end{pmatrix},$$

where Q is a square orthonormal matrix ($Q^T Q = QQ^T = I$) and R is an upper triangular matrix. In MATLAB™, it is computed by [Q,R]=qr(H). Both Q and R are local definitions in this section, and should not be confused with covariance matrices.

Now, multiply (E.12) with $Q^{-1} = Q^T$:

$$\begin{pmatrix} \mathbf{y}_0 \\ \mathbf{y}_\varepsilon \end{pmatrix} \triangleq Q^T \mathbf{y} = Q^T \mathbf{H} \hat{x} = Q^T Q \begin{pmatrix} R \\ 0 \end{pmatrix} \hat{x} = \begin{pmatrix} R \\ 0 \end{pmatrix} \hat{x}.$$

The solution is now computed by solving the triangular system of equations given by

$$R\hat{x} = \mathbf{y}_0,$$

which has good numerical properties. Technically, the condition number of $\mathbf{H}^T \mathbf{H}$ is the square of the condition number of R, which follows from

$$\mathbf{H}^T \mathbf{H} = \begin{pmatrix} R^T & 0 \end{pmatrix} Q^T Q \begin{pmatrix} R \\ 0 \end{pmatrix} = R^T R.$$

It should be noted that MATLAB™'s backslash operator performs a QR factorization in the call xhat=H\y. As a further advantage, the minimizing loss function is easily computed as

$$V(\hat{x}) = (\mathbf{y} - \mathbf{H}^T \hat{x})^T (\mathbf{y} - \mathbf{H}^T \hat{x}) = \mathbf{y}_\varepsilon^T \mathbf{y}_\varepsilon.$$

Square root implementations of sequential (recursive) algorithms also exist. For RLS, *Bierman's UD factorization* can be used. See, for instance, Friedlander (1982), Ljung and Söderström (1983), Sayed and Kailath (1994), Shynk (1989), Slock (1993) and Strobach (1991) for other fast implementations.

E.3 Comparing On-Line and Off-Line Expressions

In this section, some different ways of computing the parameters in the density function of the data are given. The following interesting equalities will turn out to be fundamental for many estimation and detection problems, and are summarized here for convenience.

Theorem E.2 (Batch versus sequential forms)
The density function for the measurements can equivalently be computed from the residuals. With a little abuse of notation, $p(y^N) = p(\varepsilon^N)$. The sum of squared on-line residuals relates to the sum of squared off-line residuals as:

$$\sum_{k=1}^{N}(y_k - H_k\hat{x}_{k-1})^T \left(H_k P_{k-1} H_k^T + R_k\right)^{-1} (y_k - H_k\hat{x}_{k-1})$$
$$= \sum_{k=1}^{N}(y_k - H_k\hat{x}_N)^T R_k^{-1}(y_k - H_k\hat{x}_N) + (x_0 - \hat{x}_N)^T P_0^{-1}(x_0 - \hat{x}_N). \quad \text{(E.13)}$$

The on-line residual covariances relate to the a posteriori parameter covariance as:

$$\sum_{k=1}^{N} \log \det \left(H_k P_{k-1} H_k^T + R_k\right) = \sum_{k=1}^{N} \log \det R_k + \log \det P_0 - \log \det P_N.$$
(E.14)

Together, this means that the likelihood given all available measurements can be computed from off-line statistics as:

$$p(y_{1:N}) \sim p(y_{1:N}|\hat{x}_N) p_x(\hat{x}_N) \left(\det P_N\right)^{1/2},$$

which holds for both a Gaussian prior $p_x(x)$ and a flat non-informative one $p_x(x) = 1$.

First, an on-line expression for the density function of the data is given. This is essentially a sum of squared prediction errors. Then, an off-line expression is given, which is interpreted as the sum of squared errors one would have obtained if the final parameter estimate was used as the true one from the beginning.

Lemma E.3 (On-line expression for the density function)
The density function of the sequence $y_{1:N}$ from (E.1) is computed recursively by

$$\log p(y_{1:N}) = \log p(y_{1:N-1}) - \frac{n_y}{2} \log 2\pi - \frac{1}{2} \log \det \left(H_N P_{N-1} H_N^T + R_N\right)$$
$$- \frac{1}{2}(y_N - H_N\hat{x}_{N-1})^T \left(H_N P_{N-1} H_N^T + R_N\right)^{-1} (y_N - H_N\hat{x}_{N-1})$$
$$= -\frac{Nn_y}{2} \log 2\pi - \sum_{k=1}^{N} \frac{1}{2} \log \det \left(H_k P_{k-1} H_k^T + R_k\right)$$
$$- \frac{1}{2} \sum_{k=1}^{N}(y_k - H_k\hat{x}_{k-1})^T \left(H_k P_{k-1} H_k^T + R_k\right)^{-1} (y_k - H_k\hat{x}_{k-1})$$
$$= \log p(\varepsilon_{1:N}).$$

E.3 Comparing On-Line and Off-Line Expressions 515

Here $\hat{x}_k$ and P_k are given by the RLS scheme in Algorithm E.1 with initial conditions x_0 and P_0.

Proof: Iteratively using Bayes' law gives

$$p(y_{1:N}) = \prod_{k=1}^{N} p(y_k|y_{1:k-1}).$$

Since $x_k|y_{1:k-1} \in \mathrm{N}(\hat{x}_{k-1}, P_{k-1})$, where $\hat{x}_{k-1}$ and P_{k-1} are the RLS estimate and covariance matrix respectively, it follows that

$$y_k|y_{1:k-1} \in \mathrm{N}(H_k\hat{x}_{k-1}, H_k P_{k-1} H_k^T + R_k),$$

and the on-line expression follows from the definition of the Gaussian PDF. ∎

Lemma E.4 (Off-line expression for the density function)
Consider the linear regression (E.1) with a prior $p_x(x)$ of the parameter vector x, which can be either Gaussian with mean x_0 and covariance $(\mathcal{I}_0)^{-1}$ or non-informative ($p_x(x) = 1$). Then the density function of the measurements can be calculated by

$$p(y_{1:N}) = p(y_{1:N}|\hat{x}_N) p_x(\hat{x}_N) (\det P_N)^{1/2} (2\pi)^{\frac{n_x}{2}} \tag{E.15}$$

if the prior is Gaussian or

$$p(y_{1:N}) = p(y_{1:N}|\hat{x}_N) (\det P_N)^{1/2} \tag{E.16}$$

if the prior is non-informative. Here $n_x = \dim x$.

Proof: Starting with the Gaussian prior we have

$$p(y_{1:N}) = \int_{-\infty}^{\infty} p(y_{1:N}|x) p(x) dx = (2\pi)^{-\frac{Nn_y + n_x}{2}} (\det \mathbf{R})^{-\frac{1}{2}} (\det \mathcal{I}_0)^{\frac{1}{2}}$$

$$\times \int_{-\infty}^{\infty} \exp\left(-\frac{1}{2}\left((\mathbf{y} - \mathbf{H}x)^T \mathbf{R}^{-1}(\mathbf{y} - \mathbf{H}x) + (x - x_0)^T \mathcal{I}_0 (x - x_0)\right)\right) dx \tag{E.17}$$

By completing the squares the exponent can be rewritten as (where $\hat{x} = \hat{x}_N$)

$$(\mathbf{y} - \mathbf{H}x)^T \mathbf{R}^{-1}(\mathbf{y} - \mathbf{H}x) + (x - x_0)^T \mathcal{I}_0 (x - x_0)$$
$$= \mathbf{y}^T \mathbf{R}^{-1}\mathbf{y} - 2x^T \mathbf{H}^T \mathbf{R}^{-1}\mathbf{y} + x^T \mathbf{H}^T \mathbf{R}^{-1}\mathbf{H}x + x^T \mathcal{I}_0 x - 2x^T \mathcal{I}_0 x_0 + x_0^T \mathcal{I}_0 x_0$$
$$= x^T (\mathcal{I}_0 + \mathcal{I}_N) x - 2x^T (\iota_0 + \iota_N) + \mathbf{y}^T \mathbf{R}^{-1}\mathbf{y} + x_0^T \mathcal{I}_0 x_0$$
$$= \left(x - (\mathcal{I}_0 + \mathcal{I}_N)^{-1}(\iota_0 + \iota_N)\right)^T (\mathcal{I}_0 + \mathcal{I}_N)\left(x - (\mathcal{I}_0 + \mathcal{I}_N)^{-1}(\iota_0 + \iota_N)\right)$$
$$- (\iota_0 + \iota_N)^T (\mathcal{I}_0 + \mathcal{I}_N)^{-1}(\iota_0 + \iota_N) + \mathbf{y}^T \mathbf{R}^{-1}\mathbf{y} + x_0^T \mathcal{I}_0 x_0$$
$$= (x - \hat{x})^T (\mathcal{I}_0 + \mathcal{I}_N)(x - \hat{x}) - \hat{x}^T(\iota_0 + \iota_N) + \mathbf{y}^T \mathbf{R}^{-1}\mathbf{y} + x_0^T \mathcal{I}_0 x_0$$
$$= (x - \hat{x})^T (\mathcal{I}_0 + \mathcal{I}_N)(x - \hat{x}) + (\mathbf{y} - \mathbf{H}\hat{x})^T \mathbf{R}^{-1}(\mathbf{y} - \mathbf{H}\hat{x}) + \hat{x}^T \iota_N - \hat{x}^T \mathcal{I}_N \hat{x}$$
$$- \hat{x}^T \iota_0 + x_0^T \mathcal{I}_0 x_0$$
$$= (x - \hat{x})^T (\mathcal{I}_0 + \mathcal{I}_N)(x - \hat{x}) + (\mathbf{y} - \mathbf{H}\hat{x})^T \mathbf{R}^{-1}(\mathbf{y} - \mathbf{H}\hat{x}) + \hat{x}^T(\iota_0 + \iota_N) - 2\hat{x}^T \iota_0$$
$$- \hat{x}^T (\mathcal{I}_0 + \mathcal{I}_N)\hat{x} + \hat{x}^T \mathcal{I}_0 \hat{x} + x_0^T \mathcal{I}_0 x_0$$
$$= (x - \hat{x})^T (\mathcal{I}_0 + \mathcal{I}_N)(x - \hat{x}) + (\mathbf{y} - \mathbf{H}\hat{x})^T \mathbf{R}^{-1}(\mathbf{y} - \mathbf{H}\hat{x})$$
$$+ (\hat{x} - x_0)^T \mathcal{I}_0 (\hat{x} - x_0).$$

Here, the relations in (E.11) are frequently used. Hence, the integral (E.17) can be evaluated by using the Gaussian PDF for $\hat{x} \in \mathcal{N}\big(x, (\mathcal{I}_0 + \mathcal{I}_N)^{-1}big\big)$,

$$p(y_{1:N}) = (2\pi)^{-\frac{Nn_y}{2}} (\det \mathbf{R})^{-\frac{1}{2}} \left(\frac{\det \mathcal{I}_0}{\det \mathcal{I}_0 + \mathcal{I}_N}\right)^{\frac{1}{2}}$$
$$\times \exp\left(-\frac{1}{2}\left((\mathbf{y} - \mathbf{H}\hat{x}_N)^T \mathbf{R}^{-1}(\mathbf{y} - \mathbf{H}\hat{x}_N) + (\hat{x}_N - x_0)^T \mathcal{I}_0(\hat{x}_N - x_0)\right)\right)$$
$$\times \int (2\pi)^{-\frac{n_x}{2}} (\det \mathcal{I}_0 + \mathcal{I}_N)^{\frac{1}{2}} \exp\left(-\frac{1}{2}(x - \hat{x}_N)^T (\mathcal{I}_0 + \mathcal{I}_N)(x - \hat{x}_N)\right) dx$$
$$= p(y_{1:N}|\hat{x}_N) \left(\frac{\det \mathcal{I}_0}{\det \mathcal{I}_0 + \mathcal{I}_N}\right)^{1/2} \exp\left(-\frac{1}{2}(\hat{x}_N - x_0)^T \mathcal{I}_0(\hat{x}_N - x_0)\right)$$
$$= p(y_{1:N}|\hat{x}_N) p_x(\hat{x}_N) (\det P_N)^{1/2} (2\pi)^{\frac{n_x}{2}}.$$

The case of non-informative prior, $p_x(x) = 1$ in (E.17), is treated by letting $\mathcal{I}_0 = 0$ in the exponential expressions, which gives

$$p(y_{1:N}) = p(y_{1:N}|\hat{x}_N) \int (2\pi)^{-\frac{n_x}{2}} \exp\left(-\frac{1}{2}(x - \hat{x}_N)^T (\mathcal{I}_N)(x - \hat{x}_N)\right) dx$$
$$= p(y_{1:N}|\hat{x}_N) (\det P_N)^{1/2}.$$

∎

It is interesting to note that the density function can be computed by just inserting the final parameter estimate into the prior densities $p(y_{1:N}|x)$ and $p_x(x)$, apart from a data independent correcting factor. The prediction errors in the on-line expression are essentially replaced by the smoothing errors.

E.3 Comparing On-Line and Off-Line Expressions

Lemma E.5
Consider the covariance matrix P_k given by RLS in Algorithm E.1. The following relation holds:

$$\sum_{k=1}^{N} \log \det \left(H_k P_{k-1} H_k^T + R_k \right)$$
$$= \sum_{k=1}^{N} \log \det R_k + \log \det P_0 - \log \det P_N.$$

Proof: By using the alternative update formula for the covariance matrix, $P_k^{-1} = P_{k-1}^{-1} + H_k^T R_k^{-1} H_k$ and the equality $\det(I + AB) = \det(I + BA)$ we have

$$\sum_{k=1}^{N} \log \det \left(H_k P_{k-1} H_k^T + R_k \right)$$
$$= \sum_{k=1}^{N} \left(\log \det R_k + \log \det \left(I + R_k^{-1} H_k P_{k-1} H_k^T \right) \right)$$
$$= \sum_{k=1}^{N} \left(\log \det R_k + \log \det \left(I + H_k^T R_k^{-1} H_k P_{k-1} \right) \right)$$
$$= \sum_{k=1}^{N} \left(\log \det R_k - \log \det P_{k-1}^{-1} + \log \det \left(P_{k-1}^{-1} + H_k^T R_k^{-1} H_k \right) \right)$$
$$= \sum_{k=1}^{N} \left(\log \det R_k - \log \det P_{k-1}^{-1} + \log \det P_k^{-1} \right)$$
$$= \sum_{k=1}^{N} \log \det R_k + \log \det P_0 - \log \det P_N$$

and the off-line expression follows. In the last equality, the telescope sum makes most of the terms cancel. ∎

By combining the two relations (E.14) and (E.13), the log likelihood can be rewritten in the following convenient form:

$$2 \log p(y_{1:N})$$
$$= -Nn_y \log 2\pi - \sum_{k=1}^{N} \log \det \left(H_k P_{k-1} H_k^T + R_k\right)$$
$$- \sum_{k=1}^{N} (y_k - H_k \hat{x}_{k-1})^T \left(H_k P_{k-1} H_k^T + R_k\right)^{-1} (y_k - H_k \hat{x}_{k-1})$$
$$= -Nn_y \log 2\pi - \sum_{k=1}^{N} (y_k - H_k \hat{x}_N)^T R_k^{-1} (y_k - H_k \hat{x}_N)$$
$$- \sum_{k=1}^{N} \log \det R_k - \log \det P_0 + \log \det P_N - (x_0 - \hat{x}_N)^T P_0^{-1} (x_0 - \hat{x}_N).$$

That is, the off-line LS method can be used to evaluate a likelihood.

E.4 Asymptotic Expressions

The next lemma explains the behavior of the other model dependent factor $\log \det P_N$ (besides the smoothing errors) in Lemma E.4.

Lemma E.6 (Asymptotic parameter covariance)
Assume that the regressors satisfy the condition that

$$\overline{\mathrm{E}} H_k^T R_k^{-1} H_k = \lim_{N \to \infty} \frac{1}{N} \sum_{k=1}^{N} \mathrm{E}[H_k^T R_k^{-1} H_k]$$

exists and is non-singular. Then

$$\frac{\log \det P_N}{\log N} \to -n_x, \quad N \to \infty$$

with probability one. Thus, for large N, $\log \det P_N$ approximately equals $-n_x \log N$.

Proof: We have from the definition of $\mathcal{I}_N$

$$-\log \det P_N = \log \det(\mathcal{I}_0 + \mathcal{I}_N)$$
$$= \log \det \left(\mathcal{I}_0 + N \frac{1}{N} \sum_{k=1}^{N} H_k^T R_k^{-1} H_k\right)$$
$$= \log \det N I_{n_x} + \log \det \left(\frac{1}{N} \mathcal{I}_0 + \frac{1}{N} \sum_{k=1}^{N} H_k^T R_k^{-1} H_k\right).$$

Here, the first term equals $n_x \log N$ and the second one tends to $\log \det Q$, and the result follows. ∎

E.5 Derivation of Marginal Densities

Throughout this section, assume a scalar linear Gaussian observation model

$$y_k = H_k x + e_k, \quad e_k \sim \mathcal{N}(0, \lambda).$$

Here, λ denotes the scalar noise variance, and x is a stochastic variable of dimension n_x. The regression vector H_k is known at time k.

This means that the analysis is conditioned on the model. The fact that all the PDF:s are conditioned on a particular model structure will be suppressed.

The simplest case is the joint likelihood for the parameter vector and the noise variance.

Theorem E.7 (No marginalization)
For given λ and x the PDF of $y_{1:N}$ is

$$p(y_{1:N}|x, \lambda) = (2\pi\lambda)^{-N/2} e^{-\frac{1}{2\lambda} V_N(x)}, \tag{E.18}$$

where $V_N(x) = \sum_{k=1}^{N} (y_k - H_k x)^2$.

Proof: The proof is trivial when $\{H_k^T\}$ is a known sequence. In the general case when H_k may contain past disturbances (through feedback), Bayes' rule gives

$$p(y_{1:N}) = p(y_N|y_{1:N-1})p(y_{1:N-1}) = \Pi_{k=1}^{N} p(y_k|y_{1:k-1})$$

and the theorem follows by noticing that

$$p(y_k|y_{1:k-1}) = \frac{1}{\sqrt{2\pi\lambda}} e^{-\frac{1}{2\lambda}(y_k - H_k x)^2}.$$

since H_k contains past information only. ∎

Next, continue with the likelihood involving only the noise variance. Since

$$p(y_{1:N}|\lambda) = \int p(y_{1:N}|x, \lambda) p(x|\lambda) dx,$$

the previous theorem can be used.

Theorem E.8 (Marginalization of x)
The PDF of $y_{1:N}$ conditioned on the noise variance λ when

$$(x|\lambda) \in N(x_0, \lambda(\mathcal{I}_0)^{-1}) \tag{E.19}$$

is given by

$$p(y_{1:N}|\lambda) = \sqrt{\frac{(2\pi)^{n_x}}{\det(\mathcal{I}_0 + \mathcal{I}_N)}} p(y_{1:N}|\hat{x}_N, \lambda) p(\hat{x}_N|\lambda) \qquad (E.20)$$

where $\hat{x}_N$ is given by

$$\hat{x}_N = (\mathcal{I}_0 + \mathcal{I}_N)^{-1} (\mathcal{I}_0 x_0 + \iota_N). \qquad (E.21)$$

Here $\mathcal{I}_N = \sum_{k=1}^{N} H_k^T H_k$ and $\iota_N = \sum_{k=1}^{N} H_k^T y_k$.

Proof: Consider the following relation

$$\sum_{k=1}^{N}(y_k - H_k x)^2 + (x - x_0)^T \mathcal{I}_0 (x - x_0) -$$
$$\sum_{k=1}^{N}(y_k - H_k \hat{x}_N)^2 - (\hat{x}_N - x_0)^T \mathcal{I}_0 (\hat{x}_N - x_0)$$
$$= -2\iota_N (\hat{x}_N - x_0)^T + x^T \mathcal{I}_N x - \hat{x}_N^T \mathcal{I}_N \hat{x}_N$$
$$+ (x - x_0)^T \mathcal{I}_0 (x - x_0) - (\hat{x}_N - x_0)^T \mathcal{I}_0 (\hat{x}_N - x_0)$$
$$= -2\left\{\hat{x}_N^T (\mathcal{I}_0 + \mathcal{I}_N) - x_0^T \mathcal{I}_0\right\}(x - \hat{x}_N)$$
$$+ x^T (\mathcal{I}_0 + \mathcal{I}_N) x - \hat{x}_N^T (\mathcal{I}_0 + \mathcal{I}_N) \hat{x}_N - 2x_0^T \mathcal{I}_0 (x - \hat{x}_N)$$
$$= -2\hat{x}_N^T (\mathcal{I}_0 + \mathcal{I}_N) x + x^T (\mathcal{I}_0 + \mathcal{I}_N) x + \hat{x}_N^T (\mathcal{I}_0 + \mathcal{I}_N) \hat{x}_N$$
$$= (x - \hat{x}_N)^T (\mathcal{I}_0 + \mathcal{I}_N) (x - \hat{x}_N). \qquad (E.22)$$

Using this together with (E.18) gives

$$p(y_{1:N}|x, \lambda) = p(y_{1:N}|\hat{x}_N, \lambda) \frac{p(y_{1:N}|x, \lambda)}{p(y_{1:N}|\hat{x}_N, \lambda)} \qquad (E.23)$$
$$= p(y_{1:N}|\hat{x}_N, \lambda) e^{-\frac{1}{2\lambda}(x - \hat{x}_N)^T (\mathcal{I}_0 + \mathcal{I}_N)(x - \hat{x}_N)} \times$$
$$e^{-\frac{1}{2\lambda}(\hat{x}_N - x_0)^T \mathcal{I}_0 (\hat{x}_N - x_0)} \times e^{\frac{1}{2\lambda}(x - x_0)^T \mathcal{I}_0 (x - x_0)}$$
$$= p(y_{1:N}|\hat{x}_N, \lambda) e^{-\frac{1}{2\lambda}(x - \hat{x}_N)^T (\mathcal{I}_0 + \mathcal{I}_N)(x - \hat{x}_N)}$$
$$e^{-\frac{1}{2\lambda}(\hat{x}_N - x_0)^T \mathcal{I}_0 (\hat{x}_N - x_0)} \cdot \frac{\sqrt{\det(\mathcal{I}_0)}}{(2\pi)^{n_x/2} \lambda^{n_x/2}} \cdot \frac{1}{p(x|\lambda)},$$

E.5 Derivation of Marginal Densities

and thus

$$p(y_{1:N}|\lambda) = \int p(y_{1:N}|x,\lambda)p(x|\lambda)dx \qquad (E.24)$$

$$= \int \frac{\sqrt{\det(\mathcal{I}_0 + \mathcal{I}_N)}}{(2\pi)^{n_x/2}\lambda^{n_x/2}} e^{-\frac{1}{2\lambda}(x-\hat{x}_N)^T(\mathcal{I}_0+\mathcal{I}_N)(x-\hat{x}_N)} dx$$

$$\times \sqrt{\frac{\det(\mathcal{I}_0)}{\det(\mathcal{I}_0+\mathcal{I}_N)}} p(y_{1:N}|\hat{x}_N,\lambda) e^{-\frac{1}{2\lambda}(\hat{x}_N-x_0)^T \mathcal{I}_0(\hat{x}_N-x_0)}$$

$$= \sqrt{\frac{\det(\mathcal{I}_0)}{\det(\mathcal{I}_0+\mathcal{I}_N)}} p(y_{1:N}|\hat{x}_N,\lambda) e^{-\frac{1}{2\lambda}(\hat{x}_N-x_0)^T \mathcal{I}_0(\hat{x}_N-x_0)}$$

$$= \sqrt{\frac{(2\pi)^{n_x}}{\det(\mathcal{I}_0+\mathcal{I}_N)}} p(y_{1:N}|\hat{x}_N,\lambda) p(\hat{x}_N|\lambda),$$

which proves the theorem. ∎

The likelihood involving the parameter vector x is given by

$$p(y_{1:N}|x) = \int p(y_{1:N}|x,\lambda) \cdot p(\lambda|x) d\lambda.$$

Computing this integral gives the next theorem.

Theorem E.9 (Marginalization of λ)
The PDF of $y_{1:N}$ conditioned on x when λ is $W^{-1}(m,\sigma)$ is given by

$$p(y_{1:N}|x) = \frac{1}{p(x)} \frac{\sigma^{m/2}}{\pi^{\frac{N+n_x}{2}}} \frac{\Gamma(\frac{N+m+n_x}{2})}{\Gamma(\frac{m}{2})} (\det(\mathcal{I}_0))^{1/2} \overline{V}_N^{-\frac{N+m+n_x}{2}}(x), \qquad (E.25)$$

where

$$\overline{V}_N(x) = V_N(x) + (x-x_0)^T \mathcal{I}_0 (x-x_0) + \sigma \qquad (E.26)$$

and where $p(y_{1:N}|x)$ is the PDF corresponding to a Student t distribution $\mathrm{ST}(x_0, \sigma(\mathcal{I}_0)^{-1}, m)$.

Proof: In order to prove (E.25), write

$$p(y_{1:N}|x) = \int p(y_{1:N}|x,\lambda) \cdot p(\lambda|x) d\lambda \qquad (E.27)$$

$$= \int p(y_{1:N}|x,\lambda) \cdot \frac{p(x|\lambda)}{p(x)} p(\lambda) d\lambda$$

$$= \frac{1}{p(x)} \int p(y_{1:N}|x,\lambda) p(x|\lambda) p(\lambda) d\lambda.$$

Then (E.25) follows by identifying the integral with a multiple of an integral of an $W^{-1}(\overline{V}_N(x), N+m+n_x)$ PDF. ∎

The distribution of the data conditioned only on the model structure is given by Theorem E.10.

Theorem E.10 (Marginalization of x and λ)
The PDF of $y_{1:N}$ when λ is $W^{-1}(m,\sigma)$ and x is distributed as in (E.19) is given by

$$p(y_{1:N}) = \frac{\sigma^{m/2}}{\pi^{N/2}} \frac{\Gamma(\frac{N+m}{2})}{\Gamma(\frac{m}{2})} \left(\frac{\det(\mathcal{I}_0)}{\det(\mathcal{I}_0 + \mathcal{I}_N)} \right)^{1/2} \overline{V}_N^{-\frac{N+m}{2}}(\hat{x}_N), \qquad (E.28)$$

where $\overline{V}_N(x)$ and $\hat{x}_N$ are defined by (E.26) and (E.21), respectively.

Remark E.11
Remember that the theorem gives the posterior distribution of the data conditioned on a certain model structure, so $p(y_{1:N}|\mathcal{M})$ is more exact.

Proof: From (E.20) and the definition of the Wishart distribution it follows

$$p(y_{1:N}) = \int p(y_{1:N}|\lambda) p(\lambda) d\lambda \qquad (E.29)$$

$$= \int \sqrt{\frac{(2\pi)^{n_x}}{\det(\mathcal{I}_0 + \mathcal{I}_N)}} p(y_{1:N}|\hat{x}_N, \lambda) p(\hat{x}_N|\lambda) p(\lambda) d\lambda \qquad (E.30)$$

$$= \int \sqrt{\frac{(2\pi)^{n_x}}{\det(\mathcal{I}_0 + \mathcal{I}_N)}} \frac{1}{(2\pi\lambda)^{N/2}} e^{-\frac{1}{2\lambda} \sum_{k=1}^{N}(y_k - H_k \hat{x}_N)^2}$$

$$\times \frac{\sqrt{\det(\mathcal{I}_0)}}{(2\pi)^{n_x/2}} e^{-\frac{1}{2\lambda}(\hat{x}_N - x_0)^T \mathcal{I}_0 (\hat{x}_N - x_0)}$$

$$\times \frac{\sigma^{m/2} e^{-\frac{\sigma}{2\lambda}}}{2^{m/2} \Gamma(m/2) \lambda^{(m+2)/2}} d\lambda$$

$$= \frac{\sigma^{m/2}}{\pi^{N/2}} \frac{\Gamma(\frac{N+m}{2})}{\Gamma(\frac{m}{2})} \left(\frac{\det(\mathcal{I}_0)}{\det(\mathcal{I}_0 + \mathcal{I}_N)} \right)^{1/2} \overline{V}_N^{-\frac{N+m}{2}}(\hat{x}_N)$$

$$\times \int \frac{|\overline{V}_N(\hat{x}_N)|^{\frac{N+m}{2}}}{\lambda^{\frac{N+m+2}{2}} \Gamma(\frac{N+m}{2}) 2^{\frac{N+m}{2}}} e^{-\frac{1}{2\lambda} \overline{V}_N(\hat{x}_N)} d\lambda$$

$$= \frac{\sigma^{m/2}}{\pi^{N/2}} \frac{\Gamma(\frac{N+m}{2})}{\Gamma(\frac{m}{2})} \left(\frac{\det(\mathcal{I}_0)}{\det(\mathcal{I}_0 + \mathcal{I}_N)} \right)^{1/2} \overline{V}_N^{-\frac{N+m}{2}}(\hat{x}_N),$$

which proves the theorem. ∎

Bibliography

3GPP. Requirements for support of radio resource management. Technical Specification TSG RAN 25.133, Section 9.1.8.2.1, 2003.

D. Ahlberg and D. Sundström. Fault diagnosis in gear boxes with sound and vibration analysis. Technical report, Dept of Elec. Eng. Linköping University, LiTH-ISY-EX-4237, 2009.

H. Akashi and H. Kumamoto. Random sampling approach to state estimation in switching environment. *Automation*, 13:429, 1977.

D.L. Alspach and H.W. Sorenson. Nonlinear Bayesian estimation using Gaussian sum approximation. *IEEE Transactions on Automatic Control*, 17:439–448, 1972.

L. Alvarez, J. Yi, and R. Horowitz. Adaptive emergency braking control with underestimation of friction coefficient. *IEEE Transactions on Control System Technology*, 2001.

B.D.O. Anderson and J.B. Moore. *Optimal filtering*. Prentice Hall, Englewood Cliffs, NJ., 1979.

P. Andersson. Long range 3D imaging using range gated laser radar images. *Optical Engineering*, 45(3), 2005.

C. Andrieu and A. Doucet. Particle filtering for partially observed Gaussian state space models. *Journal of the Royal Statistical Society*, 64(4):827–836, 2002.

I. Arasaratnam, S. Haykin, and R.J. Elliot. Discrete-time nonlinear filtering algorithms using Gauss-Hermite quadrature. *Proceedings of IEEE*, 95:953–977, 2007.

I. Arasaratnam, S. Haykin, and R.J. Elliot. Cubature Kalman filter. *IEEE Transactions on Automatic Control*, 54:1254–1269, 2009.

B. Armstrong-Hélouvry. *Control of machines with friction.* Kluwer, Boston, MA, 1991.

S. Arulampalam, S. Maskell, N. Gordon, and T. Clapp. A tutorial on particle filters for online nonlinear/non-Gaussian Bayesian tracking. *IEEE Transactions on Signal Processing*, 50(2):174–188, 2002.

A. Athalye. *Design and implementation of reconfigurable hardware for real-time particle filtering.* PhD thesis, Stody Brook University, 2007.

M. Athans, R.P. Wishner, and A. Bertolini. Suboptimal state estimation for continuous-time nonlinear systems from discrete noisy measurements. *IEEE Transactions on Automatic Control*, 1968.

T. Bailey and H. Durrant-Whyte. Simultaneous localization and mapping (SLAM): Part II. *IEEE Robotics & Automation Magazine*, 13(3):108–117, September 2006.

E. Bakker, L. Nyborg, and H.B. Pacejka. Tyre modelling for use in vehicle dynamic studies. Society of Automotive Engineers, paper 870421, 1987.

Y. Bar-Shalom and T. Fortmann. *Tracking and Data Association*, volume 179 of *Mathematics in Science and Engineering.* Academic Press, 1988.

Y. Bar-Shalom and X.R. Li. *Estimation and tracking: principles, techniques, and software.* Artech House, 1993.

Y. Bar-Shalom, X.R. Li, and T. Kirubarajan. *Estimation with Applications to Tracking and Navigation: Theory, Algorithms and Software.* John Wiley & Sons, 2001.

L.E. Baum, T. Petrie, G.Soules, and N. Weiss. A maximization technique occuring in the statistical analysis of probabilistic functions of Markov chains. *The Annals of Mathematical Statistics*, 41:164–171, 1970.

A. Benaskeur. Consistent fusion of correlated data sources. In *Proceedings of the 28th Annual Conference of the Industrial Electronics Society (IECON 02)*, volume 4, pages 2652–2656., 2002.

N. Bergman. *Recursive Bayesian Estimation: Navigation and Tracking Applications.* Dissertation nr. 579, Linköping University, Sweden, 1999.

N. Bergman and F. Gustafsson. Three statistical batch algorithms for tracking manoeuvring targets. In *Proceedings 5th European Control Conference*, Karlsruhe, Germany, 1999.

J.E. Bernard and C.L. Clover. Tire modeling for low-speed and high-speed calculations. Society of Automotive Engineers (950311), 1995.

P.J. Besl and N.D. McKay. A method for registration of 3-D shapes. *IEEE Transactions on Pattern Analysis and Machine Intelligence*, 14(2):239–256, 1992.

G.J. Bierman. *Factorization methods for discrete sequential estimation.* Academic Press, New York, 1977.

A. Blomfeldt and R. Haverstad. Sensor fusion of GPS and IMU for a racing application. Master's thesis no LiTH-ISY-EX-4143, Department of Electrical Engineering, Linköping University, 2008.

IEEE Standards Board. Standard specification format guide and test procedure for single-gyros. Technical Report IEEE Standard 962-1997. Annex C, IEEE, R2008.

K.R. Britting. *Inertial Navigation Systems Analysis.* Wiley - Interscience, 1971.

R.B. Brown and P.Y.C. Hwang. *Introduction to Random Signals and Applied Kalman Filtering.* John Wiley & Sons, 1997.

O. Cappé, S.J. Godsill, and E. Moulines. An overview of existing methods and recent advances in sequential Monte Carlo. *IEEE Proceedings*, 95:899, 2007.

G. Casella and C. P. Robert. Rao-Blackwellisation of sampling schemes. *Biometrika*, 83(1):81–94, 1996.

S. Challa, R.J. Evans, and X. Wang. A Bayesian solution and its approximations to out-of-sequence measurement problems. *Information Fusion*, 4:185–199, 2003.

R. Chen and J. S. Liu. Mixture Kalman filters. *Journal of the Royal Statistical Society*, 62(3):493–508, 2000.

D. Coppersmith and S. Winograd. Matrix multiplication via arithmetic progressions. *Journal of Symbolic Computation*, 9:251–280, 1990.

D. Crisan and A. Doucet. Convergence of sequential Monte Carlo methods. Technical Report CUED/F-INFENG/TR381, Signal Processing Group, Department of Engineering, University of Cambridge, 2000.

D. Crisan and A. Doucet. A survey of convergence results on particle filtering methods for practitioners. *IEEE Transactions on Signal Processing*, 50(3):736–746, 2002.

M. Dahlin and S. Mahl. Radar distance positioning system – with a particle filter approach. Master Thesis LiTH-ISY-EX-3998, Dept of Elec. Eng. Linköping University, S-581 83 Linköping, Sweden, 2007. In Swedish.

C. Canudas de Wit and R. Horowitz. Observers for tire/road contact friction using only angular velocity information. In *IEEE Conference on Decision and Control*, Phoenix, AZ, 1999.

C. Canudas de Wit and P. Tsiotras. Dynamic tire friction models for vehicle traction control. In *IEEE Conference on Decision and Control*, Phoenix, AZ, 1999.

C. Canudas de Wit, H. Olsson, K-J. Åström, and P. Lischinsky. A new model for control of systems with friction. *IEEE Transactions on Automatic Control*, 40(3):419–425, 1995.

J.E Dennis Jr. and B. Schnabel. *Numerical methods for unconstrained optimization and non-linear equations.* Prentice-Hall series in computational mathematics. Prentice-Hall, Englewood Cliffs, New Jersey, 1983.

P.M. Djuric, J.H. Kotecha, J. Zhang, Y. Huang, T. Ghirmai, M.F. Bugallo, and J. Miguez. Particle filtering. *IEEE Signal Processing Magazine*, 20:19, 2003.

R. Douc, A. Garivier, E. Moulines, and J. Olsson. On the forward filtering backward smoothing particle approximations of the smoothing distribution in general state spaces models. *Annals of Applied Probability*, 21(6):2109–2145, 2010.

A. Doucet, S. J. Godsill, and C. Andrieu. On sequential Monte Carlo sampling methods for Bayesian filtering. *Statistics and Computing*, 10(3):197–208, 2000a.

A. Doucet, S.J. Godsill, and C. Andrieu. On sequential simulation-based methods for Bayesian filtering. *Statistics and Computing*, 10(3):197–208, 2000b.

A. Doucet, N. de Freitas, and N. Gordon, editors. *Sequential Monte Carlo Methods in Practice*. Springer Verlag, 2001a.

A. Doucet, N. Gordon, and V. Krishnamurthy. Particle filters for state estimation of jump Markov linear systems. *IEEE Transactions on Signal Processing*, 49(3): 613–624, 2001b.

A. Doucet, M. Briers, and S. Sénécal. Efficient block sampling strategies for sequential Monte Carlo methods. *Journal of Computational and Graphical Statistics*, 15 (3):1–19, 2006.

C. Drane, M. Macnaughtan, and C. Scott. Positioning GSM telephones. *IEEE Communications Magazine*, 36(4), 1998.

H. Durrant-Whyte and T. Bailey. Simultaneous localization and mapping (SLAM): Part I. *IEEE Robotics & Automation Magazine*, 13(2):99–110, June 2006.

M. Efe and D.P. Atherton. Maneuvering target tracking with an adaptive Kalman filter. In *Proceedings of the 37th IEEE Conference on Decision and Control*, 1998.

A. Eidehall and F. Gustafsson. Combined road prediction and target tracking in collision avoidance. In *IEEE Intelligent Vehicle Symposium (IV'04)*, pages 619–624, Parma, Italy, 2004.

Y.C. Eldar. Uniformly improving the Cramér-Rao bound and maximum-likelihood estimation. *IEEE Transactions on Signal Processing*, 54(8), 2006.

K.M. Fauske, F. Gustafsson, and O.Herenaes. Estimation of AUV dynamics for sensor fusion. In *10th International Conference on Information Fusion*, Quebec, Canada, July 2007.

P. Fearnhead. *Sequential Monte Carlo methods in filter theory*. PhD thesis, University of Oxford, 1998.

W.J. Fleming. Overview of automotive sensors. *IEEE Sensors Journal*, 1:296–308, 2001.

U. Forssell, P. Hall, S. Ahlqvist, and F. Gustafsson. Novel map-aided positioning system. In *Proc. of FISITA*, number F02-1131, Helsinki, 2002.

European Maritime Radionavigation Forum. GNSS vulnerability and mitigation measures (rev. 6). Technical report, 2001.

T.I. Fossen. *Marine Control Systems. Guidance, Navigation, and Control of Ships, Righs and Underwater Vehicles*. Marine Cybernetics, Trondheim, Norway, 2002.

B. Friedlander. Lattice filters for adaptive processing. *Proceedings IEEE*, 70(8): 829–867, August 1982.

A. Gelman, J.B. Carlin, H.S. Stern, and D.B. Rubin. *Bayesian Data Analysis*. Chapman & Hall, 2004.

S. Geman and D. Geman. Stochastic relaxation, Gibbs distributions and the Bayesian restoration of images. *IEEE Trans. on Pattern Analysis and Machine Intelligence*, 6:721–741, 1984.

D.R. Gerwe and P.S. Idell. Cramer-Rao analysis of orientation estimation: Viewing geometry influences on the information conveyed by target features. *Journal of Optical Society of America A*, 20(5):797–816, May 2003.

T. Glad and L. Ljung. *Control Theory: Multivariable and Nonlinear Methods*. Taylor & Francis, 2000.

S. J. Godsill, A. Doucet, and M. West. Monte Carlo smoothing for nonlinear time series. *Journal of the American Statistical Association*, 99(465):156–168, 2004.

N.J. Gordon, D.J. Salmond, and A.F.M. Smith. A novel approach to nonlinear/non-Gaussian Bayesian state estimation. In *IEE Proceedings on Radar and Signal Processing*, volume 140, pages 107–113, 1993.

C. Grönwall, O. Steinvall, F. Gustafsson, and T. Chevalier. Influence of laser radar sensor parameters on range measurements and shape fitting uncertainties. *Optical Engineering*, 2007.

F. Gustafsson. Rotational speed sensors: Limitations, pre-processing and automotive applications. *IEEE Instrumentation and Measurement Magazine*, 13(2): 16–23.

F. Gustafsson. *Adaptive filtering and change detection*. John Wiley & Sons, Ltd, 2001.

F. Gustafsson. Automotive safety systems. *IEEE Signal Processing Magazine*, 26: 32–47, 2009.

F. Gustafsson. Particle filter theory and practice with positioning applications. *IEEE Transactions on Aerospace and Electronics Magazine Part II: Tutorials*, 7 (July):53–82, 2010.

F. Gustafsson and F. Gunnarsson. Mobile positioning using wireless networks: possibilities and fundamental limitations based on available wireless network measurements. *IEEE Signal Processing Magazine*, 22:41–53, 2005.

F. Gustafsson and G. Hendeby. On nonlinear transformations of stochastic variables and its application to nonlinear filtering. In *IEEE Conference on Acoustics, Speech and Signal Processing (ICASSP)*, Las Vegas, NV, USA, 2008.

F. Gustafsson and A. Isaksson. Best choice of coordinate system for tracking coordinated turns. In *IEEE Conference on Decision and Control*, pages 3145–3150, Kobe, Japan, 1996.

F. Gustafsson, S. Ahlqvist, U. Forssell, and N. Persson. Sensor fusion for accurate computation of yaw rate and absolute velocity. In *Society of Automotive Engineers World Congress*, number SAE 2001-01-1064, Detroit, 2001a.

F. Gustafsson, M. Drevö, U. Forssell, M. Löfgren, N. Persson, and H. Quicklund. Virtual sensors of tire pressure and road friction. In *Society of Automotive Engineers World Congress*, number SAE 2001-01-0796, Detroit, 2001b.

F. Gustafsson, F. Gunnarsson, N. Bergman, U. Forssell, J. Jansson, R. Karlsson, and P-J. Nordlund. Particle filters for positioning, navigation and tracking. *IEEE Transactions on Signal Processing*, 50(2):425–437, February 2002.

P. Hall. A Bayesian approach to map-aided vehicle positioning. Master Thesis LiTH-ISY-EX-3104, Dept of Elec. Eng. Linköping University, S-581 83 Linköping, Sweden, 2001. In Swedish.

J.M. Hammersley and K.W. Morton. Poor man's Monte Carlo. *Journal of the Royal Statistical Society, Series B*, 16:23, 1954.

J.E. Handshin. Monte Carlo techniques for prediction and filtering of nonlinear stochastic processes. 6:555, 1970.

J. Harned, L. Johnston, and G. Scharpf. Measurement of tire brake force characteristics as related to wheel slip (antilock) control system design. *SAE Transactions*, 78(690214):909–925, 1969.

M. Hata. Empirical formula for propagation loss in land mobile radio services. *IEEE Transactions on Vehicular Technology*, 29(3):317–325, 1980.

G. Hendeby. Fundamental estimation and detection limits in linear non-Gaussian systems. LIU-TEK-LIC-2005:1199, Dept. of Electrical Engineering, Linköping University, Sweden, 2005.

G. Hendeby, J.D. Hol, R. Karlsson, and F. Gustafsson. A graphics processing unit implementation of the particle filter. In *European Signal Processing Conference (EUSIPCO)*, Poznań, Poland, September 2007a.

G. Hendeby, R. Karlsson, and F. Gustafsson. A new formulation of the Rao-Blackwellized particle filter. In *Proceedings of IEEE Workshop on Statistical Signal Processing*, Madison, WI, USA, August 2007b.

Gustaf Hendeby. *Performance and implementation aspects of nonlinear filtering*. Dissertation no. 1161, Linköping University, Sweden, 2008.

D. Hochman and M. Sadok. Theory of synchronous averaging. In *Proceedings of IEEE Aerospace Conference*, pages 3636–3653, Big Sky Montana, USA, 2004.

J.D. Hol, T.B. Schön, and F.Gustafsson. On resampling algorithms for particle filters. In *Nonlinear Statistical Signal Processing Workshop*, Cambridge, United Kingdom, Sep 2006.

X.L. Hu, T.B. Schön, and L. Ljung. A basic convergence result for particle filtering. *IEEE Transactions on Signal Processing*, 56(4):1337–1348, April 2008.

M. Isard and A. Blake. Condensation - conditional density propagation for visual tracking. *International Journal of Computer Vision*, 29(1):5–28, 1998.

B. Jalving. The NDRE-AUV flight control system. *IEEE Journal of Oceanic Engineering*, 19(4):497–501, October 1994. ISSN 0364-9059. doi: 10.1109/48.338385.

A. Jazwinsky. *Stochastic Process and Filtering Theory*, volume 64 of *Mathematics in Science and Engineering*. Academic Press, New York, 1970.

A. V. Jelalian. *Laser Radar Systems*. Artech House, Norwood, MA, 1992.

A.M. Johansen and A. Doucet. A note on auxiliary particle filters. *Statistics & Probability Letters*, 78(12):1498–1504, 2008.

S. Julier and J. Uhlmann. A non-divergent estimation algorithm in the presence of unknown correlations. In *IEEE American Control Conference*, volume 4, pages 2369–2373, 1997.

S. Julier and J. Uhlmann. *Handbook of Multisensor Data Fusion, D. Hall and J. Llinas, Eds.*, chapter General decentralized data fusion with covariance intersection (CI). CRC Press, 2001.

S.J. Julier and J.K. Uhlmann. Reduced sigma point filters for the propagation of means and covariances through nonlinear transformations. In *IEEE American Control Conference*, volume 2, pages 887–892, 2002.

S.J. Julier and J.K. Uhlmann. Unscented filtering and nonlinear estimation. *Proceedings of the IEEE*, 92(3):401–422, 2004.

S.J. Julier, J.K. Uhlmann, and Hugh F. Durrant-Whyte. A new approach for filtering nonlinear systems. In *IEEE American Control Conference*, pages 1628–1632, 1995.

T. Kailath. *Linear systems*. Prentice-Hall, Englewood Cliffs, NJ, 1980.

T. Kailath, Ali H. Sayed, and Babak Hassibi. *State space estimation theory*. Course compendium at Stanford university. To be published., 1998.

T. Kailath, A.H. Sayed, and B. Hassibi. *Linear Estimation*. Information and System Sciences. Prentice-Hall, Upper Saddle River, New Jersey, 2000.

R.E. Kalman. A new approach to linear filtering and prediction problems. *J Basic Engr. Trans. ASME Series D*, 82:35–45, 1960.

G.W. Kamerman. *The Infrared and Electro-Optical Systems Handbook*, volume 6, chapter Laser radar. Infrared information analysis center and SPIE optical engineering press, 1993.

G.W. Kamerman. *Selected Papers on Laser Radar*. SPIE Press Book, 1997.

R. Karlsson and F. Gustafsson. Particle filter and Cramer-Rao lower bound for underwater navigation. In *IEEE Conference on Acoustics, Speech and Signal Processing (ICASSP)*, Hongkong, China, Apr 2003.

R. Karlsson and F. Gustafsson. Bayesian surface and underwater navigation. *IEEE Transactions on Signal Processing*, 54(11):4204–4213, November 2006. doi: 10.1109/TSP.2006.881176.

R. Karlsson, T.B. Schön, and F. Gustafsson. Complexity analysis of the marginalized particle filter. *IEEE Transactions on Signal Processing*, 53:4408–4411, 2005.

R. Karlsson, T. Schön, D. Törnqvist, G. Conte, and F. Gustafsson. Utilizing model structure for efficient simultaneous localization and mapping for a UAV application. In *IEEE Aerospace Conference*, Big Sky, Montana, 2008.

T. Karlsson. Terrain aided underwater navigation using Bayesian statistics. Master Thesis LiTH-ISY-EX-3292, Dept of Elec. Eng. Linköping University, S-581 83 Linköping, Sweden, 2002.

S.M. Kay. *Fundamentals of signal processing – estimation theory*. Prentice Hall, 1993.

S.M. Kay. *Fundamentals of signal processing – detection theory*. Prentice Hall, 1998.

U. Kiencke and A. Daiss. Estimation of tyre friction for enhanced ABS-systems. In *Proceedings of the AVEG Congress*, Tokyo, 1994.

G. Kitagawa. Monte Carlo filter and smoother for non-Gaussian nonlinear state space models. *Journal of Computational and Graphical Statistics*, 5(1):1–25, 1996.

M. Klaas. Toward practical n^2 Monte Carlo: The marginal particle filter. *Uncertainty in Artificial Intelligence*, 2005.

A. Klotz, D. Hoetzer, and J. Sparbert. Lane data fusion for driver assistance systems. In *7th International Conference on Information Fusion*, Stockholm, 2004.

C.H. Knapp and G.C. Carter. Time delay estimation. In *IEEE Conference on Acoustics, Speech and Signal Processing*, 1976.

A. Kong, J. S. Liu, and W. H. Wong. Sequential imputations and Bayesian missing data problems. *J. Amer. Stat. Assoc.*, 89(425):278–288, 1994.

J.H. Kotecha and P.M. Djuric. Gaussian particle filtering. *IEEE Transactions on Signal Processing*, 51:2592–2601, 2003a.

J.H. Kotecha and P.M. Djuric. Gaussian sum particle filtering. *IEEE Transactions on Signal Processing*, 51:2602–2611, 2003b.

S.C. Kramer and H.W. Sorenson. Recursive Bayesian estimation using piece-wise constant approximations. *Automatica*, 24:789–801, 1988.

H. Krim and M. Viberg. Two decades of array signal processing research: the parametric approach. *IEEE Signal Processing Magazine*, 13(4):67–94, 1996.

V. Krishnamurthy and J.B. Moore. On-line estimation of hidden Markov model parameters based on the Kullback-Leibler information measure. *IEEE Transactions on Signal Processing*, pages 2557–2573, 1993.

K. J. Krizman, T.E. Biedka, and T.S. Rappaport. Wireless position location: Fundamentals, implementation strategies, and sources of error. In *Proc. IEEE Vehicular Technology Conference*, June 1997.

J. Kronander. Robust vehicle positioning: Integration of GPS and motion sensors. Master Thesis LiTH-ISY-EX-3578, Dept of Elec. Eng. Linköping University, S-581 83 Linköping, Sweden, 2003.

T. Lefebvre, H. Bruyninckx, and J. DeSchutter. Comment on "A new method for the nonlinear transformation of means and covariances in filters and estimators". *IEEE Transactions on Automatic Control*, 47(8):1406–1408, 2002.

E.L. Lehmann. *Theory of point estimation*. Statistical/Probability series. Wadsworth & Brooks/Cole, 1991a.

E.L. Lehmann. *Testing statistical hypothesis*. Statistical/Probability series. Wadsworth & Brooks/Cole, 1991b.

B.C. Levy, A. Benveniste, and R. Nikoukhah. High-level primitives for recursive maximum likelihood estimation. *IEEE Transactions on Automatic Control*, 41(8):1125–1145, 1996.

X.R. Li and Y. Bar-Shalom. Design of an interactive multiple model algorithm for air traffic control tracking. *IEEE Transactions on Control Systems Technology*, 1:186–194, 1993.

X.R. Li and V.P. Jilkov. Survey of maneuvering target tracking. part I: Dynamic models. *IEEE Transactions on Aerospace and Electronic Systems*, 39(4):1333–1364, 2003.

D. Lindgren, O. Wilson, F. Gustafsson, and H. Habberstad. Shooter localization in wireless microphone networks. In *12th International Conference on Information Fusion*, 2009.

J.S. Liu. Metropolized independent sampling with comparison to rejection ampling and importance sampling. *Statistics and Computing*, 6:113–119, 1996.

J.S. Liu and R. Chen. Sequential Monte Carlo methods for dynamic systems. *Journal of the American Statistical Association*, 93, 1998.

Y. Liu and J. Sun. Target slip tracking using gain-scheduling for antilock braking systems. In *IEEE American Control Conference*, pages 1178–1182, Seattle, 1995.

L. Ljung and T. Söderström. *Theory and practice of recursive identification.* MIT Press, Cambridge, MA, 1983.

S.C. Lo, B.B. Peterson, and P.K. Enge. Loran data modulation: a primer (AESS Tutorial IV). *IEEE Aerospace and Electronic Systems Magazine*, 22:31–51, 2007.

D.G. Luenberger. Observers for multivariable systems. *IEEE Transactions on Automatic Control*, 11:190–197, 1966.

C. Lundquist and T.B. Schön. Estimation of the free space in front of a moving vehicle. *SAE Paper*, 09AE-0138, 2009.

J.M. Maciejowski. *Multivariable feedback design.* Addison Wesley, 1989.

R. Martinez-Cantin, N. de Freitas, and J.A. Castellanos. Analysis of particle methods for simultaneous robot localization and mapping and a new algorithm: Marginal-SLAM. In *Proceedings of IEEE International Conference on Robotics and Automation*, Roma, Italy, 2007.

P. S. Maybeck. *Stochastic Models, Estimation, and Control, volume 2.* Academic Press, 1982.

M. Montemerlo, S. Thrun, D. Koller, and B. Wegbreit. FastSLAM a factored solution to the simultaneous localization and mapping problem. In *Proceedings of the AAAI National Conference on Artificial Intelligence*, Edmonton, Canada, 2002.

P. Del Moral. *Feynman-Kac Formulae: Genealogical and Interacting Particle Systems with Applications.* Springer, 2004.

D. Murdin. Data fusion and fault detection in decentralized navigation systems. Master Thesis LiTH-ISY-EX-1920, Department of Electrical Engineering, Linköping University, S-581 83 Linköping, Sweden, 1998.

A. Doucet N. Bergman and N.J. Gordon. Optimal estimation and Cramer-Rao bounds for partial non-Gaussian state-space model. *Ann. Inst. Stat. Math*, 52(1): 97–112, 2001.

Lingli Ni. *Fault-Tolerant Control of Unmanned Underwater Vehicles.* PhD thesis, Virginia Polytechnic Institute and State University, Department of Mechanical Engineering, 2001.

M. Niedzwiecki and K. Cisowski. Adaptive scheme for elimination of broadband noise and impulsive disturbances from AR and ARMA signals. *IEEE Transactions on Signal Processing*, 44(3):528–537, 1996.

P-J. Nordlund and F. Gustafsson. Marginalized particle filter for accurate and reliable terrain-aided navigation. *IEEE Transactions on Aerospace and Electronic Systems*, 2009.

M. Norgaard, N.K. Poulsen, and O. Ravn. New developments in state estimation of nonlinear systems. *Automatica*, 36:1627–1638, 2000.

U. Orguner and F. Gustafsson. Particle filtering with propagation delayed measurements. In *IEEE Aerospace Conference*, 2010.

PooGyeon Park and T. Kailath. New square-root algorithms for Kalman filtering. *IEEE Transactions on Automatic Control*, 40(5):895–899, 1995.

N. Patwari, A.O. Hero III, M. Perkins, N.S. Correal, and R.J. O'Dea. Relative location estimation in wireless sensor networks. *IEEE Transactions on Signal Processing*, 51(8), August 2003.

N. Persson, F. Gustafsson, and M. Drevö. Indirect tire pressure monitoring using sensor fusion. In *Society of Automotive Engineers World Congress*, number SAE 2002-01-1250, Detroit, 2002.

M.K. Pitt and N. Shephard. Filtering via simulation: Auxiliary particle filters. *Journal of the American Statistical Association*, 94(446):590–599, June 1999.

D.N. Politis. Computer-intensive methods in statistical analysis. *IEEE Signal Processing Magazine*, 15(1):39–55, 1998.

J.E. Potter. New statistical formulas. Technical report, Memo 40, Instrumental Laboratory, MIT, 1963.

G. Poyiadjis, A. Doucet, and S.S. Singh. Maximum likelihood parameter estimation in general state-space models using particle methods. In *Proceedings of Joint Statistical Meeting*, Minneapolis, Minnesota, 2005.

G. Poyiadjis, A. Doucet, and S.S. Singh. Maximum likelihood parameter estimation using particle methods. In *IEEE Conference on Acoustics, Speech and Signal Processing*, 2006.

B. Quine. A derivative-free implementation of the extended Kalman filter. 42: 1927–1934, 2006.

H. Raiffa and R. Schlaifer. *Applied Statistical Decision Theory*. Division of Research, Graduate School of Business Administration, Harvard University, 1961.

B.D. Ripley. *Stochastic Simulation*. John Wiley, 1988.

B. Ristic, S. Arulampalam, and N. Gordon. *Beyond the Kalman filter: Particle filters for tracking applications*. Artech House, London, 2004.

C. P. Robert and G. Casella. *Monte Carlo Statistical Methods*. Springer texts in statistics. Springer, New York, 1999.

A. Rönnebjerg. A tracking and collision warning system for maritime applications. Master Thesis LiTH-ISY-EX-3709, Dept of Elec. Eng. Linköping University, S-581 83 Linköping, Sweden, 2005. In Swedish.

M.N. Rosenbluth and A.W. Rosenbluth. Monte Carlo calculation of the average extension of molecular chains. *Journal of Chemical Physics*, 23:590, 1956.

O.S. Salychev. *Inertial Systems in Navigation and Geophysics.* Bauman MSTU Press, 1998.

A.H. Sayed and T. Kailath. A state-space approach to adaptive RLS filtering. *IEEE Signal Processing Magazine*, 11(3):18–60, 1994.

A.H. Sayed, A. Tarighat, and N. Khajehnouri. Network-based wireless location. *IEEE Signal Processing Magazine*, 22:41–53, 2005.

S.F. Schmidt. Application of state-space methods to navigation problems. *Advances in Control Systems*, pages 293–340, 1966.

T.B. Schön, F. Gustafsson, and P.J. Nordlund. Marginalized particle filters for nonlinear state-space models. *IEEE Transactions on Signal Processing*, 53:2279–2289, 2005.

T.B. Schön, R. Karlsson, and F. Gustafsson. The marginalized particle filter - analysis, applications and generalizations. In *Workshop on Sequential Monte Carlo Methods: filtering and other applications*, Oxford, United Kingdom, Jul 2006.

R. Schwarz, O. Nelles, P. Scheerer, and R. Isermann. Increasing signal accuracy of automotive wheel-speed sensors by on-line learning. In *American Control Conference*, Albuquerque, New Mexico, 1997.

R. Settineri, M. Najim, and D. Ottaviani. Order statistic fast Kalman filter. In *IEEE International Symposium on Circuits and Systems, 1996. ISCAS '96*, pages 116–119, 1996.

J.H. Shapiro, B.A. Capron, and R.C. Harney. Imaging and target detection with a heterodyne-reception optical radar. *Applied Optics*, 20(19):3282–, 1981.

J.J. Shynk. Adaptive IIR filtering. *IEEE Signal Processing Magazine*, 6(2):4–21, 1989.

S. Sing, N. Kantas, B. Vo, A. Doucet, and R.J. Evans. Simulation-based optimal sensor scheduling with application to observer trajectory planning. *Automatica*, 43:817–830, 2007.

I. Skog and P. Händel. Effects of time synchronization errors in GNSS-aided INS. In *Proceedings of PLANS*, Monterey, 2008.

I. Skog and P. Händel. In-car positioning and navigation technologies – a survey. *World Congress on Intelligent Transport Systems and Services (ITS)*, 10(1):4–21, 2009.

I. Skog and P. Händel. Synchronization by two-way message exchanges: Cramér-Rao bounds, approximate maximum likelihood, and offshore submarine positioning. *IEEE Transactions on Signal Processing*, 2010.

D.T.M. Slock. On the convergence behavior of the LMS and the normalized LMS algorithms. *IEEE Transactions on Signal Processing*, 41(9):2811–2825, 1993.

G. L. Smith, S. F. Schmidt, and L. A. McGee. Application of statistical filter theory to the optimal estimation of position and velocity on board a circumlunar vehicle. Technical Report TR R-135, NASA, 1962.

SNAME. The society of naval architects and marine engineers. Nomenclature for treating the motion of a submerged body through a fluid. *Technical and Research Bulletin*, 1–5, 1950.

D.L. Snyder and A.M. Hammoud. Image recovery from data acquired with a charge-coupled-device camera. *Journal of the Optical Society of America, A*, 10(5):1014–1022, 1993.

S.Saha, P.K. Mandal, Y. Boers, H. Driessen, and A. Bagchi. Gaussian proposal density using moment matching in SMC methods. *Journal Statistics and Computing*, 2009.

W.C. Stone, M. Juberts, N. Dagalakis, J. Stone, and J. Gorman. Performance analysis of next-generation LADAR for manufacturing, construction, and mobility. Technical Report NISTIR 7117, National Institute of Standards and Technology (NIST), 2004.

P. Strobach. New forms of Levinson and Schur algorithms. *IEEE Signal Processing Magazine*, 8(1):12–36, 1991.

H.T. Szostak, R.W. Allen, and T.J. Rosenthal. Analytical modeling of driver response in crash avoidance manuevering. volume II: An interactive tire model for driver/vehicle simulation. Technical Report DOT HS 807-271, U.S. Department of Transportation, 1988.

S. Thrun, D. Fox, F. Dellaert, and W. Burgard. Particle filters for mobile robot localization. In A. Doucet, N. de Freitas, and N. Gordon, editors, *Sequential Monte Carlo Methods in Practice*. Springer-Verlag, 2001.

S. Thrun, W. Burgard, and D. Fox. *Probabilistic Robotics*. MIT Press, 2005.

P. Tichavsky, C.H. Muravchik, and A. Nehorai. Posterior Cramér-Rao bounds for discrete-time nonlinear filtering. *IEEE Transactions on Signal Processing*, 46(5):1386–1396, 1998.

D.H. Titterton and J.L. Weston. *Strapdown Inertial Navigation Technology*. IEE Radar, Sonar, Navigation and Avionics Series 5, 1997.

G. Tolt, P. Andersson, T.R. Chevalier, C.A. Grönwall, H. Larsson, and A. Wiklund. Registration and change detection techniques using 3D laser scanner data from natural environments. In *Proceedings SPIE*, volume 6396, page 63960A, October 2006.

D. Törnqvist, T.B. Schön, R. Karlsson, and F. Gustafsson. Particle filter SLAM with high dimensional vehicle model. *Journal of Intelligent and Robotic Systems*, 2009.

D. Törnqvist, A. Helmersson, and F. Gustafsson. Window based GPS integrity test using tight GPS/IMU integration applied to a sounding rocket. In *IEEE Aerospace Conference*, 2010.

David Törnqvist. *Estimation and detetection with applications to navigation*. Dissertation no. 1216, Linköping University, Sweden, 2008.

A. Urruela and J. Riba. Novel closed-form ML position estimator for hyperbolic location. In *Proc. IEEE Conference on Acoustics, Speech and Signal Processing*, Montreal, Canada, May 2004.

H.L. Van Trees. *Detection, Estimation and Modulation Theory*. Wiley, New York, 1971.

R. Velmurugan, S. Subramanian, V. Cevher, D. Abramson, K.M. Odame, J.D. Gray, H-J. Lo, M.H. McClellan, and D.V. Anderson. On low-power analog implementation of particle filters for target tracking. In *European Signal Processing Conf. EUSIPCO*, 2006.

G. Verghese and T. Kailath. A further note on backwards Markovian models. *IEEE Transactions on Information Theory*, 25:121–124, 1979.

R. Vinkvist and M. Nilsson. Sensor fusion navigation for sounding rocket applications. Master's thesis no LiTH-ISY-EX-4009, Department of Electrical Engineering, Linköping University, 2008.

J.A. Volpe. Vulnerability assessment of the transportation infrastructure relying on global positioning system final report. Technical report, National Transportation Systems Center, 2001.

M. Šimandl, J. Královec, and Söderström. Advanced point-mass method for nonlinear state estimation. *Automatica*, 42(7):1133–1145, July 2006.

J.F. Wagner and T. Wieneke. Integrating satellite and inertial navigation – conventional and new fusion approaches. *Control Engineering Practice*, 11(5):543–550, 2003.

E.A. Wan and R. van der Merwe. The unscented Kalman filter for nonlinear estimation. In *Proc. of IEEE Symposium (AS-SPCC)*, pages 153–158.

N. Yazdi, F. Ayazi, and K. Najafi. Micromachined inertial sensors. *Proceedings of the IEEE*, pages 1640–1659, 1998.

Y.Okumura, E. Ohmori, T. Kawano, and K. Fukuda. Field strength and its variability in VHF and UHF land-mobile radio service. *Review of the Electrical Communication Laboratory*, 16(9-10), 1968.

K. Zhang, X.R. Li, and Y. Zhu. Optimal update with out-of-sequence measurements. *IEEE Transactions on Signal Processing*, 53:1992–2004, 2005.

Y. Zhao. Overview of 2G LCS technologies and standards. In *Proc. 3GPP TSG SA2 LCS Workshop*, London, UK, January 2001.

Y. Zhao. Standardization of mobile phone positioning for 3G systems. *IEEE Communications Magazine*, 40(7), July 2002.

Y. Zhou and J. Li. Data fusion of unknown correlations using internal ellipsoidal approximation. In *Proceedings of the 17th IFAC World Congress*, 2008.

Index

2.5D, 382

a posteriori, 492
ABS, 388, 435
accept-reject sampling, 472
acceptance probability, 479
ADAS, 7, 433, 435
advanced driver assistance systems, 433, 435
advanced driver assistant systems, 7
algorithm
    decentralized KF, 175
    EKF1, 197
    EKF2, 197
    filter bank merging, 270
    filter bank pruning, 266
    Gibbs change detection, 275
    Gibbs–Metropolis change detection, 272
    GPB, 270
    IMM, 270
    information KF, 161
    KF, 154
    local search, 268
    RLS, 511
    smoothing KF, 178, 180
    square root KF, 183–185
aliasing, 390
Allan variance, 328

angular speed, 388
anti-locking brake systems, 388, 435
appearance signatures, 300
application specific force models., 370
assisted GPS, 402, 403
association, 5
association diversity, 306
auction algorithm, 118
auxiliary sampling proposal resampling filter, 231

batch gating, 300
Bayes estimator, 493
Bayes factor, 503
Bayes' theorem, 492
Bayesian approach, 492
Bayesian importance sampling, 477
Bayesian mean square error, 493
Bayesian risk, 495
BER, 498
best linear unbiased estimator, 25, 484
bias compensated LS, 56
bias error, 164
bias–variance trade-off, 395
Bierman's UD factorization, 513
Bierman's UD factorization algorithm, 28
bilinear interpolation, 345
bilinear transformation, 317

Index

bit-error rate, 498
BLUE, 25, 484
BMSE, 493
bootstrap PF, 223

CA, 344
calibration, 5
calibration experiments, 423, 425
Cauchy-Schwartz inequality, 485
censoring sensors, 31
central fusion, 171
certainty equivalence, 285
Chapman-Kolmogorov, 132
characteristic function, 129
clutter, 101, 102
coefficient of variation, 224
cog spectrum, 392
cog wheel, 388
coherent detection, 87, 380
compass, 385, 386
complementary Kalman filter, 198
complete sufficient statistics, 489
conjugate prior, 495
conjugate priors, 134, 457
consistent, 484
constant acceleration model, 344
constant position model, 344
constant velocity model, 344
constrained kinematics, 370
coordinated turn, 319
counter, 393
covariance intersection algorithm, 32
CP, 344
Cramér-Rao lower bound, 39, 81, 143, 484
CRLB, 39, 81, 143, 484
cut off branches, 265
CV, 344
CW radar, 375

data association, 5
data fusion, 1
dead-reckon, 433
decentralized filters, 173
decentralized fusion, 171
deep integration, 431
defusion, 31
density function

linear regression
  off-line, 515
  on-line, 514
detection, 5
detection performance, 498
DF-EKF, 199, 463
DFF, 127
DFT, 392
direct detection, 380
direction of arrival, 3, 87, 91, 405
discrete Fourier transform, 392
discretized linearization, 318
divided difference filter, 127
DOA, 3, 84, 87, 405
dynamic system, 126

effective number of samples, 224
efficient, 484
EIV, 92
EKF, 126, 139, 433
EKF1, 197
EKF2, 197
electronic stability control, 388, 433
EM, 271
emitted power, 84
equidistant sampling, 390
equivalent measurement, 30, 114
error state Kalman filter, 198
errors in variables, 92
ESC, 388, 433
Euler sampling, 331
expectation maximization, 271
exponential family, 455, 457
extended HMM filter, 139
extended Kalman filter, 126, 139, 195, 197

F-16 aircraft, 362
fast Fourier transform, 392
FFT, 141, 392
filter bank, 262
filter bank local pruning, 268
filter bank merging, 270
filter bank merging algorithm, 270
filter bank pruning, 266
filter bank pruning algorithm, 266
FIM, 39, 81
first-order hold, 317, 345

Fisher information, 485
Fisher Information Matrix, 39, 81
fixed-point smoothing, 177
FM-CW radar, 375
FOH, 317
forward filter backward simulation, 221
forward filter backward smoother, 221
forward-backward, 178
fourth order moment, 57
fusion filter, 173
fusion formula, 30, 173

gated viewing, 380, 381
Gauss' approximation formula, 61
Gauss-Newton algorithm, 50
Gaussian mixture, 139, 276
Gaussian sum Kalman filters, 127
generalized least squares, 66
generalized likelihood, 38
generalized likelihood ratio test, 103, 106, 502
generalized maximum likelihood estimate, 38
generalized pseudo-Bayesian, 269
geographical information system, 250
geophone, 385
Gibbs change detection algorithm, 275
Gibbs sampler, 272
Gibbs sampling, 480, 481
Gibbs–Metropolis change detection algorithm, 272
global positioning system, 2, 6, 431
GLRT, 103, 106, 502
GML, 38
GPB, 269, 270
GPS, 2, 6, 431
gray-box identification, 5, 334
Gripen, 363
ground truth, 416, 435
ground truth sensor, 436
GS-KF, 127
GV, 381

Hall sensor, 386, 387
higher order statistics, 491
hit or miss cost, 495
Huber's cost, 495
Hungarian algorithm, 118

hyperbolic function, 87

IMM, 269, 270
importance function, 477
importance sampling, 219, 223
importance weight, 219, 477
improper prior, 493
IMU, 2, 6, 198, 431
inclinometer, 385
incremental rotary encoders, 387
inertial measurement unit, 2, 6, 431
inertial navigation systems, 198
inference, 37
information filter, 136, 174
information form of WLS, 27
information fusion, 1
information matrix, 27
information state, 27
INS, 198
integrated sampling, 391
interacting multiple model, 269
internal ellipsoid approximation, 32
interpolation, 390
intrinsic accuracy, 40, 144
iterated Kalman filter, 170

jitter, 394
JML, 261
jump Markov linear model, 261

Kalman filter, 126, 153
    complementary, 198
    error state, 198
Kalman filter banks, 127
KF, 126, 153
kinematic models, 369

ladar, 379
landmark handling, 306
Laplace distribution, 480
largest ellipsoid algorithm, 32
laser radar, 379
law of total probability, 494
law of total variance, 264
least squares, 24, 163
lidar, 379
likelihood, 36, 483
likelihood ratio, 498

Index

likelihood ratio test, 498
line of sight, 74
linear time-invariant, 316
linearized discretization, 318
linearized Kalman filter, 198
LMP, 505
local search, 268
localization, 5, 280
locally most powerful, 505
longitudinal slip, 400
loop closure, 285, 306
loose integration, 431
LOS, 74
LRT, 498
LS, 24, 163
LTI, 316
Luenberger observer, 346

magnetometer, 386
MAP, 493
mapping, 280
marginal particle filter, 221
marginalized likelihood, 39
marginalized likelihood ratio test, 503
marginalized maximum likelihood estimate, 39
marginalized NLS, 68
marginalized particle filter, 238
Markov chain Monte Carlo, 271, 479
matrix inversion lemma, 27, 136, 296, 510
maximum a posteriori, 132, 493
maximum likelihood, 36
maximum likelihood estimator, 490
MB, 427
MCMC, 271, 479
mean square error, 416, 483, 486
measurement update, 132, 211, 265
MEMS, 384
merging, 265, 277
method of moments, 491
Metropolis step, 272
Metropolis–Hastings, 480
MHT, 300
micro-electrical-mechanical systems, 384
micromachined devices, 384
minimum variance, 132, 484
minimum variance estimator, 25

minimum variance unbiased, 484
mixture Kalman filter, 245
MLRT, 503
MML, 39
MNLS, 68
modality, 2
mode, 261
mode parameter, 261
model calibration, 334
model selection, 329
most probable branch, 266
MPF, 238
MSE, 416, 483, 486
multi-rate sampling, 6
multinomial distribution, 138
multipath, 74
multiple hypothesis test data association, 300
multiple target tracking, 5
Munkre's algorithm, 118
muzzle blast, 427
MV, 25, 484
MVU, 484

navigation, 5, 280
navigation problem, 126
nearest neighbor, 111
Neyman-Fisher factorization, 488
Neyman-Pearson's theorem, 498
NLOS, 74
NLS, 47, 84, 329
  marginalized, 68
NN, 111
non-holonomic constraints, 418
noninformative prior, 493
nonline of sight, 74
nonlinear filtering, 126, 215
nonlinear least squares, 47, 84, 329
nonlinear weighted least squares, 50
normalized parameter space, 114
nuisance parameter, 504
nuisance parameters, 494
NWLS, 50

odometry, 356, 388
optimal sampling, 226
out of sequence, 422
outlier rejection, 186

outliers, 169

parameter covariance
  asymptotic, 518
parametric CRLB, 143
parsimonious principle, 329
particle filter, 127, 211
  marginal, 221
  prediction, 222
  smoother
    ffbs, 221
    ffbsim, 221
particle predictor, 222
path loss constant, 84, 408
PD radar, 375
PDF, 483
PF, 127, 211
physical parameter space, 114
PMF, 140
point mass filter, 140, 218
pose, 279
positioning, 280
posterior, 492
posterior CRLB, 143
power of the test, 497
PRF, 375
prior, 493
probability density function, 483
probability of detection, 497
probability of false alarm, 497
propagation delayed measurements, 422
proposal density, 219
proposal distribution, 477
prune, 264
pruning, 277
pulse repetition frequency, 375

QKF, 127
QR factorization, 182, 513
quadrature Kalman filter, 127
quantization, 393
quantization noise, 394
quaternion representation, 349
quaternions, 369

RA, 145
RANSAC, 283

Rao-Blackman-Lehmann-Scheffe's theorem, 490
Rao-Blackwellized particle filter, 245
Rauch–Tung–Striebel formulas, 178
received signal strength, 75
receiver operating characteristic, 102
receiver operating characteristics, 497
recursive least squares, 509
registration, 382
regularity conditions, 484
regularization, 225
rejection sampling, 472
relative accuracy, 40, 144, 145
resampling, 211
residual, 103
Riemann sum, 392
rigid body kinematics, 369
RLS, 509
  algorithm, 511
  windowed, 511
robust filtering, 188
ROC, 102, 104, 497
rotary encoders, 387
rotational kinematics, 369
RSS, 75, 84

safe fusion, 32
sample degeneracy, 220
sample depletion, 220
sample impoverishment, 220
sampling importance resampling, 223, 478, 479
sampling importance sampling, 223
saturation effect, 395
scanning laser radar, 380
scanning systems, 380
seismometer, 385
self-occlusion, 382
sensitivity analysis, 165
sensor fusion, 1
sensor fusion formula, 29
sensor model, 10, 42, 71
sensor network, 74
separable least squares, 66, 84, 429
separable particle filter, 245
shockwave, 427
sigma points, 62
signal spectrum, 488

Index

signal to noise ratio, 225, 407
simple hypothesis test, 498
simultaneous localization and mapping, 6, 93, 279
simultaneous navigation and mapping, 280
single point sensors, 380
SIR, 223, 479
SIS, 223
situation awareness, 10
SLAM, 6, 93, 279
slip, 400
SLS, 66, 84, 429
SN, 74
SNAM, 280
SNR, 225, 407
SONAR, 383
spread of the mean, 269
square root, 181
square-root algorithms, 28
stability constraint, 157
staring laser radar, 380
staring systems, 381
state space model, 125
stochastic integration, 476
structured parameter space, 114
Student t distribution, 521
sufficient statistics, 27
super formula, 488
support, 484
SW, 427
synchronous averaging, 390
system identification, 11, 329

TDOA, 75, 84, 405
test statistic, 102, 497
tight integration, 431
time difference of arrival, 75, 405
time of arrival, 3, 51, 75, 405
time stamps, 422
time update, 132, 211, 265
tire pressure monitoring systems, 388
TLS, 92
TOA, 3, 51, 75, 84, 405
toothed wheel, 388
total least squares, 92
TPMS, 388
tracking, 5

tracking problem, 126
traction control systems, 388
transform pair, 129
translation kinematics, 369
triangulation, 5
trilateration, 5
two-filter smoothing formula, 178
type I error, 497
type II error, 497

UKF, 127
unbiased, 25, 484
unscented Kalman filter, 127, 196
unscented transform, 58, 61
UT, 58, 61

virtual measurements, 418
virtual sensor
    compass, 385, 386
    inclinometer, 385
virtual sensors, 373, 400, 436
visual triangulation, 6

weighted least squares, 24, 163
wireless sensor network, 75
WLS, 163
Wold's decomposition theorem, 488
WSN, 75

zero-order hold, 316, 345
ZOH, 316